CAMBRIDGE LIBRARY COLLECTION

Books of enduring scholarly value

Technology

The focus of this series is engineering, broadly construed. It covers technological innovation from a range of periods and cultures, but centres on the technological achievements of the industrial era in the West, particularly in the nineteenth century, as understood by their contemporaries. Infrastructure is one major focus, covering the building of railways and canals, bridges and tunnels, land drainage, the laying of submarine cables, and the construction of docks and lighthouses. Other key topics include developments in industrial and manufacturing fields such as mining technology, the production of iron and steel, the use of steam power, and chemical processes such as photography and textile dyes.

The Mechanical Principles of Engineering and Architecture

Seventh wrangler in the Cambridge mathematical tripos in 1826, Henry Moseley (1801–72) was adept at applying mathematical analysis to a wide variety of problems. Appointed professor of natural and experimental philosophy and astronomy at London's newly established King's College in 1831, he was instrumental in creating the institution's department of engineering and applied science. This 1843 textbook is based on the lectures in statics, dynamics and structures that he gave to students of engineering and architecture. Moseley draws on the latest continental work in mechanics, and the treatment of problems is mathematically sophisticated. Starting with basic statics and dynamics, Moseley covers topics of interest to both civil and military engineers, with sections on the theory of machines and on the stability of walls, arches and other structures. Notably, the American edition of this work was adopted as a textbook by the United States Military Academy at West Point.

Cambridge University Press has long been a pioneer in the reissuing of out-of-print titles from its own backlist, producing digital reprints of books that are still sought after by scholars and students but could not be reprinted economically using traditional technology. The Cambridge Library Collection extends this activity to a wider range of books which are still of importance to researchers and professionals, either for the source material they contain, or as landmarks in the history of their academic discipline.

Drawing from the world-renowned collections in the Cambridge University Library and other partner libraries, and guided by the advice of experts in each subject area, Cambridge University Press is using state-of-the-art scanning machines in its own Printing House to capture the content of each book selected for inclusion. The files are processed to give a consistently clear, crisp image, and the books finished to the high quality standard for which the Press is recognised around the world. The latest print-on-demand technology ensures that the books will remain available indefinitely, and that orders for single or multiple copies can quickly be supplied.

The Cambridge Library Collection brings back to life books of enduring scholarly value (including out-of-copyright works originally issued by other publishers) across a wide range of disciplines in the humanities and social sciences and in science and technology.

The Mechanical Principles of Engineering and Architecture

Henry Moseley

CAMBRIDGE
UNIVERSITY PRESS

CAMBRIDGE
UNIVERSITY PRESS

University Printing House, Cambridge, CB2 8BS, United Kingdom

Cambridge University Press is part of the University of Cambridge.

It furthers the University's mission by disseminating knowledge in the pursuit of
education, learning and research at the highest international levels of excellence.

www.cambridge.org
Information on this title: www.cambridge.org/9781108071970

© in this compilation Cambridge University Press 2014

This edition first published 1843
This digitally printed version 2014

ISBN 978-1-108-07197-0 Paperback

THE

MECHANICAL PRINCIPLES

OF

ENGINEERING

AND

ARCHITECTURE.

By the same Author,

ILLUSTRATIONS OF PRACTICAL MECHANICS;

AND

A TREATISE

ON

HYDROSTATICS AND HYDRODYNAMICS.

LONDON:
Printed by A. SPOTTISWOODE,
New-Street-Square.

THE

MECHANICAL PRINCIPLES

OF

ENGINEERING

AND

ARCHITECTURE.

BY

THE REV. HENRY MOSELEY, M.A. F.R.S.

LATE OF ST. JOHN'S COLLEGE, CAMBRIDGE; PROFESSOR OF NATURAL
PHILOSOPHY AND ASTRONOMY IN KING'S COLLEGE, LONDON.

WITH ILLUSTRATIONS ON WOOD.

LONDON:

PRINTED FOR
LONGMAN, BROWN, GREEN, AND LONGMANS,
PATERNOSTER-ROW.
1843.

PREFACE.

In the following work, I have proposed to myself to apply the principles of mechanics to the discussion of the most important and obvious of those questions which present themselves in the practice of the engineer and the architect ; and I have sought to include in that discussion all the circumstances on which the practical solution of such questions may be assumed to depend. It includes the substance of a course of lectures delivered to the students of King's College in the department of engineering and architecture, during the years 1840, 1841, 1842.*

In the first part I have treated of those portions of the science of STATICS which have their application in the theory of machines and the theory of construction.

In the second, of the science of DYNAMICS, and, under this head, particularly of that union of a continued pressure with a continued motion which has received from English writers the various names of

---

* The first 170 pages of the work were printed for the use of my pupils in the year 1840. Copies of them were about the same time in the possession of several of my friends in the Universities.

" dynamical effect," " efficiency," " work done," " labouring force," "work," &c.; and "moment d'activité," "quantité d'action," " puissance mécanique," "travail," from French writers.

Among the latter this variety of terms has at length given place to the most intelligible and the simplest of them, " travail." The English word " work" is the obvious translation of " travail," and the use of it appears to be recommended by the same considerations. The work of overcoming a pressure of one pound through a space of one foot has in this country been taken as the unit, in terms of which any other amount of work is estimated ; and in France the work of overcoming a pressure of one kilogramme through a space of one metre. M. Dupin has proposed the application of the term dyname to this unit.

I have gladly sheltered myself from the charge of having contributed to increase the vocabulary of scientific words by assuming the obvious term " unit of work" to represent concisely and conveniently enough the idea which is attached to it, without translation.

The work of any pressure operating through any space is evidently measured in terms of such units, by multiplying the number of pounds in the pressure by the number of feet in the space, if the direction of the pressure be continually that in which the space is described. If not, it follows, by a simple geometrical deduction, that it is measured by the product of the number of pounds in the pressure, by the number of feet in the projection of the space described*, upon the direction of the pressure ; that is, by the product of the pressure by its virtual velocity. Thus, then, we conclude, at once, by the principle of virtual velocities, that if a machine work under a constant equilibrium of the pressures applied to it, or if it work uniformly, then is the aggregate work of those pressures which tend to accelerate its motion equal to the aggregate work of those which tend to retard it ; and, by the principle of *vis viva*, that if the machine do not work under an equilibrium of the forces impressed upon it, then is the aggregate work of those which tend to accelerate the motion of the machine greater or less than the aggregate work of those which tend to retard its motion by one half the aggregate of the *vires vivæ* acquired or lost by the moving parts of the system, whilst the work is being done upon it. In no respect have the labours of the illustrious president of the Academy of Sciences more contributed to the developement of the theory of machines than in the application which he has so successfully made to it of this principle of *vis viva*.† In the elementary discussion of this principle, which is given by M. Poncelet, in the introduction to his *Mécanique Industrielle*, he has revived the term *vis inertiæ* (*vis*

---

* If the direction of the pressure remain always parallel to itself, the space described may be any finite space; if it do not, the space is understood to be so small, that the direction of the pressure may be supposed to remain parallel to itself whilst that space is described.

† See Poncelet, *Mécanique Industrielle*, troisième partie.

inertiæ, vis insita, Newton), and, associating with it
the definitive idea of a force of resistance opposed to
the acceleration or the retardation of a body's motion,
he has shown (Arts. 66. and 122.) the work expended
in overcoming this resistance through any space to be
measured by one half the *vis viva* accumulated through
the space ; so that throwing into the consideration of
the forces under which a machine works, the *vires
inertiæ* of its moving elements, and observing that one
half of their aggregate *vis viva* is equal to the aggregate work of their *vires inertiæ,* it follows, by the
principle of virtual velocities, that the difference between the aggregate work of those forces impressed
upon a machine, which tend to accelerate its motion,
and the aggregate work of those which tend to retard
the motion, is equal to the aggregate work of the
vires inertiæ of the moving parts of the machine:
under which form the principle of *vis viva* resolves
itself into the principle of virtual velocities. So many
difficulties, however, oppose themselves to the introduction of the term *vis inertiæ,* associated with the
definitive idea of an opposing force, into the discussion
of questions of mechanics, and especially of practical
and elementary mechanics, that it has appeared to the
author of this work desirable to avoid it. It is with
this view, that in the following work a new interpretation is given to that function of the velocity of a
moving body which is known as its *vis viva ;* one half
that function being interpreted to represent the number of units of work *accumulated* in the body so long
as its motion is continued, and which number of units

of work it is capable of reproducing upon any resistance which may be opposed to its motion, and bring it to rest. A very simple investigation (Art. 66.) establishes the truth of this interpretation, and gives to the principle of *vis viva* the following new and more simple enunciation :—" The difference between the aggregate work done upon the machine, during any time, by those forces which tend to accelerate the motion, and the aggregate work, during the same time, of those which tend to retard the motion, is equal to the aggregate number of units of work accumulated in the moving parts of the machine during that time if the former aggregate exceed the latter, and lost from them during that time if the former aggregate fall short of the latter." Thus, then, if the aggregate work of the forces which tend to accelerate the motion of a machine exceeds that of the forces which tend to retard it, then is the surplus work (that done upon the driving points, above that expended upon the prejudicial resistances and upon the working points) continually accumulated in the moving elements of the machine, and their motion is thereby continually accelerated. And if the former aggregate be less than the latter, then is the deficiency supplied from the work already accumulated in the moving elements, so that their motion is in this case continually retarded.

The moving power divides itself whilst it operates in a machine, first, into that which overcomes the prejudicial resistances of the machine, or those which are opposed by friction and other causes, uselessly absorbing the work in its transmission. Secondly, into that which accelerates the motion of the various moving parts of the machine, and which accumulates in them so long as the work done by the moving power upon it exceeds that expended upon the various resistances opposed to the motion of the machine. Thirdly, into that which overcomes the useful resistances, or those which are opposed to the motion of the machine at the working point, or points, by the useful work which is done by it.

Between these three elements there obtains in every machine a mathematical relation, which I have called its MODULUS. The general form of this modulus I have discussed in a memoir on the " Theory of Machines" published in the *Philosophical Transactions* for the year 1841. The determination of the particular moduli of those elements of machinery which are most commonly in use is the subject of the third part of the following work. From a combination of the moduli of any such elements there results at once the modulus of the machine compounded of them.

When a machine has acquired a state of uniform motion work ceases to accumulate in its moving elements, and its modulus assumes the form of a direct relation between the work done by the motive power upon its driving point and that yielded at its working points. I have determined by a general method* the modulus in this case, from that statical relation between the driving and working pressures upon the machine which obtains in the state bordering

---

* Art. 152. See *Phil. Trans.*, 1841, p. 290.

upon its motion, and which may be deduced from the known conditions of equilibrium and the established laws of friction. In making this deduction I have, in every case, availed myself of the following principle, first published in my paper on the theory of the arch read before the Cambridge Philosophical Society in Dec. 1833, and printed in their *Transactions* of the following year:—" In the state bordering upon motion of one body upon the surface of another, the resultant pressure upon their common surface of contact is inclined to the normal, at an angle whose tangent is equal to the coefficient of friction."

This angle I have called the limiting angle of resistance. Its values calculated, in respect to a great variety of surfaces of contact, are given in a table at the conclusion of the second part, from the admirable experiments of M. Morin*, into the mechanical details of which precautions have been introduced hitherto unknown to experiments of this class, and which have given to our knowledge of the laws of friction a precision and a certainty hitherto unhoped for.

Of the various elements of machinery those which rotate about cylindrical axes are of the most frequent occurrence and the most useful application; I have, therefore, in the first place sought to establish the general relation of the state bordering upon motion between the driving and the working pressures upon such a machine, reference being had to the weight of

* *Nouvelles Expériences sur le Frottement,* Paris, 1833.

the machine.* This relation points out the existence
of a particular direction in which the driving pressure
should be applied to any such machine, that the
amount of work expended upon the friction of the
axis may be the least possible. This direction of the
driving pressure always presents itself on the same
side of the axis with that of the working pressure,
and when the latter is vertical it becomes parallel to
it; a principle of the economy of power in machinery
which has received its application in the parallel motion of the marine engines known as the Gorgon
Engines.

I have devoted a considerable space in this portion
of my work to the determination of the modulus of a
system of toothed wheels ; this determination I have,
moreover, extended to bevil wheels, and have included
in it, with the influence of the friction of the teeth
the wheels, that of their axes and their weights.
An approximate form of this modulus applies to
any shape of the teeth under which they may be
made to work correctly; and when in this approximate form of the modulus the terms which represent
the influence of the friction of the axis and the weight
of the wheel are neglected, it resolves itself into a
well known theorem of M. Poncelet, reproduced by
M. Navier and the Rev. Dr. Whewell.† In respect

---

* In my memoir on the " Theory of Machines " (*Phil. Trans.*
1841), I have extended this relation to the case in which the number of the pressures and their directions are any whatever. The
theorem which expresses it is given in the Appendix of this work.

† In the discussion of the friction of the teeth of wheels, the

to wheels having epicycloidal and involute teeth, the
modulus assumes a character of mathematical ex-
actitude and precision, and at once establishes the
conclusion (so often disputed) that the loss of power
is greater before the teeth pass the line of centres
than *at corresponding points* afterwards; that the
contact should, nevertheless, in all cases take place
partly before and partly after the line of centres has
been passed. In the case of involute teeth, the proportion in which the arc of contact should thus be
divided by the line of centres is determined by a
simple formula; as also are the best dimensions of the
base of the involute, with a view to the most perfect
economy of power in the working of the wheels.

The greater portion of the subjects discussed in
the third part of my work I believe to be entirely
new to science. In the fourth part I have treated
of "the theory of the stability of structures," referring
its conditions, so far as they are dependent upon rotation, to the properties of a certain line which may
be conceived to traverse every structure, passing
through those points in it where its surfaces of contact are intersected by the resultant pressures upon
them. To this line, whose properties I first discussed in a memoir upon "the Stability of a System
of Bodies in Contact," printed in the sixth volume of
the *Camb. Phil. Trans.*, I have given the name of

---

direction of the mutual pressures of the teeth is determined by a
method first applied by me to that purpose in a popular treatise,
entitled *Mechanics applied to the Arts*, published in 1834.

the line of resistance; it differs essentially in its properties from a line referred to by preceding writers
under the name of the curve of equilibrium or the
line of pressure.

The distance of the line of resistance from the extrados of a structure, at the point where it most
nearly approaches it, I have taken as a measure of
the stability of a structure, and have called it the
modulus of stability*; conceiving this measure of the
stability to be of more obvious and easier application
than the coefficient of stability used by the French
writers.

That structure in respect to every independent
element of which, the modulus of stability is the
same, is evidently the structure of the greatest
stability having a given quantity of material employed in its construction; or of the greatest economy
of material having a given stability.

The application of these principles of construction
to the theory of piers, walls supported by counter-
forts and shores, buttresses, walls supporting the
thrust of roofs and the weights of the floors of dwell-
ings, and Gothic structures, has suggested to me a
class of problems never, I believe, before treated mathematically.

I have applied the well known principle of Coulomb

---

* This idea was suggested to me by a rule for the stability of
revêtement walls attributed to Vauban, to the effect, that the resultant pressure should intersect the base of such a wall at a point
whose distance from its extrados is $\frac{4}{9}$ths the distance between the
extrados at the base and the vertical through the centre of gravity.

to the determination of the pressure of earth upon revêtement walls, and a modification of that principle, suggested by M. Poncelet, to the determination of the resistance opposed to the overthrow of a wall backed by earth. This determination has an obvious application to the theory of foundations.

In the application of the principle of Coulomb I have availed myself, with great advantage, of the properties of the limiting angle of resistance. All my results have thus received a new and a simplified form.

The theory of the arch I have discussed upon principles first laid down in my memoir on "the Theory of the Stability of a System of Bodies in Contact," before referred to, and subsequently in a memoir printed in the "Treatise on Bridges" by Professor Hosking and Mr. Hann.* They differ essentially from those on which the theory of Coulomb is founded†; when, nevertheless, applied to the case treated by the French mathematicians they lead to identical results. I have inserted at the conclusion of my work the tables of the thrust of circular arches, calculated by M. Garidel from formulæ founded on the theory of Coulomb.

The fifth part of the work treats of the "strength

---

* I have made extensive use of the memoir above referred to in the following work, by the obliging permission of the publisher, Mr. Weale.

† The theory of Coulomb was unknown to me at the time of the publication of my memoirs printed in the *Camb. Phil. Trans.* For a comparison of the two methods see Mr. Hann's treatise.

of materials," and applies a new method to the determination of the deflexion of a beam under given pressures.

In the case of a beam loaded uniformly over its whole length, and supported at four different points, I have determined the several pressures upon the points of support by a method applied by M. Navier to a similar determination in respect to a beam loaded at given points.*

In treating of rupture by elongation I have been led to a discussion of the theory of the suspension bridge. This question, so complicated when reference is had to the weight of the roadway and the weights of the suspending rods, and when the suspending chains are assumed to be of uniform thickness, becomes comparatively easy when the section of the chain is assumed so to vary its dimensions as to be every where of the same strength. A suspension bridge thus constructed is obviously that which, being of a given strength, can be constructed with the least quantity of materials ; or, which is of the greatest strength having a given quantity of materials used in its construction.†

The theory of rupture by transverse strain has suggested a new class of problems, having reference to the forms of girders having wide flanges connected by

---

* As in fig. p. 521. of the following work.

† That particular case of this problem, in which the weights of the suspending rods are neglected, has been treated by Mr. Hodgkinson in the fourth vol. of the *Manchester Transactions*, with his usual ability. He has not, however, succeeded in effecting its complete solution.

slender ribs or by open frame work: the consideration
of their strongest forms leads to results of practical
importance.

In discussing the conditions of the strength of
breast-summers, my attention has been directed to the
best positions of the columns destined to support them,
and to a comparison of the strength of a beam carrying a uniform load and supported freely at its extremities, with that of a beam similarly loaded but
having its extremities firmly imbedded in masonry.

In treating of the strength of columns I have gladly
replaced the mathematical speculations upon this subject, which are so obviously founded upon false data,
by the invaluable experimental results of Mr. E.
Hodgkinson, detailed in his well known paper in the
Philosophical Transactions for 1840.

The sixth and last part of my work treats on "impact;" and the Appendix includes, together with tables
of the mechanical properties of the materials of construction, the angles of rupture and the thrusts of
arches, and complete elliptic functions, a demonstration of the admirable theorem of M. Poncelet for determining an approximate value of the square root of
the sum or difference of two squares.

In respect to the following articles of my work I
have to acknowledge my obligations to the work of
M. Poncelet, entitled *Mécanique Industrielle.* The
mode of demonstration is in some, perhaps, so far
varied as that their origin might with difficulty be
traced; the principle, however, of each demonstration

a

—all that constitutes its novelty or its value—belongs to that distinguished author.

30*, 38, 40, 45, 46, 47, 52, 58, 62, 75, 108†, 123, 202, 267‡, 268, 269, 270, 349, 354, 365.§

* The enunciation only of this theorem is given in the *Méc. Ind.*, 2me partie, Art. 38.

† Some important elements of the demonstration of this theorem are taken from the *Méc. Ind.*, Art. 79. 2me partie. The principle of the demonstration is not, however, the same as in that work.

‡ In this and the three following articles I have developed the theory of the fly-wheel, under a different form from that adopted by M. Poncelet (*Méc. Ind.*, Art. 56. 3me partie). The principle of the whole calculation is, however, taken from his work. It probably constitutes one of the most valuable of his contributions to practical science.

§ The idea of determining the work necessary to produce a given deflection of a beam from that expended upon the compression and the elongation of its component fibres was suggested by an observation in the *Méc. Ind.*, Art. 75. 3me partie. An error presents itself in the determination given by M. Poncelet in that article of the linear deflection f of a beam under a given deflecting pressure P. It consists in assuming that the work of the deflecting pressure is represented by Pf, as it would be if, in order to deflect the beam, P must always retain the same value instead of varying directly as the deflection. The true value of the work is $\frac{1}{2}Pf$; the determination of which requires a knowledge of the law of the deflection, which the demonstration does not suppose. It is due to M. Poncelet to state that the *Mécanique Industrielle* was published (uncorrected) without his concurrence or knowledge, in Belgium, from a MS. copy of his lectures lithographed for the use of the workmen at Metz to whom they were addressed.

CONTENTS.

PART III.

THE THEORY OF MACHINES.

PART IV.

THE THEORY OF THE STABILITY OF STRUCTURES.

PART V.

THE STRENGTH OF MATERIALS.

PART VI.

IMPACT.

APPENDIX.

ERRATA.

Page 36. line 2. from bottom, *for* $x_1^{\frac{2}{3}}$ *read* $x_1^{\frac{3}{2}}$.

55. line 5. from bottom, *for* ABDC *read* AEFC.

64. line 3. from bottom, *for* Y_2 *read* y_2.

64. line 5. from bottom, *for* Y_{a-0} *read* Y_{a-1}.

122. line 10. from top, *for* half *read* double.

167. line 8. from top, *for* B_1 *read* B.

172. line 3. from top, *for* 114 *read* 119.

173. line 5. from bottom, *for* $\frac{P_2 a_2^2}{a_2^2} V_1^2$ *read* $\frac{P_2 a_2^2}{a_1^2} V_1^2$.

174. line 8. from top, *for* b^2 *read* b_2.

521. line 6. from top, *for* $\frac{1}{2}$ in. square *read* $\frac{3}{16}$ in. by $\frac{7}{16}$ in.

In the table page 152. the words "without unguent" enclosed by a bracket opposite to the words "iron upon oak," belong (with the corresponding numbers) to the following bracket.

THE

MECHANICAL PRINCIPLES

OF

CIVIL ENGINEERING.

PART I.

STATICS.

1. FORCE is that which *tends* to cause or to destroy motion, or which actually causes or destroys it.

The *direction* of a force is that straight line in which it tends to cause motion in the point to which it is applied, or in which it tends to destroy the motion in it.

When more forces than one are applied to a body, and their respective tendencies to communicate motion to it counteract one another, so that the body remains at rest, these forces are said to be in EQUILIBRIUM, and are called PRESSURES.

It is found by experiment, that the effect of a pressure when applied to a solid body, is the same at whatever point in the line of its direction it is applied; so that the conditions of the equilibrium of that pressure, in respect to other pressures applied to the same body, are not altered, if, without altering the direction of the pressure, we remove its point of application, provided only the point to which we remove it be in the straight line in the direction of which it acts.

The science of STATICS is that which treats of the *equilibrium of pressures.* When two pressures *only* are applied to a body, and hold it at rest, it is found by experiment that

B

these pressures act in opposite directions, and have their directions always in the same straight line. Two such pressures are said to be *equal*.

If instead of applying two pressures which are thus equal in *opposite* directions, we apply them both in the *same* direction, the single pressure which must be applied in a direction opposite to the *two* to sustain them, is said to be *double* of either of them. If we take a third pressure equal to either of the two first, and apply the three in the same direction, the single pressure, which must be applied in a direction opposite to the three to sustain them, is said to be *triple* of either of them; and so of any number of pressures. Thus fixing upon any one pressure, and ascertaining how many pressures equal to this are necessary, when applied in an opposite direction, to sustain any other greater pressure, we arrive at a true conception of the amount of that greater pressure in terms of the first.

That single pressure, in terms of which the amount of any other greater pressure is thus ascertained, is called an UNIT of pressure.

Pressures, the amount of which are determined in terms of some known unit of pressure, are said to be *measured*.

Different pressures, the amounts of which can be determined in terms of the *same* unit, are said to be *commensurable*.

The units of pressure which it is found most convenient to use, are the weights of certain portions of matter, or the pressures with which they tend towards the centre of the earth. The units of pressure are different in different countries. With us the unit of pressure from which all the rest are derived is the weight of 22·815 * cubic inches of distilled water. This weight is one pound troy; being divided into 5760 equal parts, the weight of each is a grain troy, and 7000 such grains constitute the pound avoirdupois.

If straight lines be taken in the directions of any number

---

* This standard was fixed by Act of Parliament in 1824. The temperature of the water is supposed to be 62° Farenheit, the weight to be taken in air, and the barometer to stand at 30 inches.

of pressures, and have their lengths proportional to the numbers of units in those pressures respectively, then these lines having to one another the same proportion in length that the pressures have in magnitude, and being moreover drawn in the directions in which those pressures respectively act, are said to *represent them in magnitude and direction.*

A system of pressures being in equilibrium, let any number of them be imagined to be taken away and replaced by a single pressure, and let this single pressure be such that the equilibrium which before existed may remain, then this single pressure, producing the same effect in respect to the equilibrium that the pressures which it replaces produced, is said to be their RESULTANT.

The pressures which it replaces are said to be the COMPONENTS of this single pressure ; and the act of replacing them by such a single pressure, is called the COMPOSITION of pressures.

If, a single pressure being removed from a system in equilibrium, it be replaced by any number of other pressures, such, that whatever effect was produced by that which they replace singly, the same effect (in respect to the conditions of the equilibrium) may be produced by those pressures conjointly, then is that single pressure said to have been RESOLVED into these, and the act of making this substitution of two or more pressures for one, is called the RESOLUTION of pressures.

THE PARALLELOGRAM OF PRESSURES.

2. *The resultant of any two pressures applied to a point, is represented in direction by the diagonal of a parallelogram, whose adjacent sides represent those pressures in magnitude and direction.**

(Duchayla's Method.)

To the demonstration of this proposition, after the excellent method of Duchayla, it is necessary in the first place

---

* This proposition constitutes the foundation of the entire science of Statics.

to show, that if there be any two pressures P_2 and P_3 whose directions are in the same straight line, and a third pressure P_1 in any other direction, and if the proposition be true in respect to P_1 and P_2, and also in respect to P_1 and P_3, then it will be true in respect to P_1 and $P_2 + P_3$.

Let P_1, P_2, and P_3, form part of any system of pressures in equilibrium, and let them be applied to the point A; take AB and AC to represent, in magnitude and direction, the pressures P_1 and P_2, and CD the pressure P_3, and complete the parallelograms CB and DF. Suppose the proposition to be true with regard to P_1 and P_2, the resultant of P_1 and P_2 will then be in the direction of the diagonal AF of the parallelogram BC, whose adjacent sides AC and AB represent P_1 and P_2 in magnitude and direction. Let P_1 and P_2 be replaced by this resultant. It matters not to the equilibrium where in the line AF it is applied; let it then be applied at F. But thus applied at F it may, without affecting the conditions of the equilibrium, be in its turn replaced by (or resolved into) two other pressures acting in CF and BF, and these will manifestly be equal to P_1 and P_2, of which P_1 may be transferred without altering the conditions to C, and P_2 to E. Let this be done, and let P_3 be transferred from A to C, we shall then have P_1 and P_3 acting in the directions CF and CD at C, and P_2, in the direction FE at E, and the conditions of the equilibrium will not have been affected by the transfer of them to these points. Now suppose that the proposition is also true in respect to P_1 and P_3 as well as P_1 and P_2. Then since CF and CD represent P_1 and P_3 in magnitude and direction, therefore their resultant is in the direction of the diagonal CE. Let them be replaced by this resultant, and let it be transferred to E, and let it then be resolved into two other pressures acting in the directions DE and FE; these will evidently be P_1 and P_3. We have now then transferred all the three pressures P_1, P_2, P_3, from A to E, and they act at E in directions parallel to the directions in which they acted at A, and this has been done without affecting the conditions of the equilibrium; or, in other words, it has been shown that the pressures P_1, P_2, P_3, produce the same effect as it respects the conditions of the equilibrium, whether they be applied at A or E. The *resultant* of P_1, P_2, P_3, must therefore produce the same effect, as it regards the conditions of the equilibrium, whether it be applied at A or E. But in order that this resultant may thus produce the same effect when acting at A or E, it must act in the straight line AE, because a pressure produces the same effect when applied at two different points, only when both those points are in the line of its direction. On the supposition made therefore, the resultant of P_1, P_2, and P_3, or of P_1 and $P_2 + P_3$ acts in the direction of the diagonal AE of the parallelogram BD whose adjacent sides AD and AB represent $P_2 + P_3$ and P_1 in magnitude and direction; and it has been shown, that if the proposition be true in respect to P_1 and P_2, and also in respect to P_1 and P_3, then it is true in respect to P_1 and $P_2 + P_3$. Now this being the case for all values of P_1, P_2, P_3, it is the case when P_1, P_2, and P_3, are equal to one another. But if P_1 be equal to P_2 their resultant will manifestly have its direction as much towards one of these pressures as the other; that is, it will have its direction midway between them, and it will bisect the angle BAC: but the diagonal AF in this case also bisects the angle BAC, since P_1 being equal to P_2, AC is equal to AB; so that in this particular case the direction of the resultant *is* the direction of the diagonal, and the proposition is true, and similarly it is true of P_1 and P_3, since these pressures are equal. Since then it is true of P_1 and P_2 when they are equal, and also of P_1 and P_3, therefore it is true in this case of P_1 and $P_2 + P_3$, that is of P_1 and $2P_1$. And since it is true of P_1 and P_2, and also of P_1 and $2P_1$, therefore it is true of P_1 and $P_2 + 2P_1$, that is of P_1 and $3P_1$; and so of P_1 and mP_1, if m be any whole number; and similarly since it is true of mP_1 and P_1, therefore it is true of mP_1 and $2P_1$ &c., and of mP_1 and nP_1 where n is any whole number. Therefore the proposition is true of any two pressures mP_1 and nP_1 which are *commensurable*.

It is moreover true when the pressures are *in-commensurable*. For let AC and AB represent any two such pressures P_1 and P_2 in magnitude and direction, and complete the parallelogram ABDC, then will the direction of the resultant of P_1 and P_2 be in AD ; for if not, let its direction be AE, and draw EG parallel to CD. Divide AB into equal parts, each less than GC, and set off on AC parts equal to those from A towards C. One of the divisions of these will manifestly fall in GC. Let it be H, and complete the parallelogram AHFB. Then the pressure P_2 being conceived to be divided into as many equal units of pressure as there are equal parts in the line AB, AH may be taken to represent a pressure P_3 containing as many of these units of pressure as there are equal parts in AH, and these pressures P_2 and P_3 will be *commensurable*, being measured in terms of the same unit. Their resultant is therefore in the direction AF, and this resultant of P_3 and P_2 has its direction nearer to AC than the resultant AE of P_1 and P_2 has ; which is absurd, since P_1 is greater than P_3.

Therefore AE is not in the direction of the resultant of P_1 and P_2 ; and in the same manner it may be shown that no other than AD is in that direction. Therefore, &c.

3. *The resultant of two pressures applied in any directions to a point, is represented in magnitude as well as in direction by the diagonal of the parallelogram whose adjacent sides represent those pressures in magnitude and in direction.*

Let BA and CA represent, in magnitude and direction, any two pressures applied to the point A. Complete the parallelogram BC. Then by the last proposition AD will represent the resultant of these pressures in direction. It will also represent it in magnitude ; for, produce DA to G, and conceive a pressure to be applied in GA equal to the resultant of BA and CA, and opposite to it, and let this pressure be represented in magnitude by the line GA. Then will the pressures represented by the lines BA, CA, and GA, manifestly be pressures in equilibrium. Complete the parallelogram BG, then is the resultant of GA and BA in the direction FA; also since GA and BA are in equilibrium with CA, therefore this resultant is in equilibrium with CA, but when *two* pressures are in equilibrium, their directions are in the same straight line; therefore FAC is a straight line. But AC is parallel to BD, therefore FA is parallel to BD, and FB is, by construction, parallel to GD, therefore AFBD is a parallelogram, and AD is equal to FB and therefore to AG. But AG represents the resultant of CA and BA in magnitude, AD therefore represents it in *magnitude*. Therefore, &c.

The Principle of the Equality of Moments.

4. DEFINITION. If any number of pressures act in the same plane, and any point be taken in that plane, and perpendiculars be drawn from it upon the directions of all these pressures, produced if necessary, and if the number of units in each pressure be then multiplied by the number of units in the corresponding perpendicular, then this product is called the *moment* of that pressure *about* the point from which the perpendiculars are drawn, and these moments are said to be measured from that point.

5. *If three pressures be in equilibrium, and their moments be taken about any point in the plane in which they act, then the sum of the moments of those two pressures which tend to turn the plane in one direction about the point from which the moments are measured, is equal to the moment of that pressure which tends to turn it in the opposite direction.*

 Let P_1, P_2, P_3, acting in the directions P_1O, P_2O, P_3O, be any three pressures in equilibrium. Take any point A in the plane in which they act, and measure their moments from A, then will the sum of the moments of P_2 and P_3, which tend to turn the plane in one direction about A, equal the moment of P_1, which tends to turn it in the opposite direction.

Through A draw DAB parallel to OP_1, and produce OP_2 to meet it in D. Take OD to represent P_2, and take DB of such a length that OD may have the same proportion to DB that P_2 has to P_1. Complete the parallelogram ODBC, then will OD and OC represent P_2 and P_1 in magnitude and direction. Therefore OB will represent P_3 in magnitude and direction.

Draw AM, AN, AL, perpendiculars on OC, OD, OB, and join AO, AC. Now the triangle OBC is equal to the triangle OAC, since these triangles are upon the same base and between the same parallels.

Also, $\triangle\ ODA + \triangle\ OAB = \triangle\ ODB = \triangle\ OBC,$

$\therefore \triangle\ ODA + \triangle\ OAB = \triangle\ OAC,$

$\therefore\ \frac{1}{2}\ \overline{OD} \times \overline{AN} + \frac{1}{2}\ \overline{OB} \times \overline{AL} = \frac{1}{2}\ \overline{OC} \times \overline{AM},$

$\therefore\ P_2 \times \overline{AN} + P_3 \times \overline{AL} = P_1 \times \overline{AM}.$

Now $P_1 \times \overline{AM}$, $P_2 \times \overline{AN}$, $P_3 \times \overline{AL}$, are the moments of P_1, P_2, P_3, about A (Art. 4.)

$\therefore\ m^t\ P_2 + m^t\ P_3 = m^t\ P_1\ \dots\dots\dots$ (1).

Therefore, &c. &c.

6. If R be the resultant of P_2 and P_3, then since R is equal to P_1 and acts in the same straight line, $m^t\ R = m^t P_1$, $\therefore\ m^t\ P_2 + m^t\ P_3 = m^t\ R$. (8)

The sum of the moments therefore, about any point, of two pressures, P_2 and P_3, in the same plane, which tend to turn it in the same direction about that point, is equal to the moment of their resultant about that point.

If they had tended to turn it in opposite directions, then the *difference* of their moments would have equalled the moment of their resultant. For let R be the resultant of P_1 and P_3, which tend to turn the plane in opposite directions about A, &c. Then is R equal to P_2, and in the same straight line with it, therefore moment R is equal to moment P_2.

But by equation (1) $m^t P_1 - m^t P_3 = m^t P_2$; $\therefore m^t P_1 - m^t P_3 = m^t R$.

Generally therefore, $m^t P_1 + m^t P_2 = m^t R \ldots \ldots (2)$, *the moment therefore of the resultant of any two pressures in the same plane is equal to the sum or difference of the moments of its components, according as they act to turn the plane in the same direction about the point from which the moments are measured, or in opposite directions.*

7. *If any number of pressures in the same plane be in equilibrium, and any point be taken, in that plane, from which their moments are measured, then the sum of the moments of those pressures which tend to turn the plane in one direction about that point is equal to the sum of the moments of those which tend to turn it in the opposite direction.*

Let P_1, P_2, $P_3 \ldots \ldots P_n$ be any number of pressures in the same plane which are in equilibrium, and A any point in the plane from which their moments are measured, then will the sum of the moments of those pressures which tend to turn the plane in one direction about A equal the sum of the moments of those which tend to turn it in the opposite direction.

Let R_1 be the resultant of P_1 and P_2,
$R_2 \ldots \ldots R_1$ and P_3,
$R_3 \ldots \ldots R_2$ and P_4,
&c. $\ldots \ldots$ &c.
$R_{n-1} \ldots \ldots R_{n-2}$ and P_n.

Therefore by the last proposition, it being understood that the moments of those of the pressures P_1, P_2, which tend to turn the plane to the left of A, are to be taken negatively, we have

$$m^t R_1 = m^t P_1 + m^t P_2,$$
$$m^t R_2 = m^t R_1 + m^t P_3,$$
$$m^t R_3 = m^t R_2 + m^t P_4,$$
$$\text{&c.} = \text{&c.} \quad \text{&c.}$$
$$m^t R_{n-1} = m^t R_{n-2} + m^t P_n.$$

Adding these equations together, and striking out the terms common to both sides, we have

$$m^t R_{n-1} = m^t P_1 + m^t P_2 + m^t P_3 + \ldots \ldots + m^t P_n \ldots (3),$$

where R_{n-1} is the resultant of all the pressures P_1, P_2, $\ldots \ldots P_n$.

But these pressures are in equilibrium; they have, therefore, no resultant.

$$\therefore R_{n-1} = 0 \therefore m^t R_{n-1} = 0,$$

$$\therefore m^t P_1 + m^t P_2 + m^t P_3 + \ldots \ldots m^t P_n = 0. \ldots (4).$$

Now in this equation the moments of those pressures which tend to turn the system to the left hand are to be taken negatively. Moreover, the sum of the negative terms must equal the sum of the positive terms, otherwise the whole sum could not equal zero. It follows, therefore, that the sum of the moments of those pressures which tend to turn the system to the right must equal the sum of the moments of those which tend to turn it to the left. Therefore, &c. &c.

8. *If any number of pressures acting in the same plane be in equilibrium, and they be imagined to be moved parallel to their existing directions, and all applied to the same point, so as all to act upon that point in directions parallel to those in which they before acted upon different points, then will they be in equilibrium about that point.*

For (see the preceding figure) the pressure R_1 at whatever point in its direction it be conceived to be applied, may be resolved at that point into two pressures parallel and equal to P_1 and P_2: similarly, R_2 may be resolved, at any point in its direction, into two pressures parallel and equal to R_1 and P_3, of which R_1 may be resolved into two, parallel and equal to P_1 and P_2, so that R_2 may be resolved at any point of its direction into three pressures parallel and equal to P_1, P_2, P_3: and in like manner, R_3 may be resolved into two pressures parallel and equal to R_2 and P_4, and therefore into four pressures parallel and equal to P_1, P_2, P_3, P_4, and so of the rest. Therefore R_{n-1} may at any point of its direction be resolved into n pressures parallel and equal to $P_1, P_2, P_3, \ldots \ldots P_n$; if, therefore, n such pressures were applied to that point, they would just be held in equilibrium by a pressure equal and opposite to R_{n-1}. But $R_{n-1} = 0$; these n pressures would, therefore, be in equilibrium with one another if applied to this point.

Now it is evident, that if being thus applied to *this* point, they would be in equilibrium, they would be in equilibrium if similarly applied to any other point. Therefore, &c.

The Polygon of Pressures.

9. *The conditions of the equilibrium of any number of pressures applied to a point.*

Let OP_1, OP_2, OP_3, &c. represent in magnitude and direction pressures P_1, P_2, &c. applied to the same point O. Complete the parallelogram OP_1AP_2, and draw its diagonal OA; then will OA represent in magnitude and direction the resultant of P_1 and P_2. Complete the parallelogram $OABP_3$, then will OB represent in magnitude and direction the resultant of OA and P_3; but OA is the resultant of P_1 and P_2, therefore OB is the resultant of P_1, P_2, P_3; similarly, if the parallelogram $OBCP_4$ be completed, its diagonal OC represents the resultant of OB and P_4, that is, of P_1, P_2, P_3, P_4, and in like manner OD, the diagonal of the parallelogram $OCDP_5$, represents the resultant of P_1, P_2, P_3, P_4, P_5.

Now let it be observed, that AP_1 is equal and parallel to OP_2, AB to OP_3, BC to OP_4, CD to OP_5, so that P_1A, AB, BC, CD, represent P_2, P_3, P_4, P_5, respectively in magnitude, and are parallel to their directions. Moreover OP_1 is in the direction of P_1 and represents it in magnitude, so that the sides OP_1, P_1A, AB, BC, CD, of the polygon OP_1ABCDO, represent the pressures P_1, P_2, P_3, P_4, P_5, respectively in magnitude, and are parallel to their directions; whilst the side

OD, which completes that polygon, represents the resultant of those pressures in magnitude and direction.

If, therefore, the pressures P_1, P_2, P_3, P_4, P_5, be in equilibrium, so that they have no resultant, then the side OD of the polygon must vanish, and the point D coincide with O. Thus then if any number of pressures be applied to a point, and lines be drawn parallel to the directions of those pressures, and representing them in magnitude, so as to form sides of a polygon (care being taken to draw each line from the point where it unites with the preceding, *towards* the direction in which the corresponding pressure acts), then the line thus drawn parallel to the last pressure and representing it in magnitude, will pass through the point from which the polygon commenced, and will just complete it if the pressures be in equilibrium; and if they be *not* in equilibrium, then this last line will not complete the polygon, and if a line be drawn completing it, that line will represent the resultant of all the pressures in magnitude and direction.

This principle is that of the POLYGON OF PRESSURES; it obtains in respect to pressures applied to the same point, whether they be in the same plane or not.

10. *If any number of pressures in the same plane be in equilibrium, and each be resolved in directions parallel to any two rectangular axes, then the sum of all those resolved pressures, whose tendency is to communicate motion in one direction along either axis, is equal to the sum of those whose tendency is in the opposite direction.*

Let the polygon of pressures be formed in respect to any number of pressures, P_1, P_2, P_3, P_4, in the same plane and in equilibrium (Arts. 8, 9.), and let the sides of this polygon be *projected* on any straight line Ax in the same plane. Now it is evident, that the sum of the projections of those sides of the polygon which form that side of the figure which is nearest to Ax, is equal to the sum of the projections of those sides which form the opposite side of the polygon: moreover, that the former are those sides of the polygon which represent pressures tending to communicate motion from A towards x, or from left to right in respect to the line Ax; and the latter, those which tend to communicate motion in the opposite direction. Now each projection is equal to the corresponding side of the polygon, multiplied by the cosine of its inclination to Ax. The sum of all those sides of the polygon which represent pressures tending to communicate motion from A towards x, multiplied each by the cosine of its inclination to Ax, is equal, therefore, to the sum of all the sides representing pressures whose tendency is in the opposite direction, each being similarly multiplied by the cosine of its inclination to Ax. Now the sides of the polygon represent the pressures in magnitude, and are inclined at the same angles to Ax. Therefore each pressure being multiplied by the cosine of its inclination to Ax, the sum of all these products in respect to those which tend to communicate motion in one direction equals the sum similarly taken in respect to those which tend to communicate motion in the opposite direction; or, if in taking this sum it be understood that each term into which there enters a pressure whose tendency is from A towards x, is to be taken positively, whilst each into which there enters a pressure which tends from x towards A is to be taken negatively, then the sum of all these terms will equal zero; that is, calling the inclinations of the directions of P_1, P_2, P_3 ... P_4 to Ax, α_1, α_2, α_3 .... α_n respectively,

$$P_1 \cos. \alpha_1 + P_2 \cos. \alpha_2 + P_3 \cos. \alpha_3 + \ldots + P_n \cos. \alpha_n = 0 \ldots (5),$$

in which expression all those terms are to be taken negatively which include pressures, whose tendency is from x towards A.

This proposition being true in respect to any axis, Ax is true in respect to another axis, to which the inclinations of the directions of the pressures are represented by β_1, β_2, β_3, ...β_n, so that,

$$P_1 \cos. \beta_1 + P_2 \cos. \beta_2 + \ldots + P_n \cos. \beta_n = 0.$$

Let this second axis be at right angles to the first :

then $\beta_1 = \dfrac{\pi}{2} - \alpha_1 \;\therefore\; \cos. \; \beta_1 = \sin. \; \alpha_1, \; \beta_2 = \dfrac{\pi}{2} - \alpha_2 \;\therefore\; \cos. \; \beta_2$

$= \sin. \; \alpha_2,$ &c. $=$ &c.

$\therefore P_1 \sin. \; \alpha_1 + P_2 \sin. \; \alpha_2 + \;.\;.\;.\;. + P_n \sin. \; \alpha_n = 0 \;.\;.\;.\;. (6)$;

those terms in this equation, involving pressures which tend
to communicate motion in one direction, in respect to the
axis Ay being taken with the positive sign, and those which
tend in the opposite direction with the negative sign.

If the pressures P_1, P_2, &c. be each of them resolved
into two others, one of which is parallel to the axis Ax, and
the other to the axis Ay, it is evident that the pressures
thus resolved parallel to Ax, will be represented by $P_1 \cos \alpha_1$,
$P_2 \cos. \; \alpha_2$, &c., and those resolved parallel to Ay, by
$P_1 \sin. \; \alpha_1$, $P_2 \sin. \; \alpha_2$, &c. Thus then it follows, that if
any system of pressures in equilibrium be thus resolved
parallel to two rectangular axes, the sum of those resolved
pressures, whose tendency is in one direction along either
axis, is equal to the sum of those whose tendency is in the opposite direction.

This condition, and that of the equality of moments, are
necessary to the equilibrium of any number of pressures in
the same plane, and they are together *sufficient* to that equilibrium.

11. *To determine the resultant of any number of pressures
in the same plane.*

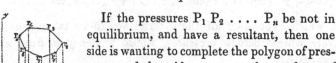

If the pressures $P_1 \; P_2 \;.\;.\;.\;.\; P_n$ be not in
equilibrium, and have a resultant, then one
side is wanting to complete the polygon of pressures, and that side represents the resultant of
all the pressures in magnitude, and is parallel to its direction
(Art. 9.). Moreover it is evident, that in this case the sum of
the projections on Ax (Art. 10.) of those lines which form one
side of the polygon, will be deficient of the sum of those of
the lines which form the other side of the polygon, by the

projection of this last deficient side ; and therefore, that the sum of the resolved pressures acting in one direction along the line Ax, will be less than the sum of the resolved pressures in the opposite direction, by the resolved part of the resultant along this line. Now if R represent this resultant, and θ its inclination to Ax, then R cos. θ is the resolved part of R in the direction of Ax. Therefore the signs of the terms being understood as before, we have

$$R \cos. \theta = P_1 \cos. \alpha_1 + P_2 \cos. \alpha_2 + \ldots + P_n \cos. \alpha_n \quad \ldots (7).$$

And reasoning similarly in respect to the axis Ay, we have

$$R \sin. \theta = P_1 \sin. \alpha_1 + P_2 \sin. \alpha_2 + \ldots + P_n \sin. \alpha_n \ldots (8).$$

Squaring these equations and adding them, and observing that $R^2 \sin.^2 \theta + R^2 \cos.^2 \theta = R^2 (\sin.^2 \theta + \cos.^2 \theta) = R^2$, we have

$$R^2 = (\Sigma P \sin. \alpha)^2 + (\Sigma P \cos. \alpha)^2 \quad \ldots \ldots \ldots (9),$$

where $\Sigma P \sin. \alpha$ is taken to represent the sum $P_1 \sin. \alpha_1 + P_2 \sin. \alpha_2 + P_3 \sin. \alpha_3 + \&c.$, and $\Sigma P \cos. \alpha$ to represent the sum $P_1 \cos. \alpha_1 + P_2 \cos. \alpha_2 + P_3 \cos. \alpha_3 + \&c.$

Dividing equation (8) by equation (7),

$$\tan. \theta = \frac{\Sigma P \sin. \alpha}{\Sigma P \cos. \alpha} \quad \ldots \ldots \ldots (10).$$

Thus then by equation (9) the magnitude of the resultant R is known, and by equation (10) its inclination θ to the axis Ax is known. In order completely to determine it, we have yet to find the perpendicular distance at which it acts from the given point A. For this we must have recourse to the condition of the equality of moments (Art. 7.).

If the sum of the moments of those of the pressures, P_1, $P_2 \ldots P_n$, which tend to turn the system in one direction about A, do not equal the sum of the moments of those which tend to turn it the other way, then a pressure being applied to the system, equal and opposite to the resultant R, will bring about the equality of these two sums, so that the moment of R must be equal to the difference of these sums.

Let then p equal the perpendicular distance of the direction of R from A.   Therefore

$$R p = m^t P_1 + m^t P_2 + m^t P_3 + \ldots + m^t P_n \ldots (11),$$

in the second member of which equation the moments of those pressures are to be taken negatively, which tend to communicate motion round A towards the left.

Dividing both sides by R we have

$$p = \frac{m^t P_1 + m^t P_2 + \ldots + m^t P_n}{R} \ldots (12).$$

Thus then by equations (9), (10), (12), the magnitude of the resultant R, its inclination to the given axis Ax, and the perpendicular distance of its direction from the point A, are known ; and thus the resultant pressure is completely determined in magnitude and direction.

The Parallelopipedon of Pressures.

12. *Three pressures,* P$_1$, P$_2$, P$_3$, *being applied to the same point* A, *in directions* xA, yA, zA, *which are not in the same plane, it is required to determine their resultant.*

Take the lines P$_1$A, P$_2$A, P$_3$A, to represent the pressures P$_1$, P$_2$, P$_3$, in magnitude and direction. Complete the parallelopipedon RP$_2$P$_3$P$_1$, of which AP$_1$, AP$_2$, AP$_3$, are adjacent edges, and draw its diagonal RA ; then will RA represent the resultant of P$_1$, P$_2$, P$_3$, in direction and magnitude.   For since P$_1$SP$_2$A is a parallelogram, whose adjacent sides P$_1$A, P$_2$A, represent the pressures P$_1$ and P$_2$ in magnitude and direction, therefore its diagonal SA represents the resultant of these two pressures.   And similarly RA, the diagonal of the parallelogram RSAP$_3$, represents in magnitude and direction the resultant of SA and P$_3$, that is, of P$_1$, P$_2$, and P$_3$, since SA is the resultant of P$_1$ and P$_2$.

It is evident that the fourth pressure necessary to produce an equilibrium with P$_1$, P$_2$, P$_3$, being equal and opposite to their resultant, is represented in magnitude and direction by AR.

13. *Three pressures,* P_1, P_2, P_3, *being in equilibrium, it is required to determine the third* P_3 *in terms of the other two, and their inclination to one another.*

Let AP_1 and AP_2 represent the pressures P_1 and P_2 in magnitude and direction, and let the inclination P_1AP_2 of P_1 to P_2 be represented by $_1\theta_2$. Complete the parallelogram AP_1RP_2, and draw its diagonal AR. Then does AR represent the resultant of P_1 and P_2 in magnitude and direction. But this resultant is in equilibrium with P_3, since P_1 and P_2 are in equilibrium with P_3. It acts, therefore, in the same straight line with P_3, but in an opposite direction, and is equal to it. Since then AR represents this resultant in magnitude and direction, therefore RA represents P_3 in magnitude and direction.

Now, $\overline{AR^2} = \overline{AP_1^2} - 2\overline{AP_1} \cdot \overline{P_1R} \cdot \cos. AP_1R + \overline{P_1R^2}$;

also, $AP_1R = \pi - P_1AP_2 = \pi - _1\theta_2$, $P_1R = AP_2$, and AP_1, AP_2, AR, represent P_1, P_2, P_3, in magnitude,

$$\therefore \quad P_3^2 = P_1^2 - 2P_1P_2 \cos. (\pi - _1\theta_2) + P_2^2.$$

Now cos. $(\pi - _1\theta_2) = -\cos. _1\theta_2$, $\therefore P_3^2 = P_1^2 + 2P_1P_2 \cos. _1\theta_2 + P_2^2$,

$$\therefore \quad P_3 = \sqrt{P_1^2 + 2P_1P_2 \cos. _1\theta_2 + P_2^2}. \quad \ldots \ldots (13).$$

14. *If three pressures,* P_1, P_2, P_3, *be in equilibrium, any two of them are to one another inversely as the sines of their inclinations to the third.*

Let the inclination of P_1 to P_3 be represented by $_1\theta_3$, and that of P_2 to P_3 by $_2\theta_3$.

Now $P_1AR = \pi - P_1AP_3 = \pi - _1\theta_3$, $\therefore$ sin. $P_1AR = $ sin. $_1\theta_3$;
$P_1RA = P_2AR = \pi - P_2AP_3 = \pi - _2\theta_3$, $\therefore$ sin. $P_1RA = $ sin. $_2\theta_3$.

c

Also, $$\frac{AP_1}{AP_2} = \frac{AP_1}{P_1R} = \frac{\sin. P_1RA}{\sin. P_1AR},$$

$$\therefore \frac{P_1}{P_2} = \frac{\sin. {}_2\theta_3}{\sin. {}_1\theta_3} \ldots \ldots (14).$$

That is, P_1 is to P_2 inversely, as the sine of the inclination of P_1 to P_3 is to the sine of the inclination of P_2 to P_3. Therefore, &c. &c.   [Q. E. D.]

OF PARALLEL PRESSURES.

15. *The principle of the equality of moments obtains in respect to pressures in the same plane whatever may be their inclinations to one another, and therefore if their inclinations be infinitely small, or if they be parallel.*

In this case of parallel pressures, the same line AB, which is drawn from a given point A, perpendicular to one of these pressures, is also perpendicular to all the rest, so that the perpendiculars are here the parts of this line AM_1, AM_2, &c. intercepted between the point A and the directions of the pressures respectively. The principle is not however in this case true only in respect to the intercepted parts of this perpendicular line AB, but in respect to the intercepted parts of *any* line AC, drawn through the point A across the directions of the pressures, since the intercepted parts Am_1, Am_2, Am_3, &c. of this second line are proportional to those, AM_1, AM_2, &c. of the first.

Thus taking the case represented in the figure, since by the principle of the equality of moments we have,

$$\overline{AM_1} \cdot P_1 + \overline{AM_4} \cdot P_4 = \overline{AM_2} \cdot P_2 + \overline{AM_3}P_3 + \overline{AM_5}P_5;$$

dividing both sides by $\overline{AM_5}$,

$$\frac{AM_1}{AM_5} \cdot P_1 + \frac{AM_4}{AM_5} \cdot P_4 = \frac{AM_2}{AM_5} \cdot P_2 + \frac{AM_3}{AM_5} \cdot P_3 + P_5.$$

But by similar triangles, $\dfrac{AM_1}{AM_5} = \dfrac{Am_1}{Am_5}$, $\dfrac{AM_2}{AM_5} = \dfrac{Am_2}{Am_5}$, &c. = &c.

$$\therefore \frac{Am_1}{Am_5} . P_1 + \frac{Am_4}{Am_5} . P_4 = \frac{Am_2}{Am_5} . P_2 + \frac{Am_3}{Am_5} . P_3 + P_5.$$

Therefore multiplying by Am_5,

$$\overline{Am_1} . P_1 + \overline{Am_4} . P_4 = \overline{Am_2} . P_2 + \overline{Am_3} . P_3 + \overline{Am_5} . P_5.$$

Therefore, &c. [Q.E.D.]

16. *To find the resultant of any number of parallel pressures in the same plane.*

It is evident that if a pressure equal and opposite to the resultant were added to the system, the whole would be in equilibrium. And being in equilibrium it has been shown (Art. 8.), that if the pressures were all moved from their present points of application, so as to remain parallel to their existing directions, and applied to the *same point*, they are such as would be in equilibrium about that point. But being thus moved, these parallel pressures would all have their directions in the same straight line. Acting therefore all in the same straight line, and being in equilibrium, the sum of those pressures whose tendency is in one direction along that line must equal the sum of those whose tendency is in the opposite direction. Now one of these sums includes the resultant R. It is evident then that before R was introduced the two sums must have been unequal, and that R equals the excess of the greater sum over the less; and generally, that if ΣP represent the sum of any number of parallel pressures, those whose tendency is in one direction being taken with the positive sign, and those whose tendency is in the opposite direction, with the negative sign; then

$$R = \Sigma P \quad \ldots \ldots (15).$$

the sign of R indicating whether it act in the direction of those pressures which are taken positively, or those which are taken negatively.

Moreover since these pressures, including R, are in equilibrium, therefore the sum of the *moments* about any point, of

c 2

those whose tendency is to communicate motion in one direction, must equal the sum of the moments of the rest, — these moments being measured on *any* line, as AC; but one of these sums includes the moment of R; these two sums must therefore, before the introduction of R, have been unequal, and the moment of R must be equal to the *excess* of the greater sum over the less, so that, representing the sum of the moments of the pressures (R not being included) by $\Sigma\, m^t\, P$, those whose tendency is to communicate motion in one direction, having the positive sign, and the rest the negative ; and representing by x the distance from A, measured along the line AC, at which R intersects that line, we have, since xR is the moment of R, xR$=\Sigma\, m^t\, P$, where the sign of xR indicates the direction in which R tends to turn the system about A, but R$=\Sigma$P,

$$\therefore\; x = \frac{\Sigma\, m^t\, P}{\Sigma P} \;\ldots\ldots\; (16).$$

Equations (15) and (16) determine completely the magnitude and the direction of the resultant of a system of parallel pressures in the same plane.

17. *To determine the resultant of any number of parallel pressures not in the same plane.*

Let P_1 and P_2 be the points of application of any two of these pressures, and let the pressures themselves be represented by P_1 and P_2. Also let their resultant R_1 intersect the line joining the points P_1 and P_2 in the point R_1; produce the line P_1, P_2, to intersect any plane given in position, in the point L. Through the points P_1, P_2, and R_1, draw P_1M_1, P_2M_2, and R_1N_1 perpendicularly to this plane: these lines will be in the same plane with one another and with P_1L; let the intersection of this last mentioned plane with the first be LM_1.

then will P_1M_1, P_2M_2, and R_1N_1 be perpendiculars to LM_1; moreover by the last proposition,

$$P_1 \overline{LP_1} + P_2 \overline{LP_2} = R_1 \overline{LR_1};$$

$$\therefore P_1 . \frac{LP_1}{LR_1} + P_2 . \frac{LP_2}{LR_1} = R_1.$$

But by similar triangles,

$$\frac{LP_1}{LR_1} = \frac{P_1M_1}{R_1N_1}, \qquad \frac{LP_2}{LR_1} = \frac{P_2M_2}{R_1N_1},$$

$$\therefore P_1 . \frac{P_1M_1}{R_1N_1} + P_2 . \frac{P_2M_2}{R_1N_1} = R_1.$$

Let now the resultant, R_2, of R_1 and P_3 intersect the line joining the points R_1 and P_3 in the point R_2, and similarly let the resultant, R_3, of R_2 and P_4 intersect the line joining the points R_2 and P_4 in the point R_3, and so on: then by the last equation,

$$P_1 . \overline{P_1M_1} + P_2 . \overline{P_2M_2} = R_1 \overline{R_1N_1}.$$

Similarly,

$$R_1 . \overline{R_1N_1} + P_3 . \overline{P_3M_3} = R_2 \overline{R_2N_2},$$

$$R_2 . \overline{R_2N_2} + P_4 . \overline{P_4M_4} = R_3 \overline{R_3N_3},$$

$$\&c. \quad + \quad \&c. \quad = \quad \&c.$$

$$R_{n-2} . \overline{R_{n-2}N_{n-2}} + P_n . \overline{P_nM_n} = R_{n-1} . \overline{R_{n-1}N_{n-1}}.$$

Adding these equations, and striking out terms common to both sides,

$$P_1 . \overline{P_1M_1} + P_2 \overline{P_2M_2} + \ldots + P_n . \overline{P_nM_n} = R_{n-1} . \overline{R_{n-1}N_{n-1}} \quad (17).$$

Now, $R_1 = P_1 + P_2,$ $R_2 = R_1 + P_3 = P_1 + P_2 + P_3,$

$$R_3 = R_2 + P_4 = P_1 + P_2 + P_3 + P_4, \&c. = \&c.$$

$$R_{n-1} = P_1 + P_2 + P_3 + \ldots + P_n;$$

$$\therefore \overline{R_{n-1}N_{n-1}} . \overline{P_1 + P_2 + P_3 + \&c. + P_n} = P_1 . \overline{P_1M_1} + P_2 . \overline{P_2M_2} + \ldots + P_n . \overline{P_nM_n};$$

$$\therefore \overline{R_{n-1}N}_{n-1} = \frac{P_1 \overline{P_1M_1} + P_2 \overline{P_2M_2} + \ldots + P_n . \overline{P_nM}_n}{P_1 + P_2 + P_3 + \ldots + P_n} \quad (18);$$

in which expression those of the parallel pressures P_1, P_2, &c. which tend in one direction, are to be taken positively, whilst those which tend in the opposite direction are to be taken negatively.

The line $R_{n-1}N_{n-1}$ represents the perpendicular distance from the given plane of a point through which the resultant of all the pressures P_1, P_2 . . . . P_n, passes. In the same manner may be determined the distance of this point from any other plane. Let this distance be thus determined in respect to three given planes at right angles to one another. Its actual position in space will then be known. Thus then we shall know a point through which the resultant of all the pressures passes, also the direction of that resultant, for it is parallel to the common direction of all the pressures, and we shall know its amount, for it is equal to the sum of all the pressure with their proper signs. Thus then the resultant pressure will be completely known. The point R_{n-1} is called the CENTRE OF PARALLEL PRESSURES.

18. The product of any pressure by its perpendicular distance from a plane (or rather the product of the number of units in the pressure by the number of units in the perpendicular), is called *the moment of the pressure, in respect to that plane.* Whence it follows from equation (17) that *the sum of the moments of any number of parallel pressures in respect to a given plane is equal to the moment of their resultant in respect to that plane.*

19. It is evident, from equation (17), that the distance $R_{n-1}\ N_{n-1}$ of the *centre of pressure* of any number of parallel pressures from a given plane, is independent of the *directions* of these parallel pressures, and is dependent wholly upon their amounts and the perpendicular distances P_1M_1, P_2M_2, &c. of their points of application from the given plane.

So that if the directions of the pressures were changed, provided that their amounts and points of application remained the same, *their centre of pressure*, determined as above, would remain unchanged; that is, the resultant, although it would alter its direction with the directions of the component pressures, would, nevertheless, always pass through the *same point.*

The *weights* of any number of different bodies or different parts of the same body, constitute a system of parallel pressures; the direction, therefore, through this system of the resultant weight may be determined by the preceding proposition; their centre of pressure is their *centre of gravity.*

THE CENTRE OF GRAVITY.

20. *The resultant of the weights of any number of bodies or parts of the same body united into a system of invariable form passes through the same point in it, into whatever position it may be turned.*

For the effect of turning it into different positions is to cause the directions of the weights of its parts to traverse the heavy body or system in different directions, at one time *lengthwise* for instance, at another *across,* at another *obliquely;* and the effect upon the direction of the resultant weight through the body, produced by thus turning it into different positions, and thereby changing the directions in which the weights of its component parts traverse its mass, is manifestly the same as would be produced, if without altering the position of the body, the direction of *gravity* could be *changed* so as, for instance, to make it at one time traverse that body longitudinally, at another obliquely, at a third transversely. But by Article 19. this *last* mentioned change, altering the common direction of the parallel pressures through the body without altering their amounts or their points of application, would not alter the position of their centre of pressure *in the body;* therefore, neither would the *first* mentioned change. Whence it follows that the *centre of pressure* of the weights of the parts of a heavy body, or of a system of invariable form, does not alter its position *in the body,* whatever may be the position into which

c 4

the body is turned; or in other words, that the resultant of the weights of its parts passes always through the same point in the body or system in whatever position it may be placed.

This point, through which the resultant of the weights of the parts of a body, or system of bodies of invariable form, passes, in whatever position it is placed; or, if it be a body or system of *variable* form, through which the resultant *would* pass, in whatever position it were placed, if it became rigid or invariable in its form, is called the CENTRE OF GRAVITY.

21. Since the weights of the parts of a body act in parallel directions, and all tend in the same direction, therefore their resultant is equal to their sum. Now, the resultant of the weights of the parts of the body would produce, singly, the same effect as it regards the conditions of the equilibrium of the body, that the weights of its parts actually do collectively, and this weight is equal to the sum of the weights of the parts, that is, to the whole weight of the body, and in every position it acts vertically downwards through the same point in the body, viz. the centre of gravity. Thus then it follows, that *in every position of the body and under every circumstance, the weights of its parts produce the same effect in respect to the conditions of its equilibrium, as though they were all collected in and acted through that one point of it — its centre of gravity.* *

* That the resultant of the weights of all the parts of a rigid body passes in all the positions of that body through the same point in it is a property of many and most important uses in the mechanism of the universe, as well as in the practice of the arts; another proof of it is therefore subjoined, which may be more satisfactory to some readers than that given in the text. The system being rigid, the distance P_1, P_2, of the points of application of any two of the pressures remains the same, into whatever position the body may be turned: the only difference produced in the circumstance under which they are applied is an alteration in the inclinations of these pressures to the line P_1, P_2: now being weights, the directions of these pressures always remain parallel to one another, whatever may be their inclination; thus

22. *To determine the position of the centre of gravity of two weights, P_1 and P_2, forming part of a rigid system.*

Let it be represented by G. Then since the resultant of P_1 and P_2 passes through G, we have by equation (16.), taking P_1 as the point from which the moments are measured,

$$\overline{P_1 + P_2} \cdot \overline{P_1 G} = P_2 \cdot \overline{P_1 P_2},$$

$$\therefore \quad P_1 G = \frac{P_2 \cdot \overline{P_1 P_2}}{P_1 + P_2};$$

whence the position of G is known.

23. *It is required to determine the centre of gravity of three weights P_1, P_2, P_3, not in the same straight line, and forming part of a rigid system.*

Find the centre of gravity G_1, of P_1 and P_2, as in the last proposition. Suppose the weights P_1 and P_2 to be collected in G_1, and find as before the common centre of gravity G_2 of this weight $P_1 + P_2$, so collected in G_1, and the third weight P_3. It is evident that this point G_2 is the centre of gravity required.

---

then it follows by the principle of the equality of moments (Art.15.), that $\overline{P_1 + P_2} \cdot \overline{P_1 R_1} = P_2 \cdot \overline{P_1 P_2}$, so that for every such inclination of the pressures to $P_1 P_2$, the line $P_1 R_1$ is of the same length, and the the point R_1 therefore the same point ; therefore, the line $P_3 R_1$ is always the same line in the body; and R_1 which equals $P_1 + P_2$, is always the same pressure, as also is P_3, and these pressures always remain parallel, therefore, for the same reason as before, R_2 is always the same point in the body in whatever position it may be turned, and so of R_3, R_4 . . . . . and R_{n-1}. That is, in every position of the body, the resultant of the weights of its parts passes through the same point R_{n-1} in it. Since the resultant of the weights of the parts of a body always passes through its centre of gravity, it is evident, that a single force applied at that point equal and opposite to this resultant, that is, equal in amount to the whole weight of the body, and in a direction vertically upwards, would in every position of the body sustain it. This *property* of the centre of gravity, viz. that it is a point in the body where a *single* force would support it, is sometimes taken as the definition of it.

Since G_2 is the centre of gravity of P_3 and $P_1 + P_2$ collected in G_1, we have by the last proposition

$$\overline{G_1 G_2} \cdot \overline{P_1 + P_2 + P_3} = \overline{G_1 P_3} \cdot P_3,$$

$$\therefore \quad \overline{G_1 \ G_2} = \frac{\overline{G_1 P_3} \cdot P_3}{P_1 + P_2 + P_3}.$$

If P_1, P_2, P_3, be all equal, then

$$G_1 G_2 = \tfrac{1}{3} \ \overline{G_1 P_3}.$$

Moreover in this case,

$$\overline{P_1 G_1} = \tfrac{1}{2} \ \overline{P_1 P_2}.$$

24. *To find the centre of gravity of four weights not in the same plane.*

Let P_1, P_2, P_3, P_4, represent these weights; find the centre of gravity G_2 of the weights P_1, P_2, P_3, as in the last proposition; suppose these three weights to be collected in G_2, and then find the centre of gravity G_3 of the weight thus collected in G_2 and P_4. G_3 will be the centre of gravity required, and since G_3 is the centre of gravity of P_4 acting at the point P_4, and of $P_1 + P_2 + P_3$ collected at G_2,

$$\therefore \quad \overline{G_2 G_3} \cdot \overline{P_1 + P_2 + P_3 + P_4} = \overline{G_2 P_4} \cdot P_4,$$

$$\therefore \quad \overline{G_2 G_3} = \frac{\overline{G_2 P_4} \cdot P_4}{P_1 + P_2 + P_3 + P_4}.$$

If all these weights be equal, then by the above equation,

$$\overline{G_2 G_3} = \tfrac{1}{4} \ \overline{G_2 P_4},$$
$$\text{also,} \quad \overline{G_1 G_2} = \tfrac{1}{3} \ \overline{G_1 P_3},$$
$$\text{and} \quad \overline{G_1 P_1} = \tfrac{1}{2} \ \overline{P_1 P_2}.$$

25. THE CENTRE OF GRAVITY OF A TRIANGLE.

Let the sides AB and BC of the triangular *lamina* ABC be bisected in E and D, and the lines CE and AD drawn to the opposite angles, then is the intersection G of these lines the centre of gravity of the triangle: for the triangle may be supposed to be made up of exceedingly narrow rectangular strips or bands, parallel to BC, each of which will be bisected by the line AD; for by similar triangles PR : DB::AR : AD::RQ : DC, therefore, alternando, PR : RQ::DB : DC; but DB=DC; therefore PR=RQ.

Therefore each of the elementary bands, or rectangles parallel to BC, which compose the triangle ABC, would separately balance on the line AD; therefore all of them joined together would balance on the line AD, therefore the centre of gravity of the triangle is in AD.

In the same manner it may be shown that the centre of gravity of the triangle is in the line CE; therefore the centre of gravity is at the intersection G of these lines.

Now $DG = \frac{1}{3} DA$: for imagine the triangle to be without weight, and three equal weights to be placed at the angles A, B, and C, then it is evident that these three weights will balance upon AD; for AD being supported, the weight A will be supported, since it is *in* that line; moreover, B and C will be supported since they are equidistant from that line.

Since, then, all three of the weights will balance upon AD, their centre of gravity is in AD. In like manner it may be shown that the centre of gravity of all three weights is in CE; therefore it is in G, and coincides with the centre of gravity of the triangle.

Now, suppose the weights B and C to be collected in their centre of gravity D, and suppose each weight to be represented in amount by A, a weight equal to 2A will then be collected in D, and a weight equal to A at A, and the centre of gravity of these is in G; therefore $DA \times A = DG \times (2A + A)$,

$\therefore DA = 3 DG$, or $DG = \frac{1}{3} DA$.   [Q.E.D.]

26. THE CENTRE OF GRAVITY OF THE PYRAMID.

 Let ABC be a pyramid, and suppose it to be made up of elementary laminæ *bcd*, parallel to the base BCD. Take G, the centre of gravity of the base BCD, and join AG; then A will pass through the centre of gravity *g* of the lamina

bcd*, therefore each of the laminæ will separately balance on
the straight line AG; therefore the laminæ when combined
will balance upon this line; therefore the whole figure will
balance on AG, and the centre of gravity of the whole is in
AG. In like manner if the centre of gravity H of the face ABD
be taken, and CH be joined, then it may be shown that the
centre of gravity of the whole is in CH; therefore the lines
AG and CH intersect, and the centre of gravity is at their
intersection K.

Now GK is one-fourth of GA; for suppose equal weights
to be placed at the angles A, B, C, and D of the pyramid
(the pyramid itself being imagined without weight), then
will these four weights balance upon the line AG, for one of
them A is *in* that line, and the line passes through the centre
of gravity G of the other three.

Since then the equal weights A, B, C, and D balance upon
the line AG, their centre of gravity is in AG; in the same
manner it may be shown that the centre of gravity of the
four weights is in CH, therefore it is in K, and coincides with
the centre of gravity of the pyramid.

Now let the number of units in each weight be represented
by A, and let the three weights B, C, and D be supposed to
be collected in their centre of gravity G; the four weights
will then be reduced to two, viz. 3A at G, and A at A,
whose common centre of gravity is K,

$$\therefore \quad GK \times \overline{3A+A} = GA \times A,$$
$$\therefore \quad 4GK = GA \text{ or } GK = \tfrac{1}{4} GA. \qquad [\text{Q.E.D.}]$$

27. *The centre of gravity of a pyramid with a polygonal base
is situated at a vertical height from the base, equal to one
fourth the whole height of the pyramid.*

For any such pyramid ABCDEF may be supposed to be

---

* For produce the plane ABG to intersect the plane ADC in AM, then
by similar triangles DM : MC :: *dm* : *mc*, but DM = MC; therefore *dm = mc*.
Also by similar triangles GM : BM :: *gm* : *bm*, but GM = $\tfrac{1}{3}$ BM; therefore
gm = $\tfrac{1}{3}$ *bm*. Since then *dm* = $\tfrac{1}{2}$ *dc* and *gm* = $\tfrac{1}{3}$ *bm*, therefore *g* is the centre
of gravity of the triangle *bdc*.

made up of triangular pyramids ABCF, ACDF, and ADEF, whose centres of gravity G, H, and K, are situated in lines AL, AM, and AN, drawn to the centres of gravity L, M, and N of their bases; LG being one fourth of LA, MH one fourth of MA, and NK one fourth of NA. The points G, H, and K, are therefore in a plane parallel to the base of the pyramid, and whose vertical distance from the base equals one fourth the vertical height of the pyramid.

Since then the centres of gravity G, H, and K of the elementary triangular pyramids which compose the whole polygonal pyramid are in this plane, therefore the centre of gravity of the whole is in this plane, *i.e.* the centre of gravity of the whole polygonal pyramid is situated at a vertical height from the base, equal to one fourth the vertical height of the whole pyramid, or at a vertical depth from the vertex, equal to three fourths of the whole. Now the above proposition is true, whatever be the number of the sides of the polygonal base, and therefore if they be infinite in number; and therefore it is true of the cone, which may be considered a pyramid having a polygonal base, of an infinite number of sides; and it is true whether the *cone* or *pyramid* be an *oblique* or a *right* *cone* or *pyramid.*

28. If a body be of a prismatic form, and symmetrical about a certain plane, then its whole weight may be supposed to be collected in the surface of that plane, and *uni-* 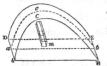 *formly* distributed through it. For let ACBEFD represent such a prismatic body, and *abc* a plane about which it is symmetrical : take *m*, an element of uniform thickness whose sides are parallel to the sides of the prism, and which is terminated by the faces ABC and DFE of the prism; it is evident that this element *m* will be bisected by the plane *abc*, and that its centre of gravity will therefore lie in that plane, so that its whole weight may be supposed collected in that plane; and this being true of every other similar element, and all these

elements being equal, it follows that the whole weight of the body may be supposed to be collected *in* and uniformly distributed *through* that plane. It is in this sense only that we can speak with accuracy of the weight and the centre of gravity of a *plane*, whereas a plane being a surface only, and having no thickness, can have no weight, and therefore no centre of gravity. In like manner when we speak of the centre of gravity of a curved surface, we mean the centre of gravity of a body, the weights of all whose parts may be supposed to be collected and uniformly distributed throughout that curved surface. It is evident that this condition is approached to whenever the body being hollow, its material is exceedingly thin. Its whole weight may then be conceived to be collected in a surface equidistant from its two external surfaces. In like manner an exceedingly thin uniform curved rod may be imagined to have its weight collected uniformly in a line passing along the centre of its thickness, and in this sense we may speak of the centre of gravity of a *line*, although a line having no breadth or thickness can have no weight, and therefore no centre of gravity.

29. THE CENTRE OF GRAVITY OF A TRAPEZOID.

Let AD and BC be the parallel sides of the trapezoid, of which AD is the less. Let AD be represented by a, BC by b, and the perpendicular distance NL of the two sides by h. Draw DE parallel to AB. Let G_1 be the intersection of the diagonals of the parallelogram ABED, then will G_1 be the centre of gravity of that parallelogram. Bisect CE in L, join DL, and take $DG_2 = \frac{2}{3} DL$, then will G_2 be the centre of gravity of the triangle DEC. Draw G_1M_1 and G_2M_2 perpendiculars to AD; then since $AG_1 = \frac{1}{2} AE$, therefore $G_1M_1 = \frac{1}{2} FE = \frac{1}{2} h$. And since $DG_2 = \frac{2}{3} DL$, therefore $G_2M_2 = \frac{2}{3} NL = \frac{2}{3} h$. Suppose the whole parallelogram to be collected in its centre of gravity G_1, and the whole triangle in its centre of gravity G_2. Let

G be the centre of gravity of the whole trapezoid, and draw GM perpendicular to AD. Then would the whole be supported by a single force equal to the weight of the trapezoid acting upwards at G. Therefore (Art. 17.),

$$\overline{MG} \cdot \overline{ABCD} = \overline{G_1M_1} \cdot \overline{ABED} + \overline{G_2M_2} \cdot \overline{CED}$$

Now, $\quad ABCD = \frac{1}{2} h (a+b),\ ABED = ha,$

$\quad CED = \frac{1}{2} h (b-a),\ G_1M_1 = \frac{1}{2} h \quad G_2M_2 = \frac{2}{3} h,$

$\therefore \overline{MG} \cdot \frac{1}{2} h (a+b) = \frac{1}{2} h \cdot ha + \frac{2}{3} h \cdot \frac{1}{2} h (b-a),$

$\therefore \overline{MG} (a+b) = ha + \frac{2}{3} h (b-a) = \frac{1}{3} h (a+2b),$

$$\therefore MG = \frac{1}{3} h \cdot \frac{a+2b}{a+b} \quad \ldots \ldots (19).$$

30. THE CENTRE OF GRAVITY OF ANY QUADRILATERAL FIGURE.

Draw the diagonals AC and BD of any quadrilateral figure ABCD, and let them intersect in E, and from the greater of the two parts, BE and DE, of either diagonal BD set off a part BF equal to the less part. Bisect the other diagonal AC in H, join HF and take HG equal to one third of HF; then will G be the centre of gravity of the whole figure.

For if not, let g be the centre of gravity, join HB and HD and take $HG_1 = \frac{1}{3} HB$ and $HG_2 = \frac{1}{3} HD$, then will G_1 and G_2 be the centres of gravity of the triangles ABC and ADC respectively (Art. 25.). Suppose these triangles to be collected in their centres of gravity G_1, G_2; it is evident that the centre of gravity g, of the whole figure, will be in the straight line joining the points G_1, G_2: let this line intersect AC in K; then since a pressure equal to the weight of the whole figure acting upwards at g, will be in equilibrium with the weights of the triangles collected in G_1 and G_2, we have, by the principle of the equality of moments. (Art. 15.)

$$\overline{Kg} \cdot \overline{ABCD} = \overline{KG_1} \cdot \overline{ABC} - \overline{KG_2} \cdot \overline{ADC}.$$

Now since $HG_1 = \frac{1}{3} HB$, and $HG_2 = \frac{1}{3} HD$, therefore G_1G_2 is parallel to DB, therefore $KG_1 = \frac{1}{3}$ BE, and $KG_2 = \frac{1}{3}$ DE. Now let the angle AED=BEC=ι. Therefore the perpendicular from B upon AC=BE sin. ι, and that from D=DE sin. , therefore area of triangle ABC=$\frac{1}{2}$ $\overline{AC}$. $\overline{BE}$ sin. ι, and area of triangle ADC=$\frac{1}{2}$ $\overline{AC}$. $\overline{DE}$ sin. ι, therefore area of quadrilateral ABCD=$\frac{1}{2}$ $\overline{AC}$. $\overline{BE}$ sin. ι + $\frac{1}{2}$ $\overline{AC}$. $\overline{DE}$ sin. $\iota = \frac{1}{2}$ (BE+DE) $\overline{AC}$ sin. ι. Substituting these values in the preceding equation,

$$\overline{Kg} \cdot \tfrac{1}{2} \, (\text{BE}+\text{DE}) \, \overline{AC} \text{ sin. } \iota = \tfrac{1}{3} \, \overline{BE} \cdot \tfrac{1}{2} \overline{AC} \cdot \overline{BE} \text{ sin. } \iota - \tfrac{1}{3} \, \overline{DE} \cdot \tfrac{1}{2} \, \overline{AC} \cdot \overline{DE} \text{ sin. } \iota,$$

$$\therefore \overline{Kg} \cdot (\text{BE}+\text{DE}) = \tfrac{1}{3} \, (\overline{\text{BE}^2} - \overline{\text{DE}^2}),$$

$$\therefore Kg = \tfrac{1}{3} \frac{\overline{\text{BE}^2} - \overline{\text{DE}^2}}{\overline{\text{BE}} + \overline{\text{DE}}} = \tfrac{1}{3} (\text{BE}-\text{DE}) = \tfrac{1}{3} (\text{BE}-\text{BF}) = \tfrac{1}{3} \text{FE}.$$

But since HG=$\frac{1}{3}$HF, $\therefore$ KG=$\frac{1}{3}$ FE, $\therefore$ Kg=KG; that is, the true centre of gravity g coincides with the point G. Therefore, &c. [Q.E.D.]

*31. In the examples hitherto given, the centre of pressure of a system of weights, or their centre of gravity, has been determined by methods which are *indirect* as compared with the direct and general method indicated in Article 17. That method supposes, however, a determination of the sum of the moments of the weights of all the various elements of the body in respect to three given planes. Now in a *continuous* body these elements are *infinite* in number, each being infinitely small; this determination supposes, therefore, the summation of an infinite number of infinitely small quantities, and requires an application of the principles of the integral calculus.

Let ΔM be taken to represent any small element of the volume M of a continuous body, and x its perpendicular distance from a given plane. Then will $x\mu\Delta$M represent the moment of the *weight* of this element about that plane, μ representing the weight of each *unit* of the volume M. Let $\mu\Sigma x\Delta$M represent the sum of all such moments, taken in respect to all the small elements, such as ΔM, which make up the volume of the body. Then if G_1 represent the distance of the centre of gravity of the body from the given plane; since $\mu\Sigma x\Delta M$ represents the sum of the *moments* of a system of parallel pressures about that plane, μM the sum of those pressures, and G_1 the distance of their centre of pressure from the plane (Art. 19.), it follows by equation (18.) that

$$G_1 = \frac{\mu\Sigma x \cdot \Delta M}{\mu M} = \frac{\Sigma x \cdot \Delta M}{M} \ \cdots \cdots \ (20).$$

Now it is proved in the theory of the integral calculus*, that a sum, such as is represented by the above expression $\Sigma x\Delta M$, whose terms are infinite in number, and each the product of a finite quantity x, and an infinitely small quantity ΔM, and in which M is, as in this case, a function of x (and therefore x a function of M), is equal to the definite integral $\int_{x_2}^{x_1} x d M$. Therefore, generally,

$$G_1 = \frac{\int_{x_2}^{x_1} x d M}{M}$$

Similarly, $\qquad G_2 = \dfrac{\int_{y_2}^{y_1} y d M}{M} \qquad \left.\right\} \ \cdots \cdots \ (21).$

$$G_3 = \frac{\int_{z_2}^{z_1} z d M}{M}$$

In the two last of which equations y and z are taken to represent, respectively, the distances of the element ΔM of the body from two other planes, as x represents its distance from

---

* Poisson, Journal de l'Ecole Polytechnique, 18me cahier, p. 320., or Art. 2. in the Treatise on Definite Integrals in the Encyclopædia Metropolitana by the author of this work. See Appendix, note A.

D

the first plane ; and G_1 and G_2 to represent the distances of its centre of gravity from those planes. The distances G_1, G_2, G_3, of the centre of gravity from three different planes being thus known, its actual position in space is fully determined. These three planes are usually taken at right angles to one another, and are then called rectangular co-ordinate planes, and their common intersections rectangular co-ordinate axes.

If the centre of gravity of the body be known to lie in a certain plane, and one of the co-ordinate planes spoken of above, as for instance that from which G_3 is measured, be taken to coincide with this plane in which the centre of gravity is known to lie, then $G_3 = 0$, and the position of the centre of gravity is determined by the two first only of the above three equations. This case occurs when the body, whose centre of gravity is to be determined, is *symmetrical* about a certain plane, since then its centre of gravity evidently lies in its plane of symmetry. If the centre of gravity of the body be known to lie in a certain *line*, and two of the co-ordinate planes, those for instance from which G_2 and G_3 are measured, be taken so as to intersect one another in that line, then the centre of gravity will be in both those planes ; therefore $G_2 = 0$ and $G_3 = 0$, and its position is determined by the first of the preceding equations alone. This case occurs when the body is *symmetrical* about a given line ; its centre of gravity is then manifestly in that line.

*32. THE CENTRE OF GRAVITY OF A CURVED LINE WHICH LIES WHOLLY IN THE SAME PLANE.

Taking M to represent the length S of such a line, we have, by equations (21),

$$G_1 = \frac{\int x dS}{S}, \qquad G_2 = \frac{\int y dS}{S} \cdots (22).$$

EXAMPLE. — *Let it be required to determine the centre of gravity of a circular arc* EF.

The centre of gravity of such an arc is evidently in the radius CA, which bisects it; since the arc is symmetrical about that radius. Take a plane Cy perpendicular to this radius, and passing through the centre, to measure the moments from. Let x represent the distance PM of any point P in this arc from this plane; also let s represent the arc PA, and S the arc EAF, a the radius CA, and C the chord EF.

$$\therefore \; x = \text{PM} = \overline{\text{CP}} \cos. \; \text{CPM} = \overline{\text{CP}} \cos. \; \text{ACP} = a \cos. \frac{s}{a};$$

$$\therefore \int x d\text{S} = a \int_{-\frac{1}{2}\text{S}}^{\frac{1}{2}\text{S}} \cos. \frac{s}{a} ds = a^2 \int_{-\frac{1}{2}\text{S}}^{\frac{1}{2}\text{S}} \cos. \frac{s}{a} d\left(\frac{s}{a}\right) = 2a^2 \sin. \left(\frac{\frac{1}{2}\text{S}}{a}\right),$$

the integral being taken between the limits $\frac{1}{2}$S and $-\frac{1}{2}$S, because these are the values of s which correspond to the *extreme* points F and E of the arc.

Now, $2a \sin.\frac{1}{2}\left(\dfrac{\text{S}}{a}\right) = $ chord of EAF $= $ C, $\therefore \int x d\text{S} = a$C,

$$\therefore \; \text{G}_1 = \frac{a\,\text{C}}{\text{S}} \; \cdots \cdots \; (23).$$

The distance of the centre of gravity of a circular arc from the centre of the circle is therefore a fourth proportional to the length of the arc, the length of the chord, and the radius of the arc.

*33. THE CENTRE OF GRAVITY OF A CURVILINEAR AREA WHICH LIES WHOLLY IN THE SAME PLANE.

Let BAC represent such an area. If x and y represent the perpendicular distances PN and PM of any point P in the curve AB from planes AC and AD, perpendicular to the plane of the given area and to one another, and M represent the area PAM, then, considering this area to be made up of rectangles parallel to PM, the width of each of which is represented by the exceedingly small quantity Δx, the volume ΔM of each such rectangle will be represented by $y\Delta x$, and its moment about AD by $\mu x y \Delta x$.

Therefore by equation (20), $G_1 = \dfrac{\Sigma x y \Delta x}{M} = \dfrac{\displaystyle\int^{x_1} x y\, dx}{M}$ . . (24).

A similar expression determines the value of G_2; but one more convenient for calculation is obtained, if we consider the weight of each of the rectangles, whose length is y, to be collected in its centre of gravity, whose distance from AC is $\frac{1}{2}y$. The moment of the weight of each rectangle about AC will then be represented $\frac{1}{2}\mu y^2 \Delta x$; whence it follows that

$$G_2 = \frac{\frac{1}{2}\mu \Sigma y^2 \Delta x}{\mu M} = \frac{1}{2}\,\frac{\displaystyle\int_{x_2}^{x_1} y^2\, dx}{M} \quad \ldots \ldots (25).$$

EXAMPLE. — *Suppose the curve* APB *to be a parabola, whose axis is* AC.

By the equation to the parabola $y^2 = 4ax$, if a be the distance of the focus from the vertex. Moreover, the limits between which the integral is to be taken are 0 and x_1 and 0 and y_1, since at A, $x=0$, $y=0$, and at C, $x=x_1$, $y=y_1$,

therefore $\displaystyle\int_{x_2}^{x_1} x y\, dx = 2\sqrt{a}\int_0^{x_1} x^{\frac{3}{2}} dx = \frac{4}{5}\sqrt{a}\, x_1^{\frac{5}{2}}$; also, $M = \displaystyle\int_0^{x_1} y\, dx =$

$2a\displaystyle\int_0^{x_1} x^{\frac{1}{2}} dx = \frac{4}{3}\sqrt{a}\, x_1^{\frac{3}{2}}$, therefore $G_1 = \frac{3}{5}\, x_1$.

Also, $\displaystyle\int_{x_2}^{x_1} y^2\, dx = 4a\int_0^{x_1} x\, dx = 2a x_1^2 = \frac{y_1^4}{8a}$, and $M = \frac{4}{3}\sqrt{a}\, x_1^{\frac{2}{3}} = \frac{y_1^3}{6a}$,

therefore $G_2 = \frac{3}{8}y$.

If, then, G be the centre of gravity of the parabolic area ACB, then $AH = \dfrac{3}{5}AC$, $HG = \dfrac{3}{8}CB$.

*34. THE CENTRE OF GRAVITY OF A SURFACE OF REVOLUTION.

Any surface of revolution BAC is evidently symmetrical about its axis of revolution AD, its centre of gravity is therefore in that axis. Let the moments be measured from a plane passing through A and perpendicular to the axis AD, and let x and y be co-ordinates of any point P in the generating curve APB of the surface, and s the length of the curve AP. Then M being taken to represent the area of the surface, and being supposed to be made up of bands parallel to PQ, the area ΔM of each such band is represented (see p. 44.)* by $2\pi y \Delta s$, and its moment by $2\pi \mu x y \Delta s$,

$$\therefore G_1 = \frac{2\pi \Sigma xy \Delta s}{M} = \frac{2\pi \displaystyle\int_{S_2}^{S_1} xy\, ds}{M} \quad \ldots \ldots (26).$$

EXAMPLE.— *To determine the centre of gravity of the surface of any zone or segment of a sphere.*

Let B_1AC_1 represent the surface of a sphere, whose centre is D, and whose radius DP is represented by a, and the arc AP by s. Then $x = DM = DP$ cos. $PDM = a \cos. \dfrac{s}{a}$, $y = PM = DP$ sin. PDM $= a$ sin.$\dfrac{s}{a}$, $\therefore 2xy = 2a^2$ sin. $\dfrac{s}{a}$ cos. $\dfrac{s}{a} = a^2$ sin.$\dfrac{2s}{a}$.

* Or Prof. Hall's Diff. Calculus, p. 168.

$$\therefore 2\pi \int_{S_2}^{S_1} xy\,ds = \pi a^2 \int_{S_2}^{S_1} \sin.\frac{2s}{a}\,ds$$

$$= \tfrac{1}{2}\,\pi a^3 \left\{ \cos.\frac{2S_2}{a} - \cos.\frac{2S_1}{a} \right\}$$

$$= \tfrac{1}{2}\pi a^3 \left\{ \left(1+\cos.\frac{2S_2}{a}\right) - \left(1+\cos.\frac{2S_1}{a}\right) \right\}$$

$$= \pi a^3 \left\{ \cos.^2\frac{S_2}{a} - \cos.^2\frac{S_1}{a} \right\} \quad \cdots \quad (27).$$

where S_1 and S_2 are the values of s at the points B_1 and B_2, where the zone is supposed to terminate.

Also, since $\dfrac{dM}{ds} = 2\pi y, \qquad \therefore M = 2\pi \int_{S_2}^{S_1} y\,ds.$

$$= 2\pi a \int_{S^2}^{S^1} \sin.\frac{s}{a}\,ds = 2\pi a^2 \left\{ \cos.\frac{S_2}{a} - \cos.\frac{S_1}{a} \right\},$$

$$\therefore G_1 = \tfrac{1}{2}a \left\{ \cos.\frac{S_2}{a} + \cos.\frac{S_1}{a} \right\}$$

$$= \tfrac{1}{2}\left\{ DE_2 + DE_1 \right\} = DE \quad \cdots \quad (28),$$

if E be the bisection of E_1E_2.

If $S_2 = 0$, or the zone commence from A, then

$$G_1 = \tfrac{1}{2}\,a\left\{ 1+\cos.\frac{S_1}{a} \right\} = a\cos.^2\frac{S_1}{2a} \quad \cdots \quad (29).$$

* 35. THE CENTRE OF GRAVITY OF A SOLID OF REVOLUTION.

Any solid of revolution BAC is evidently symmetrical

about its axis of revolution AD, its centre of gravity is therefore in that line; and taking a plane passing through A and perpendicular to that axis as the plane from which the moments are measured, we have only to determine the distance AG of the centre of gravity, from that plane.

Now, if x and y represent the co-ordinates of any point P in the generating curve, and M the volume of the portion PAQ of this solid, then, conceiving it to be made up of cylindrical laminæ parallel to PQ, the thickness of each of which is Δx, the volume of each is represented by $\pi y^2 \Delta x$, and its moment by $\pi \mu x y^2 \Delta x$.

$$\therefore \ G^1 = \frac{\pi \Sigma y^2 \Delta x}{M} = \frac{\pi \int_{x_2}^{x_1} x y^2 dx}{M} \ \ \cdots \cdots (30).$$

EXAMPLE.—*To determine the centre of gravity of any solid segment of a sphere.*

Let $B_1 A C_1$ represent any such segment of a sphere whose centre is D and its radius a. Let x and y represent the co-ordinates AM and MP of any point P, x being measured from A; then by the equation to the circle $y^2 = 2ax - x^2$,

$$\therefore \ \pi \int_{x_2}^{x_1} x y^2 dx = \pi \int_0^{x_1} x \ (2ax - x^2) dx = \pi(\tfrac{2}{3} a x_1{}^3 - \tfrac{1}{4} x_1{}^4).$$

Also, $M = \pi \int_{x_2}^{x_1} y^2 dx = \pi \int_0^{x_1} (2ax - x^2) dx = \pi(a x_1{}^2 - \tfrac{1}{3} x_1{}^3)$,

$$\therefore \ G_1 = \frac{\tfrac{2}{3} a x_1 - \tfrac{1}{4} x^2}{a - \tfrac{1}{3} x_1} = \tfrac{1}{4} x_1 \ . \ \left(\frac{8a - 3x_1}{3a - x_1} \right) \ \cdots \cdots (31).$$

If the segment become a hemisphere, $x_1 = a$, $\therefore \ G_1 = \tfrac{5}{8} a$.

36. *The centre of gravity of the sector of a circle.*

Let CAB represent such a sector; conceive the arc ADB to be a polygon of an infinite number of sides and lines, to be drawn from all the angles of the polygon to the centre C of the circle, these will divide the sector into as many triangles. Now the centre of gravity of each triangle will be at a distance from C equal to $\frac{2}{3}$ the line drawn from the vertex C of that triangle to the bisection of its base, that is equal to $\frac{2}{3}$ the radius of the circle, so that the centres of gravity of all the triangles will lie in a circular arc FE, whose centre is C and its radius CF equal to $\frac{2}{3}$CA, and the weights of the triangles may be supposed to be collected in this arc FE, and to be uniformly distributed through it, so that the centre of gravity G of the whole sector CAB is the centre of gravity of the circular arc FE. Therefore by equation (23), if S^1, C^1, and a^1, represent the arc FE, its chord FE, and its radius CF, and S, C, a, the similar arc, chord, and radius of ADB, then $CG = \dfrac{a^1 C^1}{S^1}$; but since the arcs AB and FE are similar, and that $a^1 = \frac{2}{3}a$, $\therefore$ $C^1 = \frac{2}{3}C$ and $S^1 = \frac{2}{3}S$. Substituting these values in the last equation, we have

$$CG = \tfrac{2}{3}\frac{aC}{S} \quad \cdots \cdots \quad (32).$$

37. *The centre of gravity of any portion of a circular ring or of an arch of equal voussoirs.*

Let $B_1 C_1 C_2 B_2$ represent any such portion of a circular ring whose centre is A. Let a_1 represent the radius, and C_1 the chord of the arc $B_1 C_1$, and S_1 its length, and let a_2, C_2 similarly represent the radius and chord of the arc $B_2 C_2$, and S_2 the length of that arc.

Also let G_1 represent the centre of gravity of the sector

AB_1C_1, G_2 that of the sector AB_2C_2, and G the centre of gravity of the ring. Then

$$\overline{AG_2} \times \text{sect.} AB_2C_2 + \overline{AG} \times \overline{\text{ring } B_1C_1B_2C_2} = \overline{AG_1} \times \overline{\text{sect.} AB_1C_1}.$$

Now (by equation 32), $AG_1 = \frac{2}{3}\dfrac{a_1C_1}{S_1}$, $\quad AG_2 = \frac{2}{3}\dfrac{a_2C_2}{S_2}$;

also sector $AB_1C_1 = \frac{1}{2}S_1a_1$, sector $AB_2C_2 = \frac{1}{2}S_2a_2$,

$\therefore$ ring $B_1C_1C_2B_2 = \text{sect.} AB_1C_1 - \text{sect.} AB_2C_2 = \frac{1}{2}S_1a_1 - \frac{1}{2}S_2a_2$,

$$\therefore \ \frac{2}{3}\frac{a_2C_2}{S_2} \cdot \frac{1}{2}S_2a_2 + \overline{AG}\, \frac{1}{2}(S_1a_1 - S_2a_2) = \frac{2}{3}\frac{a_1C_1}{S_1} \cdot \frac{1}{2} \cdot S_1a_1,$$

$$\therefore \ \overline{AG} \cdot (S_1a_1 - S_2'a_2) = \frac{2}{3}(C_1a_1{}^2 - C_2a_2{}^-),$$

$$\therefore \ AG = \frac{2}{3}\frac{C_1a_1{}^2 - C_2a_2{}^2}{S_1a_1 - S_2a_2} \ \cdots \cdots (33).$$

38. THE PROPERTIES OF GULDINUS.

If NL *represent any plane area, and* AB *be any axis, in the same plane, about which the area is made to revolve, so that* NL *is by this revolution made to generate a solid of revolution, then is the volume of this solid equal to that of a prism whose base is* NL, *and whose height is equal to the length of the path which the centre of gravity* G *of the area* NL *is made to describe.*

For take any rectangular area PRSQ in NL, whose sides are respectively parallel and perpendicular to AB, and let MT be the mean distance of the points P and Q, or R and S, from AB. Now it is evident that in the revolution of NL about AB, PQ will describe a superficial ring. Suppose this to be represented by QFPK, let M be the centre of the ring, and let the arc subtended by the angle QMF at distance unity from M be represented by θ, then the area FQPK equals the sector FQM

— the sector $KPM = \frac{1}{2}\overline{MQ^2} \times \theta - \frac{1}{2}\overline{MP^2} \times \theta = \frac{1}{2}\theta\,(\overline{MQ^2} -$
$\overline{MP^2}) = \theta\left(\dfrac{MQ+MP}{2}\right) \times (MQ-MP) = \theta(MT \times PQ).$

Now the solid ring generated by PRSQ is evidently equal to the superficial ring generated by PQ, multiplied by the distance PR. This solid ring equals therefore $\theta\,(\overline{MT} \times \overline{PQ}$ $\times \overline{PR})$ or $\theta \times \overline{MT} \times \overline{PRSQ}$. Now suppose the area PRSQ to be exceedingly small, and the whole area NL to be made up of such exceedingly small areas, and let them be represented by a_1, a_2, a_3, &c. and their mean distances MT by x_1, x_2, x_3, &c. then the solid annuli generated by these areas respectively will (as we have shown), be represented by $\theta x_1 a_1$, $\theta x_2 a_2$, $\theta x_3 a_3$, &c. &c.; and the sum of these annuli, or the whole solid, will be represented by $\theta x_1 a_1 + \theta x_2 a_2 + \theta x_3 a_3 + \&c.$, or by $\theta(x_1 a_1 + x_2 a_2 + x_3 a_3 + \&c.).$ Now if μ represent the weight of any superficial element of the plane NL, $x_1 a_1 \mu =$ the moment of the weight of a_1 about the axis AB, $x_2 a_2 \mu =$ that of the area a_2 about the same axis AB, and so on, therefore the sum $(x_1 a_1 + x_2 a_2 + x_3 a_3 + \&c.)\mu$ = the moment of the whole area NL about AB; but if G be the centre of gravity of NL, and GI its distance from AB, then

the moment of NL about $AB = \overline{GI} \times \overline{NL}\mu$; therefore the whole solid $= \theta \cdot \overline{GI} \cdot \overline{NL}$; but $\theta \cdot GI$ equals the length of the circular path described by G; therefore the volume of the solid equals NL multiplied by the length of the path described by G, *i.e.* it equals a *prism* NM, whose base is NL, and whose height GH is the length of the path described by G; which is the first property of GULDINUS.

39. The above proposition is applicable to finding the solid contents of the thread of a screw of variable diameter, or of the material in a spiral staircase: for it is evident that the thread of a screw may be supposed to be made up of an infinite number of small solids of revolution, arranged one above another

like the steps of a staircase, all of which (contained in one turn
of the thread) might be made to slide along the axis, so that
their surfaces should all lie in the same plane; in which case
they would manifestly form one solid of revolution, such as
that whose volume has been investigated. The principle is
moreover applicable to determine the volume of any solid
(however irregular may be its form otherwise), provided only
that it may be conceived to be generated by the motion of a
given plane area, perpendicular to a given curved line, which
 always passes through the same point in the
plane. For it is evident that whatever point in
this curved line the plane may at any instant be traversing,
it may at that instant be conceived to be revolving about
a certain fixed axis, passing through the centre of curvature of the curve at that point; and thus revolving about a
fixed axis, it is generating for an instant a solid of revolution
about that axis, the volume of which elementary solid of revolution is equal to the area of the plane multiplied by the
length of the path described by its centre of gravity; and
this being true of all such elementary solids, each being
equal to the product of the plane by the corresponding elementary path of the centre of gravity, it follows that the
whole volume of the solid is equal to the product of the area
by the whole length of the path.

40. *If* AB *represent any curved line made to revolve about the
axis* AD *so as to generate the surface
of revolution* BAC, *and* G *be the
centre of gravity of this curved line,
then is the area of this surface equal
to the product of the length of the curved line* AB, *by the
length of the path described by the point* G, *during the revolution of the curve about* AD. *This is the second property
of Guldinus.*

Let PQ be any small element of the generating curve, and
PQFK a zone of the surface generated by this element, this

zone may be considered as a portion of the surface of a cone
whose apex is M, where the tangents to the curve at T and
V, which are the middle points of PQ and FK, *meet* when
produced,   Let this band PQFK of the cone QMF be *de-
veloped**, and let PQFK represent its developement;
this figure PQFK will evidently be a circular ring,
whose centre is M; since the developement of the
whole cone is evidently a circular sector MQF whose centre
M corresponds to the apex of the cone, and its radius MQ to
the side MQ of the cone.

Now, as was shown in the last proposition, the area of this
circular ring when thus developed, and therefore of the conical
band before it was developed, is represented by $\theta \cdot \overline{MT} \cdot PQ$,
where θ represents the arc subtended by QMF at distance
unity. Now the arc whose radius is MT is represented by
$\theta \cdot \overline{MT}$; but this arc, before it was developed from the cone,
formed a complete circle whose radius was NT, and therefore
its circumference $2\pi\overline{NT}$; since then the circle has not
altered its length by its developement, we have

$$\theta\overline{MT} = 2\pi\overline{NT}.$$

Substituting this value of $\theta\overline{MT}$ in the expression for the area
of the band we have

$$\text{area of zone PQFK} = 2\pi \cdot \overline{NT} \cdot \overline{PQ}.$$

Let the surface be conceived to be divided into an infinite
number of such elementary bands, and let the lengths of the
corresponding elements of the curve AB be represented by
s_1, s_2, s_3, &c. and the corresponding values of NT by $y_1, y_2,$
y_3, &c. Then will the areas of the corresponding zones be
represented by $2\pi y_1 s_1, 2\pi y_2 s_2, 2\pi y_3 s_3$, &c. and the area of the
whole surface BAC by $2\pi y_1 s_1 + 2\pi y_2 s_2 + 2\pi y_3 s_3 + \ldots$. or by
$2\pi(y_1 s_1 + y_2 s_2 + y_3 s_3 + \ldots)$. But since G is the centre of
gravity of the curved line AB, therefore $\overline{AB} \cdot \overline{GH}\mu$ represents the moment of the weight of a uniform thread or wire
of the form of that line about AD, μ being the weight of each

* If the cone be supposed covered with a flexible sheet, and a band such
as PQFK be imagined to be cut upon it, and then unwrapped from the cone
and laid upon a plane, it is called the developement of the band.

unit in the length of the line : moreover, this moment equals the sum of the moments of the weights $s_1\mu$, $s_2\mu$, $s_3\mu$, &c. of the elements of the line.

$$\therefore \overline{AB} \cdot \overline{GH}\mu = (y_1s_1 + y_2s_2 + y_3s_3 + \ldots)\mu$$
$$\therefore \overline{AB} \cdot GH = y_1s_1 + y_2s_2 + y_3s_3 + \ldots$$

Therefore area of surface $BAC = 2\pi\overline{AB} \cdot \overline{GH} = \overline{AB} \cdot (2\pi\overline{GH})$.

But $2\pi\overline{GH}$ equals the length of the circular path described by G in its revolution about AD. Therefore, &c.

This proposition, like the last, is true not only in respect to a surface of revolution, but of any surface generated by a plane curve, which traverses perpendicularly another curve of any form whatever, and is always intersected by it in the same point. It is evident, indeed, that the same demonstration applies to both propositions. It must, however, be observed, that neither proposition applies unless the motion of the generating plane or curve be such, that no two of its consecutive positions intersect or cross one another.

41. *The volume of any truncated prismatic or cylindrical body* ABCD, *of which one extremity* CD *is perpendicular to the sides of the prism, and the other* AB *inclined to them, is equal to that of an upright prism* ABEF, *having for its base the plane* AB, *and for its height the perpendicular height* GN *of the centre of gravity* G *of the plane* DC, *above the plane of* AB.

For let ι represent the inclination of the plane DC to AB ;

take m, any small element of the plane CD, and let mr be a prism whose base is m and whose sides are parallel to AD and BC; of elementary prisms similar to which the whole solid ABCD may be supposed to be made up. Now the volume of this prism, whose base is m and its height mr, equals

$$\overline{mr} \times m = \text{sec.}\ \iota \times (\overline{mr} \cdot \cos.\ \iota) \times m = \text{sec.}\ \iota \times (\overline{mr} \cdot \sin.\ mrn)m =$$
$$\text{sec.}\ \iota \times \overline{mn} \times m.$$

Therefore the whole solid equals the sum of all such products as $mn \times m$, each such product being multiplied by the constant quantity sec. ι, or it is equal to the sum just spoken of, that sum being divided by cos. ι. Let this sum be represented by $\Sigma \overline{mn} \times m$, therefore the volume of the solid is representea by $\dfrac{\Sigma \overline{mn} \times m}{\cos. \iota}$. Now suppose CD to represent a thin lamina of uniform thickness, the weight of each square unit of which is μ, then will the weight of the element m be represented by $\mu \times m$, and its moment about the plane ABN by $\mu \times \overline{mn} \times m$, and $\mu \Sigma \overline{mn} \times m$ will represent the sum of the moments of all the elements of the lamina similar to m about that plane. Now by Art. 15. this sum equals the moment of the whole weight of the lamina $\mu \times \overline{CD}$, supposed to be collected in G, about that plane. Therefore

$$\mu \times \overline{CD} \times \overline{NG} = \mu \Sigma \overline{mn} \times m,$$
$$\therefore \overline{CD} \times \overline{NG} = \Sigma \overline{mn} \times m.$$

Substituting this value of $\Sigma \overline{mn} \times n$, we have
$$\text{volume of solid} = \text{sec.} \, \iota \times \overline{CD} \times \overline{NG}.$$

But the plane CD is the projection of AB, therefore CD $= \overline{AB} \cos. \iota$, $\therefore$ CD $\times$ sec. $\iota = $ AB ;

$\therefore$ vol. of solid $ABCD = \overline{AB} \times \overline{NG} = $ vol. of prism ABEF.

Therefore, &c.

[Q.E.D.]

PART II.

DYNAMICS.

42. MOTION is change of place.

The science of DYNAMICS is that which treats of the laws which govern the motions of material bodies, and of their relation to the forces whence those motions result.

The SPACES described by a moving body are the distances between the positions which it occupies at different successive periods of time.

UNIFORM MOTION is that in which equal spaces are described in equal successive intervals of time.

The VELOCITY of *uniform* motion is the space which a body moving uniformly describes in each *second* of time. Thus if a body move uniformly with a velocity represented by V, and during a time represented in seconds by T, then the space S described by it in those T seconds is represented by TV, or $S = TV$. Whence it follows that $V = \dfrac{S}{T}$ and $T = \dfrac{S}{V}$; so that if a body move uniformly, the space described by it is equal to the velocity multiplied by the time in seconds, the velocity is equal to the space divided by the time, and the time is equal to the space divided by the velocity.

43. It is a law of motion, established from constant observation upon the motions of the planets, and by experiment upon the motions of the bodies around us, that when once communicated to a body, it remains in that body, unaffected by the lapse of time, carrying it forward for ever with the same velocity and in the same direction in which it first began to

move, *unless some force act afterwards in a contrary direction to destroy it.* *

The velocity, at any instant, of a body moving with a VARIABLE MOTION, is the space which it would describe in one second of time if its motion were from that instant to become UNIFORM.

An ACCELERATING FORCE is that which acting continually upon a body in the direction of its motion, produces in it a continually increasing velocity of motion.

A RETARDING FORCE is that which acting upon a body in a direction opposite to that of its motion produces in it a continually diminishing velocity.

An IMPULSIVE force is that which having communicated motion to a body, ceases to act upon it after an exceedingly small time from the commencement of the motion.

'44. A UNIFORMLY accelerating or retarding force is that which produces equal increments or decrements of velocity in equal successive intervals of time. If f represent the additional velocity communicated to a body by a uniformly accelerating force in each successive second of time, and T the number of seconds during which it moves, then since by the first law of motion it retains all these increments of velocity (if its motion be unopposed), it follows that after T seconds, an additional velocity represented by fT, will have been communicated to it; and if at the *commencement* of this T seconds its velocity in the same direction was V, then this initial velocity having been retained (by the first law of motion), its *whole* velocity will have become $V + f$T.

If, on the contrary, f represent the velocity continually *taken away* from a body in each successive second of time, by a uniformly retarding force, and V the velocity with which it began to move in a direction opposite to that in which this retarding force acts, then will its remaining velocity after T

* This is the first LAW OF MOTION. For numerous illustrations of this fundamental law of motion, the reader is referred to the author's work, entitled, ILLUSTRATIONS OF MECHANICS, Art. 193.

seconds be represented by $V - fT$; so that generally the velocity V of a body acted upon by a uniformly accelerating or retarding force is represented, after T seconds, by the formula

$$v = V \pm fT \quad \ldots \ldots \quad (34).$$

The force of gravity is, in respect to the descent of bodies near the earth's surface, a constantly accelerating force, increasing the velocity of their descent by $32\frac{1}{6}$ feet in each successive second, and if they be projected upwards it is a constantly retarding force, diminishing their velocity by that quantity in each second. The symbol g is commonly used to represent this number $32\frac{1}{6}$; so that in respect to gravity the above formula becomes $v = V \pm gT$, the sign $\pm$ being taken according as the body is projected upwards or downwards.

A VARIABLE *accelerating* force is that which communicates unequal increments of velocity in equal successive intervals of time; and a variable *retarding* force that which takes away unequal decrements of velocity.

45. To DETERMINE THE RELATION BETWEEN THE VELOCITY AND THE SPACE, AND THE SPACE AND TIME OF A BODY'S MOTION.

Let AM_1, M_1M_2, M_2M_3, &c. represent the exceeding small successive periods of a body's motion, and AP the velocity with which it began to move, M_1P_1 the velocity at the expiration of the first interval of time, M_2P_2 that at the expiration of the second, M_3P_3 of the third interval of time, and so on; and instead of the body varying the velocity of its motion *continually* throughout the period AM_1, suppose it to move through that interval with a velocity which is a mean between the velocity AP at A, and that M_1P_1 at M_1, or with a velocity equal to $\frac{1}{2}(AP + M_1P_1)$.

Since on this supposition it moves with a *uniform* motion, the space it describes during the period AM_1 equals the product of that velocity by that period of time, or it equals

E

$\frac{1}{2}(AP + M_1P_1)AM_1$. Now this product represents the area of the trapezoid AM_1P_1P. The space described then in the interval AM_1, on the supposition that the body moves during that interval with a velocity which is the mean between those actually acquired at the commencement and termination of the interval, is represented by the trapezoidal area AM_1P_1P.

Similarly the areas P_1M_2, P_2M_3, &c. represent the spaces the body is made to describe in the successive intervals M_1M_2, M_2M_3, &c.; and therefore the *whole* polygonal area APCB represents the whole space the body is made to describe in the whole time. AB, on the supposition that it moves in each successively exceeding small interval of time with the mean velocity of that interval. Now the less the intervals are, the more nearly does this mean velocity of each interval approach the actual velocity of that interval; and if they be infinitely small, and therefore infinitely great in number, then the mean velocity coincides with the actual velocity of each interval, and in this case the polygonal area passes into the curvilinear area APCB.

Generally, therefore, if we represent by the *abscissæ* of a curve the times through which a body has moved, and by the corresponding *ordinates* of that curve the velocities which it has acquired after those times, then the *area* of that curve will represent the *space* through which the body has moved; or in other words, if a curve PC be taken such that the number of equal parts in any one of its abscissæ AM_3 being taken to represent the number of seconds during which a body has moved, the number of those equal parts in the corresponding ordinate M_3P_3 will represent the number of feet in the velocity then acquired; then the space which the body has described will be represented by the number of these equal parts squared which are contained in the area of that curve.

46. To DETERMINE THE SPACE DESCRIBED IN A GIVEN TIME BY A BODY WHICH IS PROJECTED WITH A GIVEN VELOCITY, AND WHOSE MOTION IS UNIFORMLY ACCELE- RATED, OR UNIFORMLY RETARDED.

Take any straight line AB to represent the whole time T,

in seconds, of the body's motion, and draw AD perpendicular to it, representing on the same scale its velocity at the commencement of its motion. Draw DE parallel to AB, and according as the motion is accelerated or retarded draw DC or DF inclined to DE, at an angle whose tangent equals f, the constant increment or decrement of the body's velocity. Then if any abscissa AM be taken to represent a number of seconds t during which the body has moved, the corresponding ordinate MP or MQ will represent the velocity then acquired by it, according as its motion is accelerated or retarded. For PR = RQ = DR tan. PDE = $\overline{\text{AM}}$ tan. PDE; but AM = t, and tan. PDE = f; therefore PR = RQ = ft. Also RM = AD = V, therefore MP = RM + PR = V + ft, and MQ = RM − RQ = V − ft; therefore by equation (34), MP or MQ represents the velocity after the time AM according as the motion is accelerated or retarded. The same being true of every other time, it follows, by the last proposition, that the whole space described in the time T or AB is represented by the area ABCD if the motion be accelerated, and by the area ABFD if it be retarded.

Now area ABCD = $\frac{1}{2}$AB(AD + BC), but AB = T, AD = V, BC = V + fT,

∴ area ABCD = $\frac{1}{2}$T(V + V + fT) = VT + $\frac{1}{2}f$T².

Also area ABFD = $\frac{1}{2}$AB(AD + BF), where AB and AD have the same values as before, and BF = V − fT,

∴ area ABFD = $\frac{1}{2}$T(V + V − fT) = VT − $\frac{1}{2}f$T².

Therefore, generally, if S represent the space described after T seconds,

$$S = VT \pm \tfrac{1}{2}f T^2 \quad \ldots \quad (35);$$

in which formula the sign ± is to be taken according as the motion is accelerated or retarded.

47. To DETERMINE A RELATION BETWEEN THE SPACE DE-
SCRIBED AND THE VELOCITY ACQUIRED BY A BODY WHICH
IS PROJECTED WITH A GIVEN VELOCITY, AND WHOSE MO-
TION IS UNIFORMLY ACCELERATED OR RETARDED.

Let v be the velocity acquired after T seconds, then by
equation (34), $v = V \pm f T$, $\therefore$ $T = \pm \dfrac{(v - V)}{f}$.

Now area $ABCD = \frac{1}{2} AB(AD + BC)$, where
$AB = T = \dfrac{(v - V)}{f}$, $AD = V$, $BC = v$,

$\therefore$ area $ABCD = \frac{1}{2} \dfrac{(v - V)}{f}(V + v) = \frac{1}{2} \dfrac{(v^2 - V^2)}{f}$,

area $ABFD = \frac{1}{2} AB(AD + BF)$, where $AB = T = -\dfrac{(v - V)}{f}$,
$AD = v$, $BF = V$,

$\therefore$ area $ABFD = -\frac{1}{2} \dfrac{(v - V)(v + V)}{f} = -\frac{1}{2} \dfrac{(v^2 - V^2)}{f}$.

Therefore generally, if S represent the space through which
the velocity v is acquired, then $S = \pm \frac{1}{2} \dfrac{(v^2 - V^2)}{f}$,

$\therefore$ $v^2 - V^2 = \pm 2f S$ . . . . . (36);

in which formula the $\pm$ sign is to be taken according as
the motion is accelerated or retarded.

If the body's motion be *retarded*, its velocity v will eventually be destroyed. Let S_1 be the space which will have been
described when v thus vanishes, then by the last equation
$0 - V^2 = -2f S_1$.

$\therefore$ $V^2 = 2f S_1$ . . . . . (37),

where V is the velocity with which the body is projected
in a direction opposite to the force, and S_1 the whole space

which by this velocity of projection it can be made to describe.

If the body's motion be *accelerated*, and it fall from *rest*, or have no velocity of projection, then $v^2 - 0 = +2fS$,

$$\therefore \quad v^2 = 2fS \ldots \ldots (38).$$

Let S_2 be the space through which it must in this case move to acquire a velocity V equal to that with which it was projected in the last case, therefore $V^2 = 2fS_2$. Whence it follows that $S_1 = S_2$, or that the whole space S_1 through which a body will move when projected with a given velocity V, and uniformly *retarded* by any force, is equal to the space S_2, through which it must move to *acquire* that velocity when uniformly *accelerated* by the same force.

In the case of bodies moving freely, and acted upon by gravity, f equals $32\frac{1}{6}$ feet, and is represented by g; and the space S_2, through which any given velocity V is acquired, is then said to be that *due* to that velocity.

WORK.

48. WORK is the union of a continued pressure with a continued motion. And a mechanical agent is thus said to WORK when a pressure is continually overcome, and a point (to which that pressure is applied) continually moved by it. Neither pressure nor motion alone is sufficient to constitute *work;* so that a man who merely supports a load upon his shoulders without moving it, no more *works,* in the sense in which that term is here used, than does a column which sustains a heavy weight upon its summit; and a stone as it falls freely *in vacuo,* no more *works* than do the planets as they wheel unresisted through space.

49. THE UNIT OF WORK. The unit of work used in this

country, in terms of which to estimate every other amount of work, is the work necessary to overcome a pressure of one pound through a distance of one foot, in a direction opposite to that in which the pressure acts. Thus, for instance, if a pound weight be raised through a vertical height of one foot, one unit of work is done; for a pressure of one pound is overcome through a distance of one foot, in a direction opposite to that in which the pressure acts.

50. *The number of units of work necessary to overcome a pressure of* M *pounds through a distance of* M *feet, is equal to the product* MN.

For since, to overcome a pressure of one pound through one foot requires *one* unit of work, it is evident that to overcome a pressure of M pounds through the same distance of one foot, will require M units. Since then M units of work are required to overcome this pressure through one foot, it is evident that N times as many units (*i. e.* NM) are required to overcome it through N feet. Thus, if we take U to represent the number of units of work done in overcoming a constant pressure of M pounds through N feet, we have

$$U = MN \ . \ . \ . \ . \ . \ . \ . \ (39).$$

51. To ESTIMATE THE WORK DONE UNDER A VARIABLE PRESSURE.

Let PC be a curved line and AB its axis, such that any one of its abscissæ AM_3, containing as many equal parts as there are units in the space through which any portion of the work has been done, the corresponding ordinate M_3P_3 may contain as many of those equal parts, as there are in the pressure under which it is *then* being done. Divide AB into exceedingly small equal parts, AM_1, M_1M_2, &c. and draw the ordinates M_1P_1, M_2P_2, &c.;

then if we conceive the work done through the space AM_1 (which is in reality done under pressures varying from AP to M_1P_1,) to be done uniformly under a pressure, which is the arithmetic mean between AP and M_1P_1, it is evident that the number of units in the work done through that small space will equal the number of square units in the trapezoid APP_1M_1 (see Art. 45.), and similarly with the other trapezoids; so that the number of units in the whole work done through the space AB will equal the number of square units in the whole polygonal area $APP_1P_2P_3$, &c. CB.

But since AM_1, M_1M_2, &c. are exceedingly small, this polygonal area passes into the curvilinear area $APCB$; the whole work done is therefore represented by the number of square equal parts in this area.

Now, generally, the area of any curve is represented by the integral $\int y dx$, where y represents the ordinate, and x the corresponding abscissa. But in this case the variable pressure P is represented by the ordinate, and the space S described under this variable pressure by the abscissa. If therefore U represent the work done between the values S_1 and S_2 of S, we have

$$U = \int_{S_1}^{S_2} P dS \quad \ldots \ldots \quad (40).$$

Mean pressure is that under which the same work would be done over the same space, provided that pressure, instead of varying throughout that space, remained the same : thus, the mean pressure in respect to an amount of work represented by the curvilinear area $AEFC$, is that under which an amount of work would be done represented by the rectilineal area $ABDC$, the area $ABDC$ being equal to the curvilinear area $ABDC$; the mean pressure in this case is represented by AB. Thus, to determine the mean pressure in any case of variable pressure, we have only to find a curvilinear area representing the work done under that variable pressure, and then to describe a rectangular parallelogram on the same base AC, which shall have an area equal to the curvilinear area.

If S represent the space described under a variable pressure, U the work done, and p the mean pressure, then $p\mathrm{S} = \mathrm{U}$, therefore $p = \dfrac{\mathrm{U}}{\mathrm{S}}$.

52. *To estimate the work of a pressure, whose direction is not that in which its point of application is made to move.*

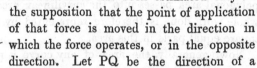

Hitherto the work of a force has been estimated only on the supposition that the point of application of that force is moved in the direction in which the force operates, or in the opposite direction. Let PQ be the direction of a pressure, whose point of application Q is made to move in the direction of the straight line AB. Suppose the pressure P to remain constant, and its direction to continue parallel to itself. It is required to estimate the work done, whilst the point of application has been moved from A to Q.

Resolve P into R and S, of which R is parallel and S perpendicular to AB. Then since no motion takes place in the direction of SQ, the pressure S does no work, and the whole work is done by R; therefore the work $= \mathrm{R} \cdot \overline{\mathrm{AQ}}$.

Now $\mathrm{R} = \mathrm{P} \cdot \cos. \mathrm{PQR}$, therefore the work $= \mathrm{P} \cdot \overline{\mathrm{AQ}} \cos.$ PQR. From the point A draw AM perpendicular to PQ, then $\overline{\mathrm{AQ}} \cos. \mathrm{PQR} = \mathrm{QM}$; therefore work $= \mathrm{P} \cdot \overline{\mathrm{QM}}$. Therefore the work of any pressure as above, not acting in the direction of the motion of the point of application of that pressure, is the same as it would have been if the point of application had been made to move in the direction of the pressure, provided that the space through which it was so moved had been the projection of the space through which it actually moves. The product $\mathrm{P} \cdot \overline{\mathrm{QM}}$ may be called the work of P resolved in the direction of P.

The above proposition which has been proved, whatever may be the distance through which the point of application is moved, in that particular case only in which the pressure remains the same in amount and always parallel to itself, is evidently true for exceedingly *small* spaces of motion, even if the pressure be variable both in amount and direction ; since for such exceedingly small variations in the positions of the points of application, the variations of the pressures themselves, both in amount and direction, arising from these variations of position, must be exceedingly small, and therefore the resulting variations in the work exceedingly small as compared with the whole work.

53. *If any number of pressures* P_1, P_2, P_3, *be applied to the same point* A, *and remain constant and parallel to themselves, whilst the point A is made to move through the straight line* AB, *then the whole work done is equal to the sum of the works of the different pressures resolved in the directions of those pressures, each being taken negatively whose point of application is made to move in an opposite direction to the pressure upon it.*

Let α_1, α_2, α_3, &c. represent the inclinations of the pressures

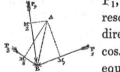

P_1, P_2, &c. to the line AB, then will the resolved parts of these pressures in the direction of that line be P_1 cos. α_1, P_2 cos. α_2, P_3 cos. α_3, &c. and they will be equivalent to a single pressure in the direction of that line represented by P_1 cos. $\alpha_1 + P_2$ cos. $\alpha_2 + P_3$ cos. α_3, &c. in which sum all those terms are to be taken negatively which involve pressures whose direction is from B towards A (since the single pressure from A towards B is manifestly equal to the difference between the sum of those resolved pressures which act in that direction, and those in the opposite direction). Therefore the whole work is equal to $\{P_1$ cos. $\alpha + P_2$ cos. $\alpha_2 + P_3$ cos. $\alpha_3 + \ldots \ldots \} . \overline{AB} = P_1 \overline{AB}$ cos. $\alpha_1 + P_2 . \overline{AB}$ cos. $\alpha_2 + P_3 \overline{AB}$ cos. $\alpha_3 + \ldots = P_1 . \overline{BM}_1 +$ $P_2 . \overline{BM}_2 + P_3 . \overline{BM}_3 + \ldots \ldots$; in which expression the successive terms are the works of the different pressures resolved in the several directions of those pressures, each being taken positively or negatively, according as the direction of the corresponding pressure is *towards* the direction of the motion or opposite to it.

Thus if U represent the whole work and U_1 and U_2 the sums of those done in opposite directions, then

$$U = U_1 - U_2 \quad \ldots \ldots \quad (41).$$

54. *If any number of pressures applied to a point be in equilibrium, and their point of application be moved, the whole work done by these pressures in the direction of the motion will equal the whole work done in the opposite direction.*

For if the pressures P_1, P_2, P_3, &c. (Art. 53.) be in equilibrium, then the sums of their resolved pressures in opposite directions along AB will be equal (Art. 10.); therefore the whole work U along AB, which by the last proposition is equal to the work of a pressure represented by the *difference* of these sums, will equal nothing, therefore $0 = U_1 - U_2$, therefore $U_1 = U_2$, that is, the whole work done in one direction along AB, by the pressures P_1, P_2, &c. is equal to the whole work done in the opposite direction.

55. *If a body be acted upon by a force whose direction is always towards a certain point S, called a centre of force, and be made to describe any given curve PA in a direction opposed to the action of that force, and Sp be measured on SA equal to SP, then will the work done in moving the body through the curve PA be equal to that which would be necessary to move it in a straight line from p to A.*

For suppose the curve PA to be a polygon of an infinite

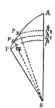 number. of sides PP_1, P_1P_2, &c. Through the points P, P_1, P_2, &c. describe circular arcs with the radii SP, SP_1, SP_2, &c. and let them intersect SA in p, p_1, p_2, &c. Then since PP_1 is exceedingly small, the force may be considered to act throughout this space always in a direction parallel to SP_1; therefore the work done through PP_1 is equal to the work which must be done to move the body through the distance mP_1 (Art. 52.), since mP_1 is the *projection* of PP_1 upon the direction SP_1 of the force. But $mP_1=pp_1$; therefore the work done through PP_1 is equal to that which would be required to move the body along the line SA through the distance pp_1; and similarly the work done through P_1P_2 is equal to that which must be done to move the body through p_1p_2, so that the work through PP_2 is equal to that through pp_2, and so of all other points in the curve. Therefore the work through PA is equal to that through pA.* Therefore, &c. [Q.E.D.]

56. *If* S *be at an exceedingly great distance as compared with* AP, *then all the lines drawn from* S *to* AP *may be considered parallel.* This is the case with the force of gravity at the surface of the earth, which tends towards a point, the earth's centre, situated at an exceedingly great distance, as compared with any of the distances through which the work of mechanical agents is usually estimated.

Thus then it follows that the work necessary to move a heavy body up any curve PA, or inclined plane, is the same as would be necessary to raise it in a *vertical* line pA to the same height.

The dimensions of the body are here supposed to be ex-

---

* Of course it is in this proposition supposed that the force, if it be not constant, is dependent for its amount only on the distance of the point at which it acts from the centre of force S; so that the distances of p and P from S being the same, the force at p is equal to that at P; similarly the force at p is equal to that at P_1, the force at p_2 equal to that at P_2, &c.

ceeding small. If it be of considerable dimensions, then whatever be the height through which its centre of gravity is
raised along the curve, the work expended is the same
(Art. 60.) as though the centre of gravity were raised vertically to that height.*

57. In the preceding propositions the work has been
estimated on the supposition that the body is made to move
so as to increase its distance from the centre S, or in a direction opposed to that of the force impelling it towards S.   It is
evident nevertheless that the work would have been precisely
the same, if instead of the body moving *from* P to A it had
moved from A to P, provided only that in this last case
there were applied to it at every point such a force as would
prevent its motion from being accelerated by the force continually impelling it towards S ; for it is evident that to prevent this acceleration, there must continually be applied to
the body a force in a direction *from* S equal to that by which
it is attracted towards it; and the work of such a force is
manifestly the same, provided the *path* be the same, whether
the body move in one direction or the other along that path,
being in the two cases the work of the same force over the
same space, but in opposite directions.

58. *If there be any number of parallel pressures,* P_1, P_2, P_3,
&c. *whose points of application are transferred, each through
any given distance from one position to another, then is the
work which would be necessary to transfer their resultant
through a space equal to that by which their centre of
pressure is displaced in this change of position, equal to the
difference between the aggregate work of those pressures
whose points of application have been moved in the directions
in which the pressures applied to them act, and those whose*

* The *only* force acting upon the body is in this proposition supposed
to be that acting towards S.   No account is taken of friction or any other
forces which oppose themselves to its motion.

points of application have been moved in the opposite directions to their pressures.

For (Art. 17.), if y_1, y_2, y_3, &c. represent the distances of the points of application of these pressures from any given plane in their first position, and h the distance of their centre of pressure from that plane, and if Y_1, Y_2, Y_3, &c. and H represent the corresponding distances in the second position, and if P_1, P_2, P_3, &c. be taken positively or negatively according as their directions are *from* or *towards* the given plane, $h\{P_1 + P_2 + P_3 + \ \ldots\} = P_1 y_1 + P_2 y_2 + P_3 y_3\ \ldots$

and $H\{P_1 + P_2 + P_3 + \ldots\} = P_1 Y_1 + P_2 Y_2 + P_3 Y_3 + \ldots$

$\therefore (H - h)\{P_1 + P_2 + P_3 + \ldots\} = P_1(Y_1 - y_1) + P_2(Y_2 - y_2)$
$+ P_3(Y_3 - y_3) + \ldots \ldots (42);$

in the second member of which equation the several terms are evidently positive or negative, according as the pressure P corresponding to each, and the difference $Y - y$ of its distances from the plane in its two positions, have the same or contrary signs. Now by supposition P is positive or negative according as it acts *from* or *towards* the plane; also $Y - y$ is evidently positive or negative according as the point of application of P is moved from or towards the plane: each term is therefore positive or negative, according as the corresponding point of application is transferred in a direction *towards* that in which its applied pressure acts, or in the opposite direction.

Now the plane from which the distances of the points of application are measured may be *any* plane whatever. Let it be a plane perpendicular to the directions of the pressures.

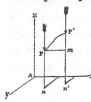
Let Axy represent this plane, and let P and P′ represent the two positions of the point of application of the pressure P (the path described by it between these two positions having been any whatever). Let MP and M′P′ represent the perpendicular distances of the points P and P′ from the plane, and draw Pm from P perpendicular to M′P′. Then $P(Y - y) = P(M'P' - MP) = P \cdot \overline{mP'}$; but, by Art. 55.,

$P . \overline{mP'}$ equals the work of P as its point of application is transferred from P to P'. Thus each term of the second member of equation (42) represents the work of the corresponding pressure, so that if Σu_1 represent the aggregate work of those pressures whose points of application are transferred *towards* the directions in which the pressures act, and Σu_2 the work of those whose points of application are moved opposite to the directions in which they severally act, then the second member of the equation is represented by $\Sigma u_1 - \Sigma u_2$. Moreover the first member of the equation is evidently the work necessary to transfer the resultant pressure $P_1 + P_2 + P_3$ &c. through the distance $H - h$, which is that by which the centre of pressure is removed *from* or *towards* the given plane, so that if U represent the quantity of work necessary to make this transfer of the centre of pressure,

$$U = \Sigma u_1 - \Sigma u_2 \ . \ . \ . \ , \ . \ . \ (43).$$

59. If the sum of those parallel pressures whose tendency is in one direction equal the sum of those whose tendency is in the opposite direction, then $P_1 + P_2 + P_3 + \ . \ . \ . \ . \ = 0$. In this case, therefore, $U = 0$, therefore $\Sigma u_1 - \Sigma u_2 = 0$, therefore $\Sigma u_1 = \Sigma u_2$; so that *when in any system of parallel pressures the sum of those whose tendency is in one direction equals the sum of those whose tendency is in the opposite direction, then the aggregate work of those whose points of application are moved in the directions of the pressures severally applied to them is equal to the aggregate work of those whose points of application are moved in the opposite directions.*

This case manifestly obtains when the parallel pressures are in EQUILIBRIUM, the sum of those whose tendency is in one direction then equalling the sum of those whose tendency is in the opposite direction, since otherwise, when applied to a point, these pressures could not be in equilibrium about that point (Art. 8.).

60. The preceding proposition is manifestly true in respect to a system of weights, these being pressures whose directions

are always parallel, wherever their points of application may be moved. Now the centre of pressure of a system of weights is its centre of gravity (Art. 19.). Thus then it follows, that if the weights composing such a system be separately moved in any directions whatever, and through any distances whatever, then the difference between the aggregate work done *upwards* in making this change of relative position and that done *downwards* is equal to the work necessary to raise the sum of all the weights through a height equal to that through which their centre of gravity is raised or depressed.* Moreover that if such a system

---

* This proposition has numerous applications. If for instance it be required to determine the aggregate expenditure of work in raising the different elements of a structure, its stone, cement, &c. to the different positions they occupy in it, we make this calculation by determining the work requisite to raise the whole weight of material at once to the height of the centre of gravity of the structure. If these materials have been carried up by labourers, and we are desirous to include the whole of their labour in the calculation, we ascertain the probable amount of each load, and conceive the weight of a labourer to be added to each load, and then all these at once to be raised to the height of the centre of gravity.

Again, if it be required to determine the expenditure of work made in raising the material excavated from a well, or in pumping the water out of it, we know that (neglecting the effect of friction, and the weight and rigidity of the cord) this expenditure of work is the same as though the whole material had been raised at one lift from the centre of gravity of the shaft to the surface. Let us take another application of this principle which offers so many practical results. The material of a railway excavation of considerable length is to be removed so as to form an embankment across a valley at some distance, and it is required to determine the expenditure of work made in this transfer of the material. Here each load of material is made to traverse a different distance, a resistance from the friction, &c. of the road being continually opposed to its motion. These resistances on the different loads constitute a system of parallel pressures, each of whose points of application is separately transferred from one given point to another given point, the directions of transfer being also parallel. Now by the preceding proposition the expenditure of work in all these separate transfers is the same as it would have been had a pressure equal to the sum of all these pressures been at once transferred from the centre of resistance of the excavation to the centre of resistance of the embankment. Now the resistances of the parts of the mass moved

of weights be supported in equilibrium by the resistance of any fixed point or points, and be put in motion, then (since the work of the resistance of each such point is *nothing*) the aggregate work of those weights which are made to descend, is equal to that of those which are made to ascend.

61. *If a plane be taken perpendicular to the directions of any number of parallel pressures and there be two different positions of the points of application of certain of these pressures in which they are at different distances from the plane, whilst the points of application of the rest of these pressures remain at the same distance from that plane, and if in both positions the system be in equilibrium, then the centre of pressure of the first mentioned pressures will be at the same distance from the plane in both positions.*

For since in both positions the system is in equilibrium, therefore in both positions $P_1 + P_2 + P_3 + \ldots = 0$,

$$\therefore (Y_1 - y_1)P_1 + (Y_2 - y_2)P_2 + (Y_3 - y_3)P_3 + \ldots + P_n(Y_n - y_n) = 0.$$

Now let P_n be any one of the pressures whose point of application is at the same distance from the given plane in both positions,

$$\therefore Y_n = y_n, \text{ and } Y_n - y_n = 0,$$

$$\therefore (Y_1 - y_1)P_1 + (Y_2 - y_2)P_2 + \ldots + (Y_{n-0} - y_{n-1})P_{n-1} = 0,$$

$$\therefore Y_1 P_1 + Y_2 P_2 + \ldots + Y_{n-1} P_{n-1} = y_1 P_1 + y_2 P_2 + \ldots + y_{n-1} P_{n-1},$$

$$\therefore \frac{Y_1 P_1 + Y_2 P_2 + \ldots + Y_{n-1} P_{n-1}}{P_1 + P_2 + \ldots + P_{n-1}} = \frac{y_1 P_1 + y_2 P_2 + \ldots + y_{n-1} P_{n-1}}{P_1 + P_2 + \ldots + P_{n-1}},$$

$$\therefore H_{n-1} = h_{n-1},$$

---

are the frictions of its elements upon the road, and these frictions are proportional to the *weights* of the elements; their centre of resistance coincides therefore with the centre of gravity of the mass, and it follows that the expenditure of work is the same as though all the material had been moved at *once* from the centre of gravity of the excavation to that of the embankment. To allow for the weight of the carriages, as many times the weight of a carriage must be added to the weight of the material as there are journeys made.

where H_{n-1} represents the distance of the centre of pressure of P_1, P_2 ... P_{n-1}, from the given plane in the first position, and h_{n-1} its distance in the second position. Its distance in the first position is therefore the same as in the second. Therefore, &c.

From this proposition it follows, that if a system of weights be supported by the resistances of one or more fixed points, and if there be any two positions whatever of the weights in both of which they are in equilibrium with the resistances of those points, then the height of the common centre of gravity of the weights is the same in both positions. And that if there be a series of positions in all of which the weights are in equilibrium about such a resisting point or points, then the centre of gravity remains continually at the same height as the system passes through this series of positions.

If all these positions of equilibrium be infinitely near to one another, then it is only during an infinitely *small* motion of the points of application that the centre of gravity ceases to ascend or descend; and, conversely, if for an infinitely small motion of the points of application the centre of gravity ceases to ascend or descend, then in two or more positions of the points of application of the system, infinitely near to one another, it is in equilibrium.

Work of Pressures applied in different Directions to a Body moveable about a fixed Axis.

62. *The work of a pressure applied to a body moveable about a fixed axis is the same at whatever point in its proper direction that pressure may be applied.*

For let AB represent the direction of a pressure applied to a body moveable about a fixed axis O; the work done by this pressure will be the same whether it be applied at A or B. For conceive the body to revolve about O, through an exceedingly small angle AOC, or BOD, so that the points A and D may describe circular arcs AC and BD. Draw Cm, Dn, and OE, perpendiculars to AB, then if P represent the pressure.

applied to AB, P . $\overline{Am}$ will represent the work done by P when applied at A (Art. 52.), and P . $\overline{Bn}$ will represent the work done by P when applied at B ; therefore the work done by P at A is the same as that done by P at B if Am is equal to Bn.

Now AC and BD being exceedingly small, they may be conceived to be straight lines. Since BD and BE are respectively perpendicular to OB and OE, therefore $\angle$ DBE $=$ $\angle$ BOE *; and because AC and AE are perpendicular to OA and OE, therefore $\angle$ CAE $= \angle$ AOE. Now A$m = \overline{CA}$. cos. CAE $=$ CA . cos. AOE $= \dfrac{CA}{OA}$. $\overline{OA}$. cos. AOE $= \dfrac{CA}{OA} \times$

OE. Similarly B$n =$ DB cos. DBE $=$ DB . cos. BOE $= \dfrac{BD}{OB}$.

$\overline{OB}$ cos. BOE $= \dfrac{BD}{OB} \times$ OE, $i. e.$ A$m =$ OE . $\dfrac{CA}{OA}$, and B$n =$

OE $\dfrac{BD}{OB}$. But $\dfrac{CA}{OA} = \dfrac{BD}{OB}$, since the $\angle$ AOC $= \angle$ BOD, therefore A$m =$ Bn.

63. *If any number of pressures be in equilibrium about a fixed axis, then the whole work of those which tend to move the system in one direction about that axis is equal to the whole work of those which tend to move it in the opposite direction about the same axis.* For let P be any one of such a system of pressures, and O a fixed axis, and OM perpendicular to the direction of P, then whatever may be the point of application of P, the work of that pressure is the same as though it were applied at M. Suppose the whole system to be moved through an exceeding small angle θ about the point O, and let OM be represented by p, then will $p\theta$ represent the space described by the point M, which will be actually in the direction of the force P, therefore

* It is a well known principle of Geometry, that if two lines be inclined at any angle, and any two others be drawn perpendicular to these, then the inclination of the last two to one another shall equal that of the first two.

the work of $P = P \cdot p \cdot \theta$. Now let P_1, P_2, P_3, &c. represent those pressures which act in the direction of the motion, and P'_1, P'_2, &c. those which act in the opposite direction, and let p_1, p_2, p_3, &c. be the perpendiculars on the first, and p'_1, p'_2, p'_3, &c. be the perpendiculars on the second; therefore by the principle of the equality of moments $P_1 p_1 + P_2 p_2 + P_3 p_3 +$ &c. $= P'_1 p'_1 + P'_2 p'_2 + P'_3 p'_3 +$ &c.; therefore multiplying both sides by θ, $P_1 p_1 \theta + P_2 p_2 \theta + P_3 p_3 \theta = P'_1 p'_1 \theta + P'_2 p'_2 \theta + P'_3 p'_3 \theta +$ &c.; but $P_1 p_1 \theta$, $P'_1 p'_1 \theta$, &c. are the works of the forces P_1, P'_1, &c.; therefore the aggregate work of those which tend to move the system in one direction is equal to the aggregate of those which tend to move it in the opposite direction.

64. THE ACCUMULATION OF WORK IN A MOVING BODY.

In every moving body there is accumulated, by the action of the forces whence its motion has resulted, a certain amount of power which it reproduces upon any resistance opposed to its motion, and which is measured by the work done by it upon that obstacle. Not to multiply terms, we shall speak of this accumulated *power* of working, thus measured by the work it is capable of producing, as ACCUMULATED WORK. It is in this sense that in a ball fired from a cannon there is understood to be accumulated the *work* it reproduces upon the obstacles which it encounters in its flight; that in the water which flows through the channel of a mill is accumulated the work which it yields up to the wheel *; and that in the carriage which is allowed rapidly to descend a hill is accumulated the work which carries it a considerable distance up the next hill. It is when the pressure under which any work is done, exceeds the resistance opposed to it, that work is thus *accumulated* in a moving body; and it will subsequently be shown (Art. 69.) that in every case the work accumulated is precisely equal to

* This remark applies more particularly to the under-shot wheel, which is carried round by the rush of the water.

the work done upon the body beyond that necessary to over-come the resistances opposed to its motion, a principle which might almost indeed be assumed as in itself evident.

65. The amount of work thus accumulated in a body moving with a given velocity, is evidently the same, whatever may have been the circumstances under which its velocity has been acquired. Whether the velocity of a ball has been communicated by projection from a steam gun, or explosion from a cannon, or by being allowed to fall freely from a sufficient height, it matters not to the result; provided the same *velocity* be communicated to it in all three cases, and it be of the same weight, the work *accumulated* in it, estimated by the effect it is capable of pro-ducing, is evidently the same.

In like manner, the whole amount of work which it is capable of yielding to overcome any resistance is the same, whatever may be the nature of that resistance.

66. TO ESTIMATE THE NUMBER OF UNITS OF WORK ACCU-MULATED IN A BODY MOVING WITH A GIVEN VELOCITY.

Let w be the weight of the body in pounds, and v its velocity in feet.

Now suppose the body to be projected with the velocity v in a direction opposite to gravity, it will ascend to the height h from which it must have fallen, to acquire that same velocity v (Art. 47.); there must then at the instant of projection have been accumulated in it an amount of work sufficient to raise it to this height h; but the number of units of work requisite to raise a weight w to a height h, is represented by wh; this then is the number of units of work accumulated in the body at the instant of projection. But since h is the height through which the body must fall to acquire the velocity v, therefore

$v^2 = 2gh$ (Art. 47.); therefore $h = \frac{1}{2}\frac{v^2}{g}$; whence it follows that if U represent the number of units of work accumulated,

$$U = \tfrac{1}{2}\frac{w}{g}v^2 \ . \ . \ . \ . \ (44).$$

Moreover it appears by the last article that this expression represents the work accumulated in a body weighing w pounds, and moving with a velocity of v feet, *whatever* may have been the circumstances under which that velocity was accumulated.

The product $\left(\dfrac{w}{g}\right)v^2$ is called the VIS VIVA of the body, so that the accumulated work is represented by half the vis viva, the quotient $\left(\dfrac{w}{g}\right)$ is called the MASS of the body.

67. *To estimate the work accumulated in a body, or lost by it, as it passes from one velocity to another.*

In a body whose weight is w, and which moves with a velocity v, there is accumulated a number of units of work represented (Art. 66.) by the formula $\tfrac{1}{2}\dfrac{w}{g}v^2$. After it has passed from this velocity to another V, there will be accumulated in it a number of units of work, represented by $\tfrac{1}{2}\dfrac{w}{g}V^2$, so that if its last velocity be greater than the first, there will have been *added* to the work accumulated in it, a number of units represented by $\tfrac{1}{2}\dfrac{w}{g}V^2 - \tfrac{1}{2}\dfrac{w}{g}v^2$; or if the second velocity be less than the first, there will have been *taken from* the work accumulated in it a number of units represented by $\tfrac{1}{2}\dfrac{w}{g}v^2 - \tfrac{1}{2}\dfrac{w}{g}V^2$. So that generally if U represent the work accumulated or lost by the body, in passing from the velocity v to the velocity V, then

$$U = \pm\tfrac{1}{2}\frac{w}{g}\{V^2 - v^2\} \ . \ . \ . \ . \ . \ (45),$$

F 3

where the $\pm$ sign is to be taken according as the motion is accelerated or retarded.

68. *The work accumulated in a body, whose motion is accelerated through any given space by given forces is equal to the work which it would be necessary to do upon the body to cause it to move back again through the same space when acted upon by the same forces.*

For it is evident that if with the velocity which a body has acquired through any space AB by the action of any forces whose direction is from A towards B, it be projected back again from B towards A, then as it returns through each successive small part or element of its path, it will be retarded by precisely the same forces as those by which it was accelerated when it *before* passed through it; so that it will, in returning through each such element, lose the same portion of its velocity as before it gained there; and when at length it has traversed the whole distance BA, and reached the point A, it will have lost between B and A a velocity, and therefore *an amount of work* (Art. 67.), precisely equal to that which before it gained between A and B.   Now the work lost between B and A is the work necessary to overcome the resistances opposed to the motion through BA.   The work accumulated from A to B is therefore equal to the work which would be necessary to overcome the resistances between B and A, or which would be necessary to move the body from a state of rest, and with a uniform motion, in opposition to these resistances, through BA.   Let this work be represented by U; also let v be the velocity with which the body started from A, and V that which it has acquired at B.   Then will $\frac{1}{2}\dfrac{W}{g}(V^2-v^2)$ represent the work accumulated between A and B,

$$\therefore \; \tfrac{1}{2}\dfrac{W}{g}(V^2-v^2)=U, \quad \therefore \; V^2-v^2=\dfrac{2gU}{W}$$

If the body, instead of being accelerated, had been *retarded,* then the work lost being that expended in overcoming the retarding forces, is evidently that necessary to move the body uniformly in opposition to these retarding forces through AB; so that if this force be represented by U, then, since $\frac{1}{2}\frac{W}{g}(v^2 - V^2)$ is in this case the work lost, we shall have $v^2 - V^2 = \frac{2gU}{W}$. Therefore, generally,

$$V^2 - v^2 = \pm \frac{2gU}{W} \quad \cdots \cdots (46),$$

where the sign $\pm$ is to be taken according as the motion is accelerated or retarded.

69. *The work accumulated in a body which has moved through any space acted upon by any force, is equal to the excess of the work which has been done upon it by those forces which tend to accelerate its motion above that which has been done upon it by those which tend to retard its motion.*

For let R be the single force which would at any point P (see last fig.) be necessary to move the body back again through an exceeding small element of the same path (the other forces impressed upon it remaining as before); then it follows by Art. 54. that the work of R over this element of the path is equal to the excess of the work over that element of the forces which are impressed upon the body in the direction of its motion above the work of those impressed in the opposite direction. Now this is true at *every* point of the path; therefore the *whole* work of the force R necessary to move the body back again from B to A is equal to the excess of the work done upon it, by the impressed forces in the direction of its motion, above the work done upon it by them in a direction opposed to its motion; whence also it follows, by the last proposition, that the *accumulated* work is equal to this excess.   Therefore, &c.

*70. If P represent the force in the direction of the motion which at a given distance S, measured along the path, acts to accelerate the motion of the body, this force being understood not to be counteracted by any other, or to be the *surplus* force in the direction of the motion over and above any resistance opposed to it, then will $\int_0^S P\,dS$ be the work which must be done in an opposite direction to overcome this force through the space S, or $U = \int_0^S P\,dS,$

$\therefore$ by equation (46), $V^2 - v'^2 = \pm \dfrac{2g\int_0^S P\,dS}{W}$ . . . . . . (47).

71. If the force P tends at first towards the direction in which the body moves, so as to *accelerate* the motion, and if after a certain space has been described it *changes* its direction so as to *retard* the motion, and U_1 represent the value of U in respect to the former motion, and V_1 the velocity acquired when that motion has terminated, whilst U_2 is the value of U in respect to the second or retarded motion, and if v be the initial and V the ultimate or actual velocity, then

$$V_1^2 - v^2 = \frac{2gU_1}{W},$$

$$V^2 - V_1^2 = -\frac{2gU_2}{W},$$

$\therefore V^2 - v^2 = \dfrac{2g(U_1 - U_2)}{W}$ . . . . . (48).

As U_2 increases, the actual velocity V of the body continually diminishes; and when at length $U_2 = U_1$, that is, when the whole work done (above the resistances) in a direction opposite to the motion, comes to equal that done, before, in the direction of the motion, then $V = v$, or the velocity of the body returns again to that which it had when the force P began to act upon it. This is that general case of reciprocating motion which is so frequently presented in the combinations of machinery, and of which the crank motion is a remarkable example.

*72. If the force which accelerates the body's motion act always *towards* the same centre S, and Sb be taken equal to SB, it has been shown (Art. 55.) that the work necessary to move the body along the curve from B to A, is equal to that which would be necessary to move it through the straight line bA. The *accumulated* work is therefore equal to that necessary to move the body through the difference bA of the two distances SA and SB (Art. 68.). If these distances be represented by R_1 and R_2, and P represent the pressure with which the body's motion along bA would be resisted at any distance R from the point S, then $\int_{R_2}^{R_1} P dR$ will represent this work. Moreover the work *accumulated* in the body between A and B is represented by $\frac{1}{2}\frac{W}{g}(V^2 - v^2)$, if V represent the velocity at B and v that at A,

$$\therefore \ \frac{1}{2}\frac{W}{g}(V^2 - v^2) = \int_{R_2}^{R_1} P dR,$$

$$\therefore \ V^2 - v^2 = \frac{2g}{W}\int_{R_2}^{R_1} P dR \ . \ . \ . \ . \ . \ (49).$$

73. The work accumulated in the body while it descends the curve AB, is the same as that which it would acquire in falling directly towards S through the distance Ab, for both of these are equal to the work which would be necessary to raise

the body from b to A. Since then the work accumulated by
the body through AB is equal to that which it would accumulate if it fell through Ab, it follows that velocity acquired by
it in falling, from rest, through AB is equal to that which it
would acquire in falling through Ab. For if V represent
the velocity acquired in the one case, and V_1 that in the other,
then the accumulated work in the first case is represented by
$\frac{1}{2}\frac{W}{g}V^2$, and that in the second case by $\frac{1}{2}\frac{W}{g}V_1^2$, therefore
$\frac{1}{2}\frac{W}{g}V^2 = \frac{1}{2}\frac{W}{g}V_1^2$, therefore $V = V_1$.

From this it follows, that if a body descend, being projected
obliquely into free space, or sliding from rest upon any curved
surface or inclined plane, and be acted upon only by the
force of gravity (that is, subject to no friction or resistance of
the air or other retarding cause), then the velocity acquired
by it in its descent is precisely the same as though it had
fallen *vertically* through the same height.

74. DEFINITION. The ANGULAR VELOCITY of a body which
rotates about a fixed axis is the arc which every particle of
the body situated at a distance unity from the axis describes
in a second of time, if the body revolves *uniformly ;* or, if the
body moves with a *variable* motion, it is the arc which it
would describe in a second of time if (from the instant when
its angular velocity is measured) its revolution were to become uniform.

75. THE ACCUMULATION OF WORK IN A BODY WHICH
ROTATES ABOUT A FIXED AXIS.

Propositions 68 and 69. apply to every case of the
motion of a heavy body. In every such case the work
accumulated or lost by the action of any moving force or
pressure, whilst the body passes from any one position to

another, is equal to the work which must be done in an opposite direction, to cause it to pass back from the second position into the first. Let us suppose U to represent this work in respect to a body of any given dimensions, which has rotated about a fixed axis from one given position into another, by the action of given forces.

Let α be taken to represent the ANGULAR VELOCITY of the body after it has passed from one of these positions into another. Then since α is the actual velocity of a particle at distance unity from the axis, therefore the velocity of a particle at any other distance ρ_1 from the axis is $\alpha\rho_1$. Let μ represent the weight of each unit of the volume of the body, and m_1 the volume of any particle whose distance from the axis is ρ_1, then will the weight of that particle be μm_1; also its velocity has been shown to be $\alpha\rho_1$, therefore the amount of work accumulated in that particle is represented by

$$\frac{1}{2}\frac{\mu m_1}{g}\alpha^2\rho_1^2, \text{ or by } \frac{1}{2}\alpha^2\frac{\mu}{g}m_1\rho_1^2.$$

Similarly the different amounts of work accumulated in the other particles or elements of the body whose distances from the axis are represented by ρ_2, ρ_3, ... and their volumes by m_2, m_3, m_4 ...., are represented by $\frac{1}{2}\alpha^2\frac{\mu}{g}m_2\rho_2^2$, $\frac{1}{2}\alpha^2\frac{\mu}{g}m_3\rho_3^2$, &c.; so that the whole work accumulated is represented by the sum $\frac{1}{2}\alpha^2\frac{\mu}{g}m_1\rho_1^2 + \frac{1}{2}\alpha^2\frac{\mu}{g}m_2\rho_2^2 + \frac{1}{2}\alpha^2\frac{\mu}{g}m_3\rho_3^2 + \ldots\ldots$, or by

$$\frac{1}{2}\alpha^2\frac{\mu}{g}\{m_1\rho_1^2 + m_2\rho_2^2 + m_3\rho_3^2 + \ldots\ldots\}.$$

The sum $m_1\rho_1^2 + m_2\rho_2^2 + m_3\rho_3^2 + \ldots$, or $\Sigma m\rho^2$ taken in respect to all the particles or elements which compose the body, is called its MOMENT OF INERTIA in respect to the particular axis about which the rotation takes place. Let it be represented by I; then will $\frac{1}{2}\alpha^2 \cdot \left(\frac{\mu}{g}\right) \cdot$ I, represent the whole amount of work accumulated in the body whilst it has been made to acquire the angular velocity α from rest. If therefore U represent the work which must be done in an opposite direction to cause the body to pass back from its last position into its first,

$$\tfrac{1}{2}\alpha^2\left(\frac{\mu}{g}\right)\mathrm{I}=\mathrm{U},$$

$$\therefore \ \alpha^2=2\left(\frac{g}{\mu}\right)\frac{\mathrm{U}}{\mathrm{I}}\ \cdot\ \cdot\ \cdot\ \cdot\ \cdot\ (50).$$

If instead of the body's first position being one of rest, it had in its first position been moving with an angular velocity α_1 which had passed, in its second position, into a velocity α; and if U represent, as before, the work which must be done in an opposite direction, to bring this body back from its second into its first position, then is $\tfrac{1}{2}\alpha^2\left(\frac{\mu}{g}\right)\mathrm{I}-\tfrac{1}{2}\alpha_1{}^2\left(\frac{\mu}{g}\right)\mathrm{I}$,

or $\tfrac{1}{2}\left(\frac{\mu}{g}\right)(\alpha^2-\alpha_1{}^2)\mathrm{I}$, the work accumulated between the first and second positions; therefore

$$\left(\frac{\mu}{g}\right)(\alpha^2-\alpha_1{}^2)\mathrm{I}=\pm\,\mathrm{U},$$

$$\therefore \ \alpha^2=\alpha_1{}^2\pm2\left(\frac{g}{\mu}\right)\frac{\mathrm{U}}{\mathrm{I}}\ \cdot\ \cdot\ \cdot\ \cdot\ \cdot\ (51),$$

where the sign $\pm$ is to be taken according as the motion is accelerated or retarded between the first and second positions, since in the one case the angular velocity increases during the motion, so that α^2 is greater than $\alpha_1{}^2$, whilst in the latter case it diminishes, so that α^2 is less than $\alpha_1{}^2$.

76. If during one part of the *motion*, the work of the impressed forces tends to accelerate, and during another to retard it, and the work in the former case be represented by U_1, and in the latter by U_2, then

$$\alpha^2=\alpha_1{}^2+2\left(\frac{g}{\mu}\right)\frac{\mathrm{U}_1}{\mathrm{I}}-2\left(\frac{g}{\mu}\right)\frac{\mathrm{U}_2}{\mathrm{I}},$$

$$\therefore \ \alpha^2=\alpha_1{}^2+2\left(\frac{g}{\mu}\right)\frac{(\mathrm{U}_1-\mathrm{U}_2)}{\mathrm{I}}\ \cdot\ \cdot\ \cdot\ \cdot\ \cdot\ (52).$$

From this equation it follows that when $U_2 = U_1$, or when the work U_2 done by the forces which tend to *resist* the motion at length, equals that done by the forces which tend to *accelerate* the motion, then $a = a_1$, or the revolving body then returns again to the angular velocity from which it set out. Whilst, if U_2 *never* becomes equal to U_1 in the course of a revolution, then the angular velocity a does not return to its original value, but is increased at each revolution; and on the other hand, if U_2 becomes at each revolution *greater* than U_1, then the angular velocity is at each revolution diminished.

The greater the moment of inertia I of the revolving mass, and the greater the weight μ of its unit of volume (that is, the heavier the material of which it is formed), the less is the variation produced in the angular velocity a by any given variation of U or $U_1 - U_2$ at different periods of the *same* revolution, or from revolution to revolution; that is, the more *steady* is the motion produced by any variable action of the impelling force. It is on this principle that the fly-wheel is used to equalize the motion of machinery under a variable operation of the moving power, or of the resistance. It is simply a contrivance for increasing the moment of inertia of the revolving mass, and thereby giving steadiness to its revolution, under the operation of variable impelling forces, on the principles stated above. This great moment of inertia is given to the fly-wheel, by collecting the greater part of its material on the rim, or about the circumference of the wheel, so that the distance ρ of each particle which composes it, from the axis about which it revolves, may be the greatest possible, and thus the sum $\Sigma m\rho^2$, or I, may be the greatest possible. At the same time the greatest value is given to the quantity μ, by constructing the wheel of the heaviest material applicable to the purpose.

What has here been said will best be understood in its application to the CRANK.

77. If we conceive a constant pressure Q to act upon the

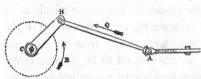

arm CB of the crank in the direction AB of the crank rod, and a constant resistance R to be opposed to the revolution of the axis C always at the same perpendicular distance from that axis, it is evident that since the perpendicular distance at which Q acts from the axis is continually varying (being at one time nothing, and at another equal to the whole length CB of the arm of the crank), the effective pressure upon the arm CB must at certain periods of each revolution *exceed* the constant resistance opposed to the motion of that arm, and at other periods *fall short* of it; so that the resultant of this pressure and this resistance, or the unbalanced pressure P upon the arm, must at one period of each revolution have its direction *in* the direction of the motion, and at another time *opposite* to it. Representing the work done upon the arm in the one case by U_1, and in the other by U_2, it follows that if $U_1 = U_2$ the arm will return in the course of each revolution, from the velocity which it had when the work U_1 began to be done, to that velocity again when the work U_2 is completed. If on the contrary U_1 exceed U_2, then the velocity will increase at each revolution; and if U_1 be less than U_2, it will diminish. It is evident from equation (52), that the greater is the moment of inertia I of the body put in motion, and the greater the weight μ of its unit of volume, the less is the variation in the value of α, produced by any given variation in the value of $U_1 - U_2$; the less therefore is the variation in the rotation of the arm of the crank, and of the machine to which it gives motion, produced by the varying action of the forces impressed upon it. Now the fly-wheel being fixed upon the same axis with the crank arm, and revolving with it, adds its own moment of inertia to that of the rest of the revolving mass, thereby increasing greatly the value of I, and therefore, on the principles stated above, equalizing the motion, whilst it does not otherwise increase the resistance to be overcome, than by the friction of its axis, and the resistance which the air opposes to its revolution.*

78. *The rotation of a body about a fixed axis when acted upon by no other moving force than its weight.*

Let U represent the work necessary to raise it from its second position into the first if it be *descending*, or from its first into its second position if it be *ascending*, and let α_1 be its angular velocity in the first position, and α in the second; then by equation (51),

$$\alpha^2 = \alpha_1^2 + 2\left(\frac{g}{\mu}\right)\left(\frac{U}{I}\right).$$

Now it has been shown (Art. 60.), that the work necessary to raise the body from its second position into the first if it be descending, or from its first into its second if it be ascending (its weight being the only force to be overcome), is the same as would be necessary to raise its whole weight collected in its centre of gravity from the one position into the other position of its centre of gravity. Let CA represent the one, and CA_1 the other position of the body, and G and G_1 the two corresponding positions of the centre of gravity, then will the work necessary to raise the body from its position CA to its position CA_1, be equal to that which is necessary to raise its whole weight W, supposed collected in G, from that point to G_1; which by Article 56. is the same as that necessary to raise it through the vertical height GM.

Let now $CG = CG_1 = h$, let CD be a vertical line through C, let $G_1 CD = \theta_1$ and $GCD = \theta$, in the case in which the body descends, and *conversely* when it ascends; therefore $GM = NN_1 = CN - CN_1 = h \cos.\ \theta - h \cos.\ \theta_1$ when the body descends, or $= h \cos.\ \theta_1 - h \cos.\ \theta$ when it ascends from the position AC to AC_1, since in this last case $GCD = \theta_1$ and $G_1 CD = \theta$. There-

---

* We shall hereafter treat fully of the crank and fly-wheel.

fore $GM = \pm h\,(\cos.\,\theta - \cos.\,\theta_1)$, the sign $\pm$ being taken according as the body ascends or descends.

Now $U = W\,.\,\overline{GM} = \pm Wh\,(\cos.\,\theta - \cos.\,\theta_1)$,

$\therefore$ by equation (51) $\alpha^2 = \alpha_1^2 + \left(\dfrac{2Wgh}{\mu I}\right)(\cos.\,\theta - \cos.\,\theta_1)$.

If M represent the volume of the revolving body $M\mu = W$,

$\therefore\ \alpha^2 = \alpha_1^2 + \left(\dfrac{2ghM}{I}\right)(\cos.\,\theta - \cos.\,\theta_1)\ \ldots\ldots\ (53)$.

When the body has descended into the vertical position, $\theta = 0$, so that $(\cos.\,\theta - \cos.\,\theta_1) = 1 - \cos.\,\theta_1 = 2\sin.^2\tfrac{1}{2}\theta_1$. When it has *ascended* into that position $\theta = \pi$, so that $(\cos.\,\theta - \cos.\,\theta_1)$ $= -(1 + \cos.\,\theta_1) = -2\cos.^2\tfrac{1}{2}\theta_1$.

In the first case, therefore,

$$\alpha^2 = \alpha_1^2 + \left(\frac{4ghM}{I}\right)\sin.^2\tfrac{1}{2}\theta_1\ \ldots\ldots\ (54).$$

In the second case,

$$\alpha^2 = \alpha_1^2 - \left(\frac{4ghM}{I}\right)\cos.^2\tfrac{1}{2}\theta_1\ \ldots\ldots\ (55).$$

When the body has descended or ascended into the horizontal position $\theta = \dfrac{\pi}{2}$, so that $(\cos.\,\theta - \cos.\,\theta_1) = -\cos.\,\theta_1$. But it is to be observed, that if the body have descended into the horizontal position, θ_1 must have been greater than $\dfrac{\pi}{2}$, and therefore $\cos.\,\theta_1$ must be negative and equal to $-\cos. BCG_1$ so that if we suppose θ_1 to be measured from CB or CD, according as the body descends or ascends, then $(\cos.\,\theta - \cos.\,\theta_1) = \pm\cos.\,\theta_1$, and we have for this case of descent or ascent to a horizontal position

$$\alpha^2 = \alpha_1^2 \pm \frac{2ghM}{I}\cos.\,\theta_1\ \ldots\ldots\ (56).$$

If the body descend from a state of rest, $\alpha_1 = 0$.

$\therefore$ by equation (53) $\alpha^2 = \dfrac{2ghM}{I}(\cos.\,\theta - \cos.\,\theta_1)\ \ldots\ (57)$.

Thus the angular velocity acquired from rest is less as the moment of inertia I is greater as compared with the volume M, or as the mass of the body is collected farther from its axis.

THE MOMENT OF INERTIA.

79. Having given the moment of inertia of a body, or system of bodies, about an axis passing through its centre of gravity, to find its moment of inertia about an axis, parallel to the first, passing through any other point in the body or system.

Let m_1 be any element of the body or system, m_1AG a plane perpendicular to the axis, about which the moments are to be measured, A the point where this plane is intersected by that axis, and G the point where it is intersected by the parallel axis passing through the centre of gravity of the body. Join AG, Am_1, Gm_1, and draw m_1M_1 perpendicular to AG. Let A$m_1 = \rho_1$, G$m_1 = r_1$, GM$_1 = x_1$, AG $= h$.

Now (Euclid, 2—12.), A$m_1^2 = \overline{AG}^2 + \overline{Gm_1}^2 + 2\overline{AG} \cdot \overline{GM_1}$,

or $\rho_1^2 = h^2 + r_1^2 + 2hx_1$.

If therefore the volume of the element be represented by m_1, and both sides of the above equation be multiplied by it,

$$\rho_1^2 m_1 = h^2 m_1 + r_1^2 m_1 + 2h x_1 m_1.$$

And if m_2, m_3, m_4, &c. represent the volumes of any other elements, and ρ_2, r_2, x_2; ρ_3, r_3, x_3, &c. be similarly taken in respect to those elements, then,

$$\rho_2^2 m_2 = h^2 m_2 + r_2^2 m_2 + 2h x_2 m_2,$$
$$\rho_3^2 m_3 = h^2 m_3 + r_3^2 m_3 + 2h x_3 m_3,$$
$$\&c. = \&c.$$

Adding these equations we have, $\rho_1^2 m_1 + \rho_2^2 m_2 + \rho_3^2 m_3 + \ldots$
$= h^2(m_1 + m_2 + m_3 + \ldots) + (r_1^2 m_1 + r_2^2 m_2 + r_3^2 m_3 + \ldots) + 2h(x_1 m_1 + x_2 m_2 + x_3 m_3 + \ldots)$,

or $\Sigma \rho^2 m = h^2 \Sigma m + \Sigma r^2 m + 2h \Sigma x m.$

G

Now Σxm is the sum of the moments of all the elements of the body about a plane perpendicular to AG, and passing through the *centre of gravity* G of the body.   Therefore (Art. 17.) $\Sigma xm = 0$,

$$\therefore \Sigma \rho^2 m = h^2 \Sigma m + \Sigma r^2 m.$$

Also $\Sigma \rho^2 m$ is the moment of inertia of the body about the given axis passing through A, and $\Sigma r^2 m$ is the moment of inertia about an axis parallel to this, passing through the centre of gravity of the body.   Let the former moment be represented by I_1, and the latter by I; and let the volume of the body Σm be represented by M,

$$\therefore \; I_1 = h^2 M + I \; \ldots \; . \; (58).$$

From which relation the moment of inertia (I_1) about any axis may be found, that (I) about an axis parallel to it, and passing through the centre of gravity of the body being known.

80. THE RADIUS OF GYRATION.   If we suppose k_1 to be the distance from the axis passing through A, at which distance, if the whole mass of the body were *collected*, the moment of inertia would remain the same, so that $k_1^2 M = I_1$, then k_1 is called the RADIUS OF GYRATION, in respect to that axis.

If k be the radius of gyration, similarly taken in respect to the axis passing through G_1, so that $k^2 M = I$, then, substituting in the preceding equation, and dividing by M,

$$k_1^2 = h^2 + k^2 \; \ldots \; . \; (59).$$

The following are examples of the determination of the moments of inertia of bodies of some of the more common geometrical forms, about the axes passing through their centres of gravity: they may thence be found about any other axes parallel to these, by equation (59).

*81. *The moment of inertia of a thin uniform rod about an axis perpendicular to its length and passing through its middle point.*

Let m represent an element of the rod contained between two plane sections perpendicular to its faces, the area of each of which is $\varkappa$, and whose distance from one another is $\Delta\rho$, and let $\varkappa$ and $\Delta\rho$ be so small that every point in this element may be considered to be at the same distance ρ from the axis A, about which the rod revolves. Then is the volume of the element represented by $\varkappa\Delta\rho$, and its moment of inertia about A by $\varkappa\rho^2\Delta\rho$. So that the whole moment of inertia I of the bar is represented by $\Sigma\varkappa\rho^2\Delta\rho$, or, since $\varkappa$ is the same throughout (the bar being uniform), by $\varkappa\Sigma\rho^2\Delta\rho$; or since $\Delta\rho$ is infinitely small, it is represented by the definite integral $\varkappa\displaystyle\int_{-\frac{1}{2}l}^{\frac{1}{2}l}\rho^2 d\rho$, where l is the whole length of the bar,

$$\therefore\ \mathrm{I}=\varkappa\{\tfrac{1}{3}(\tfrac{1}{2}l)^3-\tfrac{1}{3}(-\tfrac{1}{2}l)^3\},$$

$$\text{or } \mathrm{I}=\tfrac{1}{12}\varkappa l^3 \ . \ . \ . \ . \ . \ (60).$$

*82. *The moment of inertia of a thin rectangular lamina about an axis, passing through its centre of gravity, and parallel to one of its sides.*

It is evident that such a lamina may be conceived to be made up of an infinite number of slender rectangular rods of equal length, each of which will be bisected by the axis AB, and that the moment of inertia of the

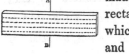

G 2

whole lamina is equal to the sum of the moments of inertia of these rods. Now if $\varkappa$ be the section of any rod, and l the length of the lamina, then the moment of inertia of that rod is, by the last proposition, represented by $\frac{1}{12}\varkappa l^3$; so that if the section of each rod be the same, and they be n in number, then the *whole* moment of inertia of the lamina is $\frac{1}{12}n\varkappa l^3$. Now $n\varkappa$ is the area of the transverse section of the lamina, which may be represented by K, so that the momentum of inertia of the lamina about the axis AB is represented by the formula

$$I = \frac{1}{12}Kl^3 \quad \cdots \quad (61).$$

*83. *The moment of inertia of a rectangular parallelopipedon about an axis, passing through its centre of gravity, and parallel to either of its edges.*

Let CD be a rectangular parallelopipedon, and AB an 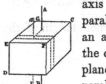 axis passing through its centre of gravity and parallel to either of its edges; also let ab be an axis parallel to the first, passing through the centre of gravity of a lamina contained by planes parallel to either of the faces of the parallelopiped. Let a, b, c, represent the three edges ED, EF, EG, of the parallelopiped, then will the momentum of inertia of the lamina about the axis ab be represented by $\frac{1}{12}Kb^3$, where K is the transverse section of the lamina (equation 61). Now let the perpendicular distance between the two axes AB and ab be represented by x. Then (by equation 58) the momentum of inertia of the lamina about the axis AB is represented by the formula $x^2M + \frac{1}{12}Kb^3$, where M represents the volume of the lamina. Let the thickness of the lamina be represented by Δx;

$\therefore$ M $= ab\Delta x$, K $= a\Delta x$; $\therefore$ m^t in^a of lam$^a = abx^2\Delta x + \frac{1}{12}ab^3\Delta x$;

$\therefore$ whole m^t in^a of parallelopiped $= ab\Sigma x^2\Delta x + \frac{1}{12}ab^3\Sigma\Delta x$; or taking Δx infinitely small, and representing the moment of inertia of the parallelopiped by I.

$$I = ab\int_{-\frac{1}{2}c}^{+\frac{1}{2}c} x^2 dx + \tfrac{1}{12}ab^3\int_{-\frac{1}{2}c}^{+\frac{1}{2}c} dx \; ;$$

or $I = ab\{\tfrac{1}{3}(\tfrac{1}{2}c)^3 - \tfrac{1}{3}(-\tfrac{1}{2}c)^3\} + \tfrac{1}{12}ab^3\{(\tfrac{1}{2}c) - (-\tfrac{1}{2}c)\}$
$$= \tfrac{1}{12}abc^3 + \tfrac{1}{12}ab^3c,$$

$$\therefore \; I = \tfrac{1}{12}abc(b^2 + c^2) \quad \cdots \cdots \; (62).$$

*84. *The moment of inertia of an upright triangular prism about a vertical axis passing through its centre of gravity.*

 Let AB be a vertical axis passing through the centre of gravity of a prism, whose horizontal section is an isosceles triangle having the equal sides ED and EF.

Let two planes be drawn parallel to the face DF of the prism, and containing between them a thin lamina pq of its volume. Let Cm, the perpendicular distance of an axis passing through the centre of gravity of this lamina from the axis AB, be represented by x; also let Δx represent the thickness of the lamina.

Let DF$=a$, DG$=b$, and let the perpendicular from the vertex E to the base DF of the triangle DEF be represented by c,

$$\therefore \; EC = \tfrac{2}{3}c, \; E m = \tfrac{2}{3}c - x; \; \text{also} \; \frac{pq}{DF} = \frac{E m}{c},$$

$$\therefore \; pq = \frac{a}{c}(\tfrac{2}{3}c - x); \; \text{also transverse section K of lamina} \; = b\Delta x.$$

$$\therefore \; \text{volume M of lamina} = \frac{ab}{c}(\tfrac{2}{3}c - x)\Delta x. \; \text{Therefore by equations (58) and (61),}$$

m^t in^a of lam^a about $AB = \dfrac{ab}{c}(\tfrac{2}{3}c - x)x^2\Delta x + \tfrac{1}{12}b\dfrac{a^3}{c^3}(\tfrac{2}{3}c - x)^3\Delta x \; ;$

$$\therefore \; \mathrm{m^t\,in^a\,of\,prism\,about\,AB} = \frac{ab}{c}\int_{-\frac{1}{2}c}^{+\frac{3}{2}c}(\tfrac{2}{3}c-x)x^2dx + \tfrac{1}{12}\frac{ba^3}{c^3}\int_{-\frac{1}{2}c}^{+\frac{3}{2}c}(\tfrac{2}{3}c-x)^3dx.$$

Performing the integrations here indicated, and representing the inertia of the prism about AB by I, we have

$$I = \tfrac{1}{12}abc(\tfrac{1}{4}a^2 + \tfrac{1}{3}c^2) \quad \cdots \cdots \quad (63).$$

*85. *The moment of inertia of a solid cylinder about its axis of symmetry.*

Let AB be the axis of such a cylinder, whose radius AC is represented by a, and its height by b. Con-
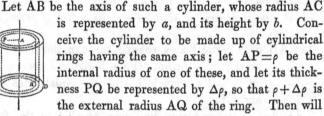
ceive the cylinder to be made up of cylindrical rings having the same axis; let $AP=\rho$ be the internal radius of one of these, and let its thickness PQ be represented by $\Delta\rho$, so that $\rho+\Delta\rho$ is the external radius AQ of the ring. Then will the *volume* of the ring be represented by $\pi b(\rho+\Delta\rho)^2 - \pi b\rho^2$, or by $\pi b[2\rho\Delta\rho+(\Delta\rho)^2]$; or if $\Delta\rho$ be taken exceedingly small, so that $(\Delta\rho)^2$ may vanish as compared with $2\rho\Delta\rho$, then is the volume of the ring represented by $2\pi b\rho\Delta\rho$.

Now this being the case, the ring may be considered as an element ΔM of the volume of the solid, every part of which element is at the same distance ρ from the axis AB, so that the whole moment of inertia $\Sigma\rho^2\Delta M$ of the cylinder $=\Sigma\rho^2(2\pi b\rho\Delta\rho)=2\pi b\Sigma\rho^3\Delta\rho$,

$$\therefore \; I = 2\pi b\int_0^a \rho^3 d\rho = \tfrac{1}{2}\pi ba^4 \quad \cdots \cdots \quad (64).$$

*86. *The moment of inertia of a hollow cylinder about its axis of symmetry.*

Let a_1 be the external radius AC, and a_2 the internal

radius AP, and b the height of the cylinder; then by the last proposition the moment of inertia of the cylinder CD, if it were solid, would be $\frac{1}{2}\pi b a_1^4$; also the moment of inertia of the cylinder PR, which is taken from this solid to form the hollow cylinder, would be $\frac{1}{2}\pi b a_2^4$. Now let I represent the moment of inertia of the hollow cylinder CP, therefore $I + \frac{1}{2}\pi b a_2^4 = \frac{1}{2}\pi b a_1^4$,

$$\therefore\ I = \frac{1}{2}\pi b(a_1^4 - a_2^4) = \frac{1}{2}\pi b(a_1^2 - a_2^2)(a_1^2 + a_2^2) = \frac{1}{2}\pi b(a_1 - a_2)(a_1 + a_2)(a_1^2 + a_2^2).$$

Let the thickness $a_1 - a_2$ of the hollow cylinder be represented by c, and its mean radius $\frac{1}{2}(a_1 + a_2)$ by R, therefore $a_1 = R + \frac{1}{2}c,\ \ a_2 = R - \frac{1}{2}c.$

Substituting these values in the preceding equation, we obtain

$$I = 2\pi bcR\{R^2 + \tfrac{1}{4}c^2\} \ \ \cdots \ \ (65).$$

*87. *The moment of inertia of a cylinder about an axis passing through its centre of gravity, and perpendicular to its axis of symmetry.*

Let AB be such an axis, and let PQ represent a lamina contained between planes perpendicular to this axis, and exceedingly near to each other. Let CD, the axis of the cylinder, be represented by b, its radius by a, and let $CM = x$. Take Δx to represent the thickness of the lamina, and let $MP = y$. Now this lamina may be considered a rectangular parallelopiped traversed through its centre of gravity by the axis AB; therefore by equation (62) its moment of inertia about that axis is represented by $\frac{1}{12}(\Delta x)b(2y)\{b^2 + (2y)^2\} = \frac{1}{6}b\{b^2 y + 4y^3\}\Delta x.$ Now the whole moment of inertia I of the cylinder about AB is evidently equal to the sum of the moments of inertia of all such laminæ;

$$\therefore\ I = \frac{1}{6}b\Sigma\{b^2 y + 4y^3\}\Delta x = \frac{1}{6}b\int_{-a}^{a}(b^2 y + 4y^3)dx.$$

Also, since x and y are the co-ordinates of a point in a circle from its centre, therefore $y=(a^2-x^2)^{\frac{1}{2}}$. Substituting this value of y, and integrating according to the well known rules of the integral calculus, we have

$$I=\tfrac{1}{4}\pi ba^2(a^2+\tfrac{1}{3}b^2) \ \ . \ . \ . \ . \ (66).$$

*88. *The moment of inertia of a cone about its axis of symmetry.*

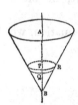

The cone may be supposed to be made up of laminæ, such as PQ, contained by planes perpendicular to the axis of symmetry AB, and each having its centre of gravity in that axis. Let BP=x, and let Δx represent the thickness of the lamina, and y its radius PR. Then, since it may be considered a cylinder of very small height, its moment of inertia about AB (equation 64) is represented by $\tfrac{1}{2}\pi y^4\Delta x$. Now the moment of inertia I of the whole cone is equal to the sum of the moments of all such elements,

$$\therefore \ I=\tfrac{1}{2}\pi\Sigma y^4\Delta x.$$

Let the radius of the base of the cone be represented by a, and its height by b; therefore $\dfrac{x}{y}=\dfrac{b}{a}$, therefore $\Delta x=\dfrac{b}{a}\Delta y$;

$$\therefore \ I=\tfrac{1}{2}\pi\frac{b}{a}\Sigma y^4\Delta y=\tfrac{1}{2}\pi\frac{b}{a}\int_0^a y^4 dy;$$

$$\therefore \ I=\tfrac{1}{10}\pi ba^4 \ \ . \ . \ . \ . \ (67).$$

89. *The momentum of inertia of a sphere about one of its diameters.*

Let C be the centre of the sphere and AB the diameter

about which its moment is to be determined. Let PQ be any lamina contained by planes perpendicular to AB ; let $CM = x$, and let Δx represent the thickness of the lamina, and y its radius; also let $CA = a$; then since this lamina, being exceedingly thin, may be considered a cylinder, its moment of inertia about the axis AB is (equation 64) $\frac{1}{2}\pi y^4 \Delta x$; and the moment of inertia I of the whole sphere is the sum of the moments of all such laminæ,

$$\therefore \ I = \frac{1}{2}\pi \Sigma y^4 \Delta x = \frac{1}{2}\pi \int_{-a}^{+a} y^4 dx.$$

Now by the equation to the circle $y^2 = a^2 - x^2$, therefore $y^4 = a^4 - 2a^2 x^2 + x^4$. If this value be substituted for y^4, and the integration be completed according to the common methods, we shall obtain the equation,

$$I = \tfrac{8}{15}\pi a^5 \ \ldots \ldots \ (68).$$

90. *The moment of inertia of a cone about an axis passing through its centre of gravity and perpendicular to its axis of symmetry.*

Let CD be an axis passing through the centre of gravity G of the cone, and perpendicular to its axis of symmetry, and let GP the distance of the lamina from G, measured along the axis, be represented by x; also let the thickness of the lamina be represented by Δx. Now this lamina may be considered a cylinder of exceedingly small thickness. If its radius be represented by y, its momentum of inertia about an axis parallel to CD passing through its centre, is therefore (equation 66) represented by $\frac{1}{4}\pi y^2 \{y^2 + \frac{1}{3}(\Delta x)^2\}\Delta x$, or if Δx be assumed exceedingly small, it is represented by $\frac{1}{4}\pi y^4 \Delta x$. Now this being the momentum of the lamina about an axis parallel to CD, passing through its centre of gravity, and the distance of this axis from CD being x, and also the volume of the lamina being $\pi y^2 \Delta x$, it follows (equation 58), that the moment of the lamina about CD is represented by $\pi y^2 x^2 \Delta x + \frac{1}{4}\pi y^4 \Delta x = \pi\{y^2 x^2 + \frac{1}{4}y^4\}\Delta x$.

Now the moment I of the whole cone about CD equals the sum of the moments of all such elements,

$$\therefore \ I = \pi\Sigma(y^2 x^2 + \frac{1}{4}y^4)\Delta x.$$

Now if a be the radius of the base of the cone and b its height, then since $BG = \frac{3}{4}b$,

$$\therefore \ \frac{\frac{3}{4}b - x}{y} = \frac{b}{a}; \ \therefore \ x = \frac{b}{a}(\tfrac{3}{4}a - y) \ \text{ and } \ \Delta x = -\frac{b}{a}\Delta y ;$$

$$\therefore \ I = -\pi\frac{b}{a}\Sigma\left\{\frac{b^2}{a^2}(\tfrac{3}{4}a - y)^2 y^2 + \tfrac{1}{4}y^4\right\}\Delta y,$$

$$\therefore \ I = -\pi\frac{b}{a}\int_a^0\left\{\frac{b^2}{a^2}(\tfrac{3}{4}a - y)^2 y^2 + \tfrac{1}{4}y^4\right\}dy,$$

$$\therefore \ I = \tfrac{1}{20}\pi a^2 b\{a^2 + \tfrac{1}{4}b^2\} \ \ . \ . \ . \ . \ (69).$$

91. *The moment of inertia of a segment of a sphere about a diameter parallel to the plane of section.*

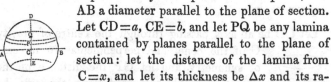

Let ADBE represent any such portion of a sphere, and AB a diameter parallel to the plane of section. Let $CD = a$, $CE = b$, and let PQ be any lamina contained by planes parallel to the plane of section: let the distance of the lamina from $C = x$, and let its thickness be Δx and its radius y. Then considering it a cylinder of exceeding small thickness, its momentum of inertia about an axis passing through its centre of gravity and parallel to AB, is represented (equation 66) by $\frac{1}{4}\pi y^2\{y^2 + \frac{1}{3}(\Delta x)^2\}\Delta x$, or (neglecting powers of Δx above the first) by $\frac{1}{4}\pi y^4\Delta x$. Hence, therefore, the moment of this lamina about the axis AB is represented (equation 58) by $\pi y''(\Delta x)x^2 + \frac{1}{4}\pi y^4\Delta x$, or by $\pi\{y^2 x^2 + \frac{1}{4}y^4\}\Delta x$; now the whole moment I of inertia of ADBE about AB is evidently equal to the sum of the moments of all such laminæ,

$$\therefore \ I = \pi\Sigma\{y^2 x^2 + \tfrac{1}{4}y^4\}\Delta x = \pi\int_{-b}^{+a}(y^2 x^2 + \tfrac{1}{4}y^4)dx.$$

Now $y^2 = a^2 - x^2$, therefore $y^2 x^2 + \frac{1}{4} y^4 = \frac{1}{4} \{2a^2 x^2 - 3x^4 + a^4\}$. Substituting this value in the integral and integrating, we have

$$I = \frac{1}{60} \pi \{16a^5 + 15a^1 b + 10a^2 b^3 - 9b^5\} \quad \ldots \ldots \quad (70).$$

THE ACCELERATION OF MOTION BY GIVEN MOVING FORCES.

92. If the forces applied to a moving body in the direction of its motion exceed those applied to it in the opposite direction (both sets of forces being resolved in the direction of a tangent to its path), the motion of the body will be *accelerated;* if they fall short of those applied in the opposite direction, the motion will be *retarded.* In either case the excess of the one set of forces above the other is called the MOVING FORCE upon the body: it is measured by that single pressure which being applied to the body in a direction opposite to the greater force, would just balance it; or which, had it been applied to the body (together with the other forces impressed upon it) when in a state of rest, would have maintained it in that state; and which therefore, if applied when its motion had commenced, would have caused it to pass from a state of *variable* to one of *uniform* motion. Thus the moving force upon a body which descends freely by gravity, is measured by its *weight,* that is, by the single force which, being applied to the body before its motion had commenced in a direction opposite to gravity, would just have supported it, and which being applied to it at any instant of its descent, would have caused its motion at that instant to pass from a state of variable to a state of uniform motion. If the resistance of the air upon its descent be taken into account, then the moving force upon the body at any instant is measured by that single pressure which, being applied upwards, would, together with the resistance of the air at that instant, just balance the weight of the body.

A moving force being thus understood to be measured by a *pressure**, being in fact the *unbalanced* pressure upon the moving body, the following relations between the amount of a moving force thus measured, and the degree of acceleration produced by it will become intelligible. These are *laws* of motion which have become known by experiment upon the motions of the bodies immediately around us, and by observation upon those of the planets.

93. When the moving force upon a body remains constantly the same in amount (as measured by the equivalent pressure) throughout the motion, or is a *uniform* moving force, it communicates to it equal additions of velocity in equal successive intervals of time. Thus the moving force upon a body descending freely by gravity (measured by its weight) being constantly the same in amount throughout its descent (the resistance of the air being neglected), the body receives from it equal additions of velocity in equal successive intervals of time, viz. $32\frac{1}{6}$ feet in each successive *second* of time (Art. 44.).

94. The increments of velocity communicated to *equal bodies* by unequal moving forces (supposed *uniform* as above) are to one another as the amounts of those moving forces (measured by their equivalent pressures).

Thus let P and P_1 be any two unequal moving forces upon two equal bodies, and let them act in the directions in which the bodies respectively move ; let them be the only forces tending to communicate motion to those bodies, and remain constantly the same in amount throughout the motion. Also let f and f_1 represent the additional velocities which these two forces respectively communicate to those two equal bodies in each successive second of time ; then it is a *law* of the motion of bodies, determined by observation and experiment, that $P : P_1 :: f : f_1$.

* Pressure and moving force are indeed but different modes of the operation of the same principle of force.

If one of the moving forces, as for instance P_1, be the *weight* W of the body moved, then the value f_1 of the increment of velocity per second corresponding to that moving force is $32\frac{1}{6}$ (Art. 44.) represented by g,

$$\therefore \ P : W :: f : g,$$

$$\therefore \ P = \frac{W}{g} f \ \ldots \ldots \ (71).$$

95. If the amount or magnitude of the moving force does not remain the same throughout the motion, or if it be a *variable* moving force, then the increments of velocity communicated by it in equal successive intervals of time are not equal; they increase continually if the moving force increases, and they diminish if it diminishes.

If two unequal moving forces, one or both of them, thus *variable* in magnitude, become the moving forces of two equal bodies, the additional velocities which they would communicate in the same interval of time to those bodies, if at any period of the motion from *variable* they become *uniform*, are to one another (Art. 94.) as the respective moving forces at that period of the motion.

Thus let f and f_1 represent the additional velocities which *would* thus be communicated to two equal bodies in one second of time, if at any instant the pressures P and P_1, which are at that instant the moving forces of those bodies, were from *variable* to become constant pressures, then (Art. 94.),

$$P : P_1 :: f : f_1.$$

This being true of any two moving forces, is evidently true, if one of them become a constant force. Let P_1 represent the weight W of the body, then will f_1 be represented by g,

$$\therefore \ P : W :: f : g.$$

Let the moving force P be supposed to remain constant during a number of seconds or parts of a second, represented by Δt, and let ΔV be the increment of velocity in the time

Δt on this supposition. Now f represents the increment of velocity in each second, and ΔV the increment of velocity in Δt seconds : moreover the force P is supposed constant during Δt, so that the motion is *uniformly* accelerated during that time (Art. 44.).

$$\therefore f\Delta t = \Delta V, \quad \therefore f = \frac{\Delta V}{\Delta t}.$$

Now this is true (if the supposition, that P remains constant during the time Δt, on which it is founded, be true), however small the time Δt may be. But if this time be infinitely small, the supposition on which it is founded is in all cases true, for P may in all cases be considered to remain the same during an infinitely small period of time, although it does not remain the same during any time which is not infinitely small. Now when Δt is infinitely small, $\frac{\Delta V}{\Delta t} = \frac{dV}{dt}$; generally therefore $f = \frac{dV}{dt}$.

If V increase as the time t increases, or if the motion be accelerated, then $\frac{dV}{dt}$ is necessarily a positive quantity. If, on the contrary, V diminishes as the time increases, then $\frac{dV}{dt}$ is negative ; so that, generally,

$$f = \pm \frac{dV}{dt} \quad \cdots \cdots (72),$$

the sign $\pm$ being taken according as the motion is accelerated or retarded. Substituting this value of f in the last proportion we have in the case, in which P represents a variable pressure,

$$P = \pm \frac{W}{g} \frac{dV}{dt} \quad \cdots \cdots (73).$$

The principles stated above constitute the fundamental relations of pressure and motion.

96. The velocity V at any instant of a body moving with a *variable* motion, being the space which it would describe in a second of time, if at that instant its motion were to become uniform, it follows, that if we represent by Δt any number of seconds or parts of a second, beginning from that instant, and by ΔS, the space which the body would describe in the time Δt, if its motion continued uniform from the commencement of that time, then,

$$V \Delta t = \Delta S, \quad \therefore V = \frac{\Delta S}{\Delta t}.$$

Now this is true if the motion remain uniform during the time Δt, however small that time may be, and therefore if it be *infinitely* small. But if the time Δt be *infinitely* small, the motion does remain uniform during that time, however variable may be the moving force; also when Δt is infinitely small, $\dfrac{\Delta S}{\Delta t} = \dfrac{dS}{dt}$. Therefore, generally,

$$V = \frac{dS}{dt} \quad \ldots \ldots \quad (74).$$

The equations (73) and (74) are the fundamental equations of dynamics: they involve those dynamical results which have been discussed on other principles in the preceding parts of this work.*

THE DESCENT OF A BODY UPON A CURVE.

*97. *If the moving force* P *upon a body varies directly as its*

---

* Thus if the latter equation be inverted, and multiplied by the former, we obtain the equation

$$P \frac{dS}{dt} = \pm \frac{W}{g} \cdot V \left(\frac{dV}{dt} \right) = \pm \frac{W}{2g} \left(\frac{dV^2}{dt} \right),$$

$$\therefore \frac{dV^2}{dS} = \pm \frac{2g}{W} P,$$

$$\therefore V^2 - v^2 = \pm \frac{2g}{W} \int_{S_2}^{S_1} P dS,$$

which is identical with equation (47).

distance at any time from a given point towards which it falls, then the whole time of the body's falling to that point will be the same, whatever may be the distance from which it falls.

Let A be the point from which the body falls, and B a point *towards* which it falls along the path APB, which may be either curved or straight; also let the body be acted upon at each point P of its path, by a force in the direction of its path at that point which varies as its distance BP, measured along the path from B; the time of falling to B will be the same, whatever may be the distance of the point A from which the body falls.

For let $BP = S$, and let the force impelling the body towards S be represented by cS, where c is a constant quantity; suppose the body, instead of falling from A towards B, to be projected with any velocity from B towards A, and let v be the velocity acquired at P, and V that at A, and let $BA = S_1$, then by equation (47),

$$V^2 - v^2 = -\frac{2g}{W}\int_S^{S_1} cS dS = -\frac{cg}{W}(S_1{}^2 - S^2).$$

Suppose now the velocity of projection from B to have been such as would only just carry the body to A, so that $V = 0$,

$$\therefore \; v^2 = \frac{cg}{W}(S_1{}^2 - S^2) \; \ldots \ldots \; (75).$$

Now by equation (74),

$$\frac{dt}{dS} = \frac{1}{v}, \quad \therefore \; t = \int\frac{dS}{v},$$

$$\therefore \; t = \int \frac{dS}{\left(\dfrac{cg}{W}\right)^{\frac{1}{2}}(S_1{}^2 - S^2)};$$

and if $\frac{1}{2}T$ represent the whole time in seconds occupied in the ascent of the body from B to A,

$$\tfrac{1}{2}T = \left(\frac{W}{cg}\right)^{\frac{1}{2}}\int_0^{S_1}\frac{dS}{(S_1{}^2 - S^2)^{\frac{1}{2}}} = -\left(\frac{W}{cg}\right)^{\frac{1}{2}}\left\{ \cos.^{-1}\frac{S_1}{S_1} - \cos.^{-1}\frac{0}{S_1} \right\},$$

$$\therefore \tfrac{1}{2}T = \left(\frac{W}{cg}\right)^{\frac{1}{2}}\frac{\pi}{2}.$$

It is clear that the time required for the body's descent from A to B is equal to that necessary for the ascent from B to A, so that the whole time required to complete the ascent and descent is equal to T, and is represented by the formula

$$T = \left(\frac{W}{cg}\right)^{\frac{1}{2}}\pi \; , \; \dots \; (76).$$

Now this expression does not contain S_1, *i.e.* the distance from which the body falls to B; the time T is the same therefore, whatever that distance may be.

THE SIMPLE PENDULUM.

98. *If a heavy particle P be imagined to be suspended from a point C by a thread without weight, and allowed to oscillate freely, but so as to deviate but little on either side of the vertical, then will its oscillations, so long as they are thus small, be performed in the same time whatever their amplitudes may be.*

For let the inclination PCB of CP to the vertical be represented by θ, and let the weight w of the particle P, which acts in the direction of the vertical VP, be resolved into two others, one of which is in the direction CP, and the other perpendicular to that direction; the former will be wholly counteracted by the tension of the thread CP, and the latter will be represented by w sin. $VPC = w$ sin. θ; and, acting in the direction in which the particle P moves, this will be the *whole* impressed *moving* force upon it (Art. 92.). Now so long as the arc θ is small, this arc does not differ sensibly from its sine, so that for *small* oscillations the impressed moving force upon P is represented by $w\theta$, or by $\frac{w(l\theta)}{l}$, or by $\frac{wS}{l}$, if l represent the length CP of the suspending thread, and S the length

H

of the arc BP. Now in this expression w and l are constant throughout the oscillation, the moving force varies therefore as S. Hence by the last proposition, the small oscillations on either side of CB are isochronous, since so long as they are thus small, the impressed moving force in the direction of the motion varies as the length of the path BP from the lowest point B. Since in the last proposition the moving force was assumed equal to cS, and that here it is represented by $\frac{w}{l}$S, therefore in this case $c = \frac{w}{l}$. Substituting this value in equation (76),

$$T = \left(\frac{l}{g}\right)^{\frac{1}{2}} \pi \ \ldots \ \ldots \ (77).$$

A single particle thus suspended by a thread without weight, is that which is meant by a SIMPLE PENDULUM. It is evident that the time of oscillation increases with the length l of the pendulum.

IMPULSIVE FORCE.

99. If any number of different moving forces be applied to as many equal bodies, the velocities communicated to them in the same exceedingly small interval of time, will be to one another as the moving forces. For let P_1, P_2, represent the moving forces, and f_1, f_2, the additional velocities they would communicate per second if each moving force remained continually of the same magnitude (Art. 93.), then would tf_1, tf_2, be the whole velocities communicated on this supposition in t seconds; let these be represented by V_1, V_2; therefore by Art. 94.

$$P_1 : P_2 :: f_1 : f_2 :: tf_1 : tf_2 :: V_1 : V_2.$$

The proposition is therefore true on the supposition that P_1 and P_2 remain constant during the interval of time t; but if t be exceedingly small, then whatever the pressures P_1 and P_2 may be, they may be considered to remain the same during that time. Therefore the proposition is true *generally*, when, as above, the moving forces are supposed to act on equal bodies, or successively on the same body, through equal exceedingly small intervals of time.

Moving forces thus acting through exceedingly small intervals of time only, are called IMPULSIVE FORCES.

THE PARALLELOGRAM OF MOTION.

100. *If two impulsive forces* P_1, P_2, *whose directions are* AB *and* AC, *be impressed at the same time upon a body at* A, *which if made to act upon it separately would cause it to move through* AB *and* AC *in the same given time, then will the body be made, by the simultaneous action of these impulsive forces, to describe in that time the diagonal* AD *of the parallelogram, of which* AB *and* AC *are adjacent sides.*

For the moving forces P_1 and P_2 acting separately upon the same body through equal infinitely small times, communicate to it velocities which are (Art. 99.) as those forces, therefore the spaces AB and AC described with these velocities in any given time are also as those forces.. Since then AB and AC are to one another as the pressures P_1 and P_2, therefore by the principle (Art. 2.) of the parallelogram of pressures, the *resultant* R of P_1 and P_2 is in the direction of the diagonal AD, and bears the same proportion to P_1 and P_2 that AD does to AB and AC.

Therefore the velocity which the resultant R of P_1 and P_2 would communicate to the body in any exceedingly small time is to the velocities which P_1 and P_2 would separately communicate to it in the same time as AD to AB and AC (Art. 99.), and therefore the spaces which the body would describe uniformly with these three velocities in *any* equal times are in the ratio of these three lines. But AB and AC are the spaces actually described in the equal times by reason of the impulses of P_1 and P_2. Therefore AD is the space de-

H 2

scribed in that time by reason of the impulse of R, that is by reason of the simultaneous impulses of P_1 and P_2.

101. THE INDEPENDENCE OF SIMULTANEOUS MOTIONS.

It is evident that if the body starting from A had been made to describe AB in a given time, and then had been made in an equal time to describe BD, it would have arrived precisely at the same point D to which the simultaneous motions AC and AB have brought it, so that the body is made to move by these *simultaneous* motions precisely to the same point to which it would have been brought by those motions, communicated to it successively, but in half the time. The following may be · taken as an illustration of this principle of the independence of simultaneous motions. Let a canal-boat be imagined to extend across the whole width of the canal, and let it be supposed that a person standing on the one bank at A is desirous to pass to a point D on the opposite bank, and that for this purpose, as the boat passes him, he steps into it, and walks across it in the direction AB, arriving at the point B in the boat precisely at the instant when the motion of the boat has carried it through BD; it is clear that he will be brought, by the joint effect of *his own* motion across the boat and the *boat's* motion along the canal, to the point D (having in reality described the diagonal AD), which point he would have reached in double the time if he had walked across a bridge from A to B in the same time that it took him to walk across the boat, and had then in an equal time walked from B to D along the opposite side.

THE POLYGON OF MOTION.

102. Let any number of impulses be communicated simul-

taneously to a body at O, one of which would cause it to move from A to O in a given time, another from B to O in the same time, a third from C to O in that time, and a fourth from D to O. Complete the parallelogram of which AO and BO are adjacent sides; then the impulses AO and BO would simultaneously cause the body to move from E to O through the diagonal EO in the time spoken of. Complete the parallelogram EOCF, and draw its diagonal OF, then would the impulses EO and CO, acting simultaneously, cause the body to move through FO in the given time: but the impulse EO produces the same effect on the body as the impulses AO and BO; therefore the impulses AO, BO, and CO, will together cause the body to move through FO in the given time. In the same manner it may be shown that the impulses AO, BO, CO, and DO, will together cause the body to move through GO in a time equal to that occupied by the body's motion through any one of these lines.

It will be observed that GD is the side which completes the polygon OAEFG, whose other sides OA, AE, EF, FG, are respectively equal and parallel to the directions OA, OB, OC, and OD, of the simultaneous impulses.

Instead of the impulses AO, &c. taking place simultaneously, if they had been received successively, the body moving first from O to A in a given time; then through AE, which is equal and parallel to OB, in an equal time; then through EF, which is equal and parallel to OC, in that time; and lastly through FG, which is equal and parallel to OD, in that time, it would have arrived at the same point G, to which these impulses have brought it simultaneously, but after a period as many times greater as there are motions, so that the principle of the independence of simultaneous motions obtains, however great may be the number of such motions.

THE PRINCIPLE OF D'ALEMBERT.

103. Let W_1, W_2, W_3, &c. represent the weights of any

number of bodies in motion, and P_1, P_2, P_3, &c. the *moving forces* (Art. 92.) upon these bodies at any given instant of the motion (*i. e.* the *unbalanced* pressures, or the pressures which are wholly employed in producing their motion, and pressures equal to which, applied in opposite directions, would bring them to rest, or to a state of uniform motion). Then (Art. 95.), $P_1 = \dfrac{W_1}{g}f_1$, $P_2 = \dfrac{W_2}{g}f_2$, $P_3 = \dfrac{W_3}{g}f_3$, &c. where f_1, f_2, f_3, &c. represent the additions of velocity which the bodies would receive in each second of time, if the moving force upon each were to become, at the instant at which it is measured, an *uniform* moving force. Suppose these bodies, whose weights are W_1, W_2, W_3, &c. to form a *system* of bodies united together by any conceivable mechanical connection, on which system are impressed, in any way, certain forces, whence result the unbalanced pressures P_1, P_2, P_3, &c. on the moving points of the system. Now conceive that to these moving points of the system there are applied pressures respectively equal to P_1, P_2, P_3, &c. but each in a direction opposite to that in which the motion of the corresponding point is accelerated or retarded. Then will the motion of each particular point evidently pass into a state of *uniform* motion, or of *rest* (Art. 92.). The whole system of bodies being thus then in a state of uniform motion, or of rest, the forces applied to its different elements must be forces in equilibrium.

Whatever, therefore, were the forces originally impressed upon the system, and causing its motion, they must, together with the pressures P_1, P_2, P_3, &c. thus applied, produce a state of equilibrium in the system ; so that these forces (originally impressed upon the system, and known in Dynamics as the IMPRESSED FORCES) have to the forces P_1, P_2, P_3, &c. when applied in directions opposite to the motions of their several points of application, the relation of forces in equilibrium. The forces P_1, P_2, P_3, &c. are known in Dynamics as the EFFECTIVE FORCES. *Thus in any system of bodies mechanically connected in any way, so that their motions may mutually influence one another, if forces equal to the effective forces were applied in directions opposite to their actual directions, these would be in equilibrium with the impressed forces, which is the principle of d'Alembert.*

104. *The work accumulated in a moving body through any space is equal to the work which must be done upon it, in an opposite direction, to overcome the effective force upon it through that space.*

This is evident from Arts. 68 and 69., since the effective force is the unbalanced pressure upon the body.

If the work of the effective force be said to be *done* upon the body*, then the work of the effective force *upon* it is equal to the work or power accumulated in it, and this work of the effective force may be all said to be *actually* accumulated in the body as in a reservoir.

MOTION OF TRANSLATION.

DEFINITION.   When a body moves forward in space, without at the same time revolving, so that all its parts move with the same velocity and in parallel directions, it is said to move with a *motion of translation* only.

105. *In order that a body may move with a motion of translation only, the resultant of the forces impressed upon it must have its direction through the centre of gravity of the body.*

For let w_1, w_2, w_3, &c. represent the weights of the parts or elements of the body, and let f represent the additional velocity per second, which any element receives or would receive if its motion were at any instant to become uniformly accelerated.   Since the motion is one of translation only, the value of f is evidently the same in respect to every other element.   The effective forces P_1, P_2, P_3, &c. on the different elements of the body are therefore represented by $\dfrac{w_1}{g}f$, $\dfrac{w_2}{g}f$, $\dfrac{w_3}{g}f$, &c. &c.

---

* This cannot perhaps be *correctly* said, since work supposes *resistance*.

Now the forces P_1, P_2, P_3, &c. are evidently *parallel* pressures. Let X be the distance of the centre (see Art. 17.) of these parallel pressures from any given plane ; and let x_1, x_2, x_3, &c. be the perpendicular distances of the elements w_1, w_2, $w_3^!$, &c. that is, of the points of application of P_1, P_2, P_3, &c. from the same plane. Therefore (by equation 18),

$$\{P_1+P_2+P_3+\ \ldots\ \}X=P_1x_1+P_2x_2+P_3x_3+\ \ldots;$$

$$\therefore\ \left\{\frac{w_1}{g}f+\frac{w_2}{g}f+\frac{w_3}{g}f+\ \ldots\ \right\}X=\frac{w_1}{g}fx_1+\frac{w_2}{g}fx_2+\frac{w_3}{g}fx_3+\ \ldots,$$

$$\therefore\ \{w_1+w_2+w_3+\ \ldots\ \}X=w_1x_1+w_2x_2+w_3x_3+\ \ldots,$$

$$\therefore\ X=\frac{w_1x_1+w_2x_2+w_3x_3+\ \ldots}{w_1+w_2+w_3+\ \ldots}.$$

But this is the expression (Art. 19.) for the distance of the centre of gravity from the given plane ; and this being true of any plane, it follows that the *centre* of the parallel pressures P_1, P_2, P_3, &c. which are the *effective* forces of the system, coincides with the centre of gravity of the system, and therefore that the resultant of the effective forces passes through the centre of gravity. Now the resultant of the effective pressures must coincide in direction with the resultant of the *impressed* pressures, since the effective pressures when applied in an opposite direction are in *equilibrium* with the impressed pressures (by d'Alembert's principle). The resultant of the impressed pressures must therefore have its direction through the centre of gravity. Therefore, &c.

MOTION OF ROTATION ABOUT A FIXED AXIS.

106. Let a rigid body or system be capable of motion about the axis A. Let m_1, m_2, m_3, &c. represent the volumes of elements of this body, and μ the weight of each unit of volume. Also let f_1, f_2, f_3, &c. represent the increments of velocity per second, communicated to these elements respectively by the action of the forces *impressed* upon the system. Let P_1, P_2, P_3, &c. represent these impressed forces, and p_1, p_2, &c. the perpendicular distances from the axis at which they are respectively applied.

Now since μm_1, μm_2, μm_3, &c. are the weights of the elements, and f_1, f_2, &c. the increments of velocity they receive per second, it follows that $\dfrac{\mu m_1}{g}f_1$, $\dfrac{\mu m_2}{g}f_2$, $\dfrac{\mu m_3}{g}f_3$, &c. are the effective forces upon them (Art. 103.). Let ρ_1, ρ_2, ρ_3, &c. represent the distances of these elements respectively from the axis of revolution, then since their effective forces are in directions perpendicular to these distances, the moments of these effective forces about the axis are $\dfrac{\mu m_1}{g}f_1\rho_1$, $\dfrac{\mu m_2}{g}f_2\rho_2$, $\dfrac{\mu m_3}{g}f_3\rho_3$, &c. Also P_1p_1, P_2p_2, P_3p_3, &c. are the moments of the impressed forces of the system about the axis. Now the impressed forces P_1, P_2, P_3, &c. together with the resistance of the axis, which is indeed one of the impressed forces, are in equilibrium with the effective forces by d'Alembert's principle. Taking then the axis as the point from which the moments are measured, the sum of the moments of P_1, P_2, &c. must equal the sum of the moments of the effective forces, or

$$\frac{\mu m_1}{g}f_1\rho_1 + \frac{\mu m_2}{g}f_2\rho_2 + \ldots = P_1p_1 + P_2p_2 + \ldots$$

Now let f represent that value of f_1, f_2, &c. which corresponds to a distance unity from the axis. Since the system is rigid, and f, f_1, f_2, &c. represent arcs described about it in the same time at the different distances 1, ρ_1, ρ_2, &c. it follows that these arcs are as their distances, and therefore that $f_1 = f\rho_1$, $f_2 = f\rho_2$, $f_3 = f\rho_3$, &c. Substituting these values in the preceding equation, we have

$$\frac{\mu}{g}m_1f\rho_1^2 + \frac{\mu}{g}m_2f\rho_2^2 + \ldots = P_1p_1 + P_2p_2 + \ldots;$$

$$\therefore f\frac{\mu}{g}\{m_1\rho_1^2 + m_2\rho_2^2 + \ldots\} = P_1p_1 + P_2p_2 + \ldots,$$

$$\text{or } f\frac{\mu}{g}\Sigma m\rho^2 = \Sigma Pp \ldots,$$

$$\therefore f = \frac{g}{\mu}\frac{\Sigma Pp}{I} \ldots \ldots (78),$$

where I represents the moment of inertia of the mass about its axis of revolution.*

107. If the impressed forces P be the *weights* of the parts of the body and θ be, in any position of the body, the inclination to the vertical Ay of the line AG, drawn from A to the centre of gravity G, then since the sum of the moments of the weights of the parts is equal to the moment of the weight of the whole mass collected in its centre of gravity (Art. 17.), we have, representing AG by G,

$$\Sigma Pp = M\mu \; . \; \overline{GG_2} = M\mu \; . \; G. \sin. \; \theta \; ;$$

Therefore (equation 78.), $f = g \dfrac{MG}{I} \sin. \; \theta \; . \; . \; . \; . \; . \; (79).$

108. *To find the resultant of the effective forces on a body which revolves about a fixed axis.*

The resultant of the effective forces upon a body which revolves about a fixed axis, is evidently equal to that single force which would just be in equilibrium with these if there were no resistance of the axis. Let R be that single force, then the moment of R about any point must equal the sum of the moments of the effective forces about that point.

---

* If a represent the angular velocity, or the velocity of an element at distance unity, then by equation (72), $f = \pm \dfrac{da}{dt}, \quad \therefore \; a\dfrac{da}{dt} = \pm \dfrac{g}{\mu I}\Sigma Pp a$;

$$\therefore \; \tfrac{1}{2}a_1{}^2 - \tfrac{1}{2}a_2{}^2 = \pm \dfrac{g}{\mu I}\Sigma \int_0^t \!\! Ppa dt.$$

Now pa is the velocity of a point at distance p, therefore Ppa is the *work* (Art. 50.) of the force P per second ; therefore $\int_0^t \!\! Ppa dt$ is the work of P (equation 40) in the time t, which is represented by U, therefore $a_1{}^2 - a_2{}^2 = \pm \dfrac{2gU}{\mu I}$, which corresponds with the result already obtained. See equation (51).

Take a point in the axis for the point about which the moments are measured, and let L be the perpendicular distance from A of the resultant R. Now, as in Art. 106. it appears that the sum of the moments of the effective forces about A is represented by $f\dfrac{\mu}{g}\Sigma m\rho^2$,

$$\therefore \mathrm{RL}=f\frac{\mu}{g}\Sigma m\rho^2 \ \ldots\ldots (80).$$

To determine the value of R let it be observed that the effective force $\dfrac{\mu}{g}fm_1\rho_1$ on any particle m_1, acting in a direction n_1m_1, perpendicular to the distance Am_1 from the axis A, may be resolved into two others, parallel to the two rectangular axes Ay and Ax, each of which is equal to the product of this effective force, whose direction is n_1m_1, and the cosine of the inclination of n_1m_1 to the corresponding axis. Now the inclination of m_1n_1 to Ax is the same as the inclination of Am_1 to Ay, since these two last lines are perpendicular to the two former. The cosine of this inclination equals therefore $\dfrac{AN_1}{Am_1}$ or $\dfrac{y_1}{\rho_1}$, if $AN_1=y_1$. Similarly the cosine of the inclination of n_1m_1 to Ay equals $\dfrac{AM_1}{Am_1}$ or $\dfrac{x_1}{\rho_1}$, if $AM_1=x_1$. The resolved parts in the directions of Ay and Ax of the effective force $\dfrac{\mu}{g}fm_1\rho_1$ are therefore $\dfrac{\mu}{g}fm_1\rho_1\dfrac{y_1}{\rho_1}$, and $\dfrac{\mu}{g}fm_1\rho_1\dfrac{x_1}{\rho_1}$, or $\dfrac{\mu}{g}fm_1y_1$ and $\dfrac{\mu}{g}fm_1x_1$.

Similarly the resolved parts in the directions of Ax and Ay of the effective force upon m_2 are $\dfrac{\mu}{g}fm_2y_2$ and $\dfrac{\mu}{g}fm_2x_2$, and so of the rest.

The sums X and Y of the resolved forces in the directions of Ax and Ay respectively (Art. 11.) are therefore

$$\frac{\mu}{g}fm_1y_1+\frac{\mu}{g}fm_2y_2+\frac{\mu}{g}fm_3y_3+ \ldots\ldots =\mathrm{X},$$

and $\dfrac{\mu}{g}fm_1x_1 + \dfrac{\mu}{g}fm_2x_2 + \dfrac{\mu}{g}fm_3x_3 + \;\cdots\; = Y \; ;$

or $\dfrac{\mu}{g}f\{m_1y_1 + m_2y_2 + m_3y_3 + \;\cdots\; \} = X,$

and $\dfrac{\mu}{g}f\{m_1x_1 + m_2x_2 + m_3x_3 + \;\cdots\; \} = Y.$

Now let G_1 and G_2 represent the distances G_2G and G_1G of the centre of gravity of the body from Ay and Ax respectively, and let the whole volume of the body be represented by M,

$\therefore$ (equation 18), $MG_2 = m_1y_1 + m_2y_2 + m_3y_3 + \;\cdots\;,$

$$MG_1 = m_1x_1 + m_2x_2 + m_3x_3 + \;\cdots\; ;$$

$$\therefore X = \frac{\mu}{g}fMG_2, \quad Y = \frac{\mu}{g}fMG_1 \;\cdots\; (81).$$

Now (Art. 11.), $R = \sqrt{X^2 + Y^2}$, therefore

$$R = \frac{\mu}{g}fM\sqrt{G_1^2 + G_2^2}.$$

Now if G be the distance AG of the centre of gravity from A, $G = \sqrt{G_1^2 + G_2^2}$,

$$\therefore R = \frac{\mu}{g}fMG \;\cdots\; (82).$$

Substituting in equation (82) the value of f from equation (78), we have

$$R = \frac{MG\Sigma Pp}{I} \;\cdots\; (83).$$

And substituting in equation (80) for R its value from equation (82),

$$f\frac{\mu}{g}MGL = f\frac{\mu}{g}I,$$

$$\therefore L = \frac{I}{MG} \;\cdots\; (84),$$

where L is the distance of the point of application of the resultant of the effective forces, from the axis.

Now let θ be the inclination of the resultant R to the axis Ax,

$\therefore$ (Art. 11.), R cos. $\theta = $X, R sin. $\theta = $Y,

$\therefore$ tan. $\theta = \dfrac{Y}{X}$; but by equations (81),

$$\frac{Y}{X} = \frac{G_1}{G_2} = \frac{AG_1}{G_1G} = \text{tan. } AGG_1,$$

$\therefore$ tan $\theta = $ tan. AGG_1, $\therefore$ $\theta = AGG_1$.

The inclination of the resultant R to Ax is therefore equal to the angle AGG_1, but the perpendicular to AG is evidently inclined to Ax at this same angle. Therefore the direction of the resultant R is perpendicular to the line AG, drawn from the axis to the centre of gravity. Moreover its magnitude and the distance of its point of application from A have been before determined by equations (83) and (84).

THE CENTRE OF PERCUSSION.

109. It is evident, that if at a point of the body through which the *resultant* of the effective forces upon it passes, there be opposed an *obstacle* to its motion, then there will be produced upon that obstacle the same effect as though the whole of the effective forces were collected in that point, and made to act there upon the obstacle, so that the whole of these forces will take effect upon the obstacle, and there will be no effect of these forces produced elsewhere, and therefore no repercussion upon the axis. It is for this reason that the point O in the resultant, where it cuts the line AG drawn from the axis to the centre of gravity, is called the CENTRE OF PERCUSSION. Its distance L from A is determined by the equation

$$L = \frac{K^2}{G} \quad \ldots \quad (85),$$

which is obtained from equation (84) by writing MK² for I (Art. 80.), K being the radius of gyration. If at the centre of percussion the body receive an impulse when at rest, then since the resultant of the effective forces thereby produced will have its direction through the point where the impulse is communicated, it follows that the whole impulse will take effect in the production of those effective forces, and no portion be expended on the axis.

THE CENTRE OF OSCILLATION.

110. It has been shown (Art. 98.) that in the simple pendulum, supposed to be a single exceedingly small element of matter suspended by a thread without weight, the time of each oscillation is dependent upon the length of this thread, or the distance of the suspended element from the axis about which it oscillates. If therefore we imagine a number of such elements to be thus suspended at *different distances* from the same axis, and if we suppose them, after having been at first united into a continuous body, placed in an inclined position, all to be released at once from this union with one another, and allowed to oscillate *freely*, it is manifest that their oscillations will all be performed in different times. Now let all these elements again be conceived united in one oscillating mass. All being then compelled to perform these oscillations in the *same* time, whilst all *tend* to perform them in different times, the motions of some are manifestly *retarded* by their connexion with the rest, and those of others *accelerated*, the former being those which lie near to the axis, and the others those more remote; so that *between* the two there must be some point in the body where the elements cease to be retarded and begin to be accelerated, and where therefore they are neither accelerated nor retarded by their connexion with the rest; an element *there* performing its oscillations precisely in the same time as it would do, if it were not connected with the rest, but suspended freely from the axis by a thread without weight. This point in the body, at the distance of which from the axis a single particle suspended freely, would perform its oscillations precisely in the same time that the body does, is called the CENTRE OF OSCILLATION.

The centre of oscillation coincides with the centre of percussion.

111. For (by equation 79) the increment of angular velocity per second f of a body revolving about an horizontal axis, the forces impressed upon it being the weights of its parts only, is represented by the formula $g\dfrac{MG}{I}$ sin. θ, where θ is the inclination to the vertical of the line AG, drawn from the axis to its centre of gravity. But (by equation 84), $L = \dfrac{I}{MG}$, where L is the distance AO of the centre of percussion from the axis,

$$\therefore f = \frac{g \text{ sin. } \theta}{L} \quad \cdots \cdots \quad (86),$$

$$\therefore fL = g \text{ sin. } \theta.$$

Now it has been shown (Art. 98.), that the impressed moving force on a particle whose weight is w, suspended from a thread without weight, inclined to the vertical at an angle θ, is represented by w sin. θ; moreover if f' represent the increment of velocity per second on this particle, then $\dfrac{w}{g}f'$ is the effective force upon it. Therefore by d'Alembert's principle,

$$w \text{ sin. } \theta = \frac{w}{g} f', \quad \therefore f' = g \text{ sin. } \theta, \quad \therefore f' = fL.$$

Now fL is the increment of velocity at the centre of percussion, and f' is that upon a single particle suspended freely at any distance from the axis. If such a particle were therefore suspended at a distance from the axis equal to that of the centre of percussion, since it would receive, *at the same distance from the axis*, the same increments of velocity per second that the centre of percussion does, it would manifestly move exactly as that point does, and perform its oscillations in the same time that the body does. Therefore, &c.

112. *The centres of suspension and oscillation are reciprocal.*

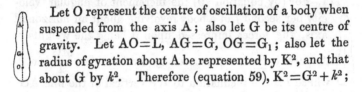

Let O represent the centre of oscillation of a body when suspended from the axis A; also let G be its centre of gravity. Let $AO=L$, $AG=G$, $OG=G_1$; also let the radius of gyration about A be represented by K^2, and that about G by k^2. Therefore (equation 59), $K^2=G^2+k^2$;

$$\therefore \text{ (equation 85), } L=\frac{G^2+k^2}{G}=G+\frac{k^2}{G} \quad\cdots\cdots (87),$$

$$\therefore \ G+G_1=G+\frac{k^2}{G},$$

$$\therefore \ G_1=\frac{k^2}{G} \quad\cdots\cdots (88).$$

Now let the body be suspended from O instead of A; when thus suspended it will have, as before, a centre of oscillation. Let the distance of this centre of oscillation from O be L_1,

$$\therefore \text{ by equation (87), } L_1=G_1+\frac{k^2}{G_1},$$

$$\therefore \text{ by equation (88), } L_1=\frac{k^2}{G}+G=L.$$

Since then the centre of oscillation in this second case is at the distance L from O, it is in A; what was before the centre of suspension has now therefore become the centre of oscillation. Thus when the centre of oscillation is converted into the centre of suspension, the centre of suspension is thereby converted into the centre of oscillation. This is what is meant, when it is said that the centres of oscillation and suspension are reciprocal.

PROJECTILES.

113. *To determine the path of a body projected obliquely in vacuo.*

Suppose the whole time, T seconds, of the flight of the body
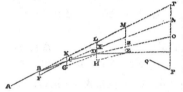
to any given point P of its path, to be divided into equal exceedingly small intervals, represented by ΔT, and conceive the whole effect of gravity upon the projectile during each one of these intervals to be collected into a single impulse at the termination of that interval, so that there may be communicated to it at *once*, by that single impulse, all the additional velocity which is in *reality* communicated to it by gravity at the different periods of the small time ΔT.

Let AB be the space which the projectile would describe, with its velocity of projection alone, in the first interval of time; then will it be projected from B at the commencement of the second interval of time in the direction ABT with a velocity which would alone carry it through the distance BK = AB in that interval of time; whilst at the same time it receives from the impulse of gravity a velocity such as would alone carry it vertically through a space in that interval of time which may be represented by BF. By reason of these two impulses communicated *together*, the body will therefore describe in the second interval of time the diagonal BC of the parallelogram of which BK and BF are adjacent sides. At the commencement of the third interval it will therefore have arrived at C, and will be projected from thence in the direction BCX, with a velocity which would alone carry it through CX = BC in the third interval; whilst at the same time it receives an impulse from gravity communicating to it a velocity which would alone carry it through a distance represented

I

by CG=BF in that interval of time. These two impulses together communicate therefore to it a velocity which carries it through CD in the third interval, and thus it is made to describe all the sides of the polygon ABCD . . . P in succession. Draw the vertical PT, and produce AB, BC, CD, &c. to meet it in T, N, O . . ., and produce GC, HD, &c. to meet BT in K, L, &c.

Now, since BC is equal to CX, and CK is parallel to XL, therefore KL is equal to BK or to AB.

Again, since CD is equal to DZ, and DL is parallel to ZM, therefore LM is equal to KL or to AB; and so of the rest.

If therefore there be n intervals of time equal to ΔT, so that there are n sides AB, BC, CD, &c. of the polygon, and n divisions AB, BK, &c. of the line AT, then $AT=n\overline{AB}$ and $BT=(n-1)\overline{AB}$,

$$\therefore TN=(n-1)\overline{KC}=(n-1)\overline{BF}.$$

Similarly $CN=(n-2)\overline{CX}$, therefore $NO=(n-2)\overline{DX}= (n-2)\overline{BF}$; and so of the remaining parts of TP.

Now these parts of TP are $(n-1)$ in number, therefore $TP=(n-1)\overline{BF}+(n-2)\overline{BF}+(n-3)\overline{BF}+\ldots\{(n-1)\text{ terms}\}$; or $TP=\{(n-1)+(n-2)+\ldots\}\overline{BF}.$

Therefore, summing the series to $(n-1)$ terms,

$$TP=\{2(n-1)-(n-2)\}\left(\frac{n-1}{2}\right)\;.\;\overline{BF},$$

$$\therefore TP=\frac{n(n-1)}{2}\overline{BF}.$$

Now g represents the additional velocity which gravity would communicate to the projectile in each second, if it acted upon it *alone*. $g\Delta T$ is therefore the velocity which it would communicate to it in each interval of ΔT seconds. $g\Delta T$ is therefore the velocity communicated to the body by each of the impulses which it has been supposed to receive from gravity.

Now BF is the space through which it would be carried in the time ΔT by this velocity,

$$\therefore\ \mathrm{BF}=(g\Delta T)\Delta T=g(\Delta T)^2,$$

$$\therefore\ \mathrm{TP}=\tfrac{1}{2}gn(n-1)(\Delta T)^2.$$

Also $\Delta T=\dfrac{T}{n}$,

$$\therefore\ \mathrm{TP}=\tfrac{1}{2}gn(n-1)\frac{T^2}{n^2}=\tfrac{1}{2}g\left(1-\frac{1}{n}\right)T^2.$$

Now this is true, however small may be the intervals of time ΔT, and therefore if they be infinitely small, that is, if the impulses of gravity be supposed to follow one another at infinitely small intervals, or if gravity be supposed to act, as it really does, *continuously*.

But if the intervals of time ΔT be infinitely small, then the number n of these intervals which make up the whole finite time T, must be infinitely great. Also when n is infinitely great, $\dfrac{1}{n}=0$.

In the actual case, therefore, of a projectile *continually* deflected by gravity, the vertical distance TP between the tangent to its path at the point of projection, and its position P after the flight has continued T seconds, is represented by the formula

$$\overline{\mathrm{TP}}=\tfrac{1}{2}gT^2\ \ .\ \ .\ \ .\ \ .\ \ (89).$$

Moreover $\mathrm{AT}=n\overline{\mathrm{AB}}$, and AB is the space which the body would describe uniformly with the velocity of projection in the time ΔT, so that $n\overline{\mathrm{AB}}$ is the space which it would describe in the time $n\ .\ \Delta T$ or T with that velocity. If therefore V equal the velocity of projection, then

$$\overline{\mathrm{AT}}=V\ .\ T\ .\ .\ .\ .\ (90);$$

so that the position of the body after the time T is the same as though it had moved through that time with the velocity of its projection *alone*, describing AT, and had then fallen through the same time by the force of gravity *alone*, describing TP (see Art. 101.).

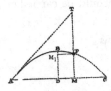

114. Let $AM = x$, $MP = y$, angle of projection $TAM = \alpha$, velocity of projection $= V$.

$$\therefore \ x \sec. \alpha = \overline{AT} = V \ . \ T, \ \therefore \ T = \frac{x \sec. \alpha}{V},$$

$$x \tan. \alpha - y = \overline{MT} - \overline{MP} = \overline{TP} = \tfrac{1}{2}gT^2. \ \ldots \ldots \ (91).$$

Substituting the value of T from the preceding equation,

$$x \tan. \alpha - y = \tfrac{1}{2}g \frac{x^2 \sec.^2 \alpha}{V^2},$$

$$\therefore \ y = x \tan. \alpha - \frac{g \sec.^2 \alpha}{2V^2} \ . \ x^2.$$

Let H be the height through which a body must fall freely by gravity to acquire the velocity V, or the height due to that velocity; then $V^2 = 2gH$ (Art. 47.), therefore $4H = \dfrac{2V^2}{g}$; therefore, by substitution,

$$y = x \tan. \alpha - \frac{\sec.^2 \alpha}{4H} x^2 \ \ldots \ldots \ (92).$$

115. *To find the time of the flight of a projectile.*

It has been shown (equation 91), that if T represent the time in seconds of the flight to a point whose co-ordinates are x and y, then

$$\tfrac{1}{2}gT^2 = x \tan. \alpha - y, \ \therefore \ T^2 = \frac{2}{g} \{x \tan. \alpha - y\},$$

$$\therefore \ T = \sqrt{\frac{2}{g}} \sqrt{x \tan. \alpha - y} \ \ldots \ldots \ (93).$$

Now, $\dfrac{2}{g} = \dfrac{2}{32\frac{1}{6}} = \dfrac{1}{16}$ nearly, $\therefore \ T = \tfrac{1}{4}\sqrt{x \tan. \alpha - y}$ nearly.

If the projectile descend again to the horizontal plane from which it was projected, and T be the whole time of its flight, and X its whole range upon the plane, then, since at the expiration of the time T, $y=0$ and $x=X$,

$$\therefore T=\sqrt{\frac{2}{g}}\sqrt{X \tan. \alpha}=\tfrac{1}{4}\sqrt{X \tan. \alpha} \text{ nearly.}$$

116. *To find the greatest horizontal distance* X, *to which a projectile ranges, having given the elevation* α *and the velocity* V *of its projection.*

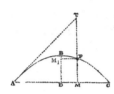

When the projectile attains its greatest horizontal range, its height y above the horizontal plane becomes 0, whilst the abscissa x of the point P, which it has then reached in its path, becomes X. Substituting these values 0 and X, for y and x in equation (92), we have $0=X \tan. \alpha - \dfrac{X^2 \sec.^2 \alpha}{4H}$,

$$\therefore X=4H \tan. \alpha \cos.^2 \alpha=4H \sin. \alpha \cos. \alpha.$$

$$\therefore X=2H \sin. 2\alpha \ldots (94).$$

If the body be projected at different angular elevations, but with the same velocity, the horizontal range will be the greatest when sin. 2α is the greatest, or when $2\alpha=\dfrac{\pi}{2}$, or $\alpha=\dfrac{\pi}{4}$.

117. *To find the greatest height which a projectile will attain in its flight if projected with a given velocity, and at a given inclination to the horizon.*

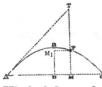

Multiplying both sides of equation (92) by 4H cos.² α, we have 4H cos.² α . $y=$ 4H cos.² α tan. α . $x-x^2=2H$ (2 cos. α sin. α) $x-x^2=2H$ sin. 2α . $x-x^2$. Subtracting both sides of this equation from H² sin.² 2α, we have

$$H^2 \sin.^2 2\alpha - 4H \cos.^2 \alpha . y=H^2 \sin.^2 2\alpha - 2H \sin. 2\alpha . x+x^2.$$

But sin.^{2} $2\alpha = 4$ sin.^{2} α cos.^{2} α,

$\therefore$ 4H cos.^{2} $\alpha\{$H sin.^{2} $\alpha - y\} = \{$H sin. $2\alpha - x\}^2$ . . . (95).

Now the second member of this equation is always a positive quantity, being a square. The first member is therefore always positive; therefore H sin.^{2} $\alpha - y$ is always positive. Whence it follows that y can never exceed H sin.^{2} α, so that it attains its greatest possible value when it equals H sin.^{2} α, a value which it manifestly attains when the second member of the above equation vanishes, or when $x =$ H sin. 2α, that is, when x becomes equal to half the greatest horizontal range, as is apparent from the last proposition; so that the greatest height BD of the projectile is represented by H sin.^{2} α, a height which it attains when AD equals half the horizontal range.

118. *The path of a projectile in vacuo is a parabola.*

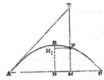

Let B be the highest point in the flight of the projectile, and BD its greatest height. Draw PM$_1$ perpendicular to BD. Let BM$_1 = x_1$, M$_1$P $= y_1$,

$\therefore$ $x_1 =$ BD $-$ M$_1$D $=$ BD $-$ PM $=$ H sin.^{2} $\alpha - y$,
   $y_1 =$ DM $=$ AM $-$ AD $= x -$ H sin. 2α.

Substituting these values in equation (95),

$$y_1{}^2 = 4\text{H cos.}^2 \alpha \cdot x_1 \quad . \quad . \quad . \quad . \quad (96),$$

which is the equation to a parabola whose vertex is in A, whose axis coincides with AD, and whose parameter is 4H cos.^{2} α.

The path of a projectile *in vacuo* is therefore a parabola, whose vertex is at the highest point attained by the projectile, and whose axis is vertical.

119. *To find the range of a projectile upon an inclined plane.*

Let R represent the range AP of a projectile upon an inclined plane AB, whose inclination is ι. Then H and α being taken to represent the same quantities as before, and x, y being the co-ordinates of P to the horizontal axis AC, we have

$$x = AM = AP \cos. PAM = R \cos. \iota,$$
$$y = PM = AP \sin. PAM = R \sin. \iota.$$

Substituting these values of x and y in the general equation (92) to the projectile, we have

$$R \sin. \iota = R \cos. \iota \tan. \alpha - \frac{R^2 \cos.^2 \iota \sec.^2 \alpha}{4H}.$$

Dividing by R, multiplying by cos. α, and transposing

$$\frac{R \cos.^2 \iota \sec. \alpha}{4H} = \cos. \iota \sin. \alpha - \sin. \iota \cos. \alpha = \sin. (\alpha - \iota),$$

$$\therefore R = 4H \frac{\sin. (\alpha - \iota) \cos. \alpha}{\cos.^2 \iota} \quad \ldots \ldots (97).$$

Now $\sin. (2\alpha - \iota) - \sin. \iota = \sin. \{\alpha + (\alpha - \iota)\} - \sin. \{\alpha - (\alpha - \iota)\} = 2 \sin. (\alpha - \iota) \cos. \alpha.$

Substituting this value of $2 \sin. (\alpha - \iota) \cos. \alpha$ in the preceding equation, we have

$$R = 2H \left\{ \frac{\sin. (2\alpha - \iota) - \sin. \iota}{\cos.^2 \iota} \right\} \quad \ldots \ldots (98).$$

Now it is evident that if α be made to vary, ι remaining the same, R will attain its greatest value when sin. $(2\alpha - \iota)$ is greatest, that is when it equals unity, or when $2\alpha - \iota = \frac{\pi}{2}$, or when $\alpha = \frac{\pi}{4} + \frac{\iota}{2}$. This, then, is the angle of elevation corresponding to the greatest range, with a given velocity upon an inclined plane whose inclination is ι.

If in the preceding expression for the range we substitute $\left\{\dfrac{\pi}{2}-(\alpha-\imath)\right\}$ for α, the value of the expression will be found to remain the same as it was before; for sin. $(2\alpha-\imath)$ will, by this substitution, become sin. $\{\pi-2(\alpha-\imath)-\imath\}=$ sin. $\{\pi-(2\alpha-\imath)\}=$ sin. $(2\alpha-\imath)$. The value of R remains therefore the same, whether the angle of elevation be α or $\dfrac{\pi}{2}-(\alpha-\imath)$.

And the projectile will range the same distance on the plane, whether it be projected at one of these angles of elevation or the other.

Let BAC be the inclination of the plane on which the

projectile ranges, and AT the direction of projection. Take DAS equal to BAT. Then BAT$=$TAC$-$BAC$=$

$\alpha-\imath$. And SAC$=$DAC$-$DAS$=\dfrac{\pi}{2}$

$-$BAT$=\dfrac{\pi}{2}-(\alpha-\imath)$. The range AP is therefore the same, whether TAC or SAC be the angle of elevation, and therefore whether AT or AS be the direction of projection.

Draw AE bisecting the angle BAD, then the angle EAC$=$

$$BAC+BAE=BAC+\tfrac{1}{2}BAD=\imath+\tfrac{1}{2}\left(\dfrac{\pi}{2}-\imath\right)=\dfrac{\pi}{4}+\dfrac{\imath}{2}.$$

The angle EAC is therefore that corresponding to the *greatest* range, and AE is the direction in which a body should be projected to range the greatest distance on the inclined plane AB.

It is evident that the directions of projection AS and AT, which correspond to equal ranges, are equally inclined to the direction AE corresponding to the greatest range.

120. *The velocity of a projectile at different points of its path.* It has been shown (Art. 56.), that if a body move in any curve acted upon by gravity, the work accumulated or lost is the same as would be accumulated or lost, provided the body, instead of moving in a curve, had moved in the direc-

tion of gravity through a space equal to the *vertical* projection of its curvilinear path.

Thus a projectile moving from A to P will accumulate or lose a quantity of work, which is equal to that which it would accumulate or lose, had it moved vertically from M to P, or from P to M, PM being the projection of its path on the direction of gravity. Now the work thus accumulated or lost equals one half the difference between the *vires vivæ* at the commencement and termination of the motion.

Let V equal the velocity at A, and v equal the velocity at P, therefore the work $=\frac{1}{2}\frac{W}{g}V^2-\frac{1}{2}\frac{W}{g}v^2$. Moreover, the work done through PM$=$W . PM, therefore $\frac{1}{2}\frac{W}{g}V^2-\frac{1}{2}\frac{W}{g}v^2 =$ W . PM, therefore $V^2-v^2=2g$MP. Let PM$=y$,

$$\therefore v^2=V^2-2gy \quad \dots \quad (99),$$

which determines the velocity at any point of the curve.

Centrifugal Force.

121. Let a body of small dimensions move in any curvilinear path AB, impelled continually towards

a given point S (called a centre of force) by a given force, whose amount, when the body has reached the point P in its path, is represented by F.* Let PQ be an exceedingly small portion of the path of the body, and conceive the force F to remain constant and parallel to itself, whilst this portion of its path is being described. Then, if PR be a tangent at P, and QR be drawn parallel to SP, PR is the space which the body would have traversed in the time of describing PQ, if it had moved with its velocity of projection from P *alone*, and had not been attracted towards

---

* The force here spoken of, and represented by F, is the *moving* force, or pressure on the body (see Art. 92.), and is therefore equal to that pressure which would just sustain its attraction towards S.

S, and RQ or PT (QT being drawn parallel to RP) is the space through which it would have fallen by its attraction towards S *alone*, or if it had not been projected at all from P.* Let v represent the velocity which it would have acquired on this last supposition, when it reached the point T. Therefore (Art. 66.), if w represents the weight of the body,

$$F \times \overline{PT} = \tfrac{1}{2}\frac{w}{g}v^2.$$

Now the velocity v, which the body would have acquired in falling through the distance PT by the action of the constant force F, is equal to *half* that which would cause it to describe the same distance *uniformly* in the same time. †

Representing therefore by V the actual velocity of the body in its path at P, we have

$$\frac{\tfrac{1}{2}v}{V} = \frac{PT}{PR}; \quad \therefore v = 2V \cdot \frac{PT}{PR}.$$

Substituting this value of v in the preceding equation,

$$F \times \overline{PT} = 2\frac{w}{g}V^2 \left(\frac{PT}{PR}\right)^2, \quad \therefore F = 2\frac{w}{g}V^2 \frac{QR}{(PR)^2}$$

Now let a circle PQV be described having a common tangent with the curve AB in the point P, and passing through the point Q. Produce PS to intersect the circumference of this circle in V, and join QV; then are the triangles PQV and QPR similar, for the angle RQP is equal to the angle QPV (QR and VP being parallel), and the angle QPR is equal to the angle QVP in the alternate segment of the circle. Therefore $\frac{QR}{PQ} = \frac{PQ}{PV}$; therefore

---

* See Art. 113. (equations 89 and 90); what is proved there of a body acted upon by the force of gravity which is constant, and whose direction is constantly parallel to itself, is evidently true of any other constant force similarly retaining a direction parallel to itself. To apply the same demonstration to any such case, we have only indeed to assume g to represent another number than $32\tfrac{1}{6}$.

† If f represent the additional velocity per second which F would communicate to the body, and t the time of describing PT, then (Art. 44.) $v = ft$; but (Art. 46.) $PT = \tfrac{1}{2}ft^2 = \left(\frac{ft}{2}\right)t = \frac{v}{2}t$; so that $\frac{v}{2}$ is the velocity with which PT would be described *uniformly* in the time t.

$QR = \dfrac{(PQ)^2}{PV}$. Substituting this value of QR in the last

equation, we have $F = 2\dfrac{w}{g}\dfrac{V^2}{PV} \cdot \left(\dfrac{PQ}{PR}\right)^2$.

Now this is true, however much PQ may be diminished. Let it be *infinitely* diminished, the supposed constant amount and parallel direction of F will then coincide with the actual case of a variable amount and inclination of that force, the ratio $\dfrac{PQ}{PR}$ will become a ratio of equality, and the circle PQV will become the circle of curvature at P, and PV that chord of the circle of curvature, which being drawn from P passes through S. Let this chord of the circle of curvature be represented by C,

$$\therefore\ F = 2\dfrac{w}{g}\dfrac{V^2}{C} \cdot \cdot \cdot \cdot \cdot (100).$$

The force or pressure F thus determined is manifestly exactly equal to that force by which the body tends in its motion continually to fly from the centre S, and may therefore be called its centrifugal force. This term is, however, generally limited in its application to the case of a body revolving in a *circle*, and to the force with which it tends to recede from the centre of that circle; or if applied to the case of motion in any other curve, then it means the force with which the body tends to recede from the centre of the circle of curvature to its path at the point through which it is, at any time, moving. When the body revolves in a circular path, the circle of curvature to the path at any one point evidently coincides with it throughout, and the chord of curvature becomes one of its diameters. Let the radius of the circle which the body thus describes be represented by R, then $C = 2R$;

$$\therefore\ F = \dfrac{w}{g}\dfrac{V^2}{R} \cdot \cdot \cdot \cdot \cdot (101).$$

Since in whatever curve a body is moving, it may be conceived at any point of its path to be revolving in the circle of curvature to the curve at that point, the force F, with which it then tends to recede from the centre of the circle of curvature is represented by the above formula, R being taken to represent the *radius of curvature* at the point of its path through which it is moving.

If a be the angular velocity of the body's revolution about the centre of its circle of curvature, then $V = aR$;

$$\therefore F = \frac{w}{g}a^2R \ . \ . \ . \ . \ . \ (102).$$

122. By transposition of equation (100) we obtain

$$V^2 = \tfrac{1}{2}\left(\frac{Fg}{w}\right)C = 2\left(\frac{Fg}{w}\right)(\tfrac{1}{4}C).$$

Now (Art. 94.) $\dfrac{Fg}{w}$ represents the additional velocity per second f, which would be communicated to a body falling towards S, if the body fell *freely* and the force F remained constant. Moreover, by Art. 47. it appears, that V is the whole velocity which the body would on this supposition acquire, whilst it fell through a distance equal to $\frac{1}{4}$C, or to one quarter of the chord of curvature. Thus, then, the velocity of a body revolving in any curve and attracted towards a centre of force is, at any point of that curve, equal to that which it would acquire in falling freely from that point towards the centre of force through one quarter of that chord of curvature which passes through the centre of force, *if the force which acted upon it at that point in the curve remained constant during its descent.* It is in this sense that the velocity of a body moving in any curve about a centre of force is said to be THAT DUE TO ONE QUARTER THE CHORD OF CURVATURE.

123. *The centrifugal force of a mass of finite dimensions.*

Let BC represent a thin lamina or slice of such a mass contained between two planes exceedingly near to one another, and both perpendicular to a given axis A, about which the mass is made to revolve.

Through A draw any two rectangular axes Ax and Ay, let m_1 be any element of the lamina whose weight is w_1, and let AM_1 and AN_1, co-ordinates of m_1, be represented by x_1 and y_1. Then by equation (91), if α represent the angular velocity of the revolution of the body, the centrifugal force on the element m_1 is represented by $\dfrac{\alpha^2}{g} w_1 \overline{Am_1}$. Let now this force, whose direction is Am_1 be resolved into two others, whose directions are Ax and Ay. The former will be represented by $\dfrac{\alpha^2}{g} w_1 \overline{Am_1}$, cos. xAm_1, or by $\dfrac{\alpha^2}{g} w_1 x_1$, and the latter by $\dfrac{\alpha^2}{g} w_1 \overline{Am_1}$ cos. yAm_1, or by $\dfrac{\alpha^2}{g} w_1 y_1$; and the centrifugal forces on all the other elements of the lamina being similarly resolved, we shall have obtained two sets of forces, those of the one set being parallel to Ax, and represented by $\dfrac{\alpha^2}{g} w_1 x_1$, $\dfrac{\alpha^2}{g} w_2 x_2$, $\dfrac{\alpha^2}{g} w_3 x_3$, &c. and those of the other set parallel to Ay represented by $\dfrac{\alpha^2}{g} w_1 y_1$, $\dfrac{\alpha^2}{g} w_2 y_2$, $\dfrac{\alpha^2}{g} w_3 y_3$, &c.

Now if X and Y represent the resolved parts parallel to the directions of Ax and Ay, of the resultant of these two sets of forces, then (Art. 11.)

$$X = \frac{\alpha^2}{g} w_1 x_1 + \frac{\alpha^2}{g} w_2 x_2 + \frac{\alpha^2}{g} w_3 x_3 + \ldots = \frac{\alpha^2}{g} \Sigma wx = \frac{\alpha^2}{g} WG_1;$$

$$Y = \frac{\alpha^2}{g} w_1 y_1 + \frac{\alpha^2}{g} w_2 y_2 + \frac{\alpha^2}{g} w_3 y_3 + \ldots = \frac{\alpha^2}{g} \Sigma wy = \frac{\alpha^2}{g} WG_2,$$

if G_1 and G_2 represent the co-ordinates AG_1 and AG_2 of the centre of gravity G of the lamina, and W its weight (Art. 18.).

Now the whole centrifugal force F on the lamina is the resultant of these two sets of forces, and is represented by $\sqrt{X^2 + Y^2}$ (Art. 11.),

$$\therefore F = \sqrt{\frac{\alpha^4}{g^2} W^2 G_1{}^2 + \frac{\alpha^4}{g^2} W^2 G_2{}^2} = \frac{\alpha^2}{g} W \sqrt{G_1{}^2 + G_2{}^2}, \text{ or}$$

$$F = \frac{\alpha^2}{g} W \cdot G \ldots \ldots (103),$$

where G is taken to represent the distance AG of the centre of gravity of the lamina from the axis of revolution.

Moreover the direction of this resultant centrifugal force is through A, since the directions of all its components are through that point.

124. From the above formula it is apparent, that if a body revolving round a fixed axis be conceived to be divided into laminæ by planes perpendicular to the axis, then the centrifugal force of each such lamina is the same as it would have been if the whole of its weight had been collected in its centre of gravity; so that if the centres of gravity of all the laminæ be in the same plane passing through the axis, then, since the centrifugal force on each lamina has its direction from the axis through the centre of gravity of that lamina, it follows that all the centrifugal forces of these laminæ are in the same plane, and that they are PARALLEL forces, so that their resultant is equal to their *sum*, those being taken with a negative sign which correspond to laminæ whose centres of gravity are on the opposite side of the axis from the rest, and whose centrifugal forces are therefore in the opposite directions to those of the rest. Thus if F′ represent the whole centrifugal force of such a mass, then $F' = \dfrac{a^2}{g} \Sigma WG$. Now let W′ represent the weight of the *whole* mass, and G′ the distance of its centre of gravity from the axis, therefore $\Sigma WG = W'G'$;

$$\therefore \ F' = \frac{a^2}{g} W'G' \ \ldots \ldots (104).$$

In the case, then, of a revolving body capable of being divided into laminæ perpendicular to the axis of revolution, the centres of gravity of all of which laminæ are in the same plane passing through the axis, the centrifugal force is the same as it would have been if the whole weight of the body had been collected in its centre of gravity, the same property obtaining in this case in respect to the *whole* body as obtains in respect to each of its individual laminæ. Since, moreover, the centrifugal forces upon the laminæ are *parallel* forces when their centres of gravity are all in the same plane passing through the axis of gravity, and since their directions are all in that plane, it follows (Art. 16.), that if we take any point O in the axis, and measure the moments of these parallel forces from that point, and call x the perpendicular distance OA of any lamina BC from that point, and H the distance of their resultant from that point, then

$$H\frac{\alpha^2}{g}\Sigma WG = \frac{\alpha^2}{g}\Sigma WGx,$$

$$\therefore \quad H = \frac{\Sigma WGx}{\Sigma WG} \quad \cdots \cdots (105).$$

The equations (93) and (94) determine the amount and the point of application of the resultant of the centrifugal forces upon the mass, upon the supposition that it can be divided into laminæ perpendicular to the axis of revolution, all of which have their centres of gravity in the same plane passing through the axis.

It is evident that this condition is satisfied, if the body be symmetrical as to a certain axis, and that axis be in the same plane with the axis of revolution, and therefore if it intersect or if it be parallel to the axis of revolution.

If, *in the case we have supposed,* $\Sigma WG = 0$, that is, if the centre of gravity be *in* the axis of revolution, then the centrifugal force vanishes. This is evidently the case where a body revolves round its axis of symmetry.

125. If the centres of gravity of the laminæ into which the body is divided by planes perpendicular to the axis of revolution be *not* in the same plane (as in the figure), then the centrifugal forces of the different laminæ will not lie in the same plane, but diverge from the axis in different directions round it. The amount and direction of their resultant cannot in this case be determined by the equations which have been given above.

THE PRINCIPLE OF VIRTUAL VELOCITIES.

126. *If any pressure* P, *whose point of application* A *is made to move through the straight line* AB, *be resolved into three others* X, Y, Z, *in the directions of the three rectangular axes,* Ox, Oy, Oz; *and if* AC, AD, *and* AE, *be the projections of* AB *upon these axes, then the work of* P *through* AB *is equal to the sum of the works of* X, Y, Z, *through* AC, AD, *and* AE *respectively, or* X . $\overline{AC}$ + Y . $\overline{AD}$ + Z . $\overline{AE}$ = P . $\overline{AM}$.

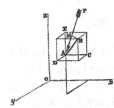

Let the inclinations of the direction of P to the axes Ox, Oy, Oz respectively, be represented by α, β, γ, and the inclinations of AB to the same axes by α_1, β_1, γ_1,

∴ (Art. 12.) X = P cos. α, Y = P cos. β, Z = P cos. γ; also AC = $\overline{AB}$ cos. α_1, AD = $\overline{AB}$ cos. β_1, AE = $\overline{AB}$ cos. γ_1,

∴ X . $\overline{AC}$ = P . $\overline{AB}$ cos. α cos. α_1, Y . $\overline{AD}$ = P . $\overline{AB}$ cos. β cos. β_1,

Z . $\overline{AE}$ = P . $\overline{AB}$ cos. γ cos. γ_1,

∴ X . $\overline{AC}$ + Y . $\overline{AD}$ + Z . $\overline{AE}$ = P . $\overline{AB}${cos. α cos. α_1 + cos. β cos. β_1 + cos. γ cos. γ_1}.

But by a well known theorem of trigonometry, cos. α cos. α_1 + cos. β cos. β_1 + cos. γ cos. γ_1 = cos. PAB,

∴ X . $\overline{AC}$ + Y . $\overline{AD}$ + Z . $\overline{AE}$ = P . $\overline{AB}$ cos. PAB ;

but $\overline{AB}$ cos. PAB = AM ;

∴ X . $\overline{AC}$ + Y . $\overline{AD}$ + Z . $\overline{AE}$ = P . $\overline{AM}$.

But (Art. 52.) the work of P through AM is equal to its work through AB. Therefore, &c.*

* This proposition may readily be deduced from Art. 53., for pressures equal and opposite to X, Y, Z, would just be in equilibrium with P, and these tending to move the point A in one direction along the line AB, P tends to move it in the opposite direction, therefore in the motion of the point A through AB, the sum of the works of X, Y, Z, must equal the work of P. But the work of X as its point of application moves through AB is equal (Art. 52.) to the work of X through the projection of AB

127. *If any number of forces be in equilibrium (being in any way mechanically connected with one another), and if, subject to that connection, their different points of application be made to move, each through any exceedingly small distance, then the aggregate of the work of those forces, whose points of application are made to move towards the directions in which the several forces applied to them act, shall equal the aggregate of the work of those forces, the motions of whose points of application are opposed to the directions of the forces applied to them.*

For let all the forces composing such a system be resolved into three sets of forces parallel to three rectangular axes, and let these three sets of parallel forces be represented by A, B, and C respectively. Then must the resultant of the parallel forces of each set equal nothing. For if any of these resultants had a finite value, then (by Art. 12.) the whole three sets of forces would have a resultant, which they cannot, since they are in equilibrium.

Now let the motion of the points of application of the forces be conceived so *small* that the *amounts* and *directions* of the forces may be made to vary, during the motion, only by an exceedingly small quantity, and so that the resolved forces upon any point of application may remain sensibly unchanged. Also let u_1, u_2, u_3, represent the *works* of these resolved forces respectively on any point, and Σu_1 the sum of all the *works* of the resolved forces of the set A, Σu_2 the sum of all the works of the forces of the set B, and Σu_3 of the set C. Now since the parallel forces of the set A have no

upon Ax, that is, through AC; similarly the work of Y, as its point of application moves through AB, is equal to its work through the projection of AB upon Ay, or through AD ; and so of Z. The sum of the works of X, Y, and Z, as their point of application is made to move through AB, is therefore equal to what would have been the sum of their works had their points of application been made to move *separately* through AC, AD, AE ; this last sum is therefore equal to the work of P through AB, which is equal to the work of P through AM, AM being the projection of AB upon the direction of P.

K

resultant, therefore (Art. 59.) the sum of the works of those forces of this set, whose points of application are moved *towards* the directions of their forces, is *equal* to the sum of the works of those whose points of application are moved *from* the directions of their forces, so that $\Sigma u_1 = 0$, if the values of u_1, which compose this sum, be taken with the positive or negative sign, according to the last mentioned condition.

Similarly, $\Sigma u_2 = 0$ and $\Sigma u_3 = 0$, $\therefore$ $\Sigma(u_1 + u_2 + u_3) = 0$.

Now let U represent the actual work of that force P_1, the works of whose components parallel to the three axes are represented by u_1, u_2, u_3; then by the last proposition,

$$u_1 + u_2 + u_3 = U,$$

$$\therefore \ \Sigma U = 0 \ . \ . \ . \ . \ . \ (106);$$

in which expression U is to be taken positively or negatively according to the same conditions as u_1, u_2, u_3; that is, according as the work at each point is done in the direction of the corresponding force, or in a direction opposite to it. Hence therefore it follows, from the above equations, that the sum of the works in one of these directions equals their sum in the opposite direction. Therefore, &c.

The projection of the line described by the point of application of any force upon the direction of that force is called its VIRTUAL VELOCITY, so that the product of the force by its virtual velocity is in fact the work of that force ; hence therefore, representing any force of the system by P, and its virtual velocity by p, we have $Pp = U$, and therefore, $\Sigma Pp = 0$, which is the principle of virtual velocities.*

128. *If there be a system of forces such that their points of application being moved through certain consecutive positions, those forces are in all such positions in equilibrium, then in respect to any finite motion of the points of application through that series of positions, the aggregate of the*

---

* This proof of the principle of virtual velocities is given here for the first time.

work of those forces, which act in the directions in which their several points of application are made to move, is equal to the aggregate of the work in the opposite direction.

This principle has been proved in the preceding proposition, only when the motions communicated to the several points of application are exceedingly small, so that the work done by each force is done only through an exceedingly small space. It extends, however, to the case in which each point of application is made to move, and the work of each force to be done, through any distance, however great, provided only that in all the different positions which the points of application of the forces of the system are thus made to take up, these forces be in equilibrium with one another; for it is evident that if there be a series of such positions immediately adjacent to one another, then the principle obtains in respect to each small motion from one of this set of positions into the adjacent one, and therefore in respect to the sum of all such small motions as may take place in the system in its passage from *any* one position into *any* other, that is, in respect to the whole motion of the system through the intervening series of positions. Therefore, &c.

The Principle of Vis Viva.

129. *If the forces of any system be not in equilibrium with one another, then the difference between the aggregate work of those whose tendency is in the direction of the motions of their several points of application, and those whose tendency is in the opposite direction, is equal to one half the aggregate vis viva of the system.*

In each of the consecutive positions which the bodies composing the system are made successively to take up, let there be applied to each body a force equal to the *effective force* (Art. 103.) upon that body, but in an opposite direction; every position will then become one of equilibrium.

Now, as the bodies which compose the system and the
various points of application of the impressed forces move
through any finite distances from one position into another,
let Σu_1 represent the aggregate work of those *impressed*
forces whose directions are *towards* the directions of the
motions of their several points of application, and let Σu_2 represent the work of those *impressed* forces which act in the
opposite directions; also let Σu_3 represent the aggregate work
of forces applied to the system equal and opposite to the *effective* forces upon it; the directions of these forces opposite to
the effective forces are manifestly opposite also to the directions
of the *motions* of their several points of application, so that
on the whole Σu_1 is the aggregate work of those forces whose
directions are *towards* the motions of their several points of
application, and $\Sigma u_2 + \Sigma u_3$ the aggregate work opposed to
them. Since therefore, by d'Alembert's principle, an equilibrium obtains in every consecutive position of the system, it
follows by the last proposition, that

$$\Sigma u_1 = \Sigma u_2 + \Sigma u_3,$$
$$\therefore \ \Sigma u_1 - \Sigma u_2 = \Sigma u_3 \ . \ . \ . \ . \ . \ (107).$$

Now u_3 is taken to represent the work of a force equal and
opposite to the effective force upon any body of the system;
but the work of such a force through any space is equal to
the work which the *effective* force (being unopposed) *accumulates* in the body through that space (Art. 69.), or it is
equal to one half the difference of the vires vivæ of the body
at the commencement and termination of the time during
which that space is described (Art. 67.). Therefore Σu_3
equals one half the aggregate difference of the *vires vivæ* of
the system at the two periods;

$$\therefore \ \Sigma u_1 - \Sigma u_2 = \frac{1}{2g}\Sigma w(v_1{}^2 - v_2{}^2) \ . \ . \ . \ . \ . \ (108).$$

Thus then it follows, that the difference between the aggregate work Σu_1 of those forces, the tendency of each of which
is towards the direction of the motion of its point of application, and that Σu_2 of those the direction of each of which is

opposed to the motion of its point of application (or, in other words, the difference between the aggregate work of the *accelerating* forces of the system and that of the *retarding forces*), is equal to one half the vis viva accumulated or lost in the system whilst the work is being done, which is the PRINCIPLE OF VIS VIVA.

130. One half the vis viva of the system measures its accumulated work ; the principle of vis viva amounts, therefore, to no more than this, that the entire difference between the work done by those forces which tend to accelerate the motions of the parts of the system to which they are applied, and those which tend to retard them, is *accumulated* in the moving parts of the system, no work whatever being lost, but all that accumulated which is done upon it by the forces which tend to accelerate its motion, above that which is expended upon the retarding forces.

This principle has been proved generally of any mechanical system; it is therefore true of the most complicated machine. The entire amount of work done by the moving power, whatever it may be, upon that machine, is yielded partly at its working points in overcoming the resistances opposed there to its motion (that is, in doing its useful work), it is partly expended in overcoming the friction and other prejudicial resistances opposed to the motion of the machine between its moving and its working points, and all the rest is *accumulated* in the moving parts of the machine, ready to be yielded up under any deficiency of the moving power, or to carry on the machine for a time, should the operation of that power be withdrawn.

131. *When the forces of any system (not in equilibrium in every position which the parts of that system may be made to take up) pass through a position of equilibrium, the vis viva of the system becomes a maximum or a minimum.*

For, as in Art. 129., let the aggregate work done in the

K 3

directions of the motions of the several parts of the system
be represented by Σu_1, and the aggregate work done in directions opposed to the motions of the several parts by Σu_2, then
(Art. 129.), one half the acquired vis viva of system $= \Sigma u_1 - \Sigma u_2$.
Now as the system passes from any one position to any other,
each of the quantities Σu_1 and Σu_2 is manifestly increased. If
Σu_1 increases by a *greater* quantity than Σu_2, then the *vis viva*
is increased in this change of position; if, on the contrary, it
is increased by a less quantity than Σu_2, then the vis viva is
diminished. Thus if $\Delta \Sigma u_1$ and $\Delta \Sigma u_2$ represent the increments of Σu and Σu_2 in this change of position, then $(\Sigma u_1 +
\Delta \Sigma u_1) - (\Sigma u_2 + \Delta \Sigma u_2)$, or $(\Sigma u_1 - \Sigma u_2) + (\Delta \Sigma u_1 - \Delta \Sigma u_2)$, repre-
senting one half the vis viva after this change of position, it
is manifest that the vis viva is increased or diminished by the
change according as $\Delta \Sigma u_1$ is greater or less than $\Delta \Sigma u_2$; and
that if $\Delta \Sigma u_1$ be equal to $\Delta \Sigma u_2$, then no change takes place in
the amount of the vis viva of the system as it passes from the
one position to the other.

Now from the principle of virtual velocities (Art. 127.), it
appears, that precisely this case occurs as the system passes
through a position of equilibrium, the aggregate work of
those forces whose tendency is to accelerate the motions of
their points of application then precisely equalling that of
the forces whose tendency is opposed to these motions. For
an exceeding small change of position therefore, passing
through a position of equilibrium, $\Delta \Sigma u_1 = \Delta \Sigma u_2$, an equality
which does not, on the other hand, obtain, unless the body do
thus pass through a position of equilibrium.

Since then the sum $\Sigma u_1 - \Sigma u_2$, and therefore the aggregate
vis viva of the system, continually increases or diminishes up
to a position of equilibrium, and then ceases (for a certain
finite space at least) to increase or diminish, it follows, that it
is in that position a maximum or a minimum. Therefore, &c.

132. *When the forces of any system pass through a position of
equilibrium, the vis viva becomes a maximum or a minimum,*

according as that position is one of stable or unstable equilibrium.

For it is clear that if the vis viva be a maximum in any position of the equilibrium of the system, so that after it has passed out of that position into another at some finite distance from it, the acquired vis viva may have become less than it was before, then the aggregate work of the forces which tend to accelerate the motion between these two positions must have been less than that of the forces which tend to retard the motion (Art. 131.). Now suppose the body to have been placed at rest in this position of equilibrium, and a small impulse to have been communicated to it, whence has resulted an aggregate amount of vis viva represented by $\Sigma m V^2$. In the transition from the first to the second position, let this vis viva have become $\Sigma m v^2$; also let the aggregate work of the forces which have tended to accelerate the motion, between the two positions, be represented by ΣU_1, and that of the forces which have tended to retard the motion by ΣU_2; then, for the reasons assigned above, it appears that ΣU_2 is greater than ΣU_1.

Moreover, by the principle of vis viva,

$$\tfrac{1}{2}\Sigma m v^2 - \tfrac{1}{2}\Sigma m V^2 = \Sigma U_1 - \Sigma U_2,$$
$$\therefore\ \Sigma m v^2 = \Sigma m V^2 - 2(\Sigma U_2 - \Sigma U_1);$$

in which equation the quantity $2(\Sigma U_2 - \Sigma U_1)$ is essentially positive, in respect to the particular range of positions through which the disturbance is supposed to take place.*

For every one of these positions there must then be a certain value of $\Sigma m V^2$, that is, a certain original impulse and disturbance of the system from its position of equilibrium, which will cause the second member of the above equation,

---

* The disturbance is of course to be limited to that particular range of positions to which the supposed position of equilibrium stands in the relation of a position of maximum vis viva. If there be other positions of equilibrium of the system, there will be other ranges of adjacent positions, in respect to each of which there obtains a similar relation of maximum or minimum vis viva.

and therefore its first member Σmv^2 to vanish. Now every term of the sum Σmv^2 is essentially positive; this sum cannot therefore vanish unless each term of it vanish, that is, unless the velocity of each body of the system vanishes, or the whole be brought to rest. This repose of the system can, however, only be instantaneous; for by supposition, the position into which it has been displaced is not one of equilibrium. Moreover, the displacement of the system cannot be continued in the direction in which it has hitherto taken place, for the negative term in the second member of the above equation would yet farther be increased so as to exceed the positive term, and the first member Σmv^2 would thus become negative, which is impossible.

The motion of the system can then only be continued by the directions of the motions of certain of the elements which compose it being changed, so that the corresponding quantities by which ΣU_1 and ΣU_2 are respectively increased may change their signs, and the whole quantity $\Sigma U_1 - \Sigma U_2$ which before *increased* continually may now continually *diminish*. This being the case, Σmv^2 will increase again until, when $\Sigma U_1 - \Sigma U_2 = 0$, it becomes again equal to ΣmV^2; that is, until the system acquires again the vis viva with which its disturbance commenced.

Thus, then, it has been shown, that in respect to every one of the supposed positions of the system * there is a certain impulse or amount of vis viva, which being communicated to the system when in equilibrium, will just cause it to oscillate as far as that position, remain for an instant at rest in it, then return again towards its position of equilibrium, and re-acquire the vis viva with which its displacement commenced. Now this being true of *every* position of the supposed range of positions, it follows that it is true of every disturbance or impulse which will not carry the system beyond this supposed range of positions; so that, having been displaced by any such disturbance or impulse, the system will constantly return

* That is, of that range of positions over which the supposed position of equilibrium holds the relation of a position of maximum vis viva.

again towards the position of equilibrium from which it set out, and is STABLE in respect to that position.

On the other hand, if the supposed position of equilibrium be one in which the vis viva is a minimum, then the aggregate work of the forces which tend to accelerate the motion must, after the system has passed through that position, exceeds that of the forces which tend to retard the motion; so that, adopting the same rotation as before, ΣU_1 must be greater than ΣU_2, and the second member of equation essentially positive. Whatever may have been the original impulse, and the communicated vis viva ΣmV^2, Σmv^2 must therefore continually increase; so that the whole system can never come to a position of instantaneous repose*; but on the contrary, the motions of its parts must continually increase, and it must deviate continually farther from its position of equilibrium, in which position it can never rest. The position is thus one of unstable equilibrium. Therefore, &c.

FRICTION.

133. It is a matter of constant experience, that a certain resistance is opposed to the motion of one body on the surface of another under any pressure, however smooth may be the surfaces of contact, not only at the first commencement, but at every subsequent period of the motion; so that, not only is the exertion of a certain force necessary to cause the one body to pass *at first* from a state of rest to a state of motion upon the surface of the other, but that a certain force is further requisite to *keep up* this state of motion. The resistance thus opposed to the motion of one body on the surface of another, when the two are *pressed* together, is

* Within that range of positions over which the supposed position of equilibrium holds the relation of minimum vis viva.

called friction; that which opposes itself to the transition
from a state of continued rest to a state of motion is called
the *friction of quiescence;* that which continually *accompanies*
the state of motion is called the *friction of motion.*

The principal experiments on friction have been made by
Coulomb *, Vince, G. Rennie †, N. Wood ‡, and recently
(at the expense of the French Government) by Morin. §
They have reference, first, to the relation of the friction
of quiescence to the friction of motion; secondly, to the
variation of the friction of the same surfaces of contact under
different pressures; thirdly, to the relation of the friction
to the *extent* of the surface of contact; fourthly, to the
relation of the amount of the friction of motion to the
velocity of the motion; fifthly, to the influence of unguents
on the laws of friction, and on its amount under the same
circumstances of pressure and contact. The following are the
principal facts which have resulted from these experiments;
they constitute the *laws* of friction.

1st. That the friction of motion is subject to the same
laws with the friction of quiescence (about to be stated), but
agrees with them more accurately. That, under the same circumstances of pressure and contact, it is nevertheless different
in amount.

2ndly. That, when no unguent is interposed, the friction
of any two surfaces (whether of quiescence or of motion) is
directly proportional to the force with which they are pressed
perpendicularly together (up to a certain *limit* of that pressure per square inch), so that, for any two given surfaces
of contact, there is a constant ratio of the friction to the
perpendicular pressure of the one surface upon the other.
Whilst this ratio is thus the same for the same surfaces of
contact, it is different for different surfaces of contact. The
particular value of it in respect to any two given surfaces
of contact is called the CO-EFFICIENT of friction in re-

---

* Mém. des Sav. Etrang. 1781.   † Phil. Trans. 1829.
‡ A Practical Treatise on Rail-roads, 3d ed. chap. 76
§ Mém. de l'Institut. 1833, 1834, 1838.

spect to those surfaces. The co-efficients of friction in respect to those surfaces of contact, which for the most part form the moving surfaces in machinery, are collected in a table, which will be found at the termination of Art. 140.

3dly. That, when no unguent is interposed, the amount of the friction is, in every case, wholly independent of the *extent* of the surfaces of contact, so that the force with which two surfaces are pressed togegether being the same, and not exceeding a certain limit (per square inch), their friction is the same whatever may be the extent of their surfaces of contact.

4thly. That the friction of motion is wholly independent of the velocity of the motion.*

5thly. That where unguents are interposed, the co-efficient of friction depends upon the nature of the unguent, and upon the greater or less abundance of the supply. In respect to the supply of the unguent, there are two extreme cases, that in which the surfaces of contact are but slightly rubbed with the unctuous matter†, and that in which, by reason of the abundant supply of the unguent, its viscous consistency, and the extent of the surfaces of contact in relation to the insistent pressure, a continuous stratum of unguent remains continually interposed between the moving surfaces, and the friction is thereby diminished, as far as it is capable of being diminished, by the interposition of the particular unguent used. In this state the amount of friction is found (as might be expected) to be dependent rather upon the nature of the unguent than upon that of the surfaces of contact; accordingly M. Morin, 'from the comparison of a great number of results, has arrived at the following remarkable conclusion, easily fixing itself in the memory, and of great practical value : — " *that with unguents, hog's lard and olive oil, interposed in a continuous stratum between them, surfaces of wood on metal, wood on wood, metal on wood, and metal on metal*

---

* This result, of so much importance in the theory of machines, is fully established by the experiments of Morin.

† As, for instance, with an oiled or a greasy cloth.

(*when in motion*), have all of them very nearly the same co-
efficient of friction, the value of that co-efficient being in all
cases included between ·07 and ·08.

" *For the unguent tallow, the co-efficient is the same as for
the other unguents in every case, except in that of metals upon
metals. This unguent appears, from the experiments of Morin, to be less suited to metallic substances than the others,
and gives for the mean value of its co-efficient under the same
circumstances* ·10."

134. Whilst there is a remarkable uniformity in the results thus obtained in respect to the friction of surfaces,
between which a perfect separation is effected throughout
their whole extent by the interposition of a continuous stratum of the unguent, there is an infinite variety in respect
to those states of unctuosity which occur between the *ex-
tremes*, of which we have spoken, of surfaces merely unctuous *
and the most perfect state of lubrication attainable by the
interposition of a given unguent. It is from this variety of
states of the unctuosity of rubbing surfaces, that so great a
discrepancy has been found in the experiments upon friction
with unguents, a discrepancy which has not probably resulted
so much from a difference in the quantity of the unguent
supplied to the rubbing surfaces in different experiments,
as in a difference of the relation of the insistent pressures
to the extent of rubbing surface. It is evident, that for every
description of unguent there must correspond a certain pressure per square inch, under which pressure a perfect separation of two surfaces is made by the interposition of a continuous stratum of that unguent between them, and which
pressure per square inch being exceeded, that perfect separation cannot be attained, however abundant may be the supply
of the unguent.

The *ingenious* experiments of Mr. Nicolas Wood,† con-

---

* Or slightly rubbed with the unguent.
† Treatise on Rail-roads, 3d ed. p. 399.

firmed by those of Mr. G. Rennie,* have fully established these important conditions of the friction of unctuous surfaces. It is much to be regretted that we are in possession of no experiments directed specially to the determination of that particular pressure per square inch, which corresponds in respect to each unguent to the state of perfect separation, and to the determination of the co-efficients of frictions in those different states of separation which correspond to pressures higher than this.

It is evident, that where the extent of the surface sustaining a given pressure is so great as to make the pressure per square inch upon that surface less than that which corresponds to the state of perfect separation, this greater extent of surface tends to increase the friction by reason of that *adhesiveness* of the unguent, dependent upon its greater or less viscosity, whose effect is proportional to the extent of the surfaces between which it is interposed. The experiments of Mr. Wood † exhibit the effects of this adhesiveness in a remarkable point of view.

---

* Trans. Royal Soc. 1829.

† It is evident that, whilst by extending the unctuous surface which sustains any given pressure, we diminish the co-efficient of friction up to a certain limit, we at the same time increase that *adhesion* of the surfaces which results from the viscosity of the unguent, so that there may be a point where the gain on the one hand begins to be exceeded by the loss on the other, and where the surface of minimum resistance under the given pressure is therefore attained.

Mr. Wood considers the pressure per square inch, which corresponds to the minimum resistance, to be 90 lbs. in the case of axles of wrought iron turning upon cast iron, with fine neat's foot oil. The experiments of Mr. Wood, whilst they place the *general* results stated above in full evidence, can scarcely however be considered satisfactory as to the particular numerical values of the constants sought in this inquiry. In those experiments, and in others of the same class, the amount of the friction is determined from the observed space or time through which a body projected with a given velocity moves before all its velocity is destroyed, that is, before its accumulated work is exhausted. This is an easy method of experiment liable to many inaccuracies. It is much to be regretted that the experiments of Morin were not extended to the friction of unctuous surfaces, reference being had to the pressure per square inch.

It is perhaps deserving of enquiry, whether in respect to those considerable pressures under which the parts of the larger machines are accustomed to move upon one another, the adhesion of the unguent to the surfaces of contact, and the opposition presented to their motion by its viscidity, are causes whose influence may be altogether neglected as compared with the ordinary friction. In the case of lighter machinery, as for instance that of clocks and watches, these considerations evidently rise into importance.

135. The experiments of M. Morin show the friction of two surfaces which have been for *a considerable time in contact*, to be not only different in its amount from the friction of surfaces *in continuous motion*, but also, especially in *this*, that the laws of friction (as stated above) are, in respect to the friction of quiescence, subject to causes of *variation* and *uncertainty* from which the friction of motion is exempt. This variation does not appear to depend upon the *extent* of the surfaces of contact, in which case it might be referred to adhesion; for with different pressures the co-efficient of the friction of quiescence was found, in certain cases, to vary exceedingly, although the surfaces of contact remained the same.* The uncertainty which would have been introduced into every question of construction by this consideration, is removed by a second very important fact developed in the course of the same experiments. It is this, that by the slightest *jar* or *shock* of two bodies in contact, their friction is made to pass from that state which accompanies quiescence to that which accompanies motion ; and as every machine or structure, of whatever kind, may be considered to be subject to such shocks or imperceptible motions of its surfaces of contact, it is evident that the state of friction to be made the basis on which all questions of statics are to be determined, should be that which accompanies continuous motion. The laws stated above have been shown, by the experiments of Morin, to

---

* Thus in the case of oak upon oak with parallel fibres, the co-efficient of the friction of quiescence varied under different pressures upon the same surface, from ·55 to ·76.

obtain, in respect to that friction which accompanies motion, with a precision and uniformity never before assigned to them; they have given to all our calculations in respect to the theory of machines (whose moving surfaces have attained their proper bearings and been worn to their natural polish) a new and unlooked-for certainty, and may probably be ranked amongst the most accurate and valuable of the constants of practical science.

It is, however, to be observed, that all these experiments were made under comparatively small insistent pressures as compared with the extent of the surface pressed (pressures not exceeding from one to two kilogrammes per square centimeter, or from about 14·3 to 28·6 lbs. per square inch). In adopting the results of M. Morin, it is of importance to bear this fact in mind, because the experiments of Coulomb, and particularly the excellent experiments of Mr. G. Rennie, carried far beyond these limits of insistent pressure *, have fully shown the co-efficient of the friction of *quiescence* to increase rapidly, from some limit attained long before the surfaces abrade. In respect to some surfaces, as, for instance, wrought iron upon wrought iron, the co-efficient nearly tripled itself as the pressure advanced to the limits of abrasion. It is greatly to be regretted that no experiments have yet been directed to a determination of the precise limit about which this change in the value of the co-efficient begins to take place. It appears, indeed, in the experiments of Mr. Rennie in respect to some of the soft metals, as, for instance, tin upon tin, and tin upon cast iron; but in respect to the harder metals, his experiments passing at once from a pressure of 32 lbs. per square inch to a pressure of 1·66 cwt. per square inch, and the co-efficient (in the case of wrought iron for instance) from about ·148 to ·25, the limit which we seek is lost in the intervening chasm. The experiments of Mr. Rennie have reference, however, only to the friction of quiescence. It seems probable that the co-efficient of the fric-

---

* Mr. Rennie's experiments were carried, in some cases, to from 5 cwt. to 7 cwt. per square inch.

tion of motion remains constant under a wider range of pressure than that of quiescence. It is moreover certain, that the limits of pressure beyond which the surfaces of contact begin to destroy one another or to abrade, are sooner reached when one of them is in motion upon the other, than when they are at rest: it is also certain that these limits are not independent of the velocity of the moving surface. The discussion of this subject, as it connects itself especially with the friction of *motion*, is of great importance ; and it is to be regretted, that, with the means so munificently placed at his disposal by the French Government, M. Morin did not extend his experiments to higher pressures, and direct them more particularly to the circumstances of pressure and velocity under which a destruction of the rubbing surfaces first begins to show itself, and to the amount of the destruction of surface or wear of the material which corresponds to the same space traversed under different pressures and different velocities. Any accurate observer who should direct his attention to these subjects would greatly promote the interests of practical science.

Summary of the Laws of Friction.

136. From what has here been stated it results, that if P represent the perpendicular or *normal* force by which one body is pressed upon the surface of another, F the friction of the two surfaces, or the force, which being applied parallel to their common surface of contact, would cause one of them to slip upon the surface of the other, and *f* the *co-efficient* of friction, then, in the case in which no unguent is interposed, *f* represents a constant quantity, and (Art. 133.)

$$F = fP \; \cdots \; (109);$$

a relation which obtains *accurately* in respect to the friction of motion, and *approximately* in respect to the friction of quiescence.

137. The same relation obtains, moreover, in respect to unctuous surfaces when merely rubbed with the unguent, or where the presence of the unguent has no other influence than to increase the smoothness of the surfaces of contact without at all separating them from one another.

In unctuous surfaces *partially* lubricated, or between which a stratum of unguent is partially interposed, the co-efficient of friction f is dependent for its amount upon the relation of the insistent pressure to the extent of the surface pressed, or upon the pressure *per square inch* of surface. This amount, corresponding to each pressure per square inch in respect to the different unguents used in machines, has not yet been made the subject of satisfactory experiments.

The amount of the resistance F opposed to the sliding of the surfaces upon one another is, moreover, as well in this case as in that of surfaces perfectly lubricated, influenced by the *adhesiveness* of the unguent, and is therefore dependent upon the extent of the adhering surface ; so that, if S represent the number of square units in this surface, and α the adherence of each square unit, then αS represents the whole adherence opposed to the sliding of the surfaces, and

$$F = f P + \alpha S \ . \ . \ . \ . \ . \ (110);$$

where f is a function of the pressure per square unit $\dfrac{P}{S}$, and α is an exceedingly small factor dependent on the viscosity of the unguent.

THE LIMITING ANGLE OF RESISTANCE.

We shall, for the present, suppose the parts of a solid body to cohere so firmly, as to be incapable of separation by the action of any force which may be impressed upon them. The limits within which this supposition is true will be discussed hereafter.

It is not to this resistance that our present inquiry has

L

reference, but to that which results from the friction of the
surface of bodies on one another, and especially to the *direction* of that resistance.

138. *Any pressure applied to the surface of an immoveable
solid body by the intervention of another body moveable
upon it, will be sustained by the resistance of the surfaces
of contact, whatever be its direction, provided only the
angle which that direction makes with the perpendicular to
the surfaces of contact do not exceed a certain angle called
the* LIMITING ANGLE OF RESISTANCE *of those* SURFACES.
*This is true, however great the pressure may be. Also, if
the inclination of the pressure to the perpendicular exceed
the limiting angle of resistance, then this pressure will not
be sustained by the resistance of the surfaces of contact ;
and this is true, however small the pressure may be.*

Let PQ represent the direction in which the surfaces of

two bodies are pressed together at Q, and let
QA be a perpendicular or *normal* to the surfaces of contact at that point, then will the pressure PQ be sustained by the resistance of the
surfaces, however great it may be, provided its
direction lie within a certain given angle AQB,
called the limiting angle of resistance ; and it will not be sustained, however small it may be, provided its direction lie
without that angle.   For let this pressure be represented by
PQ, and let it be resolved into two others AQ and RQ, of
which AQ is that by which it presses the surfaces together
perpendicularly, and RQ that by which it tends to cause
them to slide upon one another, if therefore the friction F
produced by the first of these pressures exceed the second
pressure RQ, then the one body will not be made to slip
upon the other by this pressure PQ, however great it may be;
but if the friction F, produced by the perpendicular pressure
AQ, be less than the pressure RQ, then the one body will be
made to slip upon the other however small PQ may be.   Let

the pressure in the direction PQ be represented by P, and the angle AQP by θ, the perpendicular pressure in AQ is then represented by P cos. θ, and therefore the friction of the surfaces of contact by fP cos. θ, f representing the co-efficient of friction (Art. 136.). Moreover, the resolved pressure in the direction RQ is represented by P sin. θ. The pressure P will therefore be sustained by the friction of the surfaces of contact or not, according as

P sin. θ is less or greater than fP cos. θ ;

or, dividing both sides of this inequality by P cos. θ, according as

tan. θ is less or greater than f.

Let, now, the angle AQB equal that angle whose tangent is f, and let it be represented by φ, so that tan. $\varphi = f$. Substituting this value of f in the last inequality, it appears that the pressure P will be sustained by the friction of the surfaces of contact or not, according as

tan. θ is less or greater than tan. φ,

that is, according as

θ is less or greater than φ,

or according as

AQP is less or greater than AQB.

Therefore, &c.   [Q.E.D.]

THE CONE OF RESISTANCE.

139. If the angle AQB be conceived to revolve about the axis AQ, so that BQ may generate the surface of a cone BQC, then this cone is called the CONE OF RESISTANCE: it is evident, that any pressure, however great, applied to the surfaces of contact at Q will be sustained by the resistance of the surfaces of contact, provided its direction be any where *within* the surface of this cone; and that it will

L 2

not be sustained, however small it may be, if its direction lie any where without it.

The Two States bordering upon Motion.

140. If the direction of the pressure coincide with the surface of the cone, it will be sustained by the friction of the surfaces of contact, but the body to which it is applied will be upon the point of slipping upon the other. The state of the equilibrium of this body is then said to be that BORDERING UPON MOTION. If the pressure P admit of being applied in any direction about the point Q, there are evidently an infinity of such states of the equilibrium bordering upon motion, corresponding to all the possible positions of P on the surface of the cone.

If the pressure P admit of being applied only in the same plane, there are but two such states, corresponding to those directions of P, which coincide with the two intersections of this plane with the surface of the cone : these are called the *superior* and *inferior* states bordering upon motion. In the case in which the direction of P is limited to the plane AQB, BQ and CQ represent its directions corresponding to the two states bordering upon motion. Any direction of P within the angle BQC corresponds to a state of equilibrium ; any direction, without this angle, to a state of motion.

141. Since, when the direction of the pressure P coincides with the surface of the cone of resistance, the equilibrium is in the state bordering upon motion ; it follows, conversely, and for the same reasons, that this is the direction of the pressure sustained by the surfaces of contact of two bodies whenever the state of their equilibrium is that bordering upon motion. This being, moreover, the direction of the pressure of the one body upon the other is manifestly the direction of the *resistance* opposed by the second body to the pressure of the first at their surface of contact, for this single pressure and this single resistance are forces in equilibrium, and there-

fore equal and opposite. All that has been said above of the single pressure and the single resistance sustained by two surfaces of contact, is manifestly true of the *resultant* of any number of such pressures, and of the *resultant* of any number of such resistances. Thus then it follows, that *when any number of pressures applied to a body moveable upon another which is fixed, are sustained by the resistance of the surface of contact of the two bodies, and are in the state of equilibrium bordering upon motion, then the direction of the resultant of these pressures coincides with the surface of the cone of resistance, as does that also of the resultant of the resistances of the different points of the surface of contact*, that is, they are both inclined to the perpendicular to the surface of contact (at the point where they intersect it), at an angle equal to the limiting angle of resistance.*

TABLE I.

Friction of plane surfaces, when they have been some time in contact.

Surfaces in Contact.	Disposition of the Fibres.	State of the Surfaces.	Co-efficient of Friction.	Limiting Angle of Resistance.
EXPERIMENTS OF M. MORIN.				
	parallel	without unguent	0·62	31° 48′
	ditto	rubbed with dry soap	0·44	23 45
	perpendicular	without unguent	0·54	28 22
Oak upon oak	ditto	with water	0·71	35 23
	endways of one upon the flat surface of the other	without unguent	0·43	23 16
Oak upon elm	parallel	ditto	0·38	20 49
	ditto	ditto	0·69	34 37
Elm upon oak	ditto	rubbed with dry soap	0·41	22 18
	perpendicular	without unguent	0·57	29 41

---

* The properties of the limiting angle of resistance and the cone of resistance, were first given by the author of this work in a paper published in the Cambridge Philosophical Transactions, vol. v.

Surfaces in Contact.	Disposition of the Fibres.	State of the Surfaces.	Co-efficient of Friction.	Limiting Angle of Resistance.
EXPERIMENTS OF M. MORIN —continued.				
Ash, fir, beech, service-tree, upon oak	parallel	without unguent	0·53	27° 56'
Tanned leather upon oak	the leather flat	ditto	0·61	31 23
	the leather lengthways, but sideways	ditto	0·43	23 16
		steeped in water	0·79	38 19
Black dressed leather, or strap leather. — upon a plane surface of oak	parallel	without unguent	0·74	36 30
— upon a rounded surface of oak	perpendicular	ditto	0·47	25 11
Hemp matting upon oak	parallel	ditto	0·50	26 34
	ditto	steeped in water	0·87	41 2
Hemp cords upon oak -	ditto	without unguent	0·80	38 40
		ditto	0·62	31 48
Iron upon oak -	ditto	steeped in water	0·65	33 2
Cast-iron upon oak -	ditto	ditto	0·65	33 2
Copper upon oak -	ditto	without unguent	0·62	31 48
		steeped in water	0·62	31 48
Ox-hide as a piston sheath upon cast-iron	flat or sideways	with oil, tallow, or hog's lard	0·12	6 51
Black dressed leather, or strap leather, upon a cast-iron pulley	flat	without unguent	0 28	15 39
		steeped	0·38	20 49
Cast-iron upon cast-iron -	ditto	without unguent	0·16*	9 6
Iron upon cast-iron -	ditto	ditto	0·19	10 46
Oak, elm, yoke elm, iron, cast-iron, and brass, sliding two and two, one upon another	ditto	with tallow	0·10†	5 43
		with oil, or hog's lard	0·15‡	8 32
Calcareous oolite stone upon calcareous oolite	ditto	without unguent	0·74	36 30
Hard calcareous stone, called muschelkalk, upon calcareous oolite	ditto	ditto	0·75	36 52
Brick upon calcareous oolite	ditto	ditto	0·67	33 50
Oak upon calcareous oolite	wood endways	ditto	0·63	32 13

* The surfaces retaining some unctuousness.

† When the contact has not lasted long enough to express the grease.

‡ When the contact *has* lasted long enough to express the grease, and bring back the surfaces to an unctuous state.

Surfaces in Contact.	Disposition of the Fibres.	State of the Surfaces.	Co-efficient of Friction.	Limiting Angle of Resistance.	
EXPERIMENTS OF M. MORIN — continued.					
Iron upon calcareous oolite	flat	{ without unguent }	0·49	26°	7'
Hard calcareous stone, or muschelkalk, upon muschelkalk -	ditto	ditto	0·70	35	0
Calcareous oolite stone upon muschelkalk	ditto	ditto	0·75	36	52
Brick upon muschelkalk -	ditto	ditto	0·67	33	50
Iron upon muschelkalk -	ditto	ditto	0·42	22	47
Oak upon muschelkalk -	ditto	ditto	0·64	32	38
Calcareous oolite stone upon calcareous oolite	ditto	{ with a coating of mortar, of three parts of fine sand and one part of slack lime }	0·74 *	36	30

* After a contact of from ten to fifteen minutes.

Nature of Bodies and Unguents.	Co-efficient of Friction.	Limiting Angle.	
Soft calcareous stone, well dressed, upon the same -	0·74	36°	30'
Hard calcareous stone, ditto - - -	0·75	36	52
Common brick, ditto - - - -	0·67	33	50
Oak, endways, ditto - - - -	0·63	32	13
Wrought iron, ditto - - - -	0·49	26	7
Hard calcareous stone, well dressed, upon hard calcareous stone - - - - -	0·70	35	0
Soft, ditto - - - -	0·75	36	52
Common brick, ditto - - -	0·67	33	50
Oak, endways, ditto - - - -	0·64	32	37
Wrought iron, ditto - - - -	0·42	22	47
Soft calcareous stone upon soft calcareous stone, with fresh mortar of fine sand - - -	0·74	36	30
EXPERIMENTS BY DIFFERENT OBSERVERS.			
Smooth free-stone upon smooth free-stone, dry (Rennie)	0·71	35	23
Ditto, with fresh mortar (Rennie) - - -	0·66	33	26
Hard polished calcareous stone upon hard polished calcareous stone - - - - -	0·58	30	7
Calcareous stone upon ditto, both surfaces being made rough with the chisel (Bonchardi) - -	0·78	37	58
Well dressed granite upon rough granite (Rennie) -	0·66	33	26
Ditto, with fresh mortar, ditto (Rennie) - -	0·49	26	7
Box of wood upon pavement (Regnier) - -	0·58	30	7
Ditto upon beaten earth (Herbert) - -	0·33	18	16
Libage stone upon a bed of dry clay - -	0·51	27	2
Ditto, the clay being damp and soft - -	0·34	18	47
Ditto, the clay being equally damp, but covered with thick sand (Grève) - - -	0·40	21	48

TABLE II.

Friction of plane surfaces in motion one upon the other.

Surfaces in Contact.	Disposition of the Fibres.	State of the Surfaces.	Co-efficient of Friction.	Limiting Angle of Resistance.
EXPERIMENTS OF M. MORIN.				
	parallel	without unguent	0·48	25° 39′
	ditto	rubbed with dry soap	0·16	9 6
	perpendicular	without unguent	0·34	18 47
Oak upon oak - -	ditto	steeped in water	0·25	14 3
	wood endways on wood lengthways	without unguent	0·19	10 46
	parallel	ditto	0·43	23 17
Elm upon oak - -	perpendicular	ditto	0·45	24 14
	parallel	ditto	0·25	14 3
Ash, fir, beech, wild pear-tree, and service-tree, upon oak -	ditto	ditto	0·36 to 0·40	19 48 21 49
		ditto	0·62	31 48
		with water	0·26	14 35
Iron upon oak - -	ditto	rubbed with dry soap	0·21	11 52
		without unguent	0·49	26 7
		with water	0·22	12 25
Cast-iron upon oak, -	ditto	rubbed with dry soap	0·19	10 46
Copper upon oak -	ditto	without unguent	0·62	31 48
Iron upon elm - -	ditto	ditto	0·25	14 3
Cast-iron upon elm -	ditto	ditto	0·20	11 19
Black dressed leather upon oak	ditto	ditto	0·27	15 7
Tanned leather upon oak	flat, or lengthways and edgeways	ditto	3·30 to 0·35	16 42 19 18
		with water	0·29	16 11
		without unguent	0·56	29 15
Tanned leather upon cast-iron and brass	ditto	steeped in water	0·36	19 48
		greased and steeped in water	0·23	12 58
		with oil	0·15	8 32

Surfaces in Contact.	Disposition of the Fibres.	State of the Surfaces.	Co-efficient of Friction.	Limiting Angle of Resistance.
EXPERIMENTS OF M. MORIN *— continued.*				
Hemp, in threads or in cord, upon oak	parallel	without unguent	0·52	27° 29'
	perpendicular	with water	0·33	18 16
Oak and elm upon cast-iron	parallel	without unguent	0·38	20 49
Wild pear-tree, ditto -	ditto	ditto	0·44	23 45
Iron upon iron - -	ditto	ditto	0·44	23 45 *
Iron upon cast-iron and brass	ditto	ditto	0·18 †	10 13
Cast-iron, ditto - -	ditto	ditto	0·15	8 32
Brass { upon brass -	ditto	ditto	0·20	11 19
upon cast-iron -	ditto	ditto	0·22	12 25
upon iron -	ditto	ditto	0·16 ‡	9 6
Oak, elm, yoke elm, wild pear, cast-iron, wrought iron, steel, and moving one upon another, or on themselves	ditto	greased in the usual way with tallow, hog's lard, oil, soft gom	0·07 to 0·08 §	{ 4 1 { 4 35
		slightly greasy to the touch	0·15	8 32
Calcareous oolite stone upon calcareous oolite	ditto	without unguent	0·64	32 37
Calcareous stone, called muschelkalk, upon calcareous oolite	ditto	ditto	0·67	33 50
Common brick upon calcareous oolite	ditto	ditto	0 65	33 2
Oak upon calcareous oolite	wood endways	ditto	0·38	20 49
Wrought iron, ditto -	parallel	ditto	0·69	34 37
Calcareous stone, called muschelkalk, upon muschelkalk	ditto	ditto	0·38	20 49
Calcareous oolite stone upon muschelkalk	ditto	ditto	0·65	33 2
Common brick, ditto -	ditto	ditto	0·60	30 58
Oak upon muschelkalk - {	wood endways	ditto	0·38	20 49
Iron upon muschelkalk -	parallel	ditto	0·24	
		saturated with water	0·30	16 42

* The surfaces wear when there is no grease.
† The surfaces still retaining a little unctuousness. ‡ Ibid.
§ When the grease is constantly renewed and uniformly distributed, this proportion can be reduced to 0·05.

TABLE III.

Friction of gudgeons or axle ends, in motion, upon their bearings.
(From the experiments of Morin.)

Surfaces in Contact.	State of the Surfaces.	Co-efficient of Friction when the Grease is renewed.		Limiting Angle of Resistance.
		In the usual Way.	Continuously.	
Cast-iron axles in cast-iron bearings	coated with oil of olives, with hog's lard, tallow, and soft gom	0·07 to 0·08	0·054	4° 0' 4 35 3 6
	with the same, and water	0·08	0·28	4 35
	coated with asphaltum	0·054	0·19	3 6
	greasy	0·14	- -	7 58
	greasy and wetted	0·14	- -	7 58
Cast-iron axles, ditto	coated with oil of olives, with hog's lard, tallow, and soft gom	0·07 to 0·08	0·054	4 0 4 35 3 6
	greasy	0·16	- -	9 6
	greasy and damped	0·16	*	9 6
	scarcely greasy	0·19	*	10 46
	without unguent	0·18	- -	10 12
Cast-iron axles in lignum vitæ bearings	with oil or hog's lard	- -	0·090	5 9
	greasy with ditto	0·10	- -	5 43
	greasy, with a mixture of hog's lard and molybdæna	0·14	- -	7 58
Wrought-iron axles in cast-iron bearings	coated with oil of olives, tallow, hog's lard, or soft gom	0·07 to 0·08	0·054	4 0 4 35 3 6
Iron axles in brass bearings	coated with oil of olives, hog's lard, or tallow	0·07 to 0·08	0·054	4 0 4 35 3 6
	coated with hard gom	0·09	- -	5 9
	greasy and wetted	0·19	- -	10 46
	scarcely greasy	0·25	*	14 2
Iron axles in lignum vitæ bearings	coated with oil, or hog's lard	0·11	- -	6 17
	greasy	0·19	- -	10 46
Brass axles in brass bearings	coated with oil	0·10	- -	5 43
	with hog's lard	0·09	- -	5 9
Brass axles in cast-iron bearings	Coated with oil or tallow	- -	0·045 to 0·052	2 35 2 59

* The surfaces beginning to wear.

Surfaces in Contact.	State of the Surfaces.	Co-efficient of Friction when the Grease is renewed.		Limiting Angle of Resistance.
		In the usual Way.	Continuously.	
Lignum vitæ axles, ditto	coated with hog's lard	0·12	- -	6° 51′
	greasy	0·15	- -	8 32
Lignum vitæ axles in lignum vitæ bearings	coated with hog's lard	- -	0·07	4 0

TABLE IV.

Co-efficients of friction under pressures increased continually up to the limits of abrasion. From the experiments of Mr. G. Rennie.*

Pressure per Square Inch.	Co-efficients of Friction.			
	Wrought-iron upon Wrought-iron.	Wrought-iron upon Cast-iron.	Steel upon Cast-iron.	Brass upon Cast-iron.
32·5 lb.	·140	·174	·166	·157
1·66 cwt.	·250	·275	·300	·225
2·00	·271	·292	·333	·219
2·33	·285	·321	·340	·214
2·66	·297	·329	·344	·211
3·00	·312	·333	·347	·215
3·33	·350	·351	·351	·206
3·66	·376	·353	·353	·205
4·00	·376	·365	·354	·208
4·33	·395	·366	·356	·221
4·66	·403	·366	·357	·223
5·00	·409	·367	·358	·233
5·33		·367	·359	·234
5·66		·367	·367	·235
6·00		·376	·403	·233
6·33		·434		·234
6·66				·235
7·00				·232
7·33				·273

* Phil. Trans. 1829, table 8. p. 159.

THE RIGIDITY OF CORDS.

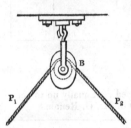

142. It is evident that, by reason of that resistance to deflexion which constitutes the rigidity of a cord, a certain force or pressure must be called into action whenever it is made to change its rectilineal direction, so as to adapt itself to the form of any curved surface over which it is made to pass; as, for instance, over the circumference of a pulley or wheel. Suppose such a cord to sustain tensions represented by P_1 and P_2, of which P_1^1 is on the point of preponderating, and let the friction of the axis of the pulley be, for the present, neglected. It is manifest that, in order to supply the force necessary to overcome the rigidity of the cord and to produce its deflexion at B, the tension P_1 must exceed P_2; whereas, if there were no rigidity, P_1 would equal P_2; so that the effect of the rigidity in increasing the tension P_1 is the same as though it had, by a certain quantity, increased the tension P_2. Now, from a very numerous series of experiments made by Coulomb upon this subject, it appears that the quantity by which the tension P_2 may thus be considered to be increased by the rigidity, is partly constant and partly dependent on the amount of P_2; so as to be represented by an algebraical formula of two terms, one of which is a constant quantity, and the other the product of a constant quantity by P_2. Thus if D represent the constant part of this formula, and E the constant factor of P_2, then is the effect of the rigidity of the cord the same as though the tension P_2 were increased by the quantity $D + E . P_2$.

When the cord, instead of being bent, under different pressures, upon circular arcs of equal radii, was bent upon circular arcs of different radii, then this quantity $D + E . P_2$, by which the tension P_2 may be considered to be increased by the rigidity, was found to vary inversely as the radii of the arcs ; so that, on the whole, it may be represented by the formula

$$\frac{D + E \cdot P_2}{R} \quad \ldots \ldots \quad (111),$$

where R represents the radius of the circular arc over which the rope is bent. Thus it appears that the yielding tension P_2 may be considered to have been increased by the rigidity of the rope, when in the state bordering upon motion, so as to become

$$P_2 + \frac{D + E \cdot P_2}{R}.$$

This formula applies only to the bending of the *same* cord under different tensions upon different circular arcs : for different cords, the constants D and E vary (within certain limits to be specified) *as the squares of the diameters or of the circumferences* of the cords, in respect to *new cords, wet or dry;* in respect to *old cords* they vary nearly as *the power $\frac{3}{2}$ of the diameters or circumferences.*

Tables have been furnished by Coulomb of the values of the constants D and E. These tables, reduced to English measures, are given on the next page.*

* The rigidity of the cord exerts its influence to increase the resistance only at that point where the cord winds upon the pulley; at the point where it leaves the pulley its elasticity favours rather, and does not perceptibly affect, the conditions of the equilibrium.

In all calculations of machines, in which the moving power is applied by the intervention of a rope passing over a pulley, *one half the diameter of rope is to be added to the radius of the pulley, or to the perpendicular on the direction of the rope from the point whence the moments are measured,* the pressure applied to the rope producing the same effect as though it were all exerted along the axis of the rope.

TABLE V. RIGIDITY OF ROPES.

Table of the values of the constants D *and* E, *according to the experiments of Coulomb (reduced to English measures). The radius* R *of the pulley is to be taken in feet.*

No. 1. New dry cords. Rigidity proportional to the square of the circumference.

Circumference of the Rope in Inches.	Value of D in lbs.	Value of E in lbs.
1	·131528	·033533
2	·526108	·023030
4	2·104451	·073175
8	8·413702	·368494

Squares of proportions of the intermediate circumferences to those of the table.

Proportions.	Squares.
1·0	1·00
1·1	1·21
1·2	1·44
1·3	1·69
1·4	1·96
1·5	2·25
1·6	2·56
1·7	2·89
1·8	3·24
1·9	3·61
2·0	4·00

No. 2. New ropes dipped in water. Rigidity proportional to the square of the circumference.

Circumference of the Rope in Inches.	Value of D in lbs.	Value of E in lbs.
1	·263053	·0057576
2	1·052217	·0230303
4	4·208902	·0731755
8	16·835606	·3684860

No. 3. Dry half-worn ropes. Rigidity proportional to the square root of the cube of the circumference.

Circumference of the Rope in Inches.	Value of D in lbs.	Value of E in lbs.
1	·146272	·0064033
2	·413656	·0180827
4	1·169641	·0512115
8	3·308787	·1448238

Square roots of the cubes of proportions of the intermediate circumferences to those of the table.

No. 4. Wetted half-worn cords. Rigidity proportional to the square root of the cube of the circumference.

Proportions.	Powers 3/2
1·0	1·000
1·1	1·154
1·2	1·315
1·3	1·482
1·4	1·657
1·5	1·837
1·6	2·024
1·7	2·217
1·8	2·415
1·9	2·619
2·0	2·828

Circumference of the Rope in Inches.	Value of D in lbs.	Value of E in lbs.
1	·292541	·006401
2	·827328	·018107
4	2·339675	·051212
8	6·616589	·144822

TABLE VI.

Tarred rope.   Rigidity proportional to the number of strands.

Number of Strands.	Value of D in lbs.	Value of E in lbs.
6	0·33390	0·009305
15	0·17212	0·021713
30	1·25294	0·044983

To determine the constants D and E for ropes whose circumferences are intermediate to those of the tables, find the ratio of the given circumference to that *nearest* to it in the tables, and seek this ratio or proportion in the first column of the auxiliary table to the right of the page. The corresponding number in the second column of this auxiliary table is a factor by which the values of D and E for the nearest circumference in the principal tables being multiplied, their values for the given circumference will be determined.

* L 8

PART III.

THE THEORY OF MACHINES.

143. THE parts of a machine are divisible into those which *receive the operation of the moving power immediately*, those which *operate immediately upon the work* to be performed, and those which *communicate between the two*, or which conduct the power or work from the *moving* to the *working* points of the machine. The first class may be called RECEIVERS, the second OPERATORS, and the third COMMUNICATORS of work.

THE TRANSMISSION OF WORK BY MACHINES.

144. The moving power divides itself whilst it operates in a machine, *first*, Into that which overcomes the *prejudicial* resistances of the machine, or those which are opposed by friction and other causes *uselessly* absorbing the work in its transmission. *Secondly*, Into that which *accelerates* the motion of the various moving parts of the machine; so long as the work done by the moving power upon it exceeds that expended upon the various resistances opposed to the motion of the machine (Art. 129.). *Thirdly*, Into that which overcomes the *useful* resistances, or those which are opposed to the motion of the machine at the working point or points by the useful work which is to be done by it. Thus, then, the work done by the moving power upon the *moving* points of the machine (as distinguished from the *working* points) divides itself in the act of transmission, first, Into the work *expended* uselessly upon the friction and other prejudicial resistances opposed to its transmission. Secondly, Into that

accumulated in the various moving elements of the machine, and *reproducible*. Thirdly, Into the *useful* work, or that done by the *operators*, whence results immediately the useful products of the machine.

145. *The aggregate number of units of useful work yielded by any machine at its working points is less than the number received upon the machine directly from the moving power, by the number of units expended upon the prejudicial resistances and by the number of units accumulated in the moving parts of the machine whilst the work is being done.*

For by the principle of vis viva (Art. 129.), if ΣU_1 represent the number of units of work received upon the machine immediately from the operation of the moving power, Σu the whole number of such units absorbed in overcoming the *prejudicial* resistances opposed to the working of the machine, ΣU_2 the whole *useful* work of the machine (or that done by its operators in producing the useful effect), and $\frac{1}{2g}\Sigma w(v_2{}^2 - v_1{}^2)$ one-half the aggregate difference of the vires vivæ of the various moving parts of the machine at the commencement and termination of the period during which the work is estimated, then, by the principle of VIS VIVA (equation 108),

$$\Sigma U_1 = \Sigma U_2 + \Sigma u + \frac{1}{2g}\Sigma w(v_2{}^2 - v_1{}^2) \ \ . \ . \ . \ . \ (112);$$

in which v_1 and v_2 represent the velocities at the commencement and termination of the period, during which the work is estimated, of that moving element of the machine whose weight is w. Now one-half the aggregate difference of the vires vivæ of the moving elements represents the work *accumulated* in them during the period in respect to which the work is estimated (Art. 130.). Therefore, &c.

M

162   THE MODULUS OF A MACHINE.

146. *If the same velocity of every part of the machine re-*
turn after any period of time, or if the motion be periodical,
then is the whole work received upon it from the moving power
during that time exactly equal to the sum of the useful work
done, and the work expended upon the prejudicial resistances.
For the velocity being in this case the same at the commencement and expiration of the period during which the
work is estimated, $\Sigma w(v_1^2 - v_2^2) = 0$, so that

$$\Sigma U_1 = \Sigma U_2 + \Sigma u \quad \ldots \ldots (113).$$

Therefore, &c.

The converse of this proposition is evidently true.

147. *If the prime mover in a machine be throughout the*
motion in equilibrium with the useful and the prejudicial
resistances, then the motion of the machine is uniform.
For in this case, by the principle of virtual velocities
(Art. 127.), $\Sigma U_1 = \Sigma U_2 + \Sigma u$; therefore (equation 112)
$\Sigma w(v_1^2 - v_2^2) = 0$; whence it follows that (in the case supposed) the velocities v_1 and v_2 of any moving element of the
machine are the same at the commencement and termination of any period of the motion however small, or that
the motion of every such element is a uniform motion.
Therefore, &c.

The converse of this proposition is evidently true.

THE MODULUS OF A MACHINE MOVING WITH A UNIFORM
OR PERIODICAL MOTION.

148. *The modulus of a machine, in the sense in which the*
term is used in this work, is the relation between the work
constantly done upon it by the moving power, and that
constantly yielded at the working points, when it has at-
tained a state of uniform motion, if it admit of such a state

*of motion ; or if the nature of its motion be periodical, then
is its modulus the relation between the work done at its
moving and at its working points in the interval of time
which it occupies in passing from any given velocity to the
same velocity again.*

The modulus is thus, in respect to any machine, the particular form *applicable to that machine* of equation 113, and
being dependent for its amount upon the amount of work Σu
expended upon the friction and other prejudicial resistances
opposed to the motion of the various elements of the machine, it measures in respect to each such machine the loss
of work due to these causes, and therefore constitutes a true
standard *for comparing the expenditure of moving power necessary to the production of the same effects by different machines :* it is thus a measure of the working qualities of machines.*

Whilst the particular modulus of every differently constructed machine is thus different, there is nevertheless a
general algebraical type or formula to which the moduli of
machines are (for the most part and with certain modifications)
referable. That form is the following,

$$U_1 = A . U_2 + B . S \dots (114),$$

where U_1 is the work done at the moving point of the
machine through the space S, U_2 the work yielded at the
working points, and A and B constants dependent for their
value upon the construction of the machine ; that is to say,
upon the dimensions and the combination of its parts, their
weights, and the co-efficients of friction at their various
rubbing surfaces.

It would not be difficult to establish *generally* this *form* of
the modulus under certain assumed conditions. As the modulus of each particular machine must however, in this work,
be discussed and determined independently, it will be better
to refer the reader to the particular moduli investigated in the

* The properties of the modulus of a machine are here, for the first
time, discussed.

following pages. He will observe that they are for the most part comprised under the form above assumed; subject to certain modifications which arise out of the discussion of each individual case, and which are treated at length.

149. There is, however, one important exception to this general form of the modulus, it occurs in the case of machines, some of whose parts move immersed in fluids. It is only when the resistances opposed to the motion of the parts of the machine upon one another are, like those of friction, proportional to the pressures, or when they are constant resistances, that this form of the modulus obtains. If there be resistances which, like those of fluids in which the moving parts are immersed (the air, for instance), vary with the velocity of the motion, and these resistances be considerable, then must other terms be added to the modulus. This subject will be further discussed when the resistances of fluids are treated of. It may here, however, be observed, that if the machine move *uniformly* subject to the resistance of a fluid during a given time T, and the resistance of the fluid be supposed to vary as the square of the velocity V, then will the work expended on this resistance vary as $V^2 . S$, or as $V^3 . T$, since $S = V . T$. If then U_1 and U_2 represent the work done at the moving and working points during the time T, then does the modulus (equation 114) assume, in this case, the form,

$$U_1 = A . U_2 + B . V . T + C . V^3 . T \quad . \quad . \quad . \quad . \quad (115).$$

THE MODULUS OF A MACHINE MOVING WITH AN ACCELE-RATED OR A RETARDED MOTION.

150. In the two last articles the work U_1, done upon the moving point or points of the machine, has been supposed to be just that necessary to overcome the useful and prejudicial resistances opposed to the motion of the machine, either continually or periodically; so that all the work may be expended upon these resistances, and none accumulate in the moving parts of the machine as the work proceeds, or else

that the accumulated work may return to the same amount from period to period. Let us now suppose this equality to cease, and the work U_1 done by the moving power to exceed that necessary to overcome the useful and prejudicial resistances; and to distinguish the work represented by U_1 in the one case from that in the other, let us suppose the former (that which is in *excess* of the resistances) to be represented by U^1; also let U_2 be the useful work of the machine, done through a given space S_2, and which is supposed the same whatever may be the velocity of the motion of the machine whilst that space is being described; moreover, let S_1 be the space described by the moving point, whilst the space S_2 is being described by the working point.

Now since U_1 is the work which must be done at the moving point just to overcome the resistances opposed to the motion of that point, and U^1 is the work actually done upon that point by the power, therefore $U^1 - U_1$ is the excess of the work done by the power over that expended on the resistances, and is therefore equal to the work *accumulated* in the machine (Art. 130.); that is, to one half of the increase of the vis viva through the space S_1 (Art. 129.); so that, if v_1 represent the velocity of any element of the machine (whose weight is w) when the work U^1 began to be done, and v_2 its velocity when that work has been completed, then (Art. 129.),

$$U^1 - U_1 = \frac{1}{2g}\Sigma w(v_2{}^2 - v_1{}^2).$$

Now by equation (114) $U_1 = AU_2 + BS_1$,

$$\therefore\ U^1 = A \cdot U_2 + B \cdot S_1 + \frac{1}{2g}\Sigma w(v_2{}^2 - v_1{}^2)\ \ \ \ \ (116).$$

If instead of the work U^1 done by the power exceeding that U_1 expended on the resistances it had been less than it, then, instead of work being accumulated continually through the space S_1, it would continually have been lost, and we should have had the relation (Art. 129.),

$$U_1 - U^1 = \frac{1}{2g}\Sigma w(v_1{}^2 - v_2{}^2);$$

M 3

so that in this case, also,

$$U^1 - U_1 = \frac{1}{2g}\Sigma w(v_2{}^2 - v_1{}^2).$$

The equation (116) applies therefore to the case of a re-tarded motion of the machine as well as to that of an acce-lerated motion, and is the *general* expression for the modulus of a machine moving with a variable motion. Whilst the co-efficients A and B of the modulus are dependent wholly upon the friction and other direct resistances to the motion of the machine, the last term of it is wholly independent of all these resistances, its amount being determined solely by the velocities of the various moving elements of the machines and their respective weights.

THE VELOCITY OF A MACHINE MOVING WITH A VARIABLE MOTION.

151. The velocities of the different parts or elements of every machine are evidently connected with one another by certain *invariable* relations, capable of being expressed by algebraical formulæ, so that, although these relations are different for different machines, they are the same for all circumstances of the motion of the same machine. In a great number of machines this relation is expressed by a constant ratio. Let the constant ratio of the velocity v_1 of any element to that V_1 of the moving point in such a machine, be repre-sented by λ, so that $v_1 = \lambda V_1$, and let v_2 and V_2 be any other values of v_1 and V_1; then $v_2 = \lambda V_2$. Substituting these values of v_1 and v_2 in equation (116), we have

$$U^1 = A \cdot U_2 + B \cdot S_1 + \frac{1}{2g}(V_2{}^2 - V_1{}^2)\Sigma w\lambda^2 \cdot \cdot \cdot \cdot \cdot (117);$$

in which expression $\Sigma w\lambda^2$ represents the sum of the weights of all the moving elements of the machine, each being mul-tiplied by the square of the ratio λ of its velocity to that of the point where the machine receives the operation of its moving power. For the *same* machine this co-efficient $\Sigma w\lambda^2$

is therefore a constant quantity. For different machines it is different. It is wholly independent of the useful or prejudicial resistances opposed to the motion of the machine, and has its value determined solely by the weights and dimensions of the moving masses, and the manner in which they are connected with one another in the machine.

Transforming this equation and reducing, we have

$$V_2{}^2 = V_1{}^2 + 2g \left\{ \frac{U^1 - A \cdot U_2 - B_1 \cdot S_1}{\Sigma w \lambda^2} \right\} \quad \ldots \ldots (118);$$

by which equation the velocity V_2 of the moving point of the machine is determined, after a given amount of work U^1 has been done upon it by the moving power, and a given amount U_2 expended on the useful resistances ; the velocity of the moving point, when this work began to be done being given and represented by V_1.

It is evident that the motion of the machine is more equable as the quantity represented by $\Sigma w \lambda^2$ is greater. This quantity, which is the same for the same machine and different for different machines, and which distinguishes machines from one another in respect to the steadiness of their motion, independently of all considerations arising out of the nature of the resistances useful or prejudicial opposed to it, may with propriety be called the CO-EFFICIENT OF EQUABLE MOTION.* The actual motion of the machine is more equable as this co-efficient and as the co-efficients A and B (supposed positive) are *greater*.

To DETERMINE THE CO-EFFICIENTS OF THE MODULUS OF A MACHINE.

152. Let that relation first be determined between the moving pressure P_1 upon the machine and its working pressure P_2, which obtains in the *state bordering upon motion* by the

---

* The co-efficient of equable motion is here, for the first time, introduced into the consideration of the theory of machines.

preponderance of P_1. This relation will, in all cases where the *constant* resistances to the motion of the machine independently of P_2 are *small* as compared with P_2, be found to be represented by formulæ of which the following is the general type or form : —

$$P_1 = P_2 . \Phi_1 + \Phi_2 \ . \ . \ . \ . \ . \ (119);$$

where Φ_1 and Φ_2 represent certain functions of the friction and other prejudicial resistances in the machine, of which the latter disappears when the resistances vanish and the former does not; so that if $\Phi_1^{(0)}$ and $\Phi_2^{(0)}$ represent the values of these functions when the prejudicial resistances vanish, then $\Phi_2^{(0)} = 0$ and $\Phi_1^{(0)} =$ a given finite quantity dependent for its amount on the composition of the machine. Let $P_1^{(0)}$ represent that value of the pressure P_1 which would be in equilibrium with the given pressure P_2, if there were no prejudicial resistances opposed to the motion of the machine. Then, by the last equation, $P_1^{(0)} = P_2 . \Phi_1^{(0)}$.

But by the principle of virtual velocities (Art. 127.), if we suppose the motion of the machine to be *uniform*, so that P_1 and P_2 are constantly in equilibrium upon it, and if we represent by S_1 any space described by the point of application of P_1, or the *projection* of that space on the direction of P_1 (Art. 52.), and by S_2 the corresponding space or projection of the space described by P_2, then $P_1^{(0)} . S_1 = P_2 . S_2$. Therefore, dividing this equation by the last, we have

$$S_1 = \frac{S_2}{\Phi_1^{(0)}} \ . \ . \ . \ . \ (120).$$

Multiplying this equation by equation (119),

$$P_1 . S_1 = P_2 . S_2 . \left\{ \frac{\Phi_1}{\Phi_1^{(0)}} \right\} + S_2 \left\{ \frac{\Phi_2}{\Phi_1^{(0)}} \right\} = P_2 . S \left\{ \frac{\Phi_1}{\Phi_1^{(0)}} \right\} + S_1 . \Phi_2 ;$$

$$\therefore \ U_1 = \left\{ \frac{\Phi}{\Phi_1^{(0)}} \right\} . U_2 + \Phi_2 . S_1 \ . \ . \ . \ . \ . \ (121),$$

which is the modulus of the machine, so that the constant A in equation (114) is represented by $\dfrac{\Phi_1}{\Phi^{(0)}}$, and the constant B by Φ_2.

The above equation has been proved for any value of S_1, provided the values of P_1 and P_2 be constant, and the motion of the machine uniform ; it evidently obtains, therefore, for an exceedingly *small* value of S_1, when the motion of the machine is *variable*.

GENERAL CONDITION OF THE STATE BORDERING UPON MOTION IN A BODY ACTED UPON BY PRESSURES IN THE SAME PLANE, AND MOVEABLE ABOUT A CYLINDRICAL AXIS.

153. *If any number of pressures* P_1, P_2, P_3, *&c. applied in the same plane to a body moveable about a cylindrical axis, be in the state bordering upon motion, then is the direction of the resistance of the axis inclined to its radius, at the point where it intersects the circumference, at an angle equal to the limiting angle of resistance.*

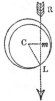 For let R represent the resultant of P_1, P_2, &c. Then, since these forces are supposed to be upon the point of causing the axis of the body to turn upon its bearings, their resultant would, if made to replace them, be also on the point of causing the axis to turn on its bearings. Hence it follows that the direction of this resultant R cannot be *through* the centre C of the axis ; for if it were, then the axis would be pressed by it in the direction of a radius, that is, *perpendicularly* upon its bearings, and could not be made to turn upon them by that pressure, or to be upon the point of turning upon them. The direction of R must then be on one side of C, so as to press the axis upon its bearings in a direction RL, *inclined* to the perpendicular CL (at the point L, where it intersects the circumference of the axis) at a certain angle RLC. Moreover, it is evident (Art. 141.), that since this force R pressing the axis upon its bearings at L is upon the point of causing it to *slip* upon them, this inclination RLC of R to the perpendicular CL, is equal to the

limiting angle of resistance of the axis and its bearings.*
Now the resistance of the axis is evidently equal and opposite to the resultant R of all the forces P_1, P_2, &c. impressed
upon the body. This resistance acts, therefore, in the direction LR, and is inclined to CL at an angle equal to the
limiting angle of resistance. Therefore, &c.

THE WHEEL AND AXLE.

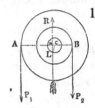

154. *The pressures P_1 and P_2 applied vertically
by means of parallel cords to a wheel and axle
are in the state bordering upon motion by the
preponderance of P_1, it is required to determine a relation between P_1 and P_2.*

The direction LR of the resistance of the axis is on that
side of the centre which is towards P_1, and is inclined to the
perpendicular CL at the point L, where it intersects the axis
at an angle CLR equal to the limiting angle of resistance.
Let this angle be represented by φ, and the radius CL of the
axis by ρ; also the radius CA of the wheel by a_1, and that
CB of the axle by a_2; and let W be the weight of the wheel
and axle, whose centre of gravity is supposed to be C. Now,
the pressures P_1, P_2, the weight W of the wheel and axle, and
the resistance R of the axis, are pressures in equilibrium.
Therefore, by the principle of the equality of moments
(Art. 7.), neglecting the rigidity of the cord, and observing
that the weight W may be supposed to act through C, we
have,

$$P_1 . \overline{CA} = P_2 . \overline{CB} + R . \overline{Cm}.$$

If, instead of P_1 preponderating, it had been on the point

---

* The *side* of C on which RL falls is manifestly determined by the
direction towards which the motion is about to take place. In this case it
is supposed about to take place to the *right* of C. If it had been to the
left, the direction of R would have been on the opposite side of C.

of yielding, or P_2 had been in the act of preponderating, then R would have fallen on the other side of C, and we should have obtained the relation $P_1 \cdot \overline{CA} = P_2 \cdot \overline{CB} - R \cdot \overline{Cm}$; so that, generally, $P_1 \cdot \overline{CA} = P_2 \cdot \overline{CB} \pm R \cdot \overline{Cm}$; the sign $\pm$ being taken according as P_1 is in the superior or inferior state bordering upon motion.

Now $CA = a_1$, $CB = a_2$, $\overline{Cm} = \overline{CL} \sin. CLR = \rho \sin. \varphi$, and $R = P_1 + P_2 \pm W$, the sign $\pm$ being taken according as the weight W of the wheel and axle acts in the same direction with the pressures P_1 and P_2, or in the opposite direction; that is, according as the pressures P_1 and P_2 act vertically *downwards* (as shown in the figure) or *upwards;*

$$\therefore \ P_1 a_1 = P_2 a_2 + (P_1 + P_2 \pm W)\rho \sin. \varphi,$$
$$\therefore \ P_1(a_1 - \rho \sin. \varphi) = P_2(a_2 + \rho \sin. \varphi) \pm W\rho \sin. \varphi.$$

Now the effect (Art. 142.) of the rigidity of the cord BP_2 is the same as though it increased the tension upon that cord from P_2 to $\left(P_2 + \dfrac{D + E \cdot P_2}{a_2}\right)$: allowing, therefore, for the rigidity of the cord, we have finally

$$P_1(a_1 - \rho \sin. \varphi) = \left(P_2 + \frac{D + E \cdot P_2}{a_2}\right)(a_2 + \rho \sin. \varphi) \pm W\rho \sin. \varphi,$$

or reducing,

$$P_1 = P_2\left(1 + \frac{E}{a_2}\right)\frac{a_2 + \rho \sin. \varphi}{a_1 - \rho \sin. \varphi} + \frac{D + \left(\dfrac{D}{a_2} \pm W\right)\rho \sin. \varphi.}{a_1 - \rho \sin. \varphi} \ \cdot\cdot \ (122),$$

which is the required relation between P_1 and P_2 in the state bordering upon motion.

$\dfrac{\rho}{a_1} \sin. \varphi$ and $\dfrac{\rho}{a_2} \sin. \varphi$ are in all cases exceedingly small;

we may therefore omit, without materially affecting the result, all terms involving powers of these quantities above the first; we shall thus obtain by reduction

$$P_1 = P_2\left(\frac{a_2 + E}{a_1}\right)\left\{1 + \left(\frac{1}{a_1} + \frac{1}{a_2}\right)\rho \sin. \varphi\right\} + \frac{D}{a_1}\left\{1 + \left(\frac{1}{a_1} + \frac{1}{a_2} \pm \frac{W}{D}\right)\rho \sin. \varphi\right\} \cdot\cdot (123).$$

155. *The modulus of uniform motion in the wheel and axle.*

It is evident from equation (122), that, in the case of the wheel and axle, the relation assumed in equation (114) obtains,

if we take $\Phi_1 = \left(1 + \dfrac{E}{a_2}\right)\dfrac{a_2 + \rho \sin. \varphi}{a_1 - \rho \sin. \varphi}$;

and $\qquad \Phi_2 = \dfrac{D + \left(\dfrac{D}{a_2} \pm W\right)\rho \sin. \varphi}{a_1 - \rho \sin. \varphi}$,

Now observing that $\Phi_1{}^{(0)}$ represents the value of Φ_1 when the prejudicial resistances vanish (or when $\varphi = 0$ and $E = 0$), we have $\Phi_1{}^{(0)} = \dfrac{a_2}{a_1}$.

$$\therefore \; \frac{\Phi_1}{\Phi^{(0)}} = \left(1 + \frac{E}{a_2}\right)\frac{a_1}{a_2}, \frac{a_2 + \rho \sin.\varphi}{a_1 - \rho \sin.\varphi} = \left(1 + \frac{E}{a_2}\right)\frac{1 + \left(\frac{\rho}{a_2}\right)\sin.\varphi}{1 - \left(\frac{\rho}{a_1}\right)\sin.\varphi} \; ;$$

Therefore by equation (121),

$$U_1 = U_2\left(1 + \frac{E}{a_2}\right)\left\{\frac{1 + \left(\frac{\rho}{a_2}\right)\sin.\varphi}{1 - \left(\frac{\rho}{a_1}\right)\sin.\varphi}\right\} + S_1\left\{\frac{D + \left(\frac{D}{a_2} \pm W\right)\rho \sin.\varphi}{a_1 - \rho \sin.\varphi}\right\} \dots (124),$$

which is the modulus of the wheel and axle.

Omitting terms involving dimensions of $\dfrac{\rho}{a_1}$ sin. φ, and $\dfrac{\rho}{a_2}$ sin. φ, and $\dfrac{E}{a_1}$ above the first, we have

$$U_1 = U_2\left\{1 + \frac{E}{a_2} + \left(\frac{1}{a_1} + \frac{1}{a_2}\right)\rho \sin.\phi\right\} + \frac{S_1 D}{a_1}\left\{1 + \left(\frac{1}{a_1} + \frac{1}{a_2} \pm \frac{W}{D}\right)\rho \sin.\phi\right\} \dots (125).$$

156. *The modulus of variable motion in the wheel and axle.*

If the relation of P_1 and P_2 be not that of either state bordering upon motion, then the motion will be continually accelerated or continually retarded, and work will continually accumulate in the moving parts of the machine, or the work already accumulated there will continually expend itself until the whole is exhausted, and the machine is brought to rest. The general expression for the modulus in this state of variable motion is (equation 116)

$$U^1 = AU_2 + BS_1 + \frac{1}{2g}\Sigma w(v_2{}^2 - v_1{}^2).$$

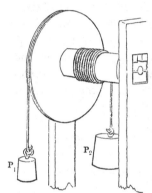

Now in this case of the wheel and axle, if V_1 and V_2 represent the velocities of P_1 at the commencement and completion of the space S_1, and α the angular velocity of the revolution of the wheel and axle; if, moreover, the pressures P_1 and P_2 be supposed to be supplied by weights suspended from the cords ; then, since the velocity of P_2 is represented by $\dfrac{a_2 V_1}{a_1}$, we have

$$\Sigma wv_1{}^2 = P_1 V_1{}^2 + P_2\left(\frac{a_2 V_1}{a_1}\right)^2 + \alpha^2 \mu_1 I_1$$

$+ \alpha^2 \mu_2 I_2$, if I_1 represent the moment of inertia of the revolving wheel, and I_2 that of the revolving axle (Art. 75.), and if μ_1 represent the weight of a unit of the wheel and μ_2 of the axle ; since $\Sigma wv_1{}^2$ represents the sum of the weights of all the moving elements of the machine, each being multiplied by the square of its velocity, and that (by Art. 75.) $\alpha^2 \mu_1 I_1$ represents this sum in respect to the wheel, and $\alpha^2 \mu_2 I_2$ in respect to the axle. Now, $V_1 = \alpha a_1$,

$$\therefore \alpha = \frac{V_1}{a_1}, \therefore \Sigma wv_1{}^2 = P_1 V_1{}^2 + \frac{P_2 a_2{}^2}{a_2} V_1{}^2 + \frac{\mu_1 I_1}{a_1{}^2} V_1{}^2 + \frac{\mu_2 I_2}{a_1{}^2} V_1{}^2 =$$

$$V_1{}^2 . \left\{ \frac{P_1 a_1{}^2 + P_2 a_2{}^2 + \mu_1 I_1 + \mu_2 I_2}{a_1{}^2} \right\}.$$

Similarly $\Sigma wv_2{}^2 = V_2{}^2 \left\{ \dfrac{P_1 a_1{}^2 + P_2 a_2{}^2 + \mu_1 I_1 + \mu_2 I_2}{a_1{}^2} \right\}$;

$$\therefore \Sigma w(v_2{}^2 - v_1{}^2) = (V_2{}^2 - V_1{}^2)\left\{ \frac{P_1 a_1{}^2 + P_2 a_2{}^2 + \mu_1 I_1 + \mu_2 I_2}{a^2} \right\}.$$

Substituting in the general expression (equation 116), we have

$$U^1 = AU_2 + BS_1 + \frac{1}{2g}(V_2{}^2 - V_1{}^2)\left\{\frac{P_1 a_1{}^2 + P_2 a_2{}^2 + \mu_1 I_1 + \mu_2 I_2}{a_1{}^2}\right\}. \ . \ (126),$$

which is the modulus of the machine in the state of variable motion, the co-efficients A and B being those already determined (equation 124), whilst the co-efficient $\dfrac{P_1 a_1{}^2 + P_2 a_2{}^2 + \mu_1 I_1 + \mu_2 I_2}{a_1{}^2}$ is the co-efficient $\Sigma w \lambda^2$ (equation 117) of *equable motion*. If the wheel and axle be each of them a solid cylinder, and the thickness of the wheel be b_1, and the length of the axle b^2, then (Art. 85.) $I_1 = \frac{1}{2}\pi b_1 a^4{}_1$, $I_2 = \frac{1}{2}\pi b_2 a_2{}^4$. Now if W_1 and W_2 represent the weights of the wheel and axle respectively, then $W_1 = \pi a_1{}^2 b_1 \mu_1$, $W_2 = \pi a_2{}^2 b_2 \mu_2$; therefore $\mu_1 I_1 = \frac{1}{2}W_1 a_1{}^2$, $\mu_2 I = \frac{1}{2}W_2 a_2{}^2$. Therefore the co-efficient of equable motion is represented by the equation

$$\Sigma w \lambda^2 = \frac{P_1 a_1{}^2 + P_2 a_2{}^2 + \frac{1}{2}W_1 a_1{}^2 + \frac{1}{2}W_2 a_2{}^2}{a_1{}^2}, \text{ or}$$

$$\Sigma w \lambda^2 = P_1 + \tfrac{1}{2}W_1 + (P_2 + \tfrac{1}{2}W_2)\left(\frac{a_2}{a_1}\right)^2 \ . \ . \ . \ . \ (127).$$

157. *To determine the velocity acquired through a given space when the relation of the weights P_1 and P_2, suspended from a wheel and axle, is not that of the state bordering upon motion.*

Let S_1 be the space through which the weight P_1 moves whilst its velocity passes from V_1 to V_2: observing that $U^1 = P_1 S_1$, and that $U_2 = P_2 S_2 = P_2 \dfrac{S_1 a_2}{a_1}$, substituting in equation 126, and solving that equation in respect to V_2, we have

$$V_2{}^2 = V_1{}^2 + 2g a_1 S_1\left\{\frac{P_1 a_1 - A \cdot P_2 a_2 - B a_1}{P_1 a_1{}^2 + P_2 a_2{}^2 + \mu_1 I_1 + \mu_2 I_2}\right\}. \ . \ . \ (128);$$

making the same suppositions as in formula 127, and representing the ratio $\dfrac{a_2}{a_1}$ by m, we have

$$V_2{}^2 = V_1{}^2 + 2g S_1\left\{\frac{P_1 - A \cdot P_2 m - B}{(P_1 + \frac{1}{2}W_1) + (P_2 + \frac{1}{2}W_2)m^2}\right\}.$$

THE PULLEY.

158. If the radius of the axle be taken equal to that of the
wheel, the wheel and axle becomes a pulley.

Assuming then in equation 122, $a_1 = a_2 = a$,
we obtain for the relation of the moving
pressures P_1 and P_2, in the state bordering
upon motion in the pulley, when the strings
are parallel,

$$P_1 = P_2\left(1 + \frac{E}{a}\right)\left\{\frac{1 + \frac{\rho}{a}\sin.\ \phi}{1 - \frac{\rho}{a}\sin.\ \phi}\right\} + \frac{D + \left(\frac{D}{|a} \pm W\right)\rho \sin.\ \phi}{a - \rho \sin.\ \phi} \quad . \ . \ (129);$$

and by equation 124 for the value of the modulus,

$$U_1 = U_2\left(1 + \frac{E}{a}\right)\left\{\frac{1 + \frac{\rho}{a}\sin.\ \phi}{1 - \frac{\rho}{a}\sin.\ \phi}\right\} + S_1\left[\frac{D + \left(\frac{D}{a} \pm W\right)\rho \sin.\ \phi}{a - \rho \sin.\ \phi}\right] \ . \ . \ (130);$$

in which the sign $\pm$ is to be taken according as the pressures
P_1' and P_2 act downwards, as in the first pulley of the preceding
figure; or upwards, as in the second. Omitting dimensions
of $\frac{\rho}{a_1}\sin.\ \phi$, $\frac{\rho}{a_2}\sin.\ \phi$, and $\frac{E}{a}$ above the first, we have by equations (123, 125)

$$P_1 = P_2\left\{1 + E + \frac{2\rho \sin.\ \phi}{a}\right\} + \frac{D}{a}\left\{1 + \left(\frac{2}{a} \pm \frac{W}{D}\right)\rho \sin.\ \phi\right\} \ . \ . \ (131),$$

$$U_1 = U_2\left\{1 + E + \frac{2\rho \sin.\ \phi}{a}\right\} + \frac{S_1 D}{a}\left\{1 + \left(\frac{2}{a} \pm \frac{W}{D}\right)\rho \sin.\ \phi\right\} . \ (132).$$

Also observing that $a_1 = a_2$, and $I_2 = 0$, the modulus of variable
motion (equation 126) becomes

$$U^1 = AU_2 + BS + \frac{1}{2g}(V_2^2 - V_1^2)\{P_1 + P_2 + \tfrac{1}{2}W\} \ . \ . \ . \ . \ (133),$$

and the velocity of variable motion (equation 118, 127) is determined by the equation

$$V_2{}^2 = V_1{}^2 + 2gS \left\{ \frac{P_1 - A \cdot P_2 - B}{P_1 + P_2^{\prime} + \frac{1}{2}W_1} \right\} \quad \cdots \cdots (134);$$

in which two last equations the values of A and B are those of the modulus of equable motion (equation 125).

SYSTEM OF ONE FIXED AND ONE MOVEABLE PULLEY.

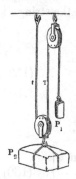

159. In the last article (equation 131) it was shown that the relation between the tensions P_1 and P_2 upon the two parts of a string passing over a pulley and parallel to one another, was, in the state bordering upon motion by the preponderance of P_1, represented by an expression of the form $P_1 = aP_2 + b$, where a and b are constants dependent upon the dimensions of the pulley and its axis, its weight, and the rigidity of the cord, and determined in terms of these elements by equation 131 ; and in which expression b has a different value according as the tension upon the cord passing over any pulley acts in the same direction with the weight of that pulley (as in the first pulley of the system shown in the figure), or in the opposite direction (as in the second pulley) : let these different values of b be represented by b and b_1. Now it is evident that before the weight P_2 can be raised by means of a system such as that shown in the figure, composed of one fixed and one moveable pulley, the state of the equilibrium of both pulleys must be that, bordering upon motion, which is described in the preceding article ; since both must be upon the point of turning upon their axes before the weight P_2 can begin to be raised.   If then T and t represent the tensions upon the two parts of the string which pass round the moveable pulley, we have

$$P_1 = aT + b, \text{ and } T = at + b_1.$$

Now the tensions T and t together support the weight P_2, and also the weight of the moveable pulley,

$$\therefore \ T + t = P_2 + W.$$

Adding $a\mathrm{T}$ to both sides of the second of the above equations, and multiplying both sides by a, we have

$$a(1+a)\mathrm{T}=a^2(\mathrm{T}+t)+ab_1=a^2(\mathrm{P}_2+\mathrm{W})+ab_1.$$

Also multiplying the first equation by $(1+a)$,

$$(1+a)\mathrm{P}_1=a(1+a)\mathrm{T}+b(1+a)=a^2(\mathrm{P}_2+\mathrm{W})+ab_1+b(1+a),$$

$$\therefore \mathrm{P}_1=\Big(\frac{a_2}{1+a}\Big)\mathrm{P}_2+\frac{a^2\mathrm{W}+b(1+a)+ab_1}{1+a}\quad\ldots\ldots (135).$$

Now if there were no friction or rigidity, a would evidently become 1 (see equation 131), and $\dfrac{a^2}{1+a}$ would become $\dfrac{1}{2}$; the co-efficients of the modulus (Art. 152.) are therefore

$$\mathrm{A}=2\Big(\frac{a^2}{1+a}\Big),\text{ and }\mathrm{B}=\frac{a^2\mathrm{W}+b(1+a)+ab_1}{1+a};$$

$$\therefore \mathrm{U}_1=2\Big(\frac{a^2}{1+a}\Big)\mathrm{U}_2+\frac{a\ \mathrm{W}+b(1+a)+ab_1}{1+a}\mathrm{S}_1\ \ldots\ldots (136),$$

which is the modulus of uniform motion to the single moveable pulley. *

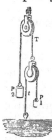

If this system of two pulleys had been arranged *thus*, with a different string passing over each, instead of with a single string as shown in the preceding figure, then, representing by t the tension upon the second part of the string to which P_1 is attached, and by T that upon the first part of the string to which P_2 is attached, we have

$$\mathrm{P}_1=at+b,\qquad \mathrm{T}=a\mathrm{P}_2+b,\qquad \mathrm{P}_1+t+\mathrm{W}=\mathrm{T}.$$

Multiplying the last of these equations by a, and adding it to the first, we have $\mathrm{P}_1(1+a)+\mathrm{W}a=\mathrm{T}a+b=a\ \mathrm{P}_2+(1+a)b$;

---

* The modulus may be determined directly from equation (135); for it is evident that if S_1 and S_2 represent the spaces described in the same time by P_1 and P_2, then $\mathrm{S}_1=2\mathrm{S}_2$. Multiplying both sides of equation (135) by this equation, we have,

$$\mathrm{P}_1\mathrm{S}_1=2\Big(\frac{a^2}{1+a}\Big)\mathrm{P}_2\mathrm{S}_2+\frac{a^2\mathrm{W}+b(1+a)+ab_1}{1+a}\,2\mathrm{S}_2;$$

now $\mathrm{P}_1\mathrm{S}_1=\mathrm{U}_1$ and $\mathrm{P}_2\mathrm{S}_2=\mathrm{U}_2$, therefore, &c.

$$\therefore P_1 = \left(\frac{a^2}{1+a}\right)P_2 + b - \frac{Wa}{1+a} \quad\ldots\ldots (137),$$

and for the modulus (equation 121),

$$U_1 = 2\left(\frac{a^?}{1+a}\right)U_2 + \left(b - \frac{Wa}{1+a}\right)S_1 \quad\ldots\ldots (138).$$

It is evident that, since the co-efficient of the second term of the modulus of this system is less than that of the first system (equation 136) (the quantities a and b being essentially positive), a given amount of work U_2 may be done by a less expense of power U_1, or a given weight P_2 may be raised to a given height with less *work*, by means of this system than the other; an advantage which is *not* due entirely to the circumstance that the weight of the moveable pulley in this case acts in *favour* of the power, whereas in the other it acts *against* it; and which advantage would exist, in a less degree, were the pulleys without weight.

A SYSTEM OF ONE FIXED AND ANY NUMBER OF MOVEABLE PULLEYS.

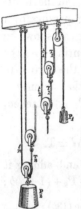

160. Let there be a system of n moveable pulleys and one fixed pulley combined as shown in the figure, a separate string passing over each moveable pulley; and let the tensions on the two parts of the string which passes over the first moveable pulley be represented by T_1 and t_1, those upon the two parts of the string which passes over the second by T_2 and t_2, &c. Also, to simplify the calculation, let all the pulleys be supposed of equal dimensions and weights, and the cords of equal rigidity;

$$\therefore T_1 = at_1 + b_1, \text{ and } T_2 + W = T_1 + t_1;$$

$$\therefore \text{ eliminating, } T_1 = \left(\frac{a}{1+a}\right)T_2 + \frac{Wa + b_1}{1+a} \quad\ldots\ldots (139).$$

Let the co-efficients of this equation be represented by α and β;

$$\therefore \; T_1 = \alpha T_2 + \beta.$$

Similarly, $T_2 = \alpha T_3 + \beta$, $T_3 = \alpha T_4 + \beta$, $T_4 = \alpha T_5 + \beta$, &c. = &c.,
$T_{n-1} = \alpha T_n + \beta$, $T_n = \alpha P_2 + \beta$.

Multiplying these equations successively, beginning from the second, by α, α^2, α^3, &c., α^{n-1}, adding them together, and striking out terms common to both sides of the resulting equation, we have

$$T_1 = \alpha^n P_2 + \beta + \alpha\beta + \alpha^2\beta + \dots + \alpha^{n-1}\beta \; ;$$

or summing the geometrical progression in the second member,

$$T_1 = \alpha^n P_2 + \beta\left(\frac{\alpha^n - 1}{\alpha - 1}\right) \; \dots \dots (140) \; ;$$

Substituting for α and β their values from equation (139), and reducing,

$$T_1 = \left(\frac{a}{1+a}\right)^n P_2 + (Wa + b_1)\left\{1 - \left(\frac{a}{1+a}\right)^n\right\}.$$

Now $P_1 = aT_1 + b$;

$$\therefore P_1 = a\left(\frac{a}{1+a}\right)^n P_2 + a(Wa + b_1)\left\{1 - \left(\frac{a}{1+a}\right)^n\right\} + b \dots (141$$

Whence observing, that, were there no friction, a would become unity, and $\left(\frac{a}{1+a}\right)^n = \left(\frac{1}{2}\right)^n$. We have (equation 121) for the modulus of this system,

$$U_1 = a\left(\frac{2a}{1+a}\right)^n U_2 + \left\{a(Wa + b_1)\{1 - \left(\frac{a}{1+a}\right)^n\} + b\right\}S_1 \dots (142$$

161. If each cord, instead of having one of its extremities attached to a fixed obstacle, had been connected by one extremity to a moveable bar carrying the weight P_2 to be raised (an arrangement which is shown in the second figure), then, adopting the same notation as before, we have

$$T_1 = at_1 + b, \qquad at_2 + b = T_2, \qquad T_2 = T_1 + t_1 + W.$$

Adding these equations together, striking out terms common to both sides, and solving in respect to T_1, we have

$$t_1 = \left(\frac{a}{a+1}\right) t_2 - \left(\frac{1}{a+1}\right) W \;;$$

in which equation it is to be observed, that the symbol b does not appear ; that element of the resistance, (which is constant,) affecting the tensions t_1 and t_2 equally, and therefore eliminating with T_1 and T_2. Let $\dfrac{a}{a+1}$ be represented by α, then

$$\left.\begin{array}{l} t_1 = \alpha t_2 - \dfrac{\alpha}{a}W. \quad \text{Similarly, } t_2 = \alpha t_3 - \dfrac{\alpha}{a}W, \\[2mm] t_3 = \alpha t_4 - \dfrac{\alpha}{a}W, \&c. = \&c., \; t_{n-1} = \alpha t_n - \dfrac{\alpha}{a}W \end{array}\right\} \;\; \ldots \;(143).$$

Eliminating between these equations precisely as between the similar equations in the preceding case (equation 140), observing only that here β is represented by $-\alpha W$, and that the equations (143) are $n-1$ in number instead of n, we have

$$t_1 = \alpha^{n-1} t_n - \frac{\alpha W}{a}\left(\frac{\alpha^{n-1}-1}{\alpha-1}\right) \;\; \ldots \ldots \;(144).$$

Also adding the preceding equations (143) together, we have

$$t_1 + t_2 + \ldots + t_{n-1} = \alpha(t_2 + t_3 + \ldots t_n) - (n-1)\frac{\alpha W}{a}.$$

Now the pressure P_2 is sustained by the tensions t_1, t_2, &c. of the different strings attached to the bar which carries it. Including in P_2, therefore, the weight of the bar, we have $t_1 + t_2 + \ldots + t_{n-1} + t_n = P_2$; $\therefore t_1 + t_2 + \ldots + t_{n-1} = P_2 - t_n$; and $t_2 + \ldots + t_n = P_2 - t_1$;

$$\therefore P_2 - t_n = \alpha(P_2 - t_1) - (n-1)\frac{\alpha W}{a}.$$

$$\therefore t_n = (1-\alpha)P_2 + \alpha t_1 + (n-1)\frac{\alpha W}{a}.$$

Substituting this value of t_n in equation (144),

$$t_1 = (1-\alpha)\alpha^{n-1}P_2 + \alpha^n t_1 + (n-1)\frac{\alpha^n W}{a} - \frac{\alpha W}{a}\frac{\alpha^{n-1}-1}{\alpha-1}.$$

Transposing and reducing,

$$(1-a^n)t_1 = (1-a)a^{n-1}P_2 + \frac{W}{a}\left\{na^n - a\frac{1-a^n}{1-a}\right\};$$

$$\therefore t_1 = \frac{(1-a)a^{n-1}}{1-a^n}P_2 + \frac{W}{a}\left\{\frac{na^n}{1-a^n} - \frac{a}{1-a}\right\}.$$

Now $a = \dfrac{a}{a+1}$, $\therefore a^{-1} = 1 + a^{-1}$; also $\dfrac{(1-a)a^{n-1}}{1-a^n} = \dfrac{a^{-1}-1}{a^{-n}-1} =$

$$\frac{a^{-1}}{(1+a^{-1})^n - 1}, \quad \frac{a^n}{1-a^n} = \frac{1}{a^{-n}-1} = \frac{1}{(1+a^{-1})^n - 1}, \text{and } \frac{a}{1-a} = a;$$

$$\therefore t_1 = \frac{a^{-1}P_2}{(1+a^{-1})^n - 1} + \frac{W}{a}\left\{\frac{n}{(1+a^{-1})^n - 1} - a\right\}.$$

Now $P_1 = at_1 + b$;

$$\therefore P_1 = \frac{P_2}{(1+a^{-1})^n - 1} + \frac{W}{a}\left\{\frac{n}{(1+a^{-1})^n - 1} - a\right\} + b \ldots (145).$$

Whence observing that when $a=1$, $\{(1+a^{-1})^n - 1 = 2^n - 1$, we obtain for the modulus of uniform motion (equation 121),

$$U_1 = \left\{\frac{2^n - 1}{(1+a^{-1})^n - 1}\right\}.U_2 + \left\{\left(\frac{W}{a}\right)\left\{\frac{n}{(1+a^{-1})^n - 1} - a\right\} + b\right\}S_1 \ldots (146).$$

A TACKLE OF ANY NUMBER OF SHEAVES.

162. If any number of pulleys (called in this case sheaves) be made to turn on as many different centres in the same block A, and if in another block B there be similarly placed as many others, the diameter of each of the last being one half that of a corresponding pulley or sheave in the first; and if the same cord attached to the first block be made to pass in succession over all the sheaves in the two blocks, as shown in the figure, it is evident that the parts of this cord 1, 2, 3, &c. passing between the two blocks, and as many in number as there are sheaves, will be parallel to each other, and will divide between them the pressure of a weight P_2 suspended from the lower block: moreover, that they would divide this pressure between them *equally* were it not

N 3

for the friction of the sheaves upon their bearings and the rigidity of the rope; so that in this case, if there were n sheaves, the tension upon each would be $\frac{1}{n}P_2$; and a pressure P_1 of that amount applied to the extremity of the cord would be sufficient to maintain the equilibrium of the state bordering upon motion. Let T_1, T_2, T_3, &c. represent the actual tensions upon the strings in the state bordering on motion by the preponderance of P_1, beginning from that which passes from P_1 over the largest sheaf; then

$$P_1 = a_1 T_1 + b_1,\quad T_1 = a_2 T_2 + b_2,\quad T_2 = a_3 T_3 + b_3,$$
$$\&c. = \&c.,\quad T_{n-1} = a_n T_n + b_n;$$

where a_1, a_2, &c., b_1, b_2, &c., represent certain constant coefficients, dependent upon the dimensions of the sheaves and the rigidity of the rope, and determined by equation (131). Moreover, since the weight P_2 is supported by the parallel tensions of the different strings, we have

$$P_1 = T_1 + T_2 + \dots + T_n.$$

It will be observed that the above equations are one more in number than the quantities T_1, T_2, T_3, &c.; the latter may therefore be eliminated among them, and we shall thus obtain a relation between the weight P_2 to be raised and that P_1 necessary to raise it, and from thence the modulus of the system.

To simplify the calculation, and to adapt it to that form of the tackle which is commonly in use, let us suppose another arrangement of the sheaves. Instead of their being of different diameters and placed all in the same plane, as shown in the last figure, let them be of equal diameter and placed side by side, as in the accompanying figure, which represents the common tackle. The inconvenience of this last mode of arrangement is, that the cord has to pass from the plane of a sheaf in one block to the plane

of the corresponding sheaf in the other *obliquely*, so that
the parts of the cords between the blocks are not truly
parallel to one another, and the sum of their tensions is not
truly equal to the weight P_2 to be raised, but somewhat
greater than it. So long, however, as the blocks are not very
near to one another, this deflection of the cord is inconsiderable, and the error resulting from it in the calculation may
be neglected. Supposing the different parts of the cord between the blocks then to be parallel, and the diameters of all
the sheaves and their axes to be equal, also neglecting the
influence of the *weight* of each sheaf in increasing the friction
of its axis, since these weights are in this case comparatively
small, the co-efficients a_1, a_2, a_3 will manifestly all be equal; as
also b_1, b_2, b_3;

$$\therefore \ P_1 = aT_1 + b, \ T_1 = aT_2 + b, \ T_2 = aT_3 + b, \\ \&c. = \&c., \ T_{n-1} = aT_n + b \quad \Big\} \ \cdots \ (147);$$

also $\qquad P_2 = T_1 + T_2 + T_3 + \cdots + T_n.$

Multiplying equations (147) successively (beginning from
the second) by a, a^2, a^3, and a^{n-1}; then adding them together, striking out the terms common to both sides, and
summing the geometric series in the second member (as in
equation 140), we have

$$P_1 = a^n T_n + b \frac{a^n - 1}{a - 1}.$$

Adding equations (147), and observing that $T_1 + T_2 + \cdots + T_n = P_2$, and that $P_1 + T_1 + T_2 + \cdots + T_{n-1} = P_1 + P_2 - T_n$, we have

$$P_1 + P_2 - T_n = aP_2 + nb.$$

Eliminating T_n between this equation and the last,

$$P_1 = a^n \{P_1 - P_2(a-1) - nb\} + b \frac{a^n - 1}{a - 1};$$

$$\therefore \ P_1 = \frac{a^n(a-1)}{a^n - 1} P_2 + \frac{nba^n}{a^n - 1} - \frac{b}{a - 1} \ \cdots \ (148).$$

To determine the modulus let it be observed, that, neglecting

N 4

friction and rigidity, a becomes unity; and that for this value of a, $\dfrac{a^n(a-1)}{a^n-1}$ becomes a vanishing fraction, whose value is determined by a well known method to be $\dfrac{1}{n}$*. Hence (Art. 152.),

$$U_1 = n\frac{a^n(a-1)}{a^n-1}U_2 + \left\{\frac{nba^n}{a^n-1} - \frac{b}{a-1}\right\}S_1 \ \ldots \ldots (149).$$

Hitherto no account has been taken of the work expended in raising the rope which ascends with the ascending weight. The correction is, however, readily made. By Art. 60. it appears that the work expended in raising this rope (different parts of which are raised different heights) is precisely the same as though the whole quantity thus raised had been raised at one lift through a height equal to that through which its centre of gravity is actually raised. Now the cord raised is that which may be conceived to lie between two positions of P_2 distant from one another by the space S_2, so that its whole length is represented by nS_2; and if μ represent the weight of each foot of it, its whole weight is represented by μnS_2: also its centre of gravity is evidently raised between the first and second positions of P_2 by the distance $\frac{1}{2}S_2$; so that the whole work expended in raising it is represented by $\frac{1}{2}\mu nS_2^2$ or by $\frac{1}{2}\dfrac{\mu S_1^2}{n}$, since $S_1 = nS_2$. Adding this work expended in raising the rope to that which would be

---

* Dividing numerator and denominator of the fraction by $(a-1)$ it becomes $\dfrac{a^n}{a^{n-1}+a^{n-2}+\ldots+1}$, which evidently equals $\dfrac{1}{n}$ when $n=1$. The modulus may readily be determined from equation (148). Let S_1 and S_2 represent the spaces described by P_1 and P_2 in any the same time; then, since when the blocks are made to approach one another by the distance S_2, each of the n portions of the cord intercepted between the two blocks is shortened by this distance S_2, it is evident that the whole length of cord intercepted between the two blocks is shortened by nS_2; but the whole of this cord must have passed over the first sheaf, therefore $S_1 = nS_2$. Multiplying equation (148) by this equation, and observing that $U_1 = P_1S^1$ and $U_2 = P_2S_2$, we obtain the modulus as given above.

necessary to raise the weight P_2, if the rope were without weight, we obtain *

$$U_1 = n\frac{a^n(a-1)}{a^n-1}U_2 + \left\{ \frac{nba^n}{a^n-1} - \frac{b}{a-1} \right\} . S_1 + \frac{\mu}{2n} . S_1{}^2 \dots (150),$$

which is the MODULUS of the tackle.

THE MODULUS OF A COMPOUND MACHINE.

163. Let the *work* of a machine be transmitted from one to another of a series of moving elements forming a compound machine, until from the *moving* it reaches the *working* point of that machine. Let P be the *pressure* under which the work is done upon the moving point, or upon the *first* moving element of the machine; P_1 that under which it is yielded from the first to the second element of the machine; P_2, from the second to the third element, &c.; and P_n the pressure under which it is yielded by the last element upon the useful product, or at the working point of the machine. Then, since each element of the compound machine is a *simple* machine, the relation between the pressures applied to that element when in the state bordering on motion will be found to present itself under the form of equation (119) (Art. 152.), in all cases where the pressure under which the work upon each element is done is *great* as compared with the *weight* of that element (see Art. 166.).

Representing, therefore, by $a_1, a_2, a_3 \dots b_1, b_2, b_3 \dots$, certain constants, which are given in terms of the forms and dimensions of the several elements and the prejudicial resistances, we have

$$P = a_1P_1 + b_1, \qquad P_1 = a_2P_2 + b_2, \qquad P_1 = a_3P_3 + b_3,$$
$$\&c. = \&c., \quad P_{n-1} = a_nP_n + b_n.$$

Eliminating the $n-1$ quantities P_1, P_2, $P_3 \dots$, P_{n-1}, between these n equations, we obtain an equation, of the form,

$$P = aP_n + b \dots \dots (151);$$

---

* A correction for the weight of the rope may be similarly applied to the modulus of each of the other systems of pulleys. The effect of the *weight of the rope* in increasing the expenditure of work on the *friction* of the pulleys is neglected as unimportant to the result.

where $a = a_1 a_2 a_3 \dots a_n$, and

$$b = a_1 a_2 \dots a_{n-1} b_n + a_1 a_2 \dots a_{n-2} b_{n-1} + \dots + a_1 b_2 + b, \left.\right\} \quad (152).$$

If the only prejudicial resistance to which each element is subjected be conceived to be friction, and the limiting angle of resistance in respect to each be represented by φ; then considering each of the quantities a_1, b_1, a_2, b_2, as a function of φ, expanding each by Maclaurin's theorem into a series ascending by powers of that variable, and neglecting terms which involve powers of it above the first, we have

$$a_1 = a_1^{(0)} + \left(\frac{da_1}{d\varphi}\right)^{(0)} \varphi, \qquad b_1 = b_1^{(0)} + \left(\frac{db_1}{d\varphi}\right)^{(0)} \varphi,$$

$$a_2 = a_2^{(0)} + \left(\frac{da_2}{d\varphi}\right)^{(0)} \varphi, \qquad b_2 = b_2^{(0)} + \left(\frac{db_2}{d\varpi}\right) d\varphi, \quad \&c. = \&c. ;$$

where, $a_1^{(0)}, b_1^{(0)}, a_2^{(0)}, b_2^{(0)}$, represent the values of a_1, b_1, a_2, b_2, &c., when $\varphi = 0$ and $\left(\frac{da_1}{d\varphi}\right)^{(0)}$, $\left(\frac{db_1}{d\varphi}\right)^{(0)}$, &c. represent the similar values of their first differential co-efficients.

Let

$$\left(\frac{da_1}{d\varphi}\right)^{(0)} \varphi = a_1^{(0)} \cdot \alpha_1, \qquad \left(\frac{db_1}{d\varphi}\right)^{(0)} \varphi = b^{(0)} \cdot \beta_1,$$

$$\left(\frac{da_2}{d\varphi}\right)^{(0)} \varphi = a_2^{(0)} \cdot \alpha_2, \qquad \left(\frac{db_2}{d\varphi}\right) \varphi = b_2^{(0)} \cdot \beta_1, \qquad \&c. = \&c.$$

Therefore $a_1 = a_1^{(0)}(1 + \alpha_1), \ b_1 = b_1^{(0)}(1 + \beta_1), \ a_2 = a_2^{(0)}(1 + \alpha_2),$ $b_2 = b_2^{(0)}(1 + \beta_2), \quad \&c. = \&c. ;$ where $\alpha_1, \beta_1, \alpha_2, \beta_2$, &c., each involving the factor φ, are exceedingly small. Substituting the values of a_1, a_2, &c. in the expression for a, and neglecting terms which involve dimensions of α_1, α_2, &c. above the first, we have

$$a = a_1^{(0)} a_2^{(0)} \dots a^{n(0)} \{1 + \alpha_1 + \alpha_2 + \alpha_3 + \dots + \alpha_n\} \dots (153).$$

Now the co-efficient of the first term of the modulus is represented (equation 121) by $\dfrac{a}{a^{(0)}}$, a representing the co-efficient of the first term of equation (119), also substituting the value of a from equation (153), and observing that

$a_{(0)} = a_1^{(0)} . a_2^{(0)} .... a_n^{(0)}$, we have $\dfrac{a}{a^{(0)}} = \{1 + \alpha_1 + ... + \alpha_n\}$;

$\therefore\ \mathrm{U} = \{1 + \alpha_1 + \alpha_2 + \alpha_3 + ... + \alpha_n\} \mathrm{U}_n + b . \mathrm{S} \ .... (154)$,

which is the modulus of a compound machine of n elements, U representing the work done at the *moving* point, U_n that at the *working* point, S the space described by the moving point, and b a constant determined by equation (152).

164. THE CONDITIONS OF THE EQUILIBRIUM OF ANY TWO PRESSURES P_1 AND P_2 APPLIED IN THE SAME PLANE TO A BODY MOVEABLE ABOUT A FIXED AXIS OF GIVEN DIMENSIONS.

In *fig.* 1. the pressures P_1 and P_2 are shown acting

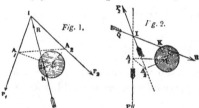

on opposite sides of the axis whose centre is C, and in *fig.* 2. upon the same side. Let the direction of the resultant of P_1 and P_2 be represented, in the first case, by IR, and in the second by RI. It is in the directions of these lines that the axis is, in the two cases, pressed upon its bearings. Suppose the relation between P_1 and P_2 to be such that the body is, in both cases, upon the point of turning in the direction in which P_1 acts. This relation obtaining between P_1 and P_2, it is evident that, if these pressures were replaced by their resultant, that resultant would also be upon the point of causing the body to turn in the direction of P_1. The direction IR of the resultant, thus acting alone upon the body, lies, therefore, in the first case, upon the *same* side of the centre C of the axis as P_1 does, and in the second case it lies upon the *opposite* side*; and, in both cases, it is inclined to the

---

* The arrows in the figure represent, not the directions of the *resultants*, but of the *resistances* of the axis, which are opposite to the resultants.

radius CK at the point K, where it intersects the axis at an angle CKR, equal to the limiting angle of resistance (see Art. 153.). Now, the resistance of the axis acts evidently in both cases in a direction opposite to the resultant of P_1 and P_2, and is equal to it; let it be represented by R. Upon the directions of P_1, P_2, and R, let fall the perpendiculars CA_1, CA_2, and CL, and let them be represented by a_1, a_2, and λ. Then, by the principle of the equality of moments, since P_1, P_2, and R are pressures in equilibrium,

$$\therefore P_1 a_1 = P_2 a_2 + \lambda R.$$

If P_1 had been upon the point of *yielding*, or P_2 on the point of preponderating, then R would have had its direction (in both cases) on the other side of C; so that the last equation would have become

$$P_1 a_1 + \lambda R = P_2 a_2.$$

According, therefore, as P_1 is in the superior or inferior state bordering upon motion,

$$P_1 a_1 - P_2 a_2 = (\pm \lambda) R.$$

And if we assume λ to be taken with the sign + or −, according as P_1 is about to preponderate or to yield, then *generally*

$$P_1 a_1 - P_2 a_2 = \lambda R \ \ . \ . \ . \ . \ . \ (155).$$

Now, since the resistance of the axis is equal to the resultant of P_1 and P_2, if we represent the angle $P_1^\prime I P_2$ by ι*, we have (Art. 13.)

$$R = \sqrt{P_1^2 + 2 P_1 P_2 \cos . \iota + P_2}.$$

Substituting this value of R in the preceding equation, and squaring both sides,

$$(P_1 a_1 - P_2 a_2)^2 = \lambda^2 (P_1^2 + 2 P_1 P_2 \cos . \iota + P_2);$$

---

* Care must be taken to measure this angle, so that P_1 and P_2 may may have their directions both *towards* or both *from* the angular point I (as shown in the figure), and not one of them *towards* that point and the other *from* it. Thus, in the second figure, the inclination ι of the pressures P_1 and P_2 is not the angle $A_2 I P_1$, but the angle $P_2 I P_1$. It is of importance to observe this distinction (see note p. 190.).

transposing and dividing by P_2^2,

$$\left(\frac{P_1}{P_2}\right)^2 (a_1^2 - \lambda^2) - 2\left(\frac{P_1}{P_2}\right)(a_1 a_2 + \lambda^2 \cos. \iota) = -(a_2^2 - \lambda^2)\ ;$$

solving this quadratic in respect to $\left(\frac{P_1}{P_2}\right)$,

$$\frac{P_1}{P_2} = \frac{(a_1 a_2 + \lambda^2 \cos. \iota) \pm \sqrt{(a_1 a_2 + \lambda^2 \cos. \iota)^2 - (a_1^2 - \lambda^2)(a_2^2 - \lambda^2)}}{a_1^2 - \lambda}\ ;$$

$$\therefore \frac{P_1}{P_2} = \frac{(a_1 a_2 + \lambda^2 \cos. \iota) \pm \lambda \sqrt{(a_1^2 + 2a_1 a_2 \cos. \iota + a_2^2) - \lambda^2 \sin.^2 \iota}}{a_1^2 - \lambda^2}.$$

Now, let the radius CK of the axis be represented by ρ, and the limiting angle of resistance CKR by φ; therefore $\lambda = \mathrm{CL} = \mathrm{CK}\ \sin.\ \mathrm{CKR} = \rho\ \sin.\ \varphi$. Also draw a straight line from A_1 to A_2 in both figures, and let it be represented by L ; $\therefore a_1^2 - 2a_1 a_2 \cos.\ A_1 CA_2 + a_2^2 = L^2$. Now, since the angles at A_1 and A_2 are right angles, therefore the angles $A_1 A_2$ and $A_1 C A_2$ are together equal to two right angles, or $A_1 C A_2 + \iota = \pi$; therefore $A_1 C A_2 = \pi - \iota$, and cos. $A_1 C A_2 = -\cos.\ \iota$; therefore $L^2 = a_1^2 + 2a_1 a_2 \cos.\ \iota + a_2^2$: substituting these values of L^2 and λ in the preceding equation,

$$P_1 = \frac{(a_1 a_2 + \rho^2 \cos. \iota \sin.^2 \varphi) \pm \rho \sin. \varphi\ (L^2 - \rho^2 \sin.^2 \iota \sin.^2 \varphi)^{\frac{1}{2}}}{(a_1^2 - \rho^2 \sin.^2 \varphi)} \cdot P_2 \ .\ .\ (156).$$

The two roots of the above equation are given by positive and negative values of λ, they correspond therefore (equation 155) to the two states bordering upon motion. These two values of λ are, moreover, given by positive and negative values of φ; assuming therefore φ to be taken positively or negatively, according as P_1 preponderates or yields, we may replace the ambiguous by the positive sign. The relation above determined between P_1 and P_2 evidently satisfies the conditions of equation (119). We obtain therefore for the *modulus* (equation 121)

$$U_1 = \left(\frac{a_1}{a_2}\right) \cdot \frac{(a_1 a_2 + \rho^2 \cos. \iota \sin.^2 \varphi) + \rho(L^2 - \rho^2 \sin.^2 \iota \sin.^2 \varphi)^{\frac{1}{2}} \sin. \varphi}{(a_1^2 - \rho^2 \sin.^2 \varphi)} U_2 \ .\ .\ .\ (157).$$

If terms involving powers of $\left(\frac{\rho}{a_1}\right) \sin.\ \varphi$ above the first be neglected, that quantity being in all cases exceedingly small, we have

$$P_1 = \left\{ \left(\frac{a_2}{a_1}\right) + \left(\frac{\rho L}{a_1{}^2}\right) \cdot \sin. \, \varphi \right\} P_2 \quad \cdots \quad (158),$$

$$U_1 = \left\{ 1 + \left(\frac{\rho L}{a_1 a_2}\right) \sin. \, \varphi \right\} U_2 \quad \cdots \quad (159).$$

165. *To determine the resultant* R *of any number of pressures* P_1, P_2, P_3 . . . ., *in terms of those pressures, and the cosines of their inclinations to one another.*

 Let α_1, α_2, α_3, &c. represent the inclinations IAC, IBC, &c. of the several pressures P_1, P_2, &c. to any given axis CA in the same plane ; and let ι_{12}, ι_{13}, ι_{23}, &c. represent the inclinations of these pressures severally to one another.

Now $\angle$ AIB $= \angle$ IBC $- \angle$ IAC (Euc. I. 32.) ;

$\therefore \iota_{12} = \alpha_2 - \alpha_1$, $\therefore$ cos. $\iota_{12} =$ cos. α_1 cos. $\alpha_2 +$ sin. α_1 sin. α_2.

Similarly, cos. $\iota_{13} =$ cos. α_1 cos. $\alpha_3 +$ sin. α_1 sin. α_3,

cos. $\iota_{23} =$ cos. α_2 cos. $\alpha_3 +$ sin. α_2 sin. α_3.

Now $R^2 = (P_1$ cos. $\alpha_1 + P_2$ cos. $\alpha_2 + P_3$ cos. $\alpha_3 + \ldots)^2 +$
$P_1 ($sin. $\alpha_1 + P_2$ sin. $\alpha_2 + P_3$ sin. $\alpha_3 + \ldots)$, (equation 9, Art. 11.).

Squaring the two terms in the second member, adding the results, and observing that cos. $^2\alpha_1 +$ sin. $^2\alpha_1 = 1$,

$R^2 = P_1{}^2 + P_2{}^2 + P_3{}^2 \ldots + 2P_1 P_2 ($cos. α_1 cos. $\alpha_2 +$ sin. α_1 sin. $\alpha_2)$
$\quad + 2P_1 P_3 ($cos. α_1 cos. $\alpha_3 +$ sin. α_1 sin. $\alpha_3) + \ldots$;

$\therefore R^{2*} = P_1{}^2 + P_2{}^2 + P_3{}^2 + \ldots + 2P_1 P_2$ cos. $\iota_{12} + 2P_1 P_3$ cos. ι_{13}
$\quad + 2P_2 P_3$ cos. $\iota_{23} +$ &c. . . . . (160).

* In which expression it is to be understood that the inclination ι_{12} of the directions of any two forces is taken on the supposition that both the forces act *from* or both act *towards* the point in which they intersect, and not one *towards* and the other *from* that point ; so that in the case represented in the accompanying figure, the inclination ι_{12} of the two forces P_1 and P_2 represented by the arrows, is not the angle $P_1 I P_2$, but the angle $Q I P_1$, since IQ and IP are directions of these two forces, both tending *from* their point of intersection ; whilst the directions $P_2 I$ and IP_1 are one of them *towards* that point, and the other *from* it.

166. THE CONDITIONS OF THE EQUILIBRIUM OF THREE PRESSURES, P_1, P_2, P_3, IN THE SAME PLANE APPLIED TO A BODY MOVEABLE ABOUT A FIXED AXIS, THE DIRECTION OF ONE OF THEM, P_3, PASSING THROUGH THE CENTRE OF THE AXIS, AND THE SYSTEM BEING IN THE STATE BORDERING UPON MOTION BY THE PREPONDERANCE OF P_1.

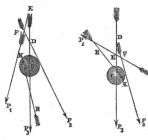

Let ι_{12}, ι_{13}, ι_{23} represent the inclinations of the directions of the pressures P_1, P_2, P_3 to one another, a_1 and a_2 the perpendiculars let fall from the centre of the axis upon P_1 and P_2', and λ the perpendicular let fall from the same point upon the resultant R of P_1, P_2, P_3. Then, since R is equal and opposite to the resistance of the axis (Art. 153.), we have, by the principle of the equality of moments, $P_1 a_1 - P_2 a_2 = \lambda R$, for P_3 passes through the centre of the axis, and its moment about that point therefore vanishes.

Substituting the value of R from equation (160),

$$P_1 a_1 - P_2 a_2 = \lambda \{ P_1{}^2 + P_2{}^2 + P_3{}^2 + 2P_1 P_2 \cos. \iota_{12} + 2P_1 P_3 \cos. \iota_{13} + 2P_2 P_3 \cos. \iota_{23}. \}^{\frac{1}{2}}$$

Squaring both sides of this equation, and transposing,

$$P_1{}^2 (a_1{}^2 - \lambda^2) - 2P_1 \{ P_2 a_1 a_2 + \lambda^2 (P_2 \cos. \iota_{12} + P_3 \cos. \iota_{13}) \} =$$
$$- P_2{}^2 a_2{}^2 + \lambda^2 \{ P_2{}^2 + P_3{}^2 + 2P_2 P_3 \cos. \iota_{23} \}.$$

If this quadratic equation be solved in respect to P_1, and terms which involve powers of λ above the first be omitted, we shall obtain the equation

P $a_1{}^2 = P_2 a_1 a_2 + \lambda \sqrt{ P_2{}^2 (a_1{}^2 + 2a_1 a_2 \cos. \iota_{12} + a_2{}^2) + P_3{}^2 a_1{}^2 + 2P_2 P_3 a_1 (a_2 \cos. \iota_{13} + a_1 \cos. \iota_{23}) }$;

or representing (as in Art. 163.) the line which joins the feet of the perpendiculars, a_1 and a_2 by L, and the function $a_1 (a_2 \cos. \iota_{13} + a_1 \cos. \iota_{23})$ by M, and substituting for λ its value $\rho \sin. \varphi$,

$$P_1 = \left(\frac{a_2}{a_1} \right) P_2 + \left(\frac{\rho \sin. \varphi}{a_1{}^2} \right) \{ P_2{}^2 L^2 + P_3{}^2 a_1{}^2 + 2P_2 P_3 M \}^{\frac{1}{2}*} \dots (161).$$

* It will be shown in the appendix, that this equation is but a particular case of a more general relation, embracing the conditions of the equilibrium of any number of pressures applied to a body movable about a cylindrical axis of given dimensions.

Representing (as in Art. 152.) the value of P_1 when the prejudicial resistances vanish, or when $\varphi = 0$, by $P_1^{(0)}$, we have $P_1^{(0)} = \left(\dfrac{a_2}{a_1}\right)P_2$. Also by the principle of virtual velocities $P_1^{(0)} . S_1 = P_2 . S_2$. Eliminating $P_1^{(0)}$ between these equations, we have $S_1 = \left(\dfrac{a_1}{a_2}\right)S_2$. Multiplying equation (161) by this,

$$P_1 S_1 = P_2 S_2 + \frac{\rho \sin . \varphi}{a_1 a_2}\{P_2^2 S_2^2 L^2 + 2P_2 P_3 S_2^2 M + P_3^2 S_2^2 a_1^2\}^{\frac{1}{2}}.$$

Substituting U_1 for $P_1 S_1$, U_2 for $P_2 S_2$, and observing that $S_2 = \dfrac{a_2}{a_1}S_1$,

$$U_1 = U_2 + \frac{\rho \sin . \varphi}{a_1 a_2}\left\{ U_2^2 L^2 + 2U_2 P_3 S_1 M\left(\frac{a_2}{a_1}\right) + P_3^2 S_1^2 a_2^2 \right\}^{\frac{1}{2}} \dots (162),$$

which is the MODULUS of the system.

If P_3 be so small as compared with P_2 that in the expansion of the binomial radical (equation 161), terms involving powers of $\dfrac{P_3}{P_2}$ above the first may be neglected; then,

$$P_1 = \left(\frac{a_2}{a_1}\right)P_2 + \left(\frac{\rho \sin . \varphi}{a_1^2}\right)\left\{ P_2 L + P_3\left(\frac{M}{L}\right) \right\} \dots \dots (163);$$

which equation may be placed under the form

$$P_1 = \left(\frac{a_2}{a_1}\right)\left\{ 1 + \frac{L\rho}{a_1 a_2}\sin . \varphi \right\} P_2 + \left(\frac{M}{a_1^2}\right)\left(\frac{\rho}{L}\right)P_3 \sin . \varphi.$$

Whence observing that the direction of P_3 being always through the centre of the axis, the point of application of that force does not *move*, so that the force P_3 does not *work* as the body is made to revolve by the preponderance of P_1; observing, moreover, that in this case the conditions of equation (119) (Art. 152.) are satisfied, we obtain for the modulus

$$U_1 = \left\{ 1 + \frac{\rho L}{a_1 a_2}\sin . \varphi \right\} U_2 + \left(\frac{M}{a_1^2}\right)\left(\frac{\rho}{L}\right)P_3 . S_1 . \sin . \varphi \dots . (164).$$

167. *The conditions of the equilibrium of two pressures* P_1 *and* P_2 *applied to a body moveable about a cylindrical axis, taking into account the weight of the body and supposing it to be symmetrical about its axis.*

The body being symmetrical about its axis, its centre of gravity is in the centre of its axis, and its weight produces the same effect as though it acted continually through the centre of its axis. In equation (161.) let then P_3 be taken to represent the weight W of the body, and ι_{13} ι_{23} the inclinations of the pressures P_1 and P_2 to the vertical. Then

$$P_1 = \left(\frac{a_2}{a_1}\right)P_2 + \left(\frac{\rho \sin.\phi}{a_1{}^2}\right)\left\{P_2{}^2L^2 + 2P_2WM + W^2a_1{}^2\right\}^{\frac{1}{2}} \dots (165.)$$

Also by the equation (162) we find for the modulus

$$U_1 = U_2 + \left(\frac{\rho \sin.\phi}{a_1 a_2}\right)\left\{U_2{}^2L^2 + 2U_2WS_1M\left(\frac{a_2}{a_1}\right) + W^2S_1{}^2a_1{}^2\right\}^{\frac{1}{2}} \dots (166.)$$

And in the case in which P_2 is considerable as compared with W, by equations (163, 164.)

$$P_1 = \left(\frac{a_2}{a_1}\right)\left\{1 + \frac{L\rho}{a_1 a_2}\sin.\phi\right\}P_2 + \left(\frac{M}{a_1{}^2}\right)\left(\frac{\rho}{L}\right)W\sin.\phi \dots (167.)$$

$$U_1 = \left\{1 + \frac{\rho L}{a_1 a_2}\sin.\phi\right\}U_2 + \left(\frac{M\rho}{a_1{}^2L}\right)WS_1\sin.\phi \dots (168.)$$

168. A MACHINE TO WHICH ARE APPLIED ANY TWO PRESSURES P_1 AND P_2, AND WHICH IS MOVEABLE ABOUT A CYLINDRICAL AXIS, IS WORKED WITH THE GREATEST ECONOMY OF POWER WHEN THE DIRECTIONS OF THE PRESSURES ARE PARALLEL, AND WHEN THEY ARE APPLIED ON THE SAME SIDE OF THE AXIS, IF THE WEIGHT OF THE MACHINE ITSELF BE SO SMALL THAT ITS INFLUENCE IN INCREASING THE FRICTION MAY BE NEGLECTED.

For, representing the weight of such a machine by W, and neglecting terms involving W sin. ϕ, it appears by equation (166.) that the modulus is

O

$$U_1 = U_2 \left\{ 1 + \frac{\rho L}{a_1 a_2} \sin. \, \varphi \right\} \, ;$$

whence it follows that the work U_1, which must be done at the moving point to yield a given amount U_2 at the working point, is less as L is less.

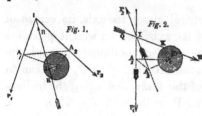

Fig. 1.   Fig. 2.

Now L represents the distance $A_1 A_2$ between the feet of the perpendiculars CA_1 and CA_2, which distance is evidently least when P_1 and P_2 act on the same side of the axis, as in *fig. 2.*, and when CA_1 and CA_2 are in the same straight line; that is, when P_1 and P_2 are parallel.

169. A MACHINE TO WHICH ARE APPLIED TWO GIVEN PRES-
SURES P_1 AND P_2, AND WHICH IS MOVEABLE ABOUT A
CYLINDRICAL AXIS, IS WORKED WITH THE GREATEST
ECONOMY OF POWER, THE INFLUENCE OF THE WEIGHT OF
THE MACHINE BEING TAKEN INTO THE ACCOUNT, WHEN
THE TWO PRESSURES ARE APPLIED ON THE SAME SIDE OF
THE AXIS, AND WHEN THE DIRECTION OF THE MOVING
PRESSURE P_1 IS INCLINED TO THE VERTICAL AT A CERTAIN
ANGLE WHICH MAY BE DETERMINED.

Let P_3 be taken to represent the weight of the machine, and let its centre of gravity coincide with the centre of its axis, then is its modulus represented (equation 166.) by

$$U_1 = U_2 + \frac{\rho \sin. \, \varphi}{a_1 a_2} \left\{ U_2^2 L^2 + 2 U_2 \, P_3 S_1 M \left(\frac{a_2}{a_1} \right) + P_3^2 S_1^2 a_2^2 \right\}^{\frac{1}{2}} \, ;$$

in which expression the work U_1, which must be done at the moving point to yield a given amount U_2 of work at the working point, is shown to be greater than that which must have been done upon the machine to yield the same amount of work if there had been *no friction* by the quantity

$$\frac{\rho \sin. \, \varphi}{a_1 a_2} \left\{ U_2^2 L^2 + 2 U_2 P_3 S_1 M \left(\frac{a_2}{a_1} \right) + P_3^2 S_1^2 a_2^2 \right\}^{\frac{1}{2}}$$

The machine is worked then with the greatest economy of power to yield a given amount of work, U_2, when this function is a *minimum*. Substituting for L^2 its value $a_1^2 + 2a_1a_2 \cos. \iota_{12} + a_2^2$, and for M its value $a_1\{a_2 \cos. \iota_{13} + a_1 \cos. \iota_{23}\}$ (see Art. 166.), also for $S_1\left(\dfrac{a_2}{a_1}\right)$ its value S_2, it becomes

$$\frac{\rho \sin. \varphi}{a_1 a_2} \left\{ U_2^2(a_1^2 + 2a_1a_2 \cos. \iota_{12} + a_2^2) + 2U_2P_3S_2a_1(a_2 \cos. \iota_{13} + a_1 \cos. \iota_{23}) + P_3^2 S_1^2 a_2^2 \right\}^{\frac{1}{2}} \cdot \cdot (169.)$$

Now let us suppose that the perpendicular distance a_2 from the centre of the axis at which the work is done, and the inclination ι_{23} of its direction to the vertical, are both *given*, as also the space S_2 *through* which it is done, so that the work is given in every respect; let also the perpendicular distance a_1 at which the power is applied, and, therefore, the space S_1 through which it is done, be *given;* and let it be required to determine that inclination ι_{12} of the power to the work which will under these circumstances give to the above function its minimum value, and which is therefore consistent with the most economical working of the machine.

Collecting all the terms in the function (169.) which contain (on the above suppositions) only *constant* quantities, and representing their sum by C, it becomes

$$\frac{\rho \sin. \varphi}{a_1 a_2} \left\{ 2a_1a_2U_2(U_2 \cos. \iota_{12} + P_3S_2 \cos. \iota_{13}) + C \right\}^{\frac{1}{2}}$$

Now C being essentially *positive*, this quantity is a minimum when $2a_1a_2U_2(U_2 \cos. \iota_{12} + P_3S_2 \cos. \iota_{13})$ is a minimum; or, observing that $U_2 = P_2S_2$ and dividing by the constant factor $2a_1a_2U_2S_2$, when

$$P_2 \cos. \iota_{12} + P_3 \cos. \iota_{13} \text{ is a minimum.}$$

From the centre of the axis C let lines Cp_1 Cp_2 be drawn parallel to the directions of the pressures P_1 P_2 respectively; and whilst Cp_2 and Cp_3 retain their positions, let the angle p_1CP_3 or ι_{13} be conceived to *increase* until P_1 attains a position in which the condition $P_2 \cos. \iota_{12} + P_3 \cos. \iota_{13} = a$ minimum is satisfied. Now $p_1CP_3 = p_1Cp_2 - p_2CP_3$, or $\iota_{13} = \iota_{12} - \iota_{23}$; substituting which value of ι_{13} this condition becomes

$$P_2 \cos. \iota_{12} + P_3 \cos. (\iota_{12} - \iota_{23}) \text{ a minimum,}$$

or $P_2 \cos. \iota_{12} + P_2 \cos. \iota_{12} \cos. \iota_{23} + P_3 \sin. \iota_{12} \sin. \iota_{23}$ a minimum,

or $(P_2 + P_3 \cos. \iota_{23}) \cos. \iota_{12} + P_3 \sin. \iota_{23} \sin. \iota_{12}$ a minimum.

Let now $\dfrac{P_3 \sin. \iota_{23}}{P_2 + P_3 \cos. \iota_{23}} = \tan. \gamma,$

$\therefore (P_2 + P_3 \cos. \iota_{23}) \cos. \iota_{12} + (P_2 + P_3 \cos. \iota_{23}) \tan. \gamma \sin. \iota_{12}$ a minimum, or dividing by the constant quantity $(P_2 + P_3 \cos. \iota_{93})$ and multiplying by cos. γ,

$$\cos. \iota_{12} \cos. \gamma + \sin. \iota_{12} \sin. \gamma = \cos. (\iota_{12} - \gamma) \text{ a minimum.}$$

$$\therefore \iota_{12} - \gamma = \pi.$$

$$\therefore \iota_{12} = \pi + \tan.^{-1} \left\{ \frac{P_3 \sin. \iota_{23}}{P_2 + P_3 \cos. \iota_{23}} \right\} \quad \ldots \text{(170.)}$$

To satisfy the condition of a minimum, the angle p_1Cp_2 must therefore be increased until it exceeds 180° by that angle γ whose tangent is represented by $\dfrac{P_3 \sin. \iota_{23}}{P_2 + P_3 \cos. \iota_{23}}$. To determine the actual direction of P_1 produce then p_2C to q, make the angle qCr equal to γ; and draw Cm perpendicular to Cr, and equal to the given perpendicular distance a_1 of the direction of P_1 from the centre of the axis. If mP_1 be then drawn through the point m parallel to Cr, it will be in the required direction of P_1; so that being applied in this direction, the moving pressure P_1 will work the machine with a greater economy of power than when applied in any other direction round the axis.

It is evident that since the value of the angle ι_{12} or p_2Cp_1, which satisfies the condition of the greatest economy of power, or of the least resistance, is essentially greater than two right angles, P_1 and P_2 must, TO SATISFY THAT CONDITION, BOTH BE APPLIED ON THE SAME SIDE OF THE AXIS. *It is then a condition necessary to the most economical working of any machine (whatever may be its weight) which is moveable about a cylindrical axis under two given pressures, that* THE MOVING PRESSURE SHOULD BE APPLIED ON THAT SIDE OF

THE AXIS OF THE MACHINE ON WHICH THE RESISTANCE IS
OVERCOME, OR THE WORK DONE. *It is a further condition
of the greatest economy of power in such a machine, that the
direction in which the moving pressure is applied should be
inclined to the vertical at an angle* ι_{13}, *whose tangent is
determined by equation* (170.).

When $\iota_{23} = 0$, or when the work is done in a *vertical* direction,
tan. $\gamma = 0$; therefore $\iota_{12} = \pi$, whence it follows that the moving
power also must in this case be applied in a vertical direction,
and on the same side of the axis as the work. When $\iota_{23} = \dfrac{\pi}{2}$

or when the work is done horizontally, tan. $\gamma = \dfrac{P_3}{P_2}$;

$$\therefore \iota_{12} = \pi + \tan.^{-1}\left(\frac{P_3}{P_2}\right).$$

The moving power must, therefore, in this case, be applied
on the same side of the axis as the work, and at an inclination to the horizon whose tangent equals the fraction obtained by dividing the weight of the machine by the working
pressure.

Since the angle ι_{12} is greater than π and less than $\dfrac{3\pi}{2}$,

cos. ι_{12} is negative; and, for a like reason, cos. ι_{13} is also
in certain cases negative. Whence it is apparent that the
function (169.) admits of a minimum value under certain
conditions, not only in respect to the *inclination* of the
moving pressure, but in respect to the *distance* a_1 of its
direction from the centre of the axis. If we suppose the
space S_1 through which the power acts whilst the given
amount of work U_2 is done to be *given*, and substitute in
that function for the product S_2a_1 its value S_1a_2, and then
assume the differential of the function in respect to a_1 to
vanish, we shall obtain by reduction

$$a_1 = -a_2 \cdot \frac{U_2^2 + 2U_2P_3S_1\cos.\iota_{13} + P_3^2S_1^2}{U_2^2\cos.\iota_{12} + U_2P_3S_1\cos.\iota_{23}} \quad \cdots (171.)$$

If we proceed in like manner assuming the space S_2 instead

o 3

of S_1 to be constant and substituting in the function (169.) for $S_1 a_2$ its value $S_2 a_1$, we shall obtain by reduction

$$a_1 = - \frac{P_2 a_2}{P_2 \cos. \iota_{12} + P_3 \cos. \iota_{23}}.$$

It is easily seen that, if when the values of ι_{12} and ι_{23} determined by equation (170.) are substituted in these equations, the resulting values of a_1 are *positive*, they correspond in the two cases to minimum values of the function (169.), and determine completely the conditions of the greatest economy of power in the machine, in respect to the direction of the moving pressure applied to it.

170. The pulley, when the tensions upon the two extremities of the cord have not vertical directions.

In the case in which the two parts of the string which pass over a pulley are not parallel to one another, the relations established in Article 158. no longer obtain; and we must have recourse to equation (167.) to establish a relation between the tensions upon them in the state bordering upon motion. Calling W the weight of the pulley, a its radius, and observing that the effect of the rigidity of the cord, in increasing the tension P_1, is the same as though it caused the tension P_2 to become $P_2\left(1 + \dfrac{E}{a}\right) + \dfrac{D}{a}$ (Art. 142.), we have

$$P_1 = \left\{ 1 + \frac{L\rho}{a^2} \sin. \phi \right\} \left\{ P_2 \left(1 + \frac{E}{a}\right) + \frac{D}{a} \right\} + \frac{M\rho}{a^2 L} W \sin. \phi.;$$

$$\therefore \; P_1 = \left(1 + \frac{E}{a}\right) \left\{ 1 + \frac{L\rho}{a^2} \sin. \phi \right\} P_2 + \frac{D}{a} + \frac{DL}{a^3} \rho \sin. \phi + \frac{MW}{a^2 L} \cdot \rho \sin. \phi;$$

or,

$$P_1 = \left(1 + \frac{E}{a}\right) \left\{ 1 + \frac{L\rho}{a^2} \sin. \phi \right\} P_2 + \frac{D}{a} \left\{ 1 + \left(\frac{L}{a^2} + \frac{MW}{LDa}\right) \rho \sin. \phi \right\} \ldots (172.)$$

where L represents the chord AB of the arc embraced by the string, and $M = a^2(\cos. \iota_{13} + \cos. \iota_{23})$, ι_{13} and ι_{23} representing the inclinations of P_1 and P_2 to the vertical: which inclinations are measured by the angles P_1EP_3 and P_2FP_3, or their supplements, according as the corresponding pressures P_1 and P_2 act downwards, as shown in the figure, or upwards (see note to Article 165.); so that if both these pressures act upwards, then the cosines of both the angles become negative, and the value of M becomes negative; whilst if one only acts upwards, then one term only of the value of M becomes negative.

Substituting this value for M, observing that $L = 2a \cos. \iota$, where 2ι represents the inclination of the two parts of the cord to one another (so that $2\iota = \iota_{13} + \iota_{23}$), and omitting terms which involve products of two of the exceedingly small quantities $\dfrac{D}{a}$, $\dfrac{E}{a}$, and $\dfrac{\rho}{a} \sin. \varphi$, we have

$$P_1 = \left\{ 1 + \frac{E}{a} + \frac{2\rho}{a} \cos. \iota \sin. \varphi \right\} P_2 + \frac{D}{a} + \frac{W\rho (\cos. \iota_{13} + \cos. \iota_{23}) \sin. \varphi}{2a \cos. \iota}$$

$$U_1 = \left\{ 1 + \frac{E}{a} + \frac{2\rho}{a} \cos. \iota \sin. \varphi \right\} U_2 + \left\{ \frac{D}{a} + \frac{W\rho(\cos. \iota_{13} + \cos. \iota_{23}) \sin. \varphi}{2a \cos. \iota} \right\} S_1 \ldots (173.);$$

which last equation is the modulus to the pulley, when the two parts of the string are inclined to the vertical and to one another.

171. THE PULLEY OF LEAST RESISTANCE.

It is evident, from an inspection of the modulus (equation 173.), that power would be transmitted the more economically by a pulley, as its radius a was greater, did not its weight W *increase* with its radius. If the thickness of the pulley be supposed to remain the same, and its other dimensions to retain the same proportions, whilst it thus increases, its weight W will vary as the square of its radius a, and there will evidently be a certain value of a, for which the value of U_1 in equation (173.), corresponding to a given value of U_2, will be a minimum; that is, there will be certain dimensions of the pulley under which it may be used with the greatest economy of power. To determine these, let W be represented by ca^2, and this value being substituted for W in equation (173.), let that equation be divided by S_1 and differentiated twice in respect to a. Then for the value of a, which corresponds to the minimum of U_1, we shall have (observing that $U_2 = P_2 S_2 = P_2 S_1$),

$$\frac{1}{S_1}\frac{dU_1}{da} = -\frac{1}{a^2}\left\{ E + 2\rho\cos.\iota\sin.\phi \right\} P_2 - \frac{D}{a^2} + \frac{c\rho(\cos.\iota_{13}+\cos.\iota_{23})\sin.\phi}{2\cos.\iota} = 0 ;$$

$$\frac{1}{S_1}\frac{d^2U_1}{da^2} = \frac{2}{a^3}\left\{ E + 2\rho\cos.\iota\sin.\phi \right\} P_2 + \frac{2D}{a^3} > 0.$$

Both these conditions are satisfied by the value

$$a = \sqrt{\frac{2\{(E + 2\rho\cos.\iota\sin.\phi)P_2 + D\}\cos.\iota}{c\rho(\cos.\iota_{13}+\cos.\iota_{23})\sin.\phi}} \quad \ldots \ldots \text{(174.)}$$

172. If both the strings be inclined at equal angles to the vertical, on opposite sides of it; or if $\iota_{13}=\iota_{23}=\iota$, so that cos. ι_{13} + cos. ι_{23} = 2 cos. ι, then equations (172.) and (173.) become

$$P_1 = \left\{ 1 + \frac{E}{a} + \frac{2\rho}{a}\cos.\iota\sin.\phi \right\} P_2 + \frac{D}{a} + \frac{W\rho}{a}\sin.\phi \ldots \text{(175.)},$$

$$U_1 = \left\{ 1 + \frac{E}{a} + \frac{2\rho}{a}\cos.\iota\sin.\phi \right\} U_2 + \left\{ \frac{D}{a} + \frac{W\rho}{a}\sin.\phi \right\} S_1 \ldots \text{(176.)}$$

The radius of the pulley of *least resistance*, or of *the greatest economy of power*, is in this case by equation (174.),

$$a = \sqrt{\frac{1}{c}\left\{ \left(\frac{E}{\rho\sin.\phi} + 2\cos.\iota\right)P_2 + \frac{D}{\rho\sin.\phi} \right\}} \quad \ldots \ldots \text{(177).}$$

173. If both parts of the cord passing over a pulley be in the same horizontal straight line, so that the pulley sustains no pressure resulting from the *tension* upon the cord, but only bears its *weight*, then $\iota = \frac{\pi}{2}$, and the term involving

cos. ι in each of the above equations vanishes. It is, however, to be observed, that the *weight* bearing upon the axis of the pulley is in this case the weight of the pulley increased by the weight of cord which it is made to support. So that if the length of cord supported by the pulley be represented by s, and the weight of each foot of cord by μ, then is the *weight* sustained by the axis of the pulley represented by $W + \mu s$. Substituting this value for W in equation (176.), and assuming cos. $\iota = 0$, we have

$$U_1 = \left(1 + \frac{E}{a}\right)U_2 + \frac{1}{a}\left\{D + (W + \mu s)\rho \sin. \varphi\right\}S_1 \ . \ . \ . \ . \ (178.)$$

The pulley of least resistance is in this case determined, as in Art. 170., by assuming $W = ca^2$, and differentiating in respect to a. The value of a which satisfies the conditions $\dfrac{dU_1}{da} = 0$ and $\dfrac{d^2U_1}{da^2} > 0$, is thus found to be

$$a = \sqrt{\frac{1}{c}\left(\frac{E \cdot P_2 + D}{\rho \sin. \varphi} + \mu s\right)}. \ . \ . \ . \ . \ . \ . \ (179.)$$

174. Let us now suppose that there are n equal pulleys sustaining each the same length s of cord, and let U_n represent the work *yielded* by the rope (through the space S_1) after it has passed over the nth, or last pulley of the system, U_1 being that *done* upon it before it passes over the first pulley; then by Art. 163., equations 152. 154. and 178., neglecting terms involving powers of $\dfrac{E}{a}, \dfrac{D}{a}, \dfrac{\rho}{a} \sin. \varphi$ above the first, and observing that $a_1 = a_2 = \&c. = 1 + \dfrac{E}{a}$, $\alpha_1 = \alpha_2 = \&c. = \dfrac{E}{a}$, $b_1 = b_2 = \&c. = \dfrac{1}{a}\left\{D + (W + \mu s)\rho \sin. \varphi\right\}$, we have

$$U_1 = \left(1 + n\frac{E}{a}\right)U_2 + \frac{n}{a}\left\{D + (W + \mu s)\rho \sin. \varphi\right\}S_1.$$

Representing the whole weight of the cord sustained by the pulleys by w, and observing that $\mu n s = w$, we have

$$U_1 = \left(1 + \frac{nE}{a}\right)U_2 + \frac{1}{a}\left\{nD + (nW + w)\rho \sin. \varphi\right\}S_1 \dots (180.)$$

In the above equations it has been supposed, that although the direction of the rope on either side of each pulley is so nearly horizontal that cos. ι may be considered $= 0$, yet that it does so far *bend* itself over each pulley as to cause the surface of the rope to adapt itself to the circumference of the pulley, and thereby to produce the whole of that resistance which is due to the *rigidity* of the cord. If the tension were so great as to cause the cord to rest upon the pulley only as a rigid rod or bar would, then must we assume $E = 0$ and $D = 0$ in the preceding equations. The radius of the pulley of least resistance (equation 179.) would in this case become

$$a = \sqrt{\frac{\mu s}{\cdot c}} \dots (181.)$$

175. If one part of the cord passing over a pulley have a horizontal, and the other a vertical, direction, as, for instance, when it passes into the shaft of a mine over the sheaf or wheel which overhangs its mouth; then one of the angles ι_{13} or ι_{23} (equation 173.) becomes $\frac{\pi}{2}$, and the other 0 or π, according as the tension on the *vertical* cord is downwards or upwards, so that cos. ι_{13} + cos. $\iota_{23} = \pm 1$, the sign $\pm$ being taken according as the tension upon the vertical cord is downwards or upwards. Moreover, in this case (art. 169.) $\iota = \frac{\pi}{4}$ and cos. $\iota = \frac{1}{\sqrt{2}}$; therefore (equation 173.)

$$P_1 = \left\{1 + \frac{E}{a} + \frac{\rho\sqrt{2}}{a}\sin. \varphi\right\}P_2 + \frac{1}{a}\left\{D \pm \frac{W\rho}{\sqrt{2}}\sin. \varphi\right\}..(182.),$$

$$U_1 = \left\{ 1 + \frac{E}{a} + \frac{\rho\sqrt{2}}{a} \sin. \varphi \right\} U_2 + \frac{1}{a} \left\{ D \pm \frac{W\rho}{\sqrt{2}} \sin. \varphi \right\} S_1 \ldots (183.)$$

The radius of the pulley of least resistance is in this case (equation 174.) determined by the equation

$$a = \sqrt{\left(\frac{1}{c}\right) \left\{ \frac{EP_2 + D}{\rho \sin. \varphi} \sqrt{2} + 2P_2 \right\}} \quad \ldots \ldots (184.)$$

176. *The modulus of a system of any number of pulleys, over one of which the rope passes vertically, and over the rest horizontally.*

Let U_1 represent the work done upon the rope through the space S_1 before it passes horizontally over the first pulley of the system, and let it pass horizontally over n such pulleys; and then, after having passed over another pulley of different dimensions, let it take a vertical direction, descending, for instance, into a shaft. Let U_2 be the work yielded by it through the space S_1 immediately that it has assumed this vertical direction: also let u_1 represent the work done upon it in the horizontal direction immediately *before* it passed over this last pulley of the system. Then by equation (183)

$$u_1 = \left\{ 1 + \frac{E}{a} + \frac{\rho\sqrt{2}}{a} \sin. \varphi \right\} U_2 + \frac{1}{a} \left\{ D + \frac{W\rho}{\sqrt{2}} \sin. \varphi \right\} S_1.$$

Also, by equation (180.), representing the radius of each of the pulleys which carry the rope horizontally by a_1, the radius of its axis by ρ_1, and its weight by W_1, and observing that U_1 is here the *power* and u_1 the *work*, we have

$$U_1 = \left(1 + \frac{nE}{a_1}\right) u_1 + \frac{1}{a_1} \left\{ nD + (nW_1 + w)\rho_1 \sin. \varphi \right\} S_1.$$

Eliminating the value of u_1 between these equations, and neglecting powers above the first in $\dfrac{E}{a}$, &c., we have

$$U_1 = \left\{ 1 + E\left(\frac{1}{a} + \frac{n}{a_1}\right) + \frac{\rho\sqrt{2}}{a}\sin.\,\phi \right\} U_2 + \left\{ D\left(\frac{1}{a} + \frac{n}{a_1}\right) + \frac{nE}{a_1} + \left\{ \frac{W\rho}{a\sqrt{2}} + \frac{(nW_1 + w)\rho_1}{a_1} \right\} \sin.\,\phi \right\} S_1 \quad \ldots \ldots (185.)$$

177. If the strings be *parallel,* and their common inclination to the vertical be represented by ι, so that $\iota_{13} = \iota_{23} = \iota$; then, since in this case $L = 2a$, we have (equation 172.), neglecting terms of more than one dimension in $\dfrac{E}{a}$ and $\dfrac{\rho}{a}$,

$$P_1 = \left\{ 1 + \frac{E}{a} + \frac{2\rho}{a}\sin.\,\phi \right\} P_2 + \frac{D}{a}\left\{ 1 + \left(\frac{2}{a} + \frac{W\cos.\,\iota}{D}\right)\rho\sin.\,\phi \right\} \ldots (186.)$$

$$U_1 = \left\{ 1 + \frac{E}{a} + \frac{2\rho}{a}\sin.\,\phi \right\} U_2 + \frac{D}{a}\left\{ 1 + \left(\frac{2}{a} + \frac{W\cos.\,\iota}{D}\right)\rho\sin.\,\phi \right\} \ldots (187.)$$

in which equation ι is to be taken greater or less than $\dfrac{\pi}{2}$, and therefore the sign of cos. ι is to be taken (as before explained) positively or negatively, according as the tensions on the cords act downwards or upwards. If the tensions are vertical, $\iota = 0$ or π, according as they act upwards or downwards, so that cos. ι $= \pm 1$. The above equations agree in this case, as they ought, with equations (131.) and (132.). If the parallel tensions are *horizontal,* then $\iota = \dfrac{\pi}{2}$, and the terms involving cos. ι in the above equations *vanish.*

178. THE FRICTION OF A PIVOT.

When an axis rests upon its bearings, not by its convex circumference, but by its extremity, as shown in the accompanying figure, it is called a pivot. Let W represent the pressure borne by such a pivot supposed to act in a direction perpendicular to its surface, and to press equally upon every part of it; also let ρ_1 represent the radius of the pivot; then will $\pi\rho_1{}^2$ represent the area of the pivot, and $\dfrac{W}{\pi\rho_1{}^2}$ the pressure sustained by each unit of that area. And if f represent the co-efficient of friction (Art. 133.), $\dfrac{Wf}{\pi\rho_1{}^2}$ will represent the force which must be applied parallel to the surface of the pivot to overcome the friction of each such unit. Now let the dotted lines in the accompanying figure represent an exceedingly narrow ring of the area of the pivot, and let ρ and $\rho+\Delta\rho$ represent the extreme radii of this ring; then will its area be represented by $\pi(\rho+\Delta\rho)^2-\pi\rho^2$, or by $\pi\{2\rho(\Delta\rho)+(\Delta\rho)^2\}$, or, since $\Delta\rho$ is exceedingly small as com-compared with ρ, by $2\pi\rho\Delta\rho$. Now the friction upon each unit of this area is represented by $\dfrac{Wf}{\pi\rho_1{}^2}$; therefore the whole friction upon the ring is represented by $\dfrac{Wf}{\pi\rho_1{}^2}\cdot 2\pi\rho\Delta\rho$, or by $\dfrac{2Wf}{\rho_1{}^2}\rho\Delta\varphi$, and the *moment* of that friction about the centre of the pivot by $\dfrac{2Wf}{\rho_1{}^2}\cdot\rho^2\Delta\rho$, and the sums of the moments of the frictions of all such rings composing the whole area of the pivot by

$$\Sigma\frac{2Wf}{\rho_1{}^2}\cdot\rho^2\Delta\rho,\ \text{or by}\ \frac{2Wf}{\rho_1{}^2}\Sigma\rho^2\Delta\rho,\ \text{or by}\ \frac{2Wf}{\rho_1{}^2}\int_0^{\rho_1}\rho^2d\rho,\ \text{or by}$$

$$\frac{2Wf}{\rho_1^2}\tfrac{1}{3}\rho_1^3, \quad \text{or by } \tfrac{2}{3}Wf\rho_1 \quad \cdots \cdots \quad (188.);$$

whence it appears that *the friction of the pivot produces the same effect to oppose the revolution of the mass which rests upon it, as though the whole pressure which it sustains were collected over a point distant by two-thirds of its radius from its centre.*

If θ represent the angle through which the pivot is made to revolve, then $\tfrac{2}{3}\rho_1\theta$ will represent the space described by the point last spoken of; so that the *work* expended upon the resistance Wf acting there, would be represented by $\tfrac{2}{3}W\rho_1 f\theta$, which therefore represents the work expended upon the friction of the pivot, whilst it revolves through the angle θ; so that the work expended on each complete revolution of the pivot is represented by

$$\tfrac{4}{3}\pi\rho_1 fW \quad \cdots \cdots \quad (189).$$

179. If the pivot be hollow, or its surface be an annular instead of a continuous circular area, then representing its internal radius by ρ_2, and observing that its area is represented by $\pi(\rho_1^2 - \rho_2^2)$, and therefore the pressure upon each unit of it by $\dfrac{W}{\pi(\rho_1^2 - \rho_2^2)}$, and the friction of each such unit by $\dfrac{Wf}{\pi(\rho_1^2 - \rho_2^2)}$, we obtain, as before, for the friction of each elementary annulus the expression $\dfrac{2Wf}{\rho_1^2 - \rho_2^2} \cdot \rho\Delta\rho$, and for the sum of the moments of the frictions of all the elements of the pivot $\dfrac{2Wf}{\rho_1^2 - \rho_2^2} \cdot \displaystyle\int_{\rho_2}^{\rho_1} \rho^2 d\rho$, or

$$\tfrac{2}{3}Wf\left(\frac{\rho_1^3 - \rho_2^3}{\rho_1^2 - \rho_2^2}\right).$$

Let r represent the mean radius of the pivot, *i. e.* let

$r = \frac{1}{2}(\rho_1 + \rho_2)$; and let l represent one half the breadth of the ring, *i. e.* let $l = \frac{1}{2}(\rho_1 - \rho_2)$; therefore $\rho_1 = r + l$ and $\rho_2 = r - l$. These values of ρ_1 and ρ_2 being substituted in the above formula, it becomes

$$\frac{2}{3}Wf\left\{\frac{(r+l)^3-(r-l)^3}{(r+l)^2-(r-l)^2}\right\}, \text{ or } \frac{2}{3}Wf\left\{\frac{6r^2l+2l^3}{4rl}\right\},$$

$$\text{or } Wfr\left\{1+\frac{1}{3}\left(\frac{l}{r}\right)^2\right\} \quad \dots \dots (190.);$$

whence it follows that the friction of an annular pivot produces the same effect as though the whole pressure were collected over a point in it distant by $r\left\{1+\frac{1}{3}\left(\frac{l}{r}\right)^2\right\}$ *from its centre, where r represents its mean radius and l one half its breadth.* From this it may be shown, as before, that the whole work expended upon each complete revolution of the annular pivot is represented by the formula,

$$2\pi fr\left\{1+\frac{1}{3}\left(\frac{l}{r}\right)^2\right\}W. \quad \dots \dots (191.)$$

180. To DETERMINE THE MODULUS OF A SYSTEM OF TWO PRESSURES APPLIED TO A BODY MOVEABLE ABOUT A FIXED AXIS, WHEN THE POINT OF APPLICATION OF ONE OF THESE PRESSURES IS MADE TO REVOLVE WITH THE BODY, THE PERPENDICULAR DISTANCE OF ITS DIRECTION FROM THE CENTRE REMAINING CONSTANTLY THE SAME.

Let the pressures P_1 and P_2, instead of retaining constantly (as we have hitherto supposed them to do) the

same relative positions, be now conceived continually to alter their relative positions by the revolution of the point of application of P_1 with the body, that pressure nevertheless retaining constantly the same perpendicular distance a from the centre of the axis, whilst the direction of P_2 and its amount remain constantly the same.

It is evident that as the point A_1 thus continually alters its position, the distance A_1A_2 or L will continually change, so that the value of P_1 (equation 158.) will continually change. Now the work done under this variable pressure during one revolution of P_1 is represented (Art. 51.) by the formula $U_1 = \int_0^{2\pi} P_1 a_1 d\theta$, if θ represent the angle A_1CA described at any time about C, by the perpendicular C_1A_1, and therefore $a_1\theta$, the space S described in the same time by the point of application A_1 of P_1 (see Art. 62.).

Substituting, therefore, for P_1 its value from equation (158.), we have

$$U_1 = \int_0^{2\pi} P_2 \left\{ \left(\frac{a_2}{a_1}\right) + \left(\frac{\rho L}{a_1^2}\right) \sin. \varphi \right\} a_1 d\theta =$$

$$\int_0^{2\pi} P_2 a_2 d\theta + \frac{\rho \sin. \varphi}{a_1} \int_0^{2\pi} P_2 . L d\theta ;$$

$$\therefore \ U_1 = U_2 + \frac{\rho \sin. \varphi}{a_1} \int_0^{2\pi} P_2 . L d\theta \ \ . \ . \ . \ . \ . \ (192.)$$

Let now P_2 be assumed a constant quantity ;

$$\therefore \ \frac{1}{a_1} \int_0^{2\pi} P_2 L d\theta = P_2 a_2 \times \frac{1}{a_1 a_2} \int_0^{2\pi} L d\theta.$$

Now $L = A_1 A_2 = \{ a_1^2 + 2a_1 a_2 \cos. \theta + a_2^2 \}^{\frac{1}{2}}$;

$$\therefore \ \frac{1}{a_1 a_2} \int_0^{2\pi} L d\theta = \frac{1}{a_1 a_2} \int_0^{2\pi} (a_1^2 + 2a_1 a_2 \cos. \theta + a_2^2)^{\frac{1}{2}} d\theta =$$

$$\frac{(a_1^2 + a_2^2)^{\frac{1}{2}}}{a_1 a_2} \int_0^{2\pi} \left\{ 1 + \frac{2a_1 a_2}{a_1^2 + a_2^2} \cos. \theta \right\}^{\frac{1}{2}} d\theta =$$

$$\left(\frac{1}{a_1^2} + \frac{1}{a_2^2} \right)^{\frac{1}{2}} \int_0^{2\pi} \left\{ 1 + 2 \left(\frac{a_1}{a_2} + \frac{a_2}{a_1} \right)^{-1} \cos. \theta \right\}^{\frac{1}{2}} d\theta =$$

$$\left(\frac{1}{a_1^2} + \frac{1}{a_2^2} \right)^{\frac{1}{2}} \int_0^{2\pi} \left\{ 1 + \left(\frac{a_1}{a_2} + \frac{a_2}{a_1} \right)^{-1} \cos. \theta \right\} d\theta \text{ nearly,}$$

neglecting powers of $\left(\dfrac{a_1}{a_2}+\dfrac{a_2}{a_1}\right)^{-1}$ above the first, since in all cases its value is less than unity. Integrating this quantity between the limits 0 and 2π the second term disappears, so that

$$\frac{1}{a_1 a_2}\int_0^{2\pi}\!\!L d\theta = \left(\frac{1}{a_1{}^2}+\frac{1}{a_2{}^2}\right)^{\frac{1}{2}} 2\pi \text{ nearly ;}$$

$$\therefore\; P_2 a_2 . \frac{1}{a_1 a_2}\int_0^{2\pi}\!\!L d\theta = P_2(2\pi a_2)\left(\frac{1}{a_1{}^2}+\frac{1}{a_2{}^2}\right)^{\frac{1}{2}} = U_2\left(\frac{1}{a_1{}^2}+\frac{1}{a_2{}^2}\right)^{\frac{1}{2}} ;$$

since $2\pi a_2$ is the space through which the point of application of the constant pressure P_2 is made to move in each revolution. Therefore by equation (192), in the case in which P_2 is constant,

$$U_1 = U_2\left\{1+\left(\frac{1}{a_1{}^2}+\frac{1}{a_2{}^2}\right)^{\frac{1}{2}}\rho\sin.\,\phi\right\}\;\ldots\ldots(193).$$

181. If the pressure P_2 be supplied by the tension of a rope winding upon a drum whose radius is a_2 (as in the capstan), then is the effect of the rigidity of the rope (Art. 142.) the same as though P_2 were increased by it so as to become

$$P_2 + \frac{D+EP_2}{a_2},\; \text{ or }\; \left(1+\frac{E}{a_2}\right)P_2 + \frac{D}{a_2}.$$

Now, assuming P_2 to be constant, and observing that $U_2 = 2\pi P_2 a_2$, we have, by equation (192),

$$U_1 = P_2 a_2\left\{2\pi + \frac{\rho\sin.\,\phi}{a_1 a_2}\int_0^{2\pi}\!\!L d\theta\right\}.$$

Substituting in this equation the *above* value for P_2,

$$U_1 = a_2\left\{\left(1+\frac{E}{a_2}\right)P_2 + \frac{D}{a_2}\right\}\left\{2\pi + \frac{\rho\sin.\,\phi}{a_1 a_2}\int_0^{2\pi}\!\!L d\theta\right\}.$$

Performing the actual multiplication of these factors, observing that $\dfrac{D}{a_2}$ is exceedingly small, and omitting the term

P

involving the product of this quantity and $\dfrac{\rho \sin. \varphi}{a_1}$, we have

$$U_1 = P_2 a_2 \left(1 + \frac{E}{a_2}\right) \left\{ 2\pi + \frac{\rho \sin. \varphi}{a_1 a_2} \int_0^{2\pi} L d\theta \right\} + 2\pi D.$$

Whence performing the integration as before, we obtain

$$U_1 = U_2 \left(1 + \frac{E}{a_2}\right) \left\{ 1 + \left(\frac{1}{a_1^2} + \frac{1}{a_2^2}\right)^{\frac{1}{2}} \rho \sin. \varphi \right\} + 2\pi D.$$

If this equation be multiplied by n, and if instead of U_1 and U_2 representing the work done during *one* complete revolution, they be taken to represent the work done through n such revolutions, then

$$U_1 = U_2 \left(1 + \frac{E}{a_2}\right) \left\{ 1 + \left(\frac{1}{a_1^2} + \frac{1}{a_2^2}\right)^{\frac{1}{2}} \rho \sin. \varphi \right\} + 2n\pi D. \; \dots (194),$$

which is the MODULUS.

182. If the quantity $\left(\dfrac{a_1}{a_2} + \dfrac{a_2}{a_1}\right)^{-1}$ be *not* so small that terms of the binomial expansion involving powers of that quantity above the first may be neglected, the value of the definite integral $\int_0^{2\pi} L d\theta$ may be determined as follows : —

$$\int_0^{2\pi} (a_1^2 + 2a_1 a_2 \cos. \theta + a_2^2)^{\frac{1}{2}} d\theta = \int_0^{2\pi} \{(a_1 + a_2)^2 - 2a_1 a_2 (1 - \cos. \theta)\}^{\frac{1}{2}} d\theta$$

$$= (a_1 + a_2) \int_0^{2\pi} \left\{ 1 - \frac{4 a_1 a_2}{(a_1 + a_2)^2} \sin.^2 \frac{\theta}{2} \right\}^{\frac{1}{2}} d\theta. \quad \text{Let } k^2 = \frac{4 a_1 a_2}{(a_1 + a_2)^n};$$

$$\therefore \int_0^{2\pi} L d\theta = (a_1 + a_2) \int_0^{2\pi} \left(1 - k^2 \sin.^2 \frac{\theta}{2}\right)^{\frac{1}{2}} d\theta = (a_1 + a_2) \int_0^{\pi} (1 - k^2 \sin.^2 \theta)^{\frac{1}{2}} d\theta$$

$$= 2(a_1 + a_2) \int_0^{\frac{\pi}{2}} (1 - k^2 \sin.^2 \theta)^{\frac{1}{2}} d\theta^* = 2(a_1 + a_2) E_1(k), \text{ where } E_1(k)$$

---

* See *Encyc. Met.* art. DEF. INT. theorem 2.

represents the complete elliptic function of the second order, whose modulus is k.* The value of this function is given for all values of k in a table which will be found at the end of this work.

Substituting in equation (192),

$$U_1 = U_2 + \frac{\rho \sin. \phi}{a_1} . 2(a_1 + a_2) . E_1(k)\dagger . P_2 = U_2 + (2\pi a_2 P_2)\frac{1}{\pi}\Big(\frac{1}{a_1} + \frac{1}{a_2}\Big)\rho \sin. \phi \ ;$$

$$\therefore \ U_1 = U_2\Big\{ 1 + \frac{1}{\pi}\Big(\frac{1}{a_1} + \frac{1}{a_2}\Big)\rho \sin. \phi \Big\} \ \ . \ . \ . \ . \ (195).$$

THE CAPSTAN.

183. The capstan, as used on shipboard, is represented in

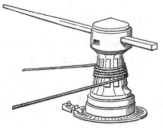

the accompanying figure. It consists of a solid timber CC, pierced through the greater part of its length by an aperture AD, which receives the upper portion of a solid shaft AB of great strength, whose lower extremity is prolonged, and strongly fixed into the timber framing of the ship. The piece CC, into the upper portion of which are

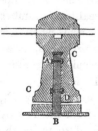

fitted the moveable arms of the capstan, *turns* upon the shaft AB, resting its weig ht upon the crown of the shaft, coiling the cable round its central portion CC, and sustaining the tension of the cable by the lateral resistance of the shaft. Thus the capstan combines the resistances of the pivot and the axis, so that the whole resistance to its motion is equal to the sum of the resistances

---

* See *Encyc. Met.* art. DEF. INT. theorem 2.

† An approximate value of $E_1(k)$ is given when k is small by the formula $E_1(k) = \frac{\pi}{2}(1 + K^{-1}$, where $K = \frac{2\sqrt{k}}{1+k}$. (See *Encyc. Met.* art. DEF. INT. equation (W′), 14.)

due separately to the axis and the pivot, and the whole work
expended in turning it equal to the whole work which would
be expended in turning it upon its pivot were there no tension of the cable upon it, added to the whole work necessary
to turn it upon its axis under the tension of the cable were
there no friction of the pivot. Now, if U_2 represent the work
to be done upon the cable in n complete revolutions, the work
which must be done upon the capstan to yield this work upon
the cable is represented (equation 194) by

$$\left(1+\frac{E}{a_2}\right)\left\{1+\left(\frac{1}{a_1{}^2}+\frac{1}{a_2{}^2}\right)^{\frac{1}{2}}\rho\sin.\varphi\right\}U_2+2n\pi D,$$

where a_1 represents the length of the arm, and a_2 the radius
of that portion of the capstan on which the cable is winding.
Moreover (Art. 177.), the work due to the friction of the
pivot in n complete revolutions is represented by $\frac{4}{3}n\pi\rho_1 f\,W$.

On the whole, therefore, it appears that the work U_1 expended upon n complete revolutions of the capstan is represented by the formula

$$U_1=\left(1+\frac{E}{a_2}\right)\left\{1+\left(\frac{1}{a_1{}^2}+\frac{1}{a_2{}^2}\right)^{\frac{1}{2}}\rho\sin.\varphi\right\}U_2+2n\pi\left\{D+\frac{2}{3}\rho_1 f\,W\right\}\dots(196).$$

which is the MODULUS of the capstan.

A single pressure P_1 applied to a single arm has been supposed to give motion to the capstan ; in reality, a number
of such pressures are applied to its different arms when it is
used to raise the anchor of a ship. These pressures, however, have in all cases, — except in one particular case about
to be described, — a single resultant. It is that single
resultant which is to be considered as represented by P_1,
and the distance of its point of application from the axis by
a_1, when more than one pressure is applied to move the
capstan.

The particular case spoken of above, in which the pressures applied to move the capstan have no resultant, or cannot be replaced by any single pressure, is that in which they
may be divided into two sets of pressure, *each* set having a

resultant, and in which these two resultants are equal, act in opposite directions, on opposite sides of the centre, perpendicular to the same straight line passing through the centre, and at equal distances from it.*

Suppose that they may thus be compounded into the equal pressures R_1 and R_2, and let them be replaced by these. The capstan will then be acted upon by four pressures, — the tension P_2 of the cable, the resistance R of the shaft or axis, and the pressures R_1 and R_2. Now these pressures are in equilibrium. If moved, therefore, parallel to their present directions, so as to be applied to a single point, they would be in equilibrium about that point (Art. 8.). But when so removed, R_1 and R_2 will act in the *same straight line* and in opposite directions. Moreover, they are equal to one another; R_1 and R_2 will therefore *separately* be in equilibrium with one another when applied to that point; and therefore P_2 and R will *separately* be in equilibrium; whence it follows, that R is equal to P_2 or the whole pressure upon the axis, equal in this case to the whole tension P_2 upon the cable. So that the friction of the axis is represented in every position of the capstan by P_2 tan. ϕ (tan. ϕ being equal to the co-efficient of friction (Art. 138.)), and the *work* expended on the friction of the axis, whilst the capstan revolves through the angle θ by $P_2 \rho \theta$ tan. ϕ, or by $P_2 a_2 \theta \left(\dfrac{\rho}{a_2} \right)$ tan. ϕ, or by $U_2 \left(\dfrac{\rho}{a_2} \right)$ tan. ϕ; so that, on the whole, introducing the correction for *rigidity* and for the friction of the pivot, the modulus (equation 196) becomes in this case

$$U_1 = U_2 \left(1 + \frac{E}{a_2} \right) \left\{ 1 + \left(\frac{\rho}{a_2} \right) \tan. \phi \right\} + 2n\pi \left\{ D + \frac{2}{3} \rho_1 f W \right\} \ldots \ldots (197).$$

This is manifestly the least possible value of the modulus,

* Two equal pressures thus placed constitute a STATICAL COUPLE. The properties of such couples have been fully discussed by M. Poinsot, and by Mr. Pritchard in his Treatise on Statical Couples; some account of them will be found in the Appendix to this work.

being very nearly that given (equation 196) by the value infinity of a_1.*

Thus, then, it appears generally from equation (196), that the loss by friction is less as a_1 is greater, or as P_1 is applied at a greater distance from the axis; but that it is least of all when the pressures are so distributed round the capstan as to be reducible to a COUPLE, that case corresponding to the value infinity of a_1. This case, in which the moving pressures upon the capstan are reducible to a *couple*, manifestly occurs when they are arranged round it in any number of pairs, the two pressures of each pair being equal to one another, acting on opposite sides of the centre, and perpendicular to the same line passing through it. This *symmetrical* distribution of the pressures about the axis of the capstan is therefore the most favourable to the working of it, as well as to the stability of the shaft which sustains the pressure upon it.

184. THE MODULUS OF A SYSTEM OF THREE PRESSURES APPLIED TO A BODY MOVEABLE ABOUT A CYLINDRICAL AXIS, TWO OF THESE PRESSURES BEING GIVEN IN DIRECTION AND PARALLEL TO ONE ANOTHER, AND THE DIRECTION OF THE THIRD CONTINUALLY REVOLVING ABOUT THE AXIS AT THE SAME PERPENDICULAR DISTANCE FROM IT.

Let P_2 and P_3 represent the parallel pressures of the system, and P_1 the revolving pressure.

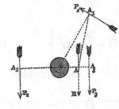

From the centre of the axis C, let fall the perpendiculars CA_1, CA_2, CA_3 upon the directions of the pressures, and let θ represent the inclination of CA_1 to CA_3 at any period of the revolution of P_1. Let P_1 be the preponderating pressure, and let P_2 act to turn the system in the same direction

* ϕ being exceedingly small, tan. ϕ is very nearly equal to sin. ϕ.

as P_1, and P_3 in the opposite direction; also let R represent the resultant of P_2 and P_3, and r the perpendicular distance CA of its direction from C. Suppose the pressures P_2 and P_3 to be replaced by R; the conditions of the equilibrium of P_1 throughout its revolution, and therefore the *work* of P_1 will remain unaltered by this change, and the system will now be a system of *two* pressures P_1 and R instead of three; of which pressures R is given in direction. The modulus of this system is therefore represented (equation 192) by the formula

$$U_1 = U_r + \frac{\rho \sin. \varphi}{a_1} \int_0^\theta R \cdot L d\theta \quad \ldots \quad (198);$$

where U_r represents the work of R, and L represents the distance AA_1 between the feet of the perpendiculars r and a_1, so that $L^2 = a_1^2 - 2a_1 r \cos. \theta + r^2 = (a_1 - r \cos. \theta)^2 + r^2 \sin.^2 \theta$;

$$\therefore R^2L^2 = (Ra_1 - Rr \cos. \theta)^2 + R^2 r^2 \sin.^2 \theta.$$

Now, $R = P_3 + P_2$, $Rr = P_3 a_3 - P_2 a_2$;

$$\therefore R^2L^2 = \{(P_3 + P_2)a_1 - (P_3 a_3 - P_2 a_2) \cos. \theta\}^2 + (P_3 a_3 - P_2 a_2)^2 \sin.^2\theta.$$

Now, if a_1 be $> a_3$, then $a_1(P_3 + P_2) > (P_3 a_3 - P_2 a_2)$,

$$\therefore a_1(P_3 + P_2) > (P_3 a_3 - P_2 a_2)(\sin.\theta + \cos. \theta);$$

for $\sin. \theta + \cos. \theta$ is never greater than unity. Whence it follows, that

$$a_1(P_3 + P_2) - (P_3 a_3 - P_2 a_2) \cos. \theta > (P_3 a_3 - P_2 a_2) \sin. \theta.$$

The value of R^2L^2 is therefore represented by the sum of the squares of two quantities, of which the first is in all cases greater than the second. Therefore, extracting the square root by Poncelet's theorem,

$$RL = \cdot 96\{(P_3 + P_2)a_1 - (P_3 a_3 - P_2 a_2) \cos. \theta\} + \cdot 4(P_3 a_3 - P_2 a_2)\sin. \theta$$

very nearly; or,

$$RL = \cdot 96 a_1(P_3 + P_2) - (P_3 a_3 - P_2 a_2)(\cdot 96 \cos. \theta - \cdot 4 \sin. \theta). \quad \ldots \ldots (199).$$

$$\therefore \int_0^\theta \mathrm{R}L d\theta = \cdot 96 a_1 \left\{ \frac{1}{a_3} \int_0^\theta \mathrm{P}_3 a_3 d\theta + \frac{1}{a_2} \int_0^\theta \mathrm{P}_2 a_2 d\theta \right\} - \int_0^\theta (\mathrm{P}_3 a_3 - \mathrm{P}_2 a_2)(\cdot 96 \cos. \theta - \cdot 4 \sin. \theta) d\theta,$$

$$\int_0^\theta \mathrm{R}L d\theta = \cdot 96 a_1 \left\{ \frac{\mathrm{U}_3}{a_3} + \frac{\mathrm{U}_2}{a_2} \right\} - \int_0^\theta (\mathrm{P}_3 a_3 - \mathrm{P}_2 a_2)(\cdot 96 \cos. \theta - \cdot 4 \sin. \theta) d\theta \, . \, (200).$$

If P_2 and P_3 be *constant*, the integral in the second member of this equation becomes $(\mathrm{P}_3 a_3 - \mathrm{P}_2 a_2)(\cdot 96 \sin. \theta + \cdot 4 \cos. \theta)$; whence observing that $\mathrm{P}_3 a_3 - \mathrm{P}_2 a_2 = \dfrac{\mathrm{P}_3 a_3 \theta - \mathrm{P}_2 a_2 \theta}{\theta} = \dfrac{\mathrm{U}_3 - \mathrm{U}_2}{\theta}$; also, that $\mathrm{U}_r = \theta \mathrm{R} r = \theta \mathrm{P}_3 a_3 - \theta \mathrm{P}_2 a_2 = \mathrm{U}_3 - \mathrm{U}_2$, and substituting in equation (198), we have

$$\mathrm{U}_1 = \mathrm{U}_3 - \mathrm{U}_2 + \rho \sin. \phi \left\{ \cdot 96 \left(\frac{\mathrm{U}_3}{a_3} + \frac{\mathrm{U}_2}{a_2} \right) - \left(\frac{\mathrm{U}_3 - \mathrm{U}_2}{a_1 \theta} \right) (\cdot 96 \sin. \theta + \cdot 4 \cos. \theta \right\} \ldots (201) ;$$

for a complete revolution making $\theta = 2\pi$, we have

$$\mathrm{U}_1 = \mathrm{U}_3 - \mathrm{U}_2 + \rho \sin. \phi \left\{ \cdot 96 \left(\frac{\mathrm{U}_3}{a_3} + \frac{\mathrm{U}_2}{a_2} \right) - \cdot 4 \left(\frac{\mathrm{U}_3 - \mathrm{U}_2}{2\pi a_1} \right) \right\} ;$$

reducing,

$$\mathrm{U}_1 = \left\{ 1 + \frac{\rho \sin. \phi}{5} \left(\frac{4 \cdot 8}{a_3} - \frac{1}{a_1 \pi} \right) \right\} \mathrm{U}_3 - \left\{ 1 - \frac{\rho \sin. \phi}{5} \left(\frac{4 \cdot 8}{a_2} + \frac{1}{a_1 \pi} \right) \right\} \mathrm{U}_2 . \ldots (202),$$

which is the modulus of the system.

185. If the pressure P_3 be supplied by the tension of a cord which winds upon a cylinder or drum at the point A_3, then allowance must be made for the rigidity of the cord, and a correction introduced into the preceding equation for that purpose. To make this correction let it be observed (Art. 142.) that the effect of the rigidity of the cord at A_3 is the same as though it increased the tension there from

$$\mathrm{P}_3 \text{ to } \mathrm{P}_3 \left(1 + \frac{\mathrm{E}}{a_3} \right) + \frac{\mathrm{D}}{a_3} ;$$

or (multiplying both sides of this inequality by a_3, and integrating in respect to θ,) as though it increased

$$\int_0^{2\pi} \mathrm{P}_3 a_3 d\theta \text{ to } \left(1 + \frac{\mathrm{E}}{a_3} \right) \int_0^{2\pi} \mathrm{P}_3 a_3 d\theta + \int_0^{2\pi} \mathrm{D} d\theta ;$$

$$\text{or, } U_3 \text{ to } \left(1 + \frac{E}{a_3}\right)U_3 + 2\pi D.$$

Thus the effect of the rigidity of the rope to which P_3 is applied upon the *work* U_3 of that force is to increase it to $\left(1 + \frac{E}{a_3}\right)U_3 + 2\pi D$. Substituting this value for U_3 in equation (202), and neglecting terms which involve products of the exceedingly small quantities $\dfrac{E}{a_3}$, $\dfrac{\rho\sin.\phi}{a_3}$, $\dfrac{\rho\sin.\phi}{a_1}$ and D, we have

$$U_1 = \left\{1 + \frac{E}{a_3} - \frac{\rho\sin.\phi}{5}\left(\frac{4\cdot8}{a_3} - \frac{1}{a_1\pi}\right)\right\}U_3 - \left\{1 - \frac{\rho\sin.\phi}{5}\left(\frac{4\cdot8}{a_2} + \frac{1}{a_1\pi}\right)\right\}U_2 + 2\pi D \ldots (203).$$

To determine the modulus for n revolutions we must substitute in this expression $n\pi$ for π.

THE CHINESE CAPSTAN.

186. This capstan is represented in the accompanying figure under an exceedingly portable and convenient form.* The axle or drum of the capstan is composed of two parts of different diameters. One extremity of the

cord is coiled upon one of these, and the other, in an opposite direction, upon the other; so that when the axle is turned, and the cord is wound *upon* one of these two parts of the drum, it is, at the same time, wound *off* the other, and the intervening cord is shortened or lengthened, at each revolution, by as much as the circumference of the one cylinder exceeds that of the other. In thus passing from one part

---

* A figure of the capstan with a double axle was seen by Dr. O. Gregory among some Chinese drawings more than a century old. It appears to have been invented under the particular form shown in the above figure by Mr. G. Eckhardt and by Mr. M'Lean of Philadelphia. (See Professor Robison's *Mech. Phil.* vol. ii. p. 255.)

of the drum to the other, the cord is made to pass round a moveable pulley which sustains the pressure to be overcome.

To determine the modulus of this machine, let u_2 and u_3 represent the work done upon the two parts of the cord respectively, whilst the work U_1 is done at the moving point of the machine, and U_2 yielded at its working point.

Then, since in this case we have a body moveable about a cylindrical axis, and acted upon by three pressures, two of which are parallel and constant, viz. the tensions of the two parts of the cord ; and the point of application of the third is made to revolve about the axis, remaining always at the same perpendicular distance from it; it follows (by equation 203), that, for n revolutions of the axis,

$$U_1 = Au_3 - Bu_2 + 2n\pi D \quad \ldots \ldots \quad (204);$$

where

$$A = \left\{ 1 + \frac{E}{a_3} + \frac{\rho \sin. \phi}{5} \left(\frac{4.8}{a_3} - \frac{1}{na_1\pi} \right) \right\}, \text{ and } B = \left\{ 1 - \frac{\rho \sin. \phi}{5} \left(\frac{4.8}{a_2} + \frac{1}{na_1\pi} \right) \right\};$$

a_2 and a_3 representing the radii of the two parts of the drum, a_1 the constant distance at which the power is applied, and ρ the radius of the axis.

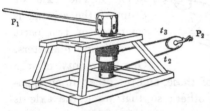

Also, since the two parts of the cord pass over a pulley, and the pulley is made to revolve under the tensions of the two parts of the cord, u_3 being the work of that tension which preponderates, we have (by equation 186),

if S represents the length of cord which passes over the pulley,

$$t_3 = A_1 t_2 + B_1;$$

where

$$A_1 = \left\{ 1 + \frac{E}{a} + \frac{2\rho_1}{a} \sin. \phi \right\}, \text{ and } B_1 = \frac{D}{a} \left\{ 1 + \left(\frac{2}{a} + \frac{W \cos. \iota}{D} \right) \rho_1 \sin. \phi \right\};$$

a representing the radius of the pulley, ρ_1 the radius of its axis, W its weight, and ι the inclination of the direction of the tensions of the two parts of the cord to the vertical, the axis of the pulley being supposed horizontal, and the two parts of the cord parallel. Now $t_3 = \dfrac{u_3}{2n\pi a_3}$, $t_2 = \dfrac{u_2}{2n\pi a_2}$. Substituting these values, and multiplying by $2n\pi a_2$, we have

$$\frac{u_3 a_2}{a_3} = A_1 u_2 + 2n\pi a_2 B_1 \quad \ldots \ldots \quad (205).$$

Since the tensions t_2 and t_3 of the two parts of the cord, and the pressure P_2 overcome by the machine, are pressures applied to the pulley and in equilibrium, and that the points of application of t_2 and P_2 are made to move in directions opposite to those in which those pressures act, whilst the point of application of t_3 is made to move in the same direction; therefore (Art. 59.)

$$U_2 + u_2 = u_3, \quad \therefore \ U_2 = u_3 - u_2.$$

Eliminating u_2 and u_3 between this equation and equation (205), we have

$$u_3 = \frac{A_1 U_2 - 2n\pi a_2 B_1}{A_1 - \dfrac{a_2}{a_3}}, \quad u_2 = \frac{\dfrac{a_2}{a_3} U_2 - 2n\pi a_2 B_1}{A_1 - \dfrac{a_2}{a_3}}.$$

Substituting these values in equation (204), and reducing,

$$U_1 = \left(\frac{A A_1 - \dfrac{a_2}{a_3} B}{A_1 - \dfrac{a_2}{a_3}} \right) U_2 - \left\{ \frac{(A - B) B_1 a_2}{A_1 - \dfrac{a_2}{a_3}} \right\} 2n\pi + 2n\pi D.$$

Substituting their values for A, A_1, B, B_1, neglecting terms involving more than one dimension of $\dfrac{\sin. \phi}{a_1}, \dfrac{E}{a_1}$, &c. and reducing, we obtain for the MODULUS of the machine

$$U_1 = \left\{ 1 + \frac{\dfrac{E}{a_3} + \dfrac{\rho \sin. \phi}{5} \left\{ \dfrac{9 \cdot 6}{a_3} - \left(1 - \dfrac{a_2}{a_3} \right) \dfrac{1}{n a_1 \pi} \right\}}{1 - \dfrac{a_2}{a_3} + \dfrac{E}{a} + \dfrac{2\rho_1}{a} \sin. \phi} \right\} U_2$$

$$- \left\{ \frac{\left\{ E \dfrac{a_2}{a_3} + \cdot 96 \left(1 + \dfrac{a_2}{a_3} \right) \rho \sin. \phi \right\} \left\{ D + W \rho_1 \cos. \iota \sin. \phi \right\}}{a \left(1 - \dfrac{a_2}{a_3} \right) + E + 2\rho_1 \sin. \phi} - D \right\} 2n\pi \ldots (206).$$

From which expression it is apparent that when the radii a_2 and a_3 of the double axle are nearly equal, a great sacrifice of power is made, in the use of this machine, by reason of the rigidity of the cord.

THE HORSE CAPSTAN, OR THE WHIM GIN.

187. The whim is a form of the capstan, used in the *first operations* of mining, for raising materials from the shaft and levels by the power of horses, before the quantity excavated is so great as to require the application of steam power, or before the valuable produce of the mine is sufficient to give a return upon the expenditure of capital necessary to the erection of a steam engine. The construction of this machine will be sufficiently understood from the accompanying figure. It will be observed that there are two ropes wound upon the drum in opposite directions, and which traverse the space

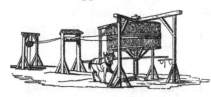

between the capstan and the mouth of the shaft. One of these carries at its extremity the descending (empty) bucket, and is con-tinually in the act of winding off the drum of the capstan as it revolves; whilst the other, from whose extremity is suspended the ascending (loaded) bucket, continually winds *on* the drum. The pressure exerted by the horses is that necessary to over-come the friction of the different bearings, and the other prejudicial resistances, and to balance the difference between the weight of the ascending load, bucket, and rope, and that of the descending bucket and rope. The rope, in passing from the capstan to the shaft, traverses (sometimes for a con-siderable distance) a series of sheaves or pulleys, such as those shown in the accompanying figure.

Let now a_2 represent the radius of the drum on which the rope is made to wind, and n the number of revolutions which it must make to wind up the whole cord; also let μ represent the weight of each foot of cord, and θ the angle which the capstan has described between the time when the ascending bucket has attained any given position in the shaft and that when it left the bottom; then does $a_2\theta$ represent the length of the ascending rope wound *on* the drum, and therefore of the descending rope wound *off* it. Also, let W represent the whole weight of the rope; then does $W - \mu a_2 \theta$ represent the weight of the *ascending* rope, and $\mu a_2 \theta$ that of the descending rope, both of which hang suspended in the shaft. Let P_2 represent the load raised at each lift in the bucket, and w the weight of the bucket; then is the tension upon the *ascending* rope at the mouth of the shaft represented by $W - \mu a_2 \theta + P_2 + w$, and that upon the *descending* rope by $\mu a_2 \theta + w$.

Let, moreover, p_3 and p_2 represent the tensions upon these ropes after they have passed from the mouth of the shaft, over the intervening pulleys, to the circumference of the capstan.

Now, since the tension upon the ascending rope, which is $W - \mu a_2 \theta + P_2 + w$ at the mouth of the shaft, is increased to p_3 at the capstan, and that the tension upon the descending rope, which is p_2 at the capstan, is increased to $\mu a_2 \theta + u$ at the mouth of the shaft, if we represent by $(1 + \alpha)$ and β the constants which enter into equation 185 (Art. 175.), we have, by that equation (observing that $U_1 = P_1 S_1$ and $U_2 = P_2 S_1$, so that S_1 disappears from both sides of it),

$$p_3 = (1 + \alpha)(W + P_2 + w - \mu a_2 \theta) + \beta, \ \ldots \ (207),$$
and
$$\mu a_2 \theta + w = (1 + \alpha) p_2 + \beta \ \ldots \ (208).$$

Multiplying the former of the above equations by $1 + \alpha$, adding them, transposing, dividing by $(1 + \alpha)$, and neglecting terms of more than one dimension in α and β,

$$p_3 - p_2 = (1 + \alpha)(W + P_2) + 2\alpha w + 2\beta - 2\mu a_2 \theta.$$

Now U_r in equation (198) represents the work of the resultant of p_3 and p_2 during n revolutions of the capstan, it therefore equals the difference between the work of p_3 and that of p_2 (see p. 216.).

$$\therefore \ \mathrm{U}_r = \int_0^{2n\pi} p_3 a_2 d\theta - \int_0^{2n\pi} p_2 a_2 d\theta = a_2 \int_0^{2n\pi} (p_3 - p_2) a_2 d\theta \ ;$$

$$\therefore \ \mathrm{U}_r = a_2 \int_0^{2n\pi} \{(1+\alpha)(\mathrm{W}+\mathrm{P}_2) + 2\alpha w + 2\beta - 2\mu a_2 \theta\} d\theta =$$

$$\{(1+\alpha)(\mathrm{W}+\mathrm{P}_2) + 2\alpha w + 2\beta\}(2n\pi a_2) - \mu(2n\pi a_2)^2 \ ;$$

$$\therefore \ \mathrm{U}_r = (1+\alpha)\mathrm{U}_2 + \{(1+\alpha)\mathrm{W} + 2\alpha w + 2\beta - \mu\mathrm{S}_2\}\mathrm{S}_2 \ . \ . \ (209);$$

observing that $2n\pi a_2 = \mathrm{S}_2$, and that $\mathrm{P}_2\mathrm{S}_2 = \mathrm{U}_2$.

Now, let it be observed that the pressures applied to the capstan are three in number; two of them, p_3 and p_2, being parallel and acting at equal distances a_2 from its axis; and the third, P_1, being made to revolve at the constant distance a_1 from the axis (P_1 representing the pressure of the horses, or the *resultant* of the pressures of the horses, if there be more than one, and a_1 the distance at which it is applied); so that equation 198 (Art. 184.) *obtains* in respect to the pressures P_1, p_3, p_2; a_3 being assumed equal to a_2.

Substituting p_2 and p_3 for P_2 and P_3 in equation (199),

$$\mathrm{RL} = \cdot 96 a_1 (p_3 + p_2) - a_2(p_3 - p_2) \ (\cdot 96 \cos. \ \theta - \cdot 4 \sin. \ \theta) \ ;$$

$$\therefore \int_0^{2n\pi} \mathrm{RL} d\theta = \cdot 96 a_1 \int_0^{2n\pi} (p_3 + p_2) d\theta - a_2 \int_0^{2n\pi} (p_3 - p_2) \ (\cdot 96 \cos. \ \theta - \cdot 4 \sin. \ \theta) d\theta.$$

Now, the terms of equation (185), represented in the above equations by α and β, are all of one dimension in the exceedingly small quantities D, E, sin. φ. If, therefore, the values of p_2 and p_3 given by these equations be substituted in the value of $\dfrac{\rho \sin. \ \varphi}{a_1} \displaystyle\int_0^{2n\pi} \mathrm{RL} d\theta$ (equation 198), then all the terms of that expression which involve the quantities α and β will be at least of *two* dimensions in D, E, sin. φ, and may be neglected. Neglecting, therefore, the values of α and β in equations (207, 208), we obtain

$$p_3 + p_2 = \mathrm{W} + \mathrm{P}_2 + 2w, \text{ and } p_3 - p_2 = \mathrm{W} + \mathrm{P}_2 - 2\mu a_2 \theta \ ;$$

$$\therefore \ a_1 \int_0^{2n\pi} (p_3 + p\ )d\theta = a_1\{W + P_2 + 2w\}2n\pi = \left(\frac{a_1}{a_2}\right)\{(2n\pi a_2)P_2 + (2n\pi a_2)\,(W + 2w)\}$$

$$= \left(\frac{a_1}{a_2}\right)\{S_2 P_2 + S_2(W + 2w)\} = \left(\frac{a_1}{a_2}\right)\{U_2 + S_2(W + 2w)\};$$

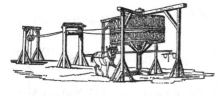

representing by S_2 the space described by the load, and by U_2 the useful work done upon it, during n revolutions of the capstan.

Similarly,

$$a_2 \int_0^{2n\pi}(p_3 - p_2)(\cdot 96\cos.\ \theta - \cdot 4\sin.\ \theta)d\theta = a_2\int_0^{2n\pi}\{W + P_2 - 2\mu a_2\theta\}(\cdot 96\cos.\ \theta - \cdot 4\sin.\ \theta)d\theta$$

$$= a_2(W + P_2)\int_0^{2n\pi}(\cdot 96\cos.\ \theta - \cdot 4\sin.\ \theta)d\theta - 2\mu a_2{}^2\int(\cdot 96\cos.\ \theta - \cdot 4\sin.\ \theta)\theta d\theta.$$

Now $\int_0^{2n\pi}(\cdot 96\cos.\ \theta - \cdot 4\sin.\ \theta)d\theta = \cdot 4$, and $\int_0^{2n\pi}(\cdot 96\cos.\ \theta - \cdot 4\sin.\ \theta)\theta d\theta = \cdot 8n\pi$ *;

$$\therefore \ a_2\int_0^{2n\pi}(p_3 - p_2)(\cdot 96\cos.\ \theta - \cdot 4\sin.\ \theta)d\theta = \cdot 4a_2(W + P_2) - \cdot 8\mu a_2(2n\pi a_2)$$

$$= \cdot 4a_2\frac{U_2}{S_2} + \cdot 4a_2(W - 2\mu S_2^i);\ \text{observing that } P_2 = \frac{U_2}{S_2},$$

$$\therefore \int_0^{2n\pi}RLd\theta = \cdot 96\left(\frac{a_1}{a_2}\right)\{U_2 + S_2(W + 2w)\} - \cdot 4a_2\frac{U_2}{S_2} - \cdot 4a_2(W - 2\mu S_2);$$

$$\frac{\rho\sin.\ \phi}{a_1}\int_0^{2n\pi}RLd\theta = \frac{\cdot 4\rho\sin.\ \phi}{a_1}\left\{\left(2\cdot 4\frac{a_1}{a_2} - \frac{a_2}{S_2}\right)U_2 + 2\{1\cdot 2(W + 2w)\left(\frac{a_1}{a_2}\right) + \mu a_2\}S\ - Wa_2\right\}.$$

---

* For $\int_0^\theta \theta\cos.\ \theta d\theta = \theta\sin.\ \theta - \int_0^\theta \sin.\ \theta d\theta = \theta\sin.\ \theta - \text{vers.}\ \theta;$ also $\int_0^\theta \theta\sin.\ \theta d\theta$

$= -\theta\cos.\ \theta + \int_0^\theta \cos.\ \theta d\theta = -\theta\cos.\ \theta + \sin.\ \theta.$ Now, substituting $2n\pi$ for θ, these integrals become respectively 0 and $-2n\pi.$

Substituting this value, and also that of U_r (equation 209) in equation (198), and assuming

$$C_1 = (1+a)W + 2aw + 2\beta \text{ and } C_2 = 1\cdot2(W+2w)\left(\frac{a_1}{a_2}\right) + \mu a_2,$$

we have

$$U_1 = (1+a)U_2 + C_1 S_2 - \mu S_2{}^2 + \frac{\cdot4\rho \sin.\, \phi}{a_1}\left\{\left(2\cdot4\frac{a_1}{a_2} - \frac{a_2}{S_2}\right)U_2 + 2C_2 S_2 - Wa_2\right\};$$

$$\therefore U_1 = \left\{1 + a + \frac{\cdot4\rho \sin.\, \phi}{a_1}\left(2\cdot4\frac{a_1}{a_2} - \frac{a_2}{S_2}\right)\right\}U_2 - \mu S_2{}^2 + \left(C_1 + \frac{\cdot8\rho C_2 \sin.\, \phi}{a_1}\right)S_2 - \frac{\cdot4 W \rho a_2 \sin.\, \phi}{a_1},$$

which is the MODULUS of the machine, all the various elements, whence a sacrifice of power may arise in the working of it, being taken into account.

THE FRICTION OF CORDS.

188. Let the polygonal line ABC ... YZ, of an infinite

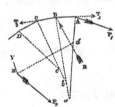

number of sides, be taken to represent the curved portion of a cord embracing any arc of a cylindrical surface (whether circular or not), in a plane perpendicular to the axis of the cylinder; also let Aa, Bb, Cc, &c. be normals or perpendiculars to the curve, inclined to one another at equal angles, each represented by $\Delta\theta$. Imagine the surface of the cylinder to be removed between each two of the points A, B, &c. in succession, so that the cord may be supported by a small portion only of the surface remaining at each of those points, whilst in the intermediate space it assumes the direction of a straight line joining them, and does not touch the surface of the cylinder. Let P_1 represent the tension upon the cord before it has passed over the point A ; T_1 the tension upon it after it has passed over that point, or before it passes over the point B ; T_2 the tension upon it after it has passed over the point B, or before it passes over C ; T_3 that after it has passed over C ; and let P_2

represent the tension upon the cord after it has passed over the nth or last point Z.

Now, any point B of the cord is held at rest by the tensions T_1 and T_2 upon it at that point, in the directions BC and BA, and by the resistance R of the surface of the cylinder there; and, if we conceive the cord to be there in the state bordering upon motion, then (Art. 138.) the direction of this resistance R is inclined to the perpendicular bB to the surface of the cylinder at an angle Rbb equal to the limiting angle of resistance ϕ.

Now, T_1, T_2, and R are pressures in equilibrium; therefore (Art. 14.)

$$\frac{T_1}{T_2} = \frac{\sin. T_2 BR}{\sin. T_1 BR};$$

but $T_1 BR = AB b - R B b = \frac{1}{2}(\pi - A a B) - R B b = \frac{\pi}{2} - \frac{\Delta\theta}{2} - \phi,$

$T_2 BR = CB b + R B b = \frac{1}{2}(\pi - B b C) + R B b = \frac{\pi}{2} - \frac{\Delta\theta}{2} + \phi;$

$$\therefore \frac{T_1}{T_2} = \frac{\sin.\left\{\frac{\pi}{2} - \left(\frac{\Delta\theta}{2} - \phi\right)\right\}}{\sin.\left\{\frac{\pi}{2} - \left(\frac{\Delta\theta}{2} + \phi\right)\right\}} = \frac{\cos.\left(\frac{\Delta\theta}{2} - \phi\right)}{\cos.\left(\frac{\Delta\theta}{2} + \phi\right)};$$

$$\therefore \frac{T_1 - T_2}{T_2} = \frac{\cos.\left(\frac{\Delta\theta}{2} - \phi\right) - \cos.\left(\frac{\Delta\theta}{2} + \phi\right)}{\cos.\left(\frac{\Delta\theta}{2} + \phi\right)} = \frac{2\sin.\frac{\Delta\theta}{2}\sin.\phi}{\cos.\frac{\Delta\theta}{2}\cos.\phi - \sin.\frac{\Delta\theta}{2}\sin.\phi};$$

or dividing numerator and denominator of the fraction in the second member by $\cos.\frac{\Delta\theta}{2} . \cos.\phi,$

$$\frac{T_1 - T_2}{T_2} = \frac{2\tan.\frac{\Delta\theta}{2}\tan.\phi}{1 - \tan.\frac{\Delta\theta}{2}\tan.\phi}.$$

Suppose now the angles AaB, BbC, &c. each of which equals $\Delta\theta$, to be exceedingly small, and therefore the points A, B, C, &c. to be exceedingly near to one another, and

Q

exceedingly numerous. By this supposition we shall manifestly approach exceedingly near to the actual case of an *infinite* number of such points and a *continuous* surface ; and if
we suppose $\Delta\theta$ infinitely small, our supposition will *coincide*
with that case. Now, on the supposition that $\Delta\theta$ is exceedingly small, $\tan.\frac{\Delta\theta}{2}$. $\tan.\varphi$ is exceedingly small, and may
be neglected as compared with unity ; it may therefore be ne-
glected in the denominator of the above fraction. Moreover,
$\Delta\theta$ being exceedingly small, $\tan.\frac{\Delta\theta}{2}=\frac{\Delta\theta}{2}$

$$\therefore \frac{T_1-T}{T_2}=\tan.\varphi\,.\,\Delta\theta* \;;\; \therefore\; T_1=T_2(1+\tan.\varphi\,.\,\Delta\theta).$$

Now the number of the points A, B, C, &c. being represented by n, and the whole angle AdZ between the *extreme* nor-
mals at A and Z by θ, it follows (Euclid, i. 32.) that $\theta=n\,.\,\Delta\theta$;
therefore $\Delta\theta=\frac{\theta}{n}$;

$$\therefore\; T_1=T_2\left(1+\frac{\theta}{n}\tan.\varphi\right).$$

Similarly, $$P_1=T_1\left(1+\frac{\theta}{n}\tan.\varphi\right),$$

$$T_2=T_3\left(1+\frac{\theta}{n}\tan.\varphi\right),$$

$$\&c.=\&c.=\&c.$$

$$T_{n-1}=P_2\left(1+\frac{\theta}{n}\tan.\varphi\right).$$

* If we consider the tension T as a function of θ, of which any consecutive values are represented by T_1 and T_2, and their difference or the
increment of T by ΔT, then $\frac{-\Delta T}{T}=\tan.\varphi.\Delta\theta$; therefore $\frac{1}{T}\cdot\frac{\Delta T}{\Delta\theta}=$
$-\tan.\varphi$; therefore, passing to the *limit* $\frac{1}{T}\frac{dT}{d\theta}=-\tan.\varphi$, and *integrating*
between the limits 0 and θ, observing that at the latter limit $T=P_2$,
and that at the former it equals P_1, we have $\log._e\left(\frac{P_2}{P_1}\right)=-\theta\tan.\varphi$;
therefore $P_1=P_2\varepsilon^{\theta\tan.\varphi}$

Multiplying these equations together, and striking out factors common to both sides of their product, we have

$$P_1 = P_2 \left(1 + \frac{\theta}{n}\tan. \, \varphi\right)^n ;$$

or $P_1 = P_2 \left\{ 1 + n\frac{\theta}{n}\tan. \, \phi + n\frac{n-1}{2}\frac{\theta^2}{n^2}\tan.\,^2\phi + n\frac{n-1}{2}\frac{n-2}{3}\frac{\theta^3}{n}\tan.\,^3\phi + \&c. \right\} ;$

or $P_1 = P_2 \left\{ 1 + \theta\tan. \, \phi + \frac{1-\frac{1}{n}}{2}\theta^2\tan.\,^3\phi + \frac{\left(1-\frac{1}{n}\right)\left(1-\frac{2}{n}\right)}{3}\theta^3\tan.\,^3\phi + \ldots \right\}.$

Now this relation of P_1 and P_2 obtains however *small* $\Delta\theta$ be taken, or however *great* n be taken. Let n be taken *infinitely* great, so that the points A, B, C, &c. may be infinitely numerous and infinitely near to each other. The *supposed* case thus passes into the *actual* case of a continuous surface, the fractions $\frac{1}{n}, \frac{2}{n}, \frac{3}{n}$, &c. *vanish*, and the above equation becomes

$$P_1 = P_2 \left\{ 1 + \frac{\theta\tan. \, \varphi}{1} + \frac{\theta^2\tan.\,^2\varphi}{1.2} + \frac{\theta^3\tan.\,^3\varphi}{1.2.3} + \ldots. \right\}.$$

But the quantity within the brackets is the well known expansion (by the exponential theorem) of the function $\varepsilon^{\,\theta\tan. \, \varphi}$,

$$\therefore \; P_1 = P_2 \varepsilon^{\,\theta\tan. \, \varphi}. \; . \; . \; . \; . \; (210).$$

Since the length of cord S_1, which passes over the point A, is the same with that S_2 which passes over the point Z, it follows that the *modulus* (Art. 152.) of such a cylindrical surface considered as a machine, and supposed to be *fixed* and to have a rope pulled and made to slip over it, is

$$U_1 = U_2 \varepsilon^{\,\theta\tan. \, \varphi}. \; . \; . \; . \; (211).$$

It is remarkable that these expressions are wholly independent of the form and dimensions of the surface sustaining the tension of the rope, and that they depend exclusively

Q 2

upon the inclination θ or AdZ of the normals to the points
A and Z, where the cord leaves the surface, and upon
the co-efficient of friction (tan. φ), of the material of which
the rope is composed and the material of which the surface
is composed. It matters not, for instance, so far as the *friction* of the rope or band is concerned, whether it passes over
a large pulley or *drum*, or a small one, provided the angle
subtended by the arc which it embraces is the same, and the
materials of the pulley and rope the same.

In the case in which a cord is made to pass m times round
such a surface, $\theta = 2m\pi$;

$$\therefore\ P_1 = P_2 \varepsilon^{2\,m\,\pi\,\tan.\varphi}.$$

And this is true whatever be the *form* of the surface, so that
the pressure necessary to cause a cord to slip when wound
completely round such a cylindrical surface a given number of
times is the same (and is always represented by this quantity),
whatever may be the form or dimension of the surface, provided that its material be the same. It matters not whether
it be square, or circular, or elliptical.

189. If P_1', P_1'', P_1''', &c. represent the pressures which must
be applied to one extremity of a rope to cause it to slip when
wound once, twice, three times, &c. round any such surface,
the *same* tension P_2 being in each case supposed to be applied
to the other extremity of it, we have

$$P_1' = P_2 \varepsilon^{2\pi \tan.\varphi},\ P_1'' = P_2 \varepsilon^{4\pi \tan.\varphi},\ P_1''' = P_2 \varepsilon^{6\pi \tan.\varphi}, \&c. = \&c.$$

So that the pressures P_1', P_1'', P_1''', &c. are in a geometrical progression, whose common ratio is $\varepsilon^{2\pi \tan.\varphi}$, which ratio
is always greater than unity. Thus it appears by the expe-
riments of M. Morin (p. 153.), that the co-efficient of friction
between hempen rope and oak free from unguent is ·33, when
the rope is wetted. In this case tan. $\varphi = ·33$ and 2π tan. φ
$= 2 \times 3·14159 \times ·33 = 2·07345$. The common ratio of the
progression is therefore in this case $\varepsilon^{2·07345}$, or it is the number

whose hyperbolic logarithm is 2·07345. This number is
7·95; so that each additional coil increases the friction nearly
eight times. Had the rope been dry, this proportion would
have been much greater. If an additional *half* coil had been
supposed continually to be put upon the rope instead of a
whole coil, the friction would have been found in the same
way to increase in geometrical progression, but the common
ratio would in this case have been $\varepsilon^{\pi \tan. \varphi}$ instead of $\varepsilon^{2\pi \tan. \varphi}$.
In the above example the value of this ratio would for each
half coil have been 2·82.

The enormous increase of friction which results from
each additional turn of the cord upon a capstan or drum,
may from these results be understood.

190. We may, from what has been stated above, readily explain the reason why a knot connecting the two extremities of
a cord effectually resists the action of any force tending
to separate them. If a wetted cord be wound round a
cylinder of oak as in
fig. 1., and its extremities be acted upon by
two forces P and R., it
has been shown that P
will not overcome R., un-
less it be equal to some
where about eight times that force. Now if the string to which
R is attached be brought underneath the other string so
as to be pressed by it against the surface of the cylinder,
as at *m*, *fig.* 2.; then, provided the friction produced by this
pressure be not less than one eighth of P, the string will not
move even although the force R cease to act. And if both
extremities of the string be thus made to pass between the
coil and the cylinder, as in *fig.* 3., a still less *pressure* upon
each will be requisite. Now by diminishing the radius of
the cylinder, this pressure can be increased to any extent,
since, by a known property of funicular curves, it varies in-

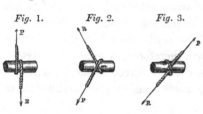

Fig. 1.   Fig. 2.   Fig. 3.

versely as the radius.* We may, therefore, so far diminish the radius of a cylinder, as that no force, however great, shall be able to pull away a rope coiled upon it, as represented in *fig. 3.*, even although one extremity were loose, and acted upon by no force.

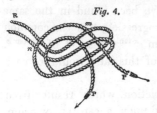

Fig. 4.

Let us suppose the rope to be doubled as in *fig.* 4., and coiled as before. Then it is apparent, from what has been said, that the cylinder may be made so small, that no forces P and P applied to the extremities of either of the double cords will be sufficient to pull them from it, in whatever directions these are applied.

Now let the cylinder be removed. The cord then being drawn tight, instead of being coiled round the cylinder, will be coiled round portions of itself, at the points *m* and *n;* and instead of being pressed at those points upon the cylinder by a force acting on one portion of its circumference, it will be pressed by a greater force acting all round its circumference. All that has been proved before, with regard to the impossibility of pulling either of the cords away from the coil when the cylinder is inserted, will therefore now obtain in a greater degree; whence it follows that no forces P and P′ acting to pull the extremities of the cords asunder, may be sufficient to separate the knot.

THE FRICTION BREAK.

191. There are certain machines whose motion *tends*, at certain stages, to a destructive acceleration; as, for instance, a crane, which, having raised a heavy weight in one position of its beam, allows it to descend by the action of gravity in another; or a railway train, which, on a certain portion of its line of transit, descends a gradient, having an inclination greater

* This property will be proved in that portion of the work which treats of the THEORY OF CONSTRUCTION.

than the limiting angle of resistance. In each of these cases, the work done by gravity on the descending weight exceeds the work expended on the ordinary resistance due to the friction of the machine ; and if some other resistance were not, under these circumstances, opposed to its motion, this excess (of the work done by gravity upon it over that expended upon the friction of its rubbing surfaces) would be accumulated in it (Art. 130.) under the form of *vis viva*, and be accompanied by a rapid acceleration and a destructive velocity of its moving parts. The extraordinary resistance required to take up this excess of work, and to prevent this accumulation, is sometimes supplied in the crane by the *work* of the labourer, who, to let the weight down gradually, exerts upon the revolving crank a pressure in a direction opposite to that which he used in raising it. It is more commonly supplied in the crane, and always in the railway train, without any *work* at all of the labourer, by a simple *pressure* of his hand or foot on the lever of the friction break, which useful instrument is represented in the accompanying figure under the form in which it is commonly applied to the crane,—a form of it which may serve to illustrate the principle of its application under

every other. BC represents a wheel fixed commonly upon that axis of the machine to which the crank is attached, and which axis is carried round by it with greater velocity than any other. The periphery of this wheel, which is usually of cast iron, is embraced by a strong band ABCE of wrought iron, fixed firmly by its extremity A to the frame of the machine, and by its extremity E to the short arm AE of a bent lever PAE, which turns upon a fixed axis or fulcrum at A, and whose arm PA, being prolonged, carries a counterpoise D just sufficient to overbalance the weight of the arm AP, and to relieve the point E of all tension, and loosen the strap from the periphery of the wheel, when no force P is applied to the extremity of the arm AP, or when the break is out of action.

It is evident that a pressure P applied to the extremity of the lever will produce a pressure upon the point E, and a tension upon the band in the direction ABCE, and that being fixed at its extremity A, the band will thus be tightened upon the wheel, producing by its friction a certain resistance upon the circumference of the wheel.

Moreover, it is evident that this resistance of friction upon the circumference of the wheel is precisely equal to the tension upon the extremity A of the band, being, indeed, wholly borne by that tension ; and that it is the same whether the wheel move, as in this case it does, under the band at rest, or whether the band move (under the same tensions upon its extremities, but in the opposite direction) over the wheel at rest. Let R and Q represent the tensions upon the extremities A and E of the band ; then if we suppose the wheel to be at rest, and the band to be drawn over it in the direction ECB by the tension R, and θ to represent the angle subtended at the centre of the wheel by that part of its circumference which the band embraces, we have (equation 210)

$$R = Q \varepsilon^{\theta \tan \varphi}.$$

Let a_1 represent the length of the arm AP, and a_2 the length of the perpendicular let fall from A upon the direction of a tangent to that point in the circumference of the wheel where the end EC of the band leaves it.

Then, neglecting the friction of the axis A, we have (Art. 5.)

$$P \cdot a_1 = Q \cdot a_2 ;$$

$$\therefore R = \frac{P a_1}{a_2} \varepsilon^{\theta \tan \varphi} \ \ \cdot \ \cdot \ \cdot \ \cdot \ \cdot \ (212).$$

If P_1 represent any pressure applied to the circumference of the break wheel, and P_2 a pressure applied to the working point of the machine, whatever it may be, to which the break is applied, and if $P_1 = a P_2 + b$ (Art. 152.) represent the relation between P_1 and P_2 in the *inferior* state bordering upon motion by the preponderance of P_2 ; then, when P_2 is taken in this expression to represent the pressure W, whose action upon the working point of the machine the break is intended to control, P_1 will represent that value R of the friction upon the break which must be produced by the intervention of the lever to control the action of the pressure W upon the machine ; so that taking R to represent the same quantity as in equation (212), we have

$$R = aW + b.$$

Eliminating R between this equation and equation (212), and solving in respect to P,

$$P = \frac{a_2}{a_1}(aW + b)\varepsilon^{-\theta \tan. \varphi}. \quad \ldots \quad (213).$$

The Band.

192. When the circular motion of any shaft in a machine, and the pressure which accompanies that motion, constituting together with it the *work* of the shaft, are to be communicated to any other distant shaft, this communication is usually

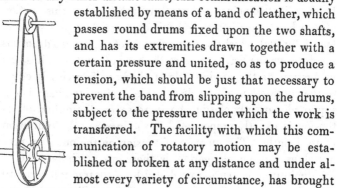

established by means of a band of leather, which passes round drums fixed upon the two shafts, and has its extremities drawn together with a certain pressure and united, so as to produce a tension, which should be just that necessary to prevent the band from slipping upon the drums, subject to the pressure under which the work is transferred. The facility with which this communication of rotatory motion may be established or broken at any distance and under almost every variety of circumstance, has brought the band so extensively into use in machinery, that it may be considered as a principal channel through which work is made to flow in its distribution to the successive stages of every process of mechanism, carried on in the same workshop or manufactory.

193. *The sum of the tensions upon the two parts of a band is the same, whatever be the pressure under which the band is*

driven, or the resistance overcome, the tension of the driving part of the band being always increased by just so much as that of the driven part is diminished.

This principle was first given by M. Poncelet; it has since been amply confirmed by the experiments of M. Morin.* It may be proved as follows†: — In the *very commencement* of the motion of that drum to which the *driving* pressure is applied, no motion is communicated by it to the other drum. Before any such motion can be communicated to the latter, a *difference* must be produced between the tensions of the two parts of the band sufficient to overcome the resistance, whatever it may be, which is opposed to the revolution of the *driven* drum. Now, an increase of the tension on the driving side of the band must be followed by an *elongation* of that side of the band (since the band is elastic), and by the revolution of the circumference of the *driving* drum through a space precisely equal to this elongation. Supposing, then, the other, or driven side of the band, to remain extended, as before, in a straight line between its two points of contact with the drums, this portion of the band must evidently have *contracted* by precisely the length through which the circumference of the *driving* drum has revolved, or the driving side of the band *elongated*. Thus, the elongation of the driving side of the band is precisely equal to the contraction of the driven side. Now, the band being supposed perfectly elastic, the increase or diminution of its tension is exactly proportional to the increase or diminution of its length. The increase of tension on the one side, produced by a given elongation, is therefore precisely equal to the diminution of tension produced by a contraction equal to that elongation on the other side. Thus, if T represent the tension upon each side of the band before the driving pressure, whatever it may be, was applied, and if T_1 and T_2 represent the tensions upon the driving and the driven sides of the band after that pressure is applied; then, since $T_1 - T$ re-

---

* Nouvelles Expériences sur le Frottement, &c.   Metz.
† No demonstration appears to have been given of it by M. Poncelet.

presents the increase of tension on the one side, and $T - T_2$ the diminution of tension on the other, $T_1 - T = T - T_2$;

$$\therefore\ T_1 + T_2 = 2T\ \ \ .\ .\ .\ .\ .\ (214).$$

It is a great principle of the economy of power in the use of the band to adjust this initial tension T, so that it may just be sufficient to prevent the band from slipping upon the drum under any pressure which it is required to transmit. The means of making this adjustment will be explained hereafter.

THE MODULUS OF THE BAND.

194. For simplifying the consideration of this important element in machinery, we shall first consider a particular case of its application. Let the two *drums*, whose axes are C_1 and C_2, be supposed *equal* to one another, so that the two parts of the band which pass round them may be parallel. Let,

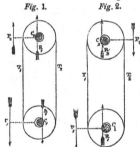

Fig. 1.   *Fig.* 2.   moreover, the centres of the two drums be in the same vertical straight line, so that the two parts of the band may be vertical.

Let P_1 and P_2 be pressures applied, in vertical directions, to turn the drums, and at perpendicular distances from their centres, represented by $C_1 P_1$ and $C_2 P_2$; of which pressures P_2 is the *working* or *driven* pressure, or that which is upon the point of yielding by the preponderance of the other P_1. In *fig.* 1. P_2 is seen applied on the same side of the centre of the drums as P_1, and in *fig.* 2. on the opposite side. Let T_1 and T_2 represent the tensions upon the two parts of the band, T_1 being that on the *driving*, and T_2 that on the *driven* side.

$$a_1 = C_1 P_1,\ \ a_2 = C_2 P_2,$$

$r =$ radius of each drum,

$W =$ weight of each drum,

$\rho =$ radius of axis of each drum,

R_1 and R_2 =resistances of axes of drums,

φ =limiting angle of resistance.

Now, the parallel pressures P_1, W, T_1, T_2, R_1, applied to the lower drum, are in *equilibrium;* therefore (Art. 16.),

$$R_1 = \pm(T_1 + T_2 - P_1 - W);$$

or substituting for $T_1 + T_2$ its value 2T (equation 214),

$$R_1 = \pm(2T - P_1 - W) \quad \ldots \ldots \text{ (215)}.$$

The sign $\pm$ being taken according as 2T is greater or less than $P_1 + W_1$, or according as the axis of the lower drum presses upon the *upper* surface of its bearings, as shown in *fig.* 1., or upon the *lower* surface, as shown in *fig.* 2. In like manner, the pressures P_2, W, T_1, T_2, R_2, applied to the upper drum, being in equilibrium,

$$R_2 = T_1 + T_2 \mp P_2 + W,$$

or (equation 214) $R_2 = 2T \mp P_2 + W \ldots$ (216),

where the sign $\mp$ is to be taken according as P_2 is applied on the same side of the axis as P_1, or on the opposite side.

Since, moreover, R_1 and R_2 act, in the state bordering upon motion, at perpendicular distances from the centre of the axis, which are each represented by $\rho \sin. \phi$ (Art. 153.), we have, by the principle of the equality of moments,

$$\left. \begin{array}{l} P_1 a_1 + T_2 r = T_1 r + R_1 \rho \sin. \phi \\ P_2 a_2 + T_2 r + R_2 \rho \sin. \varphi = T_1 r \end{array} \right\} \quad \ldots \ldots \text{ (217)},$$

observing that the resultant of all the pressures applied to each drum (excepting only the resistance of its axis) must be such as would *alone* communicate motion to it in the direction in which it actually moves, and therefore that the resistance of the axis, which is opposite to this resultant, must tend to communicate motion to the drum in a direction *opposite* to that in which it actually moves.

Subtracting the above equations, and transposing,

$$P_1 a_1 - P_2 a_2 = (R_1 + R_2) \rho \sin. \varphi.$$

Substituting the values of R_1 and R_2 from equations (215) and

(216), we obtain, in the case in which the negative sign of R_1 is to be taken, or in which $2T$ is less than $P_1 + W$, the axis C_1 resting upon the *lower* surface of its collar as shown in *fig. 2.*,

$$P_1 a_1 - P_2 a_2 = (P_1 \mp P_2 + 2W)\rho \sin. \phi\;;$$

and in the case in which the positive sign of R_1 is to be taken, $2T$ being greater than $P_1 + W$, and the axis C_1 pressing against the *upper* surface of its collar, as shown in *fig. 1.*,

$$P_1 a_1 - P_2 a_2 = (4T - P_1 \mp P_2)\rho \sin. \phi.$$

Transposing and reducing, we obtain for the relation between the driving and driven pressures in these two cases respectively,

$$P_1 = P_2\left(\frac{a_2 \mp \rho \sin. \phi}{a_1 - \rho \sin. \phi}\right) + \frac{2W\rho \sin. \phi}{a_1 - \rho \sin. \phi} \quad \cdots \cdots (218),$$

$$P_1 = P_2\left(\frac{a_2 \mp \rho \sin. \phi}{a_1 + \rho \sin. \phi}\right) + \frac{4T\rho \sin. \phi}{a_1 + \rho \sin. \phi} \quad \cdots \cdots (219),$$

and therefore (by equation 121), for the moduli in the two cases,

$$U_1 = U_2\left\{\frac{1 \mp \left(\frac{\rho}{a_2}\right) \sin. \phi}{1 - \left(\frac{\rho}{a_1}\right) \sin. \phi}\right\} + \frac{2S_1 W\rho \sin. \phi}{a_1 - \rho \sin. \phi} \quad \cdots \cdots (220),$$

$$U_1 = U_2\left\{\frac{1 \mp \left(\frac{\rho}{a_2}\right) \sin. \phi}{1 + \left(\frac{\rho}{a_1}\right) \sin. \phi}\right\} + \frac{4S_1 T\rho \sin. \phi}{a_1 + \rho \sin. \phi} \quad \cdots \cdots (221).$$

In all which equations the sign $\mp$ is to be taken according as P_2 is applied on the same side of the line C_1C_2, joining the axis as P_1, or on the opposite side.

195. *To determine the initial tension* T *upon the band, so that it may not slip upon the surface of the drum when subjected to the given resistance opposed to its motion by the work.*

Suppose the maximum resistance which may, during the

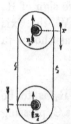

action of the machine, be opposed to the motion of the drum to be represented by a pressure P applied at a given distance a from its centre C_2. Suppose, moreover, that the band has received such an initial tension T as shall just cause it to be on the point of slipping when the motion of the drum is subjected to this maximum resistance; and let t_1 and t_2 be the tensions upon the two parts of the band when it is thus just in the act of slipping and of overcoming the resistance P. Now, the two parts of the band being parallel, it embraces one half of the circumference of each drum; the relation between t_1 and t_2 is therefore expressed (equation 210) by the equation

$t_1 = t_2 \varepsilon^{\pi \tan. \phi}$, whence we obtain $\dfrac{t_1 - t_2}{t_1 + t_2} = \dfrac{\varepsilon^{\pi \tan. \phi} - 1}{\varepsilon^{\pi \tan. \phi} + 1}$. But $t_1 + t_2 = $ 2T (equation 214),

$$\therefore t_1 - t_2 = 2T \left(\frac{\varepsilon^{\pi \tan. \phi} - 1}{\varepsilon^{\pi \tan. \phi} + 1} \right).$$

Also, the relation between the resistance P, opposed to the motion of the upper drum, and the tensions t_1 and t_2 upon the two parts of the band, when this resistance is on the the point of being overcome, is expressed (equation 217) by the equation

$$Pa + t_2 r + R_2 \rho \sin. \phi = t_1 r;$$

or substituting the value of R_2 (equation 216), and transposing,

$$Pa + (2T \mp P + W)\rho \sin. \phi = (t_1 - t_2)r;$$

whence, substituting the value of $t_1 - t_2$, determined above, and transposing, we have

$$P(a \mp \rho \sin. \phi) + W \rho \sin. \phi = 2T \left\{ \left(\frac{\varepsilon^{\pi \tan. \phi} - 1}{\varepsilon^{\pi \tan. \phi} + 1} \right) r - \rho \sin. \phi \right\};$$

$$\therefore T = \tfrac{1}{2}\left\{ \frac{\dfrac{P(a \mp \rho \sin.\ \varphi) + W\rho \sin.\ \varphi}{\pi \tan.\ \varphi}}{\left(\dfrac{\varepsilon - 1}{\varepsilon + 1}\right) r - \sin.\ \varphi} \right\} \quad \ldots \ldots (222).$$

196. *The modulus of the band under its most general form.*

The accompanying figure represents an elastic band passing

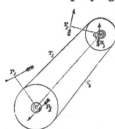

round drums of unequal radii, the line joining whose centres C_1 and C_2 is inclined at any angle to the vertical, and which are acted upon by any given pressures P_1 and P_2, P_1 being supposed to be upon the point of giving motion to the system.

Let T_1 and T_2 represent the tensions upon the two parts of the band, T_1 being that on the driving side.

$a_1\ a_2$ perpendiculars upon the directions of P_1 and P_2 respectively.

$\theta_1\ \theta_2$ the inclinations of the directions of P_1 and P_2 to the line C_1C_2.

$r_1\ r_2$ the radii of the drums.

$W_1\ W_2$ the weights of the drums.

ι the inclination of the line C_1C_2 to the vertical, and α the inclination of the two parts of the band to one another.

$\rho_1\ \rho_2$ the radii of the axes of the drums.

φ the limiting angle of resistance between the axis of the drum and its collar.

$R_1\ R_2$ the resistances of the collars in which the axes of the drums turn in the state bordering upon motion, or the resultants of the pressures upon these axes. The perpendicular distances at which these resistances act from the centres of the axes are (Art. 153.) $\rho_1 \sin.\ \varphi$ and $\rho_2 \sin.\ \varphi$. Since the pressures acting upon the lower drum are T_1, T_2, P_1, W_1, and R_1, and that these pressures are in equilibrium, W_1 acting through the centre of the axis, and T_1 and R_1 acting to turn the drum in one direction about the axis, and

P_1 and T_2 to turn it in the opposite direction; we have, by the principle of the equality of moments (Art. 153.),

$$P_1 a_1 + T_2 r_1 = T_1 r_1 + R_1 \rho_1 \sin. \, \varphi.$$

And since T_1, T_2, P_2, W_2, R_2 are similarly in equilibrium on the upper drum, W_2 acting through the centre, and P_2, R_2, T_2 acting to turn it in one direction, whilst T_1 acts to turn it in the opposite direction,

$$\therefore \quad P_2 a_2 + T_2 r_2 + R_2 \rho_2 \sin. \, \varphi = T_1 r_2;$$

$$\left. \begin{array}{l} \therefore \quad P_1 a_1 - (T_1 - T_2) r_1 = R_1 \rho_1 \sin. \, \varphi \\ \quad\quad P_2 a_2 - (T_1 - T_2) r_2 = -R_2 \rho_2 \sin. \, \varphi \end{array} \right\}.$$

Let $T_1 - T_2 = 2t$, and $T_1 + T_2 = 2T$,

$$\left. \begin{array}{l} \therefore \quad P_1 a_1 - 2t r_1 = R_1 \rho_1 \sin. \, \varphi \\ \quad\quad P_2 a_2 - 2t r_2 = -R_2 \rho_2 \sin. \, \varphi \end{array} \right\} \dots \dots (223).$$

To determine the values of R_1 and R_2 let the pressures applied to each drum be resolved (Art. 11.) in directions parallel and perpendicular to the line $C_1 C_2$; those applied to the lower drum which, being thus resolved, are *parallel* to $C_1 C_2$, are

$$+ T_1 \cos. \, \alpha, \; + T_2 \cos. \, a, \; - P_1 \cos. \, \theta_1, \; - W_1 \cos. \, \iota,$$

those pressures being taken positively which tend to move the axis of the drum from C_1 towards C_2, and those negatively whose tendency is in the opposite direction.

In like manner the pressures resolved perpendicular to $C_1 C_2$ are

$$- T_1 \sin. \, \alpha, \; + T_2 \sin. \, \alpha, \; + P_1 \sin. \, \theta_1, \; - W_1 \sin. \, \iota,$$

those pressures being taken negatively whose tendency when thus resolved perpendicular to $C_1 C_2$ is to bring that line nearer to a vertical direction, and those *positively* whose tendency is in the opposite direction.

Observing that R_1 is the *resultant* of all these pressures, we have (Art. 11.)

$$R^2 = \{(T_1 + T_2) \cos. \, a - P_1 \cos. \, \theta_1 - W_1 \cos. \, \iota\}^2 + \{P_1 \sin. \, \theta_1 - (T_1 - T_2) \sin. \, \alpha - W_1 \sin. \, \iota\}^2.$$

Proceeding similarly in respect to the pressures applied to the upper drum, we shall obtain

$$R_2{}^2 = \{(T_1+T_2)\cos. a - P_2 \cos. \theta_2 + W_2 \cos. \iota\}^2 + \{P_2 \sin. \theta_2 + (T_1-T_2)\sin. a - W_2 \sin. \iota\}^2;$$

or substituting $2T$ for $T_1 + T_2$, and $2t$ for $T_1 - T_2$,

$$\left.\begin{array}{l}R_1{}^2 = \{2T \cos. a - P_1 \cos. \theta_1 - W_1 \cos. \iota\}^2 + \{P_1 \sin. \theta_1 - 2t \sin. a - W_1 \sin. \iota\}^2 \\ R_2{}^2 = \{2T \cos. a - P_2 \cos. \theta_2 + W_2 \cos. \iota\}^2 + \{P_2 \sin. \theta_2 + 2t \sin. a - W_2 \sin. \iota\}^2\end{array}\right\} \ldots (224).$$

By eliminating R_1, R_2, and t between the four equations (223) and (224), a relation is determined between the three quantities P_1, P_2, T. To simplify this elimination let us suppose that the preceding hypothesis in respect to the directions in which the pressures are to be taken *positively* and *negatively* is so made, that the expressions enclosed within the brackets in the above equations (224) and squared may, each of them, represent a *positive* quantity. Let us, moreover, suppose the *first* of the two quantities squared in each equation to be considerably greater than the second, or the pressure upon the axis of each drum in the direction of the line $C_1 C_2$ joining their centres, greatly to exceed the pressure upon it in a direction perpendicular to that line; an hypothesis which will in every practical case be realised. These suppositions being made, we obtain, with a sufficient degree of approximation, by Poncelet's Theorem*,

$$R_1 = \cdot 96\{2T \cos. a - P_1 \cos. \theta_1 - W_1 \cos. \iota\} + \cdot 4\{P_1 \sin. \theta_1 - 2t \sin. a - W_1 \sin. \iota\},$$
$$R_2 = \cdot 96\{2T \cos. a - P_2 \cos. \theta_2 + W_2 \cos. \iota\} + \cdot 4\{P_2 \sin. \theta_2 + 2t \sin. a - W_2 \sin. \iota\}.$$

Substituting these values of R_1 and R_2 in equation (223), and reducing, we have

$$\left.\begin{array}{l}P_1 a_1 - 2t(r_1 - \cdot 4\rho_1 \sin. a \sin. \phi) = \rho_1\{1\cdot 92T \cos. a - P_1\beta_1 - W_1\gamma_1\}\sin. \phi \\ P_2 a^2 - 2t(r_2 - \cdot 4\rho_2 \sin. a \sin. \phi) = -\rho_2\{1\cdot 92T \cos. a - P_2\beta_2 + W_2\gamma_2\}\sin. \phi\end{array}\right\} \cdot \cdot \cdot (225),$$

where $\beta_1 = (\cdot 96 \cos. \theta_1 - \cdot 4 \sin. \theta_1)$,
$\beta_2 = (\cdot 96 \cos. \theta_2 - \cdot 4 \sin. \theta_2)$,
$\gamma_1 = (\cdot 96 \cos. \iota + \cdot 4 \sin. \iota)$,
$\gamma_2 = (\cdot 96 \cos. \iota - \cdot 4 \sin. \iota)$.

* See Appendix.

R

Eliminating t between these equations, and neglecting terms above the first dimension in $\rho_1 \sin. \phi$ and $\rho_2 \sin. \phi$,

$$\left.\begin{array}{l} +P_1 a_1(r_2 - \cdot 4\rho_2 \sin. \alpha \sin. \phi) \\ -P_2 a_2(r_1 - \cdot 4\rho_1 \sin. \alpha \sin. \phi) \end{array}\right\} = \left\{\begin{array}{l} +\rho_1 r_2(1\cdot 92T \cos. \alpha - P_1\beta_1 - W_1\gamma_1) \\ +\rho_2 r_1(1\cdot 92T \cos. \alpha - P_2\beta_2 + W_2\gamma_2) \end{array}\right\} \sin. \phi \ldots (226),$$

α being for the most part exceedingly small, the terms $\cdot 4\rho_1 \sin. \alpha \sin. \phi$ and $\cdot 4\rho_2 \sin. \alpha \sin. \phi$ may be neglected; we shall then obtain by transposition and reduction

$$\left.\begin{array}{l} +P_1 a_1 r_2(1+\dfrac{\rho_1\beta_1}{a_1} \sin. \phi) \\ -P_2 a_2 r_1(1-\dfrac{\rho_2\beta_2}{a_2} \sin. \phi) \end{array}\right\} = \left[\begin{array}{l} +1\cdot 92T(\rho_1 r_2 + \rho_2 r_1) \cos. \alpha \\ -(W_1\rho_1\gamma_1 r_2 - W_2\rho_2\gamma_2 r_1) \end{array}\right] \sin. \phi \ldots (227).$$

If this equation be compared with equation (219), it will be found to agree with it, *mutatis mutandis*, except that the coefficient $1\cdot 92$ is in that equation 2. This difference manifestly results from the *approximate* character of the theorem of Poncelet.

Substituting the latter co-efficient for the former, multiplying both sides of the equation by $(1 - \dfrac{\rho_1\beta_1}{a_1^2} \sin. \phi)$, neglecting terms of more than two dimensions in $\dfrac{\rho_1}{a_1}$, $\dfrac{\rho_2}{a_2}$, and sin. ϕ, and reducing,

$$P_1 = \left(\dfrac{a_2 r_1}{a_1 r_2}\right)\{1-\left(\dfrac{\rho_1\beta_1}{a_1}+\dfrac{\rho_2\beta_2}{a_2}\right) \sin. \phi\}P_2 + \left\{\begin{array}{l} +2T(\rho_1 r_2 + \rho_2 r_1) \cos. \alpha \\ -(W_1\rho_1\gamma_1 r_2 - W_2\rho_2\gamma_2 r_1) \end{array}\right\} \sin. \phi \ldots (228),$$

which is the relation between the moving and working pressures in the state bordering upon motion. From this relation we obtain for the MODULUS of the band (equation 121)

$$U_1 = \{1-\left(\dfrac{\rho_1\beta_1}{a_1}+\dfrac{\rho_2\beta_2}{a_2}\right) \sin. \phi\}U_2 + S_1 \left\{\begin{array}{l} +2T(\rho_1 r_2 + \rho_2 r_1) \cos. \alpha \\ -(W_1\rho_1\gamma_1 r_2 - W_2\rho_2\gamma_2 r_1) \end{array}\right\} \sin. \phi \ldots (229).$$

If the angle θ_2 be conceived to increase until it exceed $\dfrac{\pi}{2}$, P_2 will pass to the opposite side of $C_1 C_2$, and β_2 will become *negative;* whence it is apparent, that equation (229) agrees with equation (219) in other respects, and in the condition of the ambiguous sign. It is moreover apparent, from the form assumed by the modulus in this case and in that of that of the preceding article, *that the greatest economy of power is obtained by applying the moving and the working pressures on the same side of the line* C_1C_2 *joining the axes of the drums.* This is in fact but a particular case of the general principle established in Art. 168.

197. The initial tension T of the band may be determined precisely as in the former case (equation 222). Re-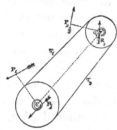presenting by θ the angle subtended by the circumference which the band embraces on the second or driven drum, by P the maximum resistance opposed to its motion at the distance a, by Φ the limiting angle of resistance between the band and the surface of the drum, and by t_1 and t_2 the tensions upon the two parts of the band, when its maximum resistance being opposed, it is upon the point of slipping; observing, moreover, that in this case $2(t_1-t_2)$ or $2t$ is repre-

sented (Art. 195.) by $2T\dfrac{\varepsilon^{\theta\tan.\Phi}-1}{\varepsilon^{\theta\tan.\Phi}+1}$; then substituting in the

second of equations (225) this value for $2t$, and P and a for P_2 and a_2, and neglecting the exceedingly small term which involves the product sin. α sin. φ, we have

$$Pa-2T\left(\frac{\varepsilon^{\theta\tan.\Phi}-1}{\varepsilon^{\theta\tan.\Phi}+1}\right)r_2=-\rho_2\{2T\cos.\alpha-P\beta_2+W_2\gamma_2\}\sin.\varphi.$$

Also, since α represents the inclination of the two parts of the band to one another; since, moreover, these touch the surfaces of the drums, and that θ represents the inclination of the radii drawn from the centre of the lesser drum to the touching points, therefore $\theta=\pi-\alpha$. Substituting this value of θ in the above equation, and solving it in respect to T, we have

$$T = \frac{1}{2}\left\{\frac{P(a - \rho_2\beta_2 \sin. \varphi) + W_2\rho_2\gamma_2 \sin. \varphi}{\left(\dfrac{\varepsilon^{\frac{1}{(\pi-\alpha)\,\tan.\,\Phi}} - 1}{\varepsilon^{\frac{1}{(\pi-\alpha)\,\tan.\,\Phi}} + 1}\right) r_2 - \rho_2 \cos. \alpha \sin. \varphi}\right\} \ \ldots \ldots (230).$$

198. *The modulus of the band, when the two parts of it, which intervene between the drums, are made to cross one another.*

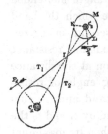

If the directions of the two parts of the band be made to cross, as shown in the accompanying figure, the moving pressure T_1 upon the second drum is applied to it on the side opposite to that on which it is applied when the bands do not cross; so that in this case, in order that the greatest economy of power may be attained (Art. 168.), the working pressure or resistance P_2 should be applied to it on the side opposite to that in which it was applied in the other case, and therefore on the side of the line C_1C_2, opposite to that on which the moving pressure P_1 upon the first drum is applied. This disposition of the moving and working pressures being supposed, and this case being investigated by the same steps as the preceding, we shall arrive at precisely the same expressions (equations 228 and 229) for the relation of the moving and the working pressures, and for the modulus.

In estimating the value of the *initial* tension T (equation 230) it will, however, be found, that the angle θ, subtended at the centre C_2 of the second drum by the arc KML, which is embraced by the band, is no longer in this case represented by $\pi - \alpha$ but by $\pi + \alpha$. This will be evident if we consider that the four angles of the quadrilateral figure C_2KIL being equal to four right angles, and its angles at K and L being right angles, the remaining angles KIL and KC$_2$L are equal to two right angles, so that KC$_2$L $= \pi - \alpha$; but the angle subtended by KML equals $2\pi - $ KC$_2$L; it equals therefore $\pi + \alpha$. If this value be substituted for θ in equation (230) it becomes

$$T = \frac{1}{2} \left\{ \frac{P(a - \rho_2 \beta_2 \sin. \phi) + W_2 \rho_2 \sin. \phi}{\left(\dfrac{\varepsilon^{\frac{\pi}{(\pi+\alpha)\tan. \phi}} - 1}{\varepsilon^{\frac{\pi}{(\pi+\alpha)\tan. \phi}} + 1} \right) r_2 - \rho_2 \cos. \alpha \sin. \phi} \right\} \quad \ldots \quad (231).$$

Now the fraction in the denominator of this expression being essentially greater in value than that in the denominator of the preceding (equation 230), it follows that the initial tension T, which must be given to the band in order that it may transmit the work from the one drum to the other under a given resistance P, is less when the two parts of the band cross than when they do not, and, therefore, that the modulus (equation 229) is less; so that *the band is worked with the greatest economy of power (other things being the same) when the two parts of it which intervene between the drums are made to cross one another.* Indeed it is evident, that since in this case the arc embraced by the band on each drum subtends a greater angle than in the other case, a less tension of the band in this case than in the other is required (Art. 187.) to prevent it from slipping under a given resistance, so that the friction upon the axis of the drums which results from the tension of the band is less in this case than the other, and therefore the work expended on that friction less in the same proportion.

The Teeth of Wheels.

199. Let A, B represent two circles in contact at D, and moveable about fixed centres at C_1 and C_2. It is evident, that if by reason of the friction of these two circles upon one another at D any motion of rotation given to A be communicated to B, the angles PC_1D and QC_2D described in the same time by these two circles will be such as will make the arcs PD and QD which they subtend at the circumfer-

ences of the circles equal to one another. Let the angle
PC_1D* be represented by θ_1, and the angle QC_2D by θ_2;
also let the radii C_1D and C_2D of the circles be represented
by r_1 and r_2. Now, arc $PD = r_1\theta_1$, arc $QD = r_2\theta_2$; and since
$PD = QD$, therefore $r_1\theta_1 = r_2\theta_2$;

$$\therefore \frac{\theta_1}{\theta_2} = \frac{r_2}{r_1} \quad \ldots \ldots \quad (232).$$

The angles described, in any the same time, by two circles
which revolve in contact are therefore inversely proportional
to the radii of the circles, so that their angular velocities
(Art. 74.) bear a constant proportion to one another; and
if one revolves uniformly, then the other revolves uniformly;
if the angular revolution of the one varies in any proportion,
then that of the other varies in like proportion.

When the *resistance* opposed to the rotation of the *driven*
circle or wheel B is considerable, it is no longer possible to
give motion to that circle by the friction on its circumference
of the driving circle. It becomes therefore necessary in the
great majority of cases to cause the rotation of the driven
wheel by some other means than the friction of the circumference of the driving wheel.

One expedient is the band already described, by means of
which the wheels may be made to drive one another at any
distances of their centres, and under a far greater resistance
than they could by their mutual contact. When, however,
the pressure is considerable, and the wheels may be brought,
into actual contact, the common and the more certain method

is to transfer the motion from one
to the other by means of projections
on the one wheel called TEETH,
which engage in similar projections
on the other.

In the construction of these teeth
the problem to be solved is, to give
such shapes to their surfaces of

---

* Or rather the arc which this angle subtends to radius unity.

mutual contact, as that the wheels shall be made to turn by the intervention of their teeth precisely as they would by the friction of their circumferences.

200. That it is *possible* to construct teeth which shall answer this condition may thus be shown.

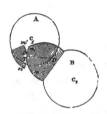

Let *mn* and *m'n'* be two curves, the one described on the plane of the circle A, and the other on the plane of the circle B; and let them be such that as the circle A revolves, carrying round with it the circle B, by their mutual contact at D, these two curves *mn* and *m'n'* may *continually touch one another*, altering of course, as they will do continually, their relative positions and their point of contact T.

It is evident that the two circles would be made to revolve by the contact of teeth whose edges were of the forms of these two curves *mn* and *m'n'* precisely as they would by their friction upon the circumferences of one another at the point D; for in the former case a certain series of points of contact of the circles (infinitely near to one another) at D, brings about another given series of points of contact (infinitely near to one another) of the curves *mn* and *m'n'* at T; and in the latter case the same series of points in the curves *mn* and *m'n'* brought into contact necessarily produces the contact of the same series of points in the two circumferences of the two circles at D.

To construct teeth whose surfaces of contact shall possess the properties here assigned to the curves *mn* and *m'n'* is the problem to be solved. Of the solution of this problem the following is the fundamental principle : —

201. *In order that two circles* A *and* B *may be made to revolve by the contact of the surfaces mn and m'n' of their teeth*

R 4

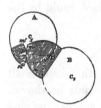

*precisely as they would by the friction of their circumfer-
ences, it is necessary, and it is sufficient, that
a line drawn from the point of contact* T *of
the teeth to the point of contact* D *of the
circumferences should, in every position of
the point* T, *be perpendicular to the surfaces
in contact there,* i. e. *a normal to both the
curves mn and m′n′.*

To prove this principle, we must first establish the following
LEMMA : — If two circles M and N be made to revolve about

the fixed centres E and F by their mutual
contact at L, and if the planes of these
circles be conceived to be carried round with
them in this revolution, and a point P on the
plane of M to trace out a curve PQ on the
plane of N whilst thus revolving, then is
this curved line PQ precisely the same as would have been
described on the plane of N by the same point P, if the
latter plane, instead of revolving, had remained at rest, and
the centre E of the circle M having been released from its
axis, that circle had been made to roll (carrying its plane
with it) on the circumference of N. For conceive O to represent a third plane on which the centres of E and F are
fixed. It is evident that if, whilst the circles M and N are
revolving by their mutual contact, the plane O, to which
their centres are both fixed, be in any way moved, no change
will thereby be produced in form of the curve PQ, which the
point P in the plane of M is describing upon the plane of N,
such a motion being *common* to both the planes M and N.*

Now let the direction in which the circle N is revolving be
that shown by the arrow, and its angular velocity uniform ;
and conceive the plane O to be made to revolve about F with

* Thus, for instance, if the circles M and N continue to revolve, we
may evidently place the whole machine in a ship under sail, in a moving
carriage, or upon a revolving wheel, without in the least altering the form
of the curve, which the point P, revolving with the plane of the circle M, is
made to trace on the plane of N, because the motion we have communicated is common to both these circles.

an angular velocity (Art. 74.) which is equal to that of N, but

in an opposite direction, communicating this angular velocity to M and N, these revolving meantime in respect to one another, and by their mutual contact, precisely as they did before.*

It is clear that the circle N being carried round by its own proper motion in one direction, and by the motion common to it and the plane O with the *same angular velocity* in the opposite direction, will, in reality, *rest* in space; whilst the centre E of the circle M, having no motion proper to itself, will revolve with the angular velocity of the plane O, and the various other points in that circle with angular velocities, compounded of their proper velocities, and those which they receive in common with the plane O, these velocities neutralising one another at the point L of the circle, by which point it is in contact with the circle N. So that whilst M revolves round N, the point L, by which the former circle at any time touches the other, is at rest; this quiescent point of the circle M nevertheless continually varying its position on the circumferences of both circles, and the circle M being in fact made to *roll* on the circle N at rest.

Thus then it appears, that by communicating a certain common angular velocity to both the circles M and N about the centre F, the former circle is made to *roll* upon the other at rest; and moreover, that this common angular velocity does not alter the form of the curve PQ, which a point P in the plane of the one circle is made to trace upon the plane of the other, or, in other words, that the curve traced under these circumstances is the same, whether the circles revolve round fixed centres by their mutual contact, or whether the centre of one circle be released, and it be made to roll upon the circumference of the other at rest.

This lemma being established, the truth of the proposition stated at the head of this article becomes evident; for if M

---

* M and N may be imagined to be placed upon a horizontal wheel O, first at rest, and then made to revolve *backwards* in respect to the motion of N.

roll on the circumference of N, it is evident that P will, at any instant, be describing a circle about their point of contact L.*

Since then P is describing, at every instant, a circle about L when M rolls upon N, N being fixed, and since the curve described by P upon this supposition is precisely the same as would have been traced by it if the centres of both circles had been fixed, and they had turned by their mutual contact, it follows that in this last case (when the circles revolve about fixed centres by their mutual contact) the point P is at any instant of the revolution describing, during that instant, an exceedingly small circular arc about the point L; whence it follows that PL is always a *perpendicular* to the curve PQ at the point P, or a *normal* to it.

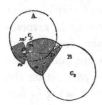

Now let *p* be a point exceedingly near to T in the curve *m′n′*, which curve is fixed upon the plane of the circle A. It is evident that, as the point *p* passes through its contact with the curve *mn* at T (see Art. 200.), it will be made to describe, on the plane of the circle B, an exceedingly small portion of that curve *mn*.

But the curve which it is (under these circumstances) at any instant describing upon the plane of B has been shown to be always perpendicular to the line DT; the curve *mn* is therefore at the point T perpendicular to the line DT; whence it follows that the curve *m′n′* is also perpendicular to that line, and *that* DT *is a normal to both those curves at* T. This is the characteristic property of the curves *mn* and *m′n′*, so that they may satisfy the condition of a continual contact with one another, whilst the circles revolve by the contact of their circumferences at D, and therefore conversely, so that these curves may, by their mutual contact, give to the circles the same motion as they would receive from the contact of their circumferences.

* For either circle may be imagined to be a polygon of an infinite number of sides, on one of the angles of which tne rolling circle will, at any instant, be in the act of turning.

202. *To describe, by means of circular arcs, the form of a tooth on one wheel which shall work truly with a tooth of any given form on another wheel.*

Let the wheels be required to revolve by the action of their

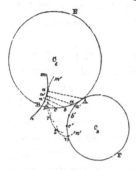

teeth, as they would by the contact of the circles ABE and ADF, called their *primitive* or *pitch* circles. Let AB represent an arc of the pitch circle ABE, included between any two *similar points* A and B of consecutive teeth, and let AD represent an arc of the pitch circle ADF equal to the arc AB, so that the points D and B may come simultaneously to A, when the circles are made to revolve by their mutual contact. AB and AD are called the *pitches* of the teeth of the two wheels. Divide each of these pitches into the same number of equal parts in the points a, b, &c., a', b', &c.; the points a and a', b and b', &c. will then be brought *simultaneously* to the point A. Let mn represent the form of the face of a tooth on the wheel, whose centre is C_1, with which tooth a corresponding tooth on the other wheel is to work truly; that is to say, the tooth on the other wheel, whose centre is C_2, is to be cut, so that, driving the surface mn, or being driven by it, the wheels shall revolve precisely as they would by the contact of their pitch circles ABE and ADF at A. From A measure the *least distance* Aα to the curve mn, and with radius Aα and centre A describe a circular arc $\alpha\beta$ on the plane of the circle whose centre is C_2. From a measure, in like manner, the least distance $a\alpha'$ to the curve mn, and with this distance $a\alpha'$ and the centre a' describe a circular arc $\beta\gamma$, intersecting the arc $\alpha\beta$ in β. From the point b measure similarly the shortest distance $b\alpha''$ to mn, and with the centre b' and this distance $b\alpha''$ describe a circular arc $\gamma\delta$, intersecting $\beta\gamma$ in γ, and so with the other points of division.

A curve touching these circular arcs $\alpha\beta$, $\beta\gamma$, $\gamma\delta$, &c. will give the true surface or boundary of the tooth.*

In order to prove this let it be observed, that the shortest distance $a\alpha'$ from a given point a to a given curve mn is a normal to the curve at the point α' in which it meets it ; and therefore, that if a circle be struck from this point a with this least distance as a radius, then this circle must touch the curve in the point α', and the curve and circle have a common normal in that point.

Now the points a and α' will be brought by the revolution of the pitch circles simultaneously to the point of contact A, and the least distance of the curve mn from the point A will then be $a\alpha'$, so that the arc $\beta\gamma$ will then be an arc struck from the centre A, with this last distance for its radius. This circular arc $\beta\gamma$ will therefore *touch* the curve mn in the point α' and the line $a\alpha'$, which will then be a line drawn from the point of contact A of the two pitch circles to the point of contact α' of the two curves mn and $m'n'$, will also be a normal to both curves at that point. The circles will therefore at that instant drive one another (Art. 201.) by the contact of the surfaces mn and $m'n'$, precisely as they would by the contact of their circumferences. And as every circular arc of the curve $m'n'$ similar to $\beta\gamma$ becomes in its turn the acting surface of the tooth, it will, in like manner, at *one point* work truly with a corresponding point of mn, so that the circles will thus drive one another *truly* at as many points of the surfaces of their teeth, as there have been taken points of division a, b, &c. and arcs $\alpha\beta$, $\beta\gamma$, &c.†

* This method of describing, geometrically, the forms of teeth is given, without demonstration, by M. Poncelet in his *Mécanique Industrielle*, 3me partie, Art. 60.

† The greater the number of these points of division, the more accurate the form of the tooth. It appears, however, to be sufficient, in most cases, to take three points of division, or even two, where no great accuracy is required. M. Poncelet (*Méc. Indust.* 3me partie, Art. 60.) has given the following, yet easier, method by which the true form of tooth may be *approximated* to with sufficient accuracy in most cases. Suppose the *given* tooth N upon the one wheel to be placed in the position in which

INVOLUTE TEETH.

203. *The teeth of two wheels will work truly together if they be bounded by curves of the form traced out by the extremity of a flexible line, unwinding from the circumference of a circle, and called the involute of a circle, provided that the circles of which these are the involutes be concentric with the pitch circles of the wheels, and have their radii in the same proportion with the radii of the pitch circles.*

Let OE and OF represent the pitch circles of two wheels,

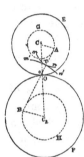

AG and BH two circles concentric with them and having their radii C_1A and C_2B in the same proportion with the radii C_1O and C_2O of the pitch circles. Also let mn and $m'n'$ represent the edges of teeth on the two wheels struck by the extremities of flexible lines unwinding from the circumferences of the circles AG and BH respectively. Let these teeth be in contact, in any position of the wheels, in the point T, and from the point T draw TA and TB tangents to the generating circles GA and BH in the points A and B. Then does AT evidently represent the position of the flexible line when its extremity was in the act of generating the point T in the curve mn; whence it follows, that AT is a *normal* to the curve mn at the point T *; and in like manner that BT is

---

it is first to engage or disengage from the *required* tooth on the other

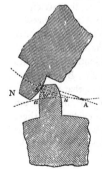

wheel, and let Aa and Ab be equal arcs of the pitch circles of the two wheels whose point of contact is A. Draw Aa the shortest distance between A and the face of the tooth N; join aa; bisect that line in m, and draw mn perpendicular to aa intersecting the circumference Aa in n. If from the centre n a circular arc be described passing through the points a and a, it will give the required form of the tooth nearly.

* For if the circle be conceived a polygon of an infinite number of sides, it is evident that the line, when in the act of unwinding from it at A, is

a normal to the curve $m'n'$ at the same point T. Now the two curves have a common tangent at T ;

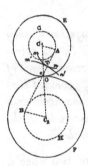

therefore their normals TA and TB at that point are in the same straight line, being both perpendicular to their tangent there. Since then ATB is a straight line, and that the vertical angles at the point o where AB and C_1C_2 intersect are equal, as also the right angles at A and B, it follows that the triangles AoC_1 and BoC_2 are similar, and that $C_1o : C_2o :: C_1A : C_2B$. But $C_1A : C_2B :: C_1O : C_2O$; $\therefore C_1o : C_2o :: C_1O : C_2O$; therefore the points O and o coincide, and the straight line AB, which passes through the point of contact T of the two teeth, and is perpendicular to the surfaces of both at that point, passes also through the point of contact O of the pitch circles of the wheels. Now this is true, whatever be the positions of the wheels, and whatever, therefore, be the points of contact of the teeth. Thus then the condition established in Art. 201. as that necessary and sufficient to the true action of the teeth of wheels, viz. " that a line drawn from the point of contact of the pitch circles to the point of contact of the teeth should be a normal to their surfaces at that point, in all the different positions of the teeth," obtains in regard to involute teeth. *

---

turning upon one of the angles of that polygon, and therefore that its extremity is, through an infinitely small angle, describing a circular arc about that point.

* The author proposes the following illustration of the action of involute teeth, which he believes to be new. Conceive AB to represent a *band* passing round the circles AG and BH, the wheels would evidently be driven by this band precisely as they would by the contact of their pitch circles, since the radii of AG and BH are to one another as the radii of the pitch circles. Conceive, moreover, that the circles BH and AG carry round with them their *planes* as they revolve, and that a tracer is fixed at any point T of the band, tracing, at the same time, lines mn and $m'n'$, upon both planes, as they revolve beneath it. It is evident that these curves, being traced by the same point, must be in contact in all positions of the circles when driven by the band, and therefore when driven by

The point of contact T of the teeth *moves along* the straight line AB, which is drawn touching the generating circles BH and AG of the involutes; this line is what is called the *locus* of the different points of contact. Moreover, this property obtains, whatever may be the number of teeth in contact at once, so that all the points of contact of the teeth, if there be more than one tooth in contact at once, lie always in this line; which is a characteristic, and a most

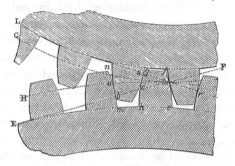

important property of teeth of the involute form. Thus in the above figure, which represents part of two wheels with involute teeth, it will be seen that the points *r s* of contact of the teeth are in the same straight line touching the base * of one of the involutes, and passing through the point of contact A of the pitch circles, as also the points A and *b* in that touching the base of the other.

EPICYCLOIDAL AND HYPOCYCLOIDAL TEETH.

204. If one circle be made to roll externally on the cir-

---

their mutual contact. The wheels would therefore be driven by the contact of teeth of the forms *mn* and *m′n′* thus traced by the point T of the band precisely as they would by the contact of their pitch circles. Now it is easily seen, that the curves *mn* and *m′n′*, thus described by the point T of the band, are *involutes* of the circles AG and BH.

* The circles from which the involutes are described are called their *bases*. This cut and that at page 257. are copied from Mr. Hawkins' edition of Camus on the Teeth of Wheels.

cumference of another, and if, whilst this motion is taking
place, a point in the circumference of the rolling circle be

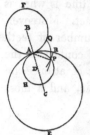

made to trace out a curve upon the plane of
the fixed circle, the curve so generated is
called an EPICYCLOID, the rolling circle being
called the *generating* circle of the epicycloid, and the circle upon which it rolls, its
base.

If the generating circle, instead of rolling
on the outside or convex circumference of
its base, roll on its inside or concave circumference, the curve generated is called the HYPOCYCLOID.

Let PQ and PR be respectively an epicycloid and a hypocycloid, having the same generating circle APH, and having
for their bases the pitch circles AF and AE of two wheels. If
teeth be cut upon these wheels, whose edges coincide with
the curves PQ and PR, they will work truly with one
another; for let them be in contact at P, and let their common generating circle APH be placed so as to touch the
pitch circles of both wheels at A, then will its circumference
evidently pass through the point of contact P of the teeth:
for if it be made to roll through an exceedingly small angle
upon the point A, rolling there upon the circumference of
both circles, its generating point will traverse exceedingly
small portions of both curves; since then a given point in the
circumference of the circle APH is thus shown to be at one
and the same time in the perimeters of both the curves PQ and
PR, that point must of necessity be the point of contact P of
the curves; since, moreover, when the circle APH rolls upon
the point A, its generating point *traverses* a small portion of
the perimeter of each of the curves PQ and PR at P, it
follows that the line AP is a normal to both curves at that
point; for whilst the circle APH is rolling through an exceedingly small angle upon A, the point P in it, is describing
a circle about that point whose radius is AP.* Teeth,

---

* The circle APH may be considered a polygon of an infinite number
of sides, on one of the angles of which polygon it may at any instant
be conceived to be turning.

therefore, whose edges are of the forms PQ and PR satisfy the condition that the line AP drawn from the point of contact of the pitch circles to any point of contact of the teeth is a normal to the surfaces of both at that point, which condition has been shown (Art. 201.) to be that necessary and sufficient to the correct working of the teeth.*

Thus then it appears, that if an *epicycloid* be described on the plane of one of the wheels with any generating circle, and with the pitch circle of that wheel for its base ; and if a hypocycloid be described on the plane of the other wheel with the pitch circle of *that* wheel for its base ; and if the

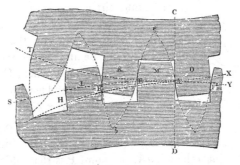

faces or acting surfaces of the teeth on the two wheels be cut so as to coincide with this epicycloid and this hypocycloid

---

* The entire demonstration by which it has been here shown that the curves generated by a point in the circumference of a given generating circle APH rolling upon the convex circumference of one of the pitch circles, and upon the concave circumference of the other are proper to form the edges of contact of the teeth, is evidently applicable if any other generating curve be substituted for APH. It may be shown precisely in the same manner, that the curves PQ and PR generated by the rolling of any such curve (not being a circle) upon the pitch circles, possess this property, that the line PA drawn from any point of their contact to the point of contact of their pitch circles is a *normal* to both, which property is necessary and sufficient to their correct action as teeth.   This was first demonstrated as a general principle of the construction of the teeth of wheels by Mr. Airy, in the Cambridge Phil. Trans. vol. ii. He has farther shown, that a tooth of any form whatever being cut upon a wheel, it is possible to find a curve which, rolling upon the pitch circle of that wheel, shall by a certain generating point traverse the edge of the given tooth.

respectively, then will the wheels be driven correctly by the intervention of these teeth. Parts of two wheels having epicycloidal teeth are represented in the preceding figure.

205. LEMMA. — *If the diameter of the generating circle of a hypocycloid equal the radius of its base, the hypocycloid becomes a straight line having the direction of a radius of its base.*

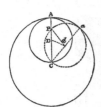 Let D and d represent two positions of the centre of such a generating circle, and suppose the generating point to have been at A in the first position, and join AC; then will the generating point be at P in the second position, *i. e.* at the point where CA intersects the circle in its second position; for join Ca and Pd, then $\angle \mathrm{P}da = \angle \mathrm{PC}d + \angle \mathrm{CP}d$ $= 2\mathrm{AC}a$. Also $2da = \mathrm{CA}$; $\therefore 2\overline{da} \times \overline{\mathrm{P}da} = 2\overline{\mathrm{CA}} \times \overline{\mathrm{AC}a}$; $\therefore \overline{da} \times \overline{\mathrm{P}da} = \overline{\mathrm{CA}} \times \overline{\mathrm{AC}a}$; $\therefore$ arc A$a =$ arc Pa. Since then the arc aP equals the arc aA, the point P is that which in the first position coincided with A, *i. e.* P is the generating point; and this is true for all positions of the generating circle; the generating point is therefore always in the straight line AC. The edge, therefore, of a hypocycloidal tooth, the diameter of whose generating circle equals half the diameter of the pitch circle of its wheel, is a straight line whose direction is towards the centre of the wheel.*

The curve thus found being made to roll on the circumference of the pitch circle of a second wheel, will therefore trace out the form of a tooth which will work truly with the first. This beautiful property involves the theoretical solution of the problem which Poncelet has solved by the geometrical construction given in Article 202. If the rolling curve be a logarithmic spiral, the involute form of tooth will be generated.

* The following very ingenious application has been made of this property of the hypocycloid to convert a circular into an alternate rectilinear

To set out the Teeth of Wheels.

206. All the teeth of the same wheel are constructed of the same form and of equal dimensions: it would, indeed, evidently be impossible to construct two wheels with different numbers of teeth, which should work truly with one another, if all the teeth on each wheel were not thus alike.

motion. AB represents a ring of metal, fixed in position, and having teeth cut upon its concave circumference. C is

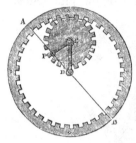

the centre of a wheel, having teeth cut in its circumference to work with those upon the circumference of the ring, and having the diameter of its pitch circle equal to half that of the pitch circle of the teeth of the ring. This being the case, it is evident, that if the pitch circle of the wheel C were made to roll upon that of the ring, any point in its circumference would describe a straight line passing through the centre D of the ring; but the circle C would roll upon the ring by the mutual action of their teeth as it would by the contact of their pitch circles; if the circle C then be made to roll upon the ring by the intervention of teeth cut upon both, any point in the circumference of C will describe a straight line passing through D. Now, conceive C to be thus made to roll round the ring by means of a double or forked link CD, between the two branches of which the wheel is received, being perforated at their extremities by circular apertures, which serve as bearings to the solid axis of the wheel. At its other extremity D, this forked link is rigidly connected with an axis passing through the centre of the ring, to which axis is communicated the circular motion to be converted by the instrument into an alternating rectilinear motion. This circular motion will thus be made to carry the centre C of the wheel round the point D, and, at the same time, cause it to roll upon the circumference of the ring. Now, conceive the axis C of the wheel, which forms part of the wheel itself, to be *prolonged* beyond the collar in which it turns, and to have rigidly fixed upon its extremity a bar CP. It is evident that a point P in this bar, whose distance from the axis C of the wheel equals the radius of its pitch circle, will move precisely as a point in the pitch circle of the wheel moves, and therefore that it will describe continually a straight line passing through the centre D of the ring. This point P receives, therefore, the alternating rectilinear motion which it was required to communicate.

All the teeth of a wheel are therefore set out by the workman from the same pattern or model, and it is in determining
the form and dimensions of this single pattern or model of
one or more teeth in reference to the mechanical effects which
the wheel is to produce, when all its teeth are cut out upon
it and it receives its proper place in the mechanical combination of which it is to form a part, that consists the art of
the description of the teeth of wheels.

The mechanical function usually assigned to toothed wheels
is the transmission of work under an increased or diminished
velocity.   If CD, DE, &c. represent arcs of the pitch circle

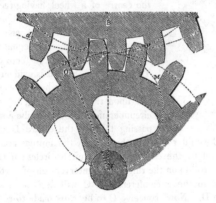

of a wheel intercepted between similar points of consecutive teeth (the chords of which arcs are called the *pitches*
of the teeth), it is evident that all these arcs must be equal,
since the teeth are all equal and similarly placed; so that
each tooth of either wheel, as it passes through its contact
with a corresponding tooth of the other, carries its pitch line
through the same space CD, over the point of contact C of the
pitch lines.   Since, therefore, the pitch line of the one wheel
is carried over a space equal to CD, and that of the other
over a space equal to cd by the contact of any two of their
teeth, and since the wheels revolve by the contact of their
teeth as they would by the contact of their pitch circles at C,
it follows that the arcs CD and cd are equal.   Now let r_1 and
r_2 represent the radii of the pitch circles of the two wheels,

then will $2\pi r_1$ and $2\pi r_2$ represent the circumferences of their pitch circles; and if n_1 and n_2 represent the numbers of teeth cut on them respectively, then $\mathrm{CD} = \dfrac{2\pi r_1}{n_1}$ and $cd = \dfrac{2\pi r_2}{n_2}$, therefore $\dfrac{2\pi r_1}{n_1} = \dfrac{2\pi r_2}{n_2}$;

$$\therefore \frac{r_1}{r_2} = \frac{n_1}{n_2} \ \cdots \cdots (232);$$

Therefore the radii of the pitch circles of the two wheels must be to one another as the numbers of teeth to be cut upon them respectively.

Again, let m_1 represent the number of revolutions made by the first wheel, whilst m_2 revolutions are made by the second; then will $2\pi r_1 m_1$ represent the space described by the circumference of the pitch circle of the first wheel while these revolutions are made, and $2\pi r_2 m_2$ that described by the circumference of the pitch circle of the second; but the wheels revolve as though their pitch circles were in contact, therefore the circumferences of these circles revolve through equal spaces, therefore $2\pi r_1 m_1 = 2\pi r_2 m_2$;

$$\therefore \frac{r_1}{r_2} = \frac{m_2}{m_1} \ \cdots \cdots (233).$$

The radii of the pitch circles of the wheels are therefore inversely as the numbers of revolutions made in the same time by them.

Equating the second members of equations (232) and (233),

$$\frac{m_2}{m_1} = \frac{n_1}{n_2} \ \cdots \cdots (234).$$

The numbers of revolutions made by the wheels in the same time are therefore to one another inversely as the numbers of teeth.

207. *In a train of wheels, to determine how many revolutions the last wheel makes whilst the first is making any given number of revolutions.*

When a wheel, driven 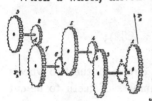 by another, carries its axis round with it, on which axis a *third* wheel is fixed, engaging with and giving motion to a *fourth*, which, in like manner, is fixed upon its axis, and carries round with it a *fifth* wheel fixed upon the same axis, which fifth wheel engages with a sixth upon another axis, and so on as shown in the above figure, the combination forms a *train* of wheels. Let n_1, n_2, n_3 ... n_{2p} represent the numbers of teeth in the successive wheels forming such a train of p pairs of wheels; and whilst the first wheel is making m revolutions, let the second and third (which revolve together, being fixed on the same axis) make m_1 revolutions; the fourth and fifth (which, in like manner, revolve together) m_2 revolutions, the sixth and seventh m_3, and so on; and let the last or $2p^{th}$ wheel thus be made to revolve m_p times whilst the first revolves m times. Then, since the first wheel which has n_1 teeth gives motion to the second which has n_2 teeth, and that whilst the former makes m revolutions the latter makes m_1 revolutions, therefore (equation 234), $\dfrac{m_1}{m} = \dfrac{n_1}{n_2}$; and since, while the third wheel (which revolves with the second) makes m_1 revolutions, the fourth makes m_2 revolutions; therefore $\dfrac{m_2}{m_1} = \dfrac{n_3}{n_4}$. Similarly, since while the fifth wheel, which has n_5 teeth, makes m_2 revolutions (revolving with the fourth), the sixth, which has n_6 teeth, makes m_3 revolutions; therefore $\dfrac{m_3}{m_2} = \dfrac{n_5}{n_6}$. In like manner $\dfrac{m_4}{m_3} = \dfrac{n_7}{n_8}$, &c. &c.

$\dfrac{m_p}{m_{p-1}} = \dfrac{n_{2p-1}}{n_{2p}}$. Multiplying these equations together, and striking out factors common to the numerator and denomi-

nator of the first member of the equation which results from
their multiplication, we obtain

$$\frac{m_p}{m} = \frac{n_1 \cdot n_3 \cdot n_5 \cdot \ldots \cdot n_{2p-1}}{n_2 \cdot n_4 \cdot n_6 \cdot \ldots \cdot n_{2p}} \ldots \ldots (235).$$

The factors in the numerator of this fraction represent the
numbers of teeth in all the *driving* wheels of this train,
and those in the denominator the numbers of teeth in the
driven wheels, or *followers* as they are more commonly
called.

If the numbers of teeth in the former be all equal and
represented by n_1, and the numbers of teeth in the latter
also equal and represented by n_2, then

$$\frac{m_p}{m} = \left(\frac{n_1}{n_2}\right)^p \cdot \cdot \cdot \cdot \cdot \cdot (236).$$

Having determined what should be the number of teeth
in each of the wheels which enter into any mechanical
combination, with a reference to that particular modification
of the velocity of the revolving parts of the machine which
is to be produced by that wheel*, it remains next to consider,
what must be the dimensions of each tooth of the wheel, so
that it may be of sufficient strength to transmit the work
which is destined to pass through it, under that velocity, or
to bear the pressure which accompanies the transmission of
that work at that particular velocity; and it remains further
to determine, what must be the dimensions of the wheel
itself consequent upon these dimensions of each tooth, and
this given number of its teeth.

208. *To determine the pitch of the teeth of a wheel, knowing
the work to be transmitted by the wheel.*

Let U represent the number of units of work to be transmitted by the wheel per minute, m the number of revolutions

---

* The reader is referred for a more complete discussion of this subject
(which belongs more particularly to descriptive mechanics) to Professor
Willis's Principles of Mechanism, chap. vii., or to Camus on the Teeth of
Wheels, by Hawkins, p. 90.

to be made by it per minute, n the number of the teeth to be cut in it, T the *pitch* of each tooth in feet, P the pressure upon each tooth in pounds.

Therefore nT represents the circumference of the pitch circle of the wheel, and mnT represents the space in feet described by it per minute. Now U represents the work transmitted by it through this space per minute, therefore $\frac{U}{mnT}$ represents the *mean* pressure under which this work is transmitted (Art. 50.);

$$\therefore P = \frac{U}{mnT} \dots \dots (237).$$

The pitch T of the teeth would evidently equal twice the breadth of each tooth, if the spaces between the teeth were equal in width to the teeth. In order that the teeth of wheels which act together may engage with one another and extricate themselves, with facility, it is however necessary that the pitch should exceed twice the breadth of the tooth by a quantity which varies according to the accuracy of the construction of the wheel from $\frac{1}{10}$th to $\frac{1}{15}$th of the breadth.*

Since the pitch T of the tooth is dependant upon its breadth, and that the breadth of the tooth is dependant, by the theory of the strength of materials, upon the pressure P which it sustains, it is evident that the quantity P in the above equation is a function of T. This function† may be assumed of the form

$$T = c\sqrt{P} \dots \dots (238);$$

where c is a constant dependant for its amount upon the nature of the material out of which the tooth is formed. Eliminating P between this equation and the last, and solving in respect to T,

$$T = \sqrt[3]{\frac{c^2 U}{mn}}.$$

* For a full discussion of this subject see Professor Willis's Principles of Mechanism, Arts. 107—112.

† See Appendix, on the dimensions of wheels.

The number of units of work transmitted by any machine per minute is usually represented in *horses' power*, one horse's power being estimated at 33,000 units, so that the number of horses' power transmitted by the machine means the number of times 33,000 units of work are transmitted by it every minute, or the number of times 33,000 must be taken to equal the number of units of work transmitted by it every minute. If therefore H represent the number of horses' power transmitted by the wheel, then U $=$ 33,000H. Substituting this value in the preceding equation, and representing the constant $33,000c^2$ by C^3, we have

$$T = C \sqrt[3]{\frac{H}{mn}} \ \ldots \ldots (239).$$

The values of the constant C for teeth of different materials are given in the Appendix.

209. *To determine the radius of the pitch circle of a wheel which shall contain n teeth of a given pitch.*

Let AB represent the pitch T of a tooth, and let it be supposed to coincide with its chord AMB. Let R represent the radius AC of the pitch circle, and n the number of teeth to be cut upon the wheel.

Now there are as many pitches in the circumference as teeth, therefore the angle ABC subtended by each pitch is represented by $\frac{2\pi}{n}$.

Also $T = 2\overline{AM} = 2\overline{AC} \sin. \tfrac{1}{2}ACB = 2R \sin. \dfrac{\pi}{n}$;

$$\therefore \ R = \tfrac{1}{2}T \cosec.\frac{\pi}{n} \ \ldots \ldots (240).$$

210. *To make the pattern of an epicycloidal tooth.*

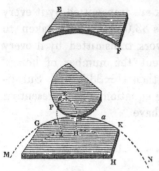

Having determined as above, the pitch of the teeth, and the radius of the pitch circle, strike an arc of the pitch circle on a thin piece of oak board or metal plate, and, with a fine saw, cut the board through along the circumference of this circle, so as to divide it into two parts, one having a *convex* and the other a corresponding *concave* circular edge. Let EF represent one of these portions of the board, and GH another.

Describe an arc of the pitch circle upon a second board or plate from which the pattern is to be cut. Let MN represent this arc. Fix the piece GH upon this board, so that its circular edge may accurately coincide with the circumference of the arc MN. Take, then, a circular plate D of wood or metal, of the dimensions which it is proposed to give to the generating circle of the epicycloid; and let a small point of steel P be fixed in it, so that this point may project slightly from its inferior surface, and accurately coincide with its circumference. Having set off the width AB of the tooth, so that twice this width increased by from $\frac{1}{10}$th to $\frac{1}{15}$th of that width (according to the accuracy of workmanship to be attained) may equal the pitch, cause the circle D to roll upon the convex edge GK of the board GH, pressing it, at the same time, slightly upon the surface of the board on which the arc MN is described, and from which the pattern is to be cut, having caused the steel point in its circumference first of all to coincide with the point A; an epicycloidal arc AP will thus be described by the point P upon the surface MN. Describe, similarly, an epicycloidal arc BE through the point B, and let them meet in E.

Let the board GH now be removed, and let EF be applied and fixed, so that its concave edge may accurately coincide with the circular arc MN. With the same circular plate D pressed upon the concave edge of EF, and made to roll upon it, cause the point in its circumference to describe in like manner, upon the surface of the board from which the pattern is to be cut, a hypocycloidal arc BH passing through the point B, and another AI passing through the point A. HEI will then represent the form of a tooth, which will work correctly (Art. 204.) with the teeth *similarly cut* upon any other wheel; provided that the pitch of the teeth so cut upon the other wheel be equal to the pitch of the teeth upon this, and *provided that the same generating circle D be used to strike the curves upon the two wheels.*

211. *To determine the proper lengths of epicycloidal teeth.*

The general forms of the teeth of wheels being determined by the method explained in the preceding article, it remains to cut them off of such lengths as may cause them successively to take up the work from one another, and transmit it under the circumstances most favourable to the economy of its transmission, and to the durability of the teeth.

In respect to the economy of the power in its transmission, it is customary, for reasons to be assigned hereafter, to provide that no tooth of the one wheel should come into action with a tooth of the other until both are in the act of passing through the line of centres. This condition may be satisfied in all cases where the numbers of teeth on neither of the wheels is exceedingly small, by properly adjusting the lengths of the teeth. Let two of the teeth of the wheels be in contact at the point A in the line CD, joining the centres of the two wheels; and let the wheel whose centre is C be the driving wheel. Let AH be a portion of the circumference of the generating circle of the teeth, then will the points A and B, where this circle intersects the edges of the teeth O and K of the driving wheel, be points of contact with the edges of the teeth M and L of the driven wheel (Art. 204.). Now, since each tooth is to come into action only when it comes into the line of centres, it is clear that the tooth L must have been driven by K from the time when their con-

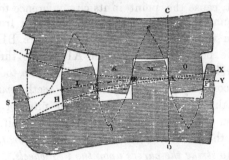

tact was in the line of centres, until they have come into the position shown in the figure, when the point of contact of the anterior face of the next tooth O of the driving wheel with the flank * of the next tooth M of the driven wheel has just passed *into* the line of centres; and since the tooth O is now to take up the task of impelling the driven wheel, and the tooth K to yield it, all that portion of the last-mentioned tooth which lies beyond the point B may evidently be removed; and if it *be* thus removed, then the tooth K, passing out of contact, will manifestly, at that period of the motion, yield all the driving strain to the tooth O, as it is required to do. In order to cut the pattern tooth of the proper length, so as to satisfy the proposed condition, we

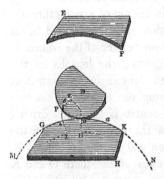

have only then to take A*a* (see the accompanying figure) equal to the pitch of the tooth, and to bring the convex circumference of the generating circle, so as to touch the convex circumference of the arc MN in that point *a*; the point of intersection *e* of this circle with the face AE of the tooth will be the last *acting* point of the

---

* That portion of the edge of the tooth which is *without* the pitch circle is called its *face*, that *within* it its *flank*.

tooth ; and if a circle be struck from the centre of the pitch circle passing through that point, all that portion of the tooth which lies beyond this circle may be cut off.*

The length of the tooth on the wheel intended to act with this, may be determined in like manner.

212. In the preceding article we have supposed the same generating circle to be used in striking, the entire surfaces of the teeth on both wheels. It is not however necessary to the correct working of the teeth, that the same circle should thus be used in striking the *entire* surfaces of two teeth which act together, but only that the generating circle of every two portions of the two teeth which come into actual contact should be the same. Thus the *flank* of the driving tooth and the *face* of the driven tooth being in contact at

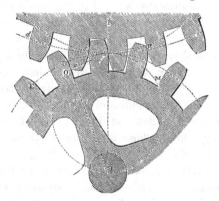

P in the accompanying figure†, this face of the one tooth and flank of the other must be respectively an epicycloid and a hypocycloid struck with the same generating circle. Again, the *face* of a driving tooth and the *flank* of a driven tooth being in contact at Q, these, too, must be struck by the

---

* The point *e* thus determined will, in some cases, fall beyond the extremity E of the tooth. In such cases it is therefore impossible to cut the tooth of such a' length as to satisfy the required conditions, viz. that it shall drive only after it has passed the line of centres. A full discussion of these impossible cases will be found in Professor Willis's work (Arts. 102—104.).

† The upper wheel is here supposed to drive the lower.

same generating circle. But it is evidently unnecessary that
the generating circle used in the second case should be the
same as that used in the first. Any generating circle will
satisfy the conditions in either case (Art. 204.), provided it be
the same for the epicycloid as for the hypocycloid which is to
act with it.

According to a general (almost a universal) custom among
mechanics, two different generating circles are thus used for
striking the teeth on two wheels which are to act together,
the *diameter* of the generating circle for striking the *faces* of
the teeth on the one wheel being equal to the *radius* of the
pitch circle of the other wheel. Thus if we call the wheels
A and B, then the epicycloidal faces of the teeth on A, and
the corresponding hypocycloidal flanks on B, are generated
by a circle whose *diameter* is equal to the *radius* of the pitch
circle of B. The hypocycloidal flanks of the teeth on B
thus become straight lines (Art. 205.), whose directions are
those of radii of that wheel. In like manner the epicycloidal
faces of the teeth on B, and the corresponding hypocycloidal
flanks of the teeth on A, are struck by a circle whose
diameter is equal to the radius of the pitch circle of A; so
that the hypocycloidal flanks of the teeth of A become in like
manner straight lines, whose directions are those of radii of
the wheel A. By this expedient of using two different
generating circles, the flanks of the teeth on both wheels
become straight lines, and the faces only are curved. The
teeth shown in the above figure are of this form. The
motive for giving this particular value to the generating
circle appears to be no other than that saving of trouble
which is offered by the substitution of a *straight* for a *curved*
flank of the tooth. A more careful consideration of the subject, however, shows that there is no real economy of labour
in this. In the first place, it renders necessary the use of
two different generating circles or templets for striking the
teeth of any given wheel or pinion, the curved portions of
the teeth of the wheel being struck with a circle whose
diameter equals half the diameter of the pinion, and the
curved portions of the teeth of the pinion with a circle whose

diameter equals half that of the wheel. Now, one generating circle would have done for both, had the workman been contented to make the flanks of his teeth of the hypocycloidal forms corresponding to it. But there is a yet greater practical inconvenience in this method. A wheel and pinion thus constructed will only work with *one another;* neither will work truly any *third* wheel or pinion of a different number of teeth, although it have the same pitch. Thus the wheels A and B having each a given number of teeth, and being made to work with one another, will neither of them work truly with C of a different number of teeth of the same pitch. For that A may work truly with C, the face of its teeth must be struck with a generating circle, whose diameter is half that of C : but they are struck with a circle whose diameter is half that of B ; the condition of uniform action is not therefore satisfied. Now let us suppose that the epicycloidal faces, and the hypocycloidal flanks of all the teeth A, B, and C had been struck with the *same* generating circle, and that all three had been of the same pitch, it is clear that any one of them would then have worked truly with any other, and that this would have been equally true of any number of teeth of the same pitch. Thus, then, the mechanist may, by the use of the same generating circle, for all his pattern wheels of the same pitch, so construct them, as that any one wheel of that pitch shall work with any other. This offers, under many circumstances, great advantages, especially in the very great reduction of the number of patterns which he will be required to keep. There are, moreover, many cases in which some arrangement similar to this is indispensable to the true working of the wheels, as when one wheel is required (which is often the case) to work with two or three others, of different numbers of teeth, A for instance to turn B and C ; by the ordinary method of construction this combination would be impracticable, so that the wheels should work truly. Any generating circle common to a whole set of the same pitch, satisfying the above condition, it may be asked whether there is any other consideration determining the best dimensions of this circle. There is such a consideration arising out of a limitation of the dimensions of the generating circle of the hypocycloidal portion of the tooth to a diameter *not greater* than half that of its base. As long as it remains within these limits, the hypocycloid generated by it is of that concave form by which the flank of the tooth is made to *spread* itself, and the base of the tooth to *widen;* when it exceeds these limits, the flank of the tooth takes the convex form, the base of the tooth is thus contracted, and its strength diminished. Since, then, the generating circle should not have a diameter greater than half that of any of the wheels of the set for which it is used, it will manifestly be the greatest which will satisfy this condition when its diameter is equal to half that of the least wheel of the set. Now no pinion should have less than twelve or fourteen teeth. Half the diameter of a wheel of the proposed pitch, which has twelve or fourteen teeth, is then the true diameter for the generating circle of the set. The above suggestions are due to Professor Willis.*

213. To describe involute teeth.

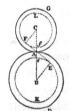

Let AD and AG represent the pitch circles of two wheels intended to work together. Draw a straight line FE through the point of contact A of the pitch circles and inclined to the line of centres CAB of these wheels at a certain angle FAC, the influence of the dimensions of which on the action of the teeth will hereafter be explained, but which appears usually to be taken not less than 80°.† Describe two circles eEK and fFL from the centres B

---

* Professor Willis has suggested a new and very ingenious method of striking the teeth of wheels by means of circular arcs. A detailed description of this method has been given by him in the Transactions of the Institution of Civil Engineers, vol. ii., accompanied by tables, &c. which render its practical application exceedingly simple and easy.

† See Camus on the Teeth of Wheels, by Hawkins, p. 168.

and C, each touching the straight line EF. These circles are to be taken as the *bases* from which the involute faces of the teeth are to be struck. It is evident (by the similar triangles ACF and AEB) that their radii CF and BE will be to one another as the radii CA and BA of the pitch circles, so that the condition necessary (Art. 203.) to the correct action of the teeth of the wheels will be satisfied, provided their faces be involutes to

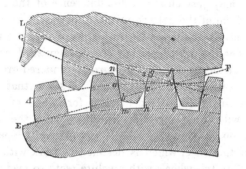

these two circles. Let AG and AH in the above figure represent arcs of the pitch circles of the wheels on an enlarged scale, and *e*E, *f*L, corresponding portions of the circles *e*EK and *f*FL of the preceding figure. Also let A*a* represent the pitch of one of the teeth of either wheel. Through the points A and *a* describe involutes *ef* and *mn**. Let *b*

* Mr. Hawkins recommends the following as a convenient method of striking involute teeth in his edition of " Camus on the Teeth of Wheels," p. 166. Take a thin board, or a plate of metal, and reduce its edge MN

so as accurately to coincide with the circular arc *e*E, and let a piece of thin watch-spring OR, having two projecting points upon it as shown at P, and which is of a width equal to the thickness of the plate, be fixed upon its edge by means of a screw O. Let the edge of the plate be then made to coincide with the arc *e*E in such a position, that when the spring is stretched, the point P in it may coincide with the point from which the tooth is to be struck; and the spring being kept continually stretched, and wound or unwound from the circle, the involute arc is thus to be described by the point P upon the face of the board from which the pattern is to be cut.

be the point where the line EF intersects the involute *mn* ;
then if the teeth on the two wheels are to be nearly of the
same thickness at their bases, bisect the line A*b* in *c* ; or if
they are to be of different thicknesses, divide the line A*b* in *c*
in the same proportion *, and strike through the point *c* an
involute curve *hg*, similar to *ef*, but inclined in the opposite
direction. If the extremity *fg* of the tooth be then cut off
so that it may just clear the circumference of the circle *f*L,
the true form of the pattern involute tooth will be obtained.

There are two remarkable properties of involute teeth, by
the combination of which they are distinguished from teeth
of all other forms, and *cæteris paribus* rendered greatly pre-
ferable to all others. The first of these is, that any two
wheels having teeth of the involute form, and of the same
pitch †, will work correctly together, since the forms of
the teeth on any one such wheel are entirely independent of
those on the wheel which is destined to work with it (Art.
203.). Any two wheels with involute teeth so made to work
together will revolve precisely as they would by the actual
contact of two circles, whose radii may be found by dividing
the line joining their centres in the proportion of the radii of
the generating circles of the involutes. This property involute
teeth possess, however, in common with the epicycloidal teeth
of different wheels, all of which are struck with the same
generating circle (Art. 212.). The second no less important
property of involute teeth — a property which distinguishes
them from teeth of all other forms — is this, *that they work
equally well, however far the centres of the wheels are removed
asunder from one another;* so that the action of the teeth of
two wheels is not impaired when their axes are displaced by

---

* This rule is given by Mr. Hawkins (p. 170.) ; it can only be an approximation, but may be sufficiently near to the truth for practical pur-
poses. It is to be observed that the teeth may have their bases in any
other circles than those, *f*L and *e*E, from which the involutes are
struck.

† The teeth being also of equal thicknesses at their bases, the method
of ensuring which condition has been explained above.

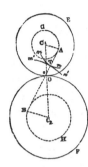

that wearing of their brasses or collars, which soon results from a continued and a considerable strain. The existence of this property will readily be admitted, if we conceive AG and BH to represent the generating circles or bases of the teeth, and these to be placed with their centres C_1 and C_2 any distance asunder, a band AB (p. 254., note) passing round both, and a point T in this band generating a curve mn, $m'n'$ on the plane of each of the circles as they are made to revolve under it. It has been shown that these curves mn and $m'n'$ will represent the faces of two teeth which will work truly with one another; moreover, that these curves are respectively involutes of the two circles AG and BH, and are therefore wholly independent in respect to their forms of the distances of the centres of the circles from one another, depending only on the dimensions of the circles. Since then the circles would drive at any distance correctly by means of the band; since, moreover, at every such distance they would be driven by the curves mn and $m'n'$ precisely as by the band; and since these curves would in every such position be the same curves, viz. involutes of the two circles, it follows that the same involute curves mn and $m'n'$ would drive the circles correctly at whatever distances their centres were placed; and, therefore, that involute teeth would drive these wheels correctly at whatever distances the axes of those wheels were placed.

THE TEETH OF A RACK AND PINION.

214. *To determine the pitch circle of the pinion.* Let H represent the distance through which the rack is to be moved by each tooth of the pinion, and let these teeth be N in number; then will the rack be moved through the space N . H during one complete revolution of the wheel. Now the rack and pinion are to be driven by the action of their teeth, as they would by the contact of the circumference of

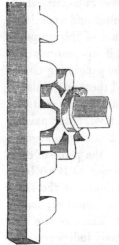

the pitch circle of the pinion with the plane face of the rack, so that the space moved through by the rack during one complete revolution of the pinion must precisely equal the circumference of the pitch circle of the pinion. If, therefore, we call R the radius of the pitch circle of the pinion, then

$$2\pi R = N . H ; \therefore R = \frac{1}{2\pi} N . H.$$

215. *To describe the teeth of the pinion, those of the rack being straight.* The properties which have been shown to belong to involute teeth (Art. 203.) manifestly obtain, however great may be the dimensions of

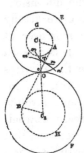

the pitch circle of their wheels, or whatever disproportion may exist between them. Of two wheels OF and OE with involute teeth which work together, let then the radius of the pitch circle of one OF become *infinite*, its circumference will then become a straight line represented by the face of a rack. Whilst the radius C_2O of the pitch circle OF thus becomes infinite, that C_2B of the circle from which its involute teeth are struck (bearing a constant ratio to the first) will also become infinite, so that the involute $m'n'$ will become a straight line * perpendicular to the line AB given in position. The involute teeth on the wheel OF will thus become straight teeth (see *fig.* 1.), having their faces perpendicular to the line AB de-

---

* For it is evident that the extremity of a line of infinite length un- winding itself from the circumference of a circle of infinite diameter will describe, through a finite space, a straight line perpendicular to the cir- cumference of the circle. The idea of giving an oblique position to the straight faces of the teeth of a rack appears first to have occurred to Professor Willis.

termined by drawing through the point O a tangent to the circle AC, from which the involute teeth of the pinion are struck. If the circle AC from which the involute teeth of the pinion are struck coincide with its *pitch* circle, the line AB becomes parallel to the face of the rack, and the edges of the teeth of the rack perpendicular to its face (*fig. 2.*).

Now, the involute teeth of the one wheel have remained unaltered, and the truth of their action with teeth of the other wheel has not been influenced by that change in the dimensions of the pitch circle of the last, which has converted it into a rack, and its curved into straight teeth. Thus, then, it follows, that straight teeth upon a rack, work truly with involute teeth upon a pinion. Indeed it is evident,

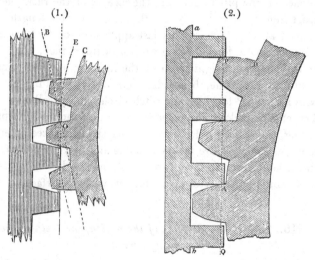

(1.)   (2.)

that if from the point of contact P (*fig. 2.*) of such an involute tooth of the pinion with the straight tooth of a rack we draw a straight line PQ parallel to the face *ab* of the rack, that straight line will be perpendicular to the surfaces of both the teeth at their point of contact P, and that being perpendicular to the face of the involute tooth, it will also touch the circle of which this tooth is the involute in the point A, at which the face *ab* of the rack would touch that circle if they revolved by mutual contact. Thus, then, the condition shown in Art. 201. to be necessary and sufficient to the correct action of the teeth, namely, that a line drawn from their point of contact, at any time, to the point of contact of their pitch circles, is satisfied in respect to these teeth. Divide, then, the circumference of the pitch circle, determined as above (Art. 214.), into N equal parts, and describe (Art. 213.) a pattern involute tooth from the circumference of the pitch circle, limiting the length of the face of the tooth to a little more than the length BP of the involute curve generated by unwinding a length AP of the flexible line equal to the distance H through which the rack is to be moved by each tooth of the pinion. The straight teeth of the rack are to be cut of the same length, and the circumference of the pitch circle and the face *ab* of the rack placed apart from one another by a little more than this length.

It is an objection to this last application of the involute form of tooth for a pinion working with a rack, that the point P of the straight tooth of the rack upon which it acts is always the same, being determined by its intersection with a line AP touching the pitch circle, and parallel to the face of the rack. The objection does not apply to the preceding, the case (*fig.* 1.) in which the straight faces of each tooth of the rack are inclined to one another. By the continual action upon a single point of the tooth of the rack, it is liable to an excessive wearing away of its surface.

216. *To describe the teeth of the pinion, the teeth of the rack being curved.*

This may be done by giving to the face of the tooth of

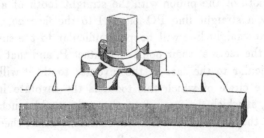

the rack a cycloidal form, and making the face of the tooth of the pinion an epicycloid, as will be apparent if we conceive 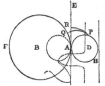 the diameter of the circle whose centre is C (see *fig.* p. 256.) to become *infinite*, the other two circles remaining unaltered. Any finite portion of the circumference of this *infinite* circle will then become a straight line. Let AE in the accompanying figure represent such a portion, and let PQ and PR represent, as before, curves generated by a point P in the circle whose centre is D, when all three circles revolve by their mutual contact at A. Then are PR and PQ the true forms of the teeth which would drive the circles as they are driven by their mutual contact at A (Art. 204.). Moreover, the curve PQ is the same (Art. 201.) as would be generated by the point P in the circumference of APH; if that circle rolled upon the circumference AQF, it is therefore an *epicycloid*; and the curve PR is the same as would be generated by the point P, if the circle APH rolled upon the circumference or straight line AE, it is therefore a *cycloid*. Thus then it appears, that after the teeth have passed the line of centres, when the face of the tooth of the pinion is driving the flank of the tooth of the rack, the former must have an epicycloidal, and the latter a cycloidal form. In like manner, by transferring the circle APH to the opposite side of AE, it may be shown, that before the teeth have passed the line of centres when the flank of the tooth of the pinion is driving the face of the tooth of the wheel, the former must have a hypo-cycloidal, and the latter a cycloidal form, the cycloid having its curvature in opposite directions on the flank and the face of the tooth. The generating circle will be of the most convenient dimensions for the description of the teeth when its diameter equals the radius of the pitch circle of the pinion. The hypocycloidal flank of the tooth of the pinion will then pass into a straight flank. The radius of the pitch circle of the pinion is determined as in Art. 214., and the method of describing its teeth is explained in Art. 210.

217. The teeth of a wheel working with a lantern
OR TRUNDLE.

In some descriptions of mill work the ordinary form of the
toothed wheel is replaced by a contrivance called a lantern or

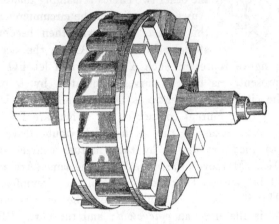

trundle, formed by two circular discs, which are connected
with one another by cylindrical columns called staves, engaging, like the teeth of a pinion, with the teeth of a wheel
which the lantern is intended to drive. This combination is
shown in the above figure.

It is evident that the teeth on the wheel which works with
the lantern have their shape determined by the cylindrical
shape of the staves. Their forms may readily be found by
the method explained in Art. 202.

Having determined upon the dimensions of the staves in
reference to the strain they are to be subjected to, and
upon the diameters of the pitch circles of the lantern and
wheel, and also upon the pitch of the teeth; strike arcs

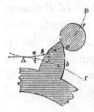

AB and AC of these circles, and set off
upon them the pitches Aa and Ab from the
point of contact A of the pitch circles (if
the teeth are *first* to come into contact in
the line of centres, if not, set them off from
the points behind the line of centres where
the teeth are first to come into contact).

Describe a circle *ae*, having its centre in AB, passing through *a*, and having its diameter equal to that of the stave, and divide each of the pitches A*a* and A*b* into the same number of equal parts (say three). From the points of division A, α, β in the pitch A*a*, measure the shortest distances to the circle *ae*, and with these shortest distances, respectively, describe from the points of division γ, δ of the pitch A*b*, circular arcs intersecting one another; a curve *ab* touching all these circular arcs will give the true face of the tooth (Art. 202.). The opposite face of the tooth must be struck from similar centres, and the base of the tooth must be cut so far within the pitch circle as to admit one half of the stave *ae* when that stave passes the line of centres.

218. THE RELATION BETWEEN TWO PRESSURES P_1 AND P_2 APPLIED TO TWO TOOTHED WHEELS IN THE STATE BORDERING UPON MOTION BY THE PREPONDERANCE OF P_1.

Let the influence of the weights of the wheels be in the first place neglected. Let B and C represent the centres of the pitch circles of the wheels, A their point of contact, P the point of contact of the driving and driven teeth at any period of the motion, RP the direction of the whole resultant pressure upon the teeth at their point of contact, which resultant pressure is equal and opposite to the resistance R of the follower to the driver, BM and CN perpendiculars from the centres of the axes of the wheels upon RP ; and BD and CE upon the directions of P_1 and P_2.

$BD = a_1$, $CE = a_2$, $BM = m_1$, $CN = m_2$.
$BA = r_1$, $CA = r_2$.
ρ_1, $\rho_2 =$ radii of axes of wheels.
φ_1, $\varphi_2 =$ limiting angles of resistance between the axes of the wheels and their bearings.

Then, since P_1 and R applied to the wheel whose centre is B are in the state bordering upon motion by the preponderance of P_1, and since a_1 and m_1 are the perpendiculars on

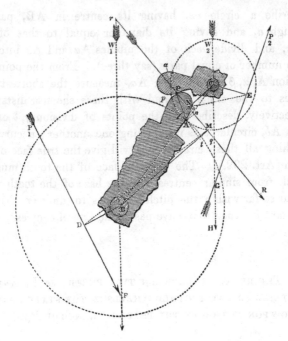

the directions of these pressures respectively, we have (equation 158)

$$P_1 = \left\{ \frac{m_1}{a_1} + \left(\frac{\rho_1 L_1}{a_1'^2}\right) \sin.\varphi_1 \right\} R = \frac{'1}{a_1} \left\{ m_1 + \left(\frac{\rho_1 L_1}{a_1}\right) \sin.\varphi_1 \right\} R \ldots (241),$$

where L_1 represents the length of the line DM joining the feet of the perpendiculars BM and BD.

Again, since R and P_2, applied to the wheel whose centre is C are in the state bordering upon motion by the *yielding* of P_2 (Art. 163.),

$$\therefore P_2 = \left\{ \frac{m_2}{a_2} - \left(\frac{\rho_2 L_2}{a_2^2}\right) \sin.\varphi_2 \right\} R = \frac{1}{a_2} \left\{ m_2 - \left(\frac{\rho_2 L_2}{a_2}\right) \sin.\varphi_2 \right\} R \ldots (242),$$

where L_2 represents the distance NE between the feet of the perpendiculars CE and CN. Eliminating R between these equations, we have

$$P_1 = \left(\frac{a_2}{a_1}\right) \left\{ \frac{m_1 + \left(\frac{\rho_1 L_1}{a_1}\right)\sin.\phi_1}{m_2 - \left(\frac{\rho_2 L_2}{a_2}\right)\sin.\phi_2} \right\} P_2 \ . \ . \ . \ (243).$$

Now let it be observed, that the line AP drawn from the point of contact A of the pitch circles to the point of contact P of the teeth is perpendicular to their surfaces at that point P, whatever may be the forms of the teeth, provided that they act truly with one another (Art. 201.); moreover, that when the point of contact P has passed the line of centres, as shown in the figure, that point is in the act of moving on the driven surface Pp *from* the centre C, or from P towards p, so that the friction of that surface is exerted in the opposite direction, or from p towards P; whence it follows that the resultant of this friction, and the perpendicular resistance aP of the driven tooth upon the driver, has its direction rP within the angle aPp and that it is inclined (Art. 141.) to the perpendicular aP at an angle aPr equal to the limiting angle of resistance. Now this resistance is evidently equal and opposite to the resultant pressure upon the surfaces of the teeth in the state bordering upon motion; whence it follows that the angle RPA is equal to the limiting angle of resistance between the surfaces of contact of the teeth. Let this angle be represented by φ, and let AP$=\lambda$. Also let the inclination PAC of AP to the line of centres BC be represented by θ. Through A draw An perpendicular to RP, and sAt parallel to it. Then,

$m_1 = BM=Bt$$+$$tM=Bt$$+An = \overline{\text{BA}}$ sin. BA$t + \overline{\text{AP}}$ sin. APR.

Also BA$t=$BOR$=$PAC$+$APR$=\theta + \varphi$;

$$\therefore \ m_1 = r_1 \sin.(\theta + \phi) + \lambda \sin.\varphi \ . \ . \ . \ . \ (244);$$

$m_2 = CN=Cs$$-$$sN=Cs$$-An=$CA sin. CA$s$$-$AP sin. APR.

But As is parallel to PR, therefore CA$s=$BOR$=\theta + \varphi$;

$$\therefore \ m_2 = r_2 \sin.(\theta + \varphi) - \lambda \sin.\varphi. \ . \ . \ . \ (245).$$

Substituting these values of m_1 and m_2 in the preceding equation,

$$P_1 = \left(\frac{a_2}{a_1}\right) \left[\frac{r_1 \sin.(\theta + \varphi) + \lambda \sin. \varphi + \left(\frac{\rho_1 L_1}{a_1}\right) \sin. \varphi_1}{r_2 \sin.(\theta + \varphi) - \lambda \sin. \varphi - \left(\frac{\rho_2 L_2}{a_2}\right) \sin. \varphi_2}\right] P_2 \ .. (246).$$

219. In the preceding investigation the point of contact P of the teeth of the driving and driven wheels is supposed to have passed the line of centres, or to be *behind* that line ; let us now suppose it not to have passed the line of centres, or to be *before* that line.

It is evident that in this case the point of contact P is in the act of moving upon the surface pPq of the driven tooth

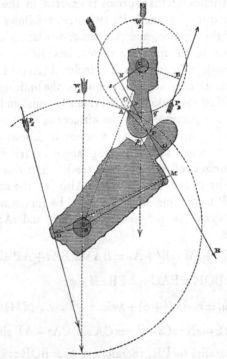

towards the centre C, or from P towards q, as in the other case it is *from* the centre, or from P towards p. In this case, therefore, the friction of the driven surface is exerted in

the direction qP ; whence it follows, that in this state bordering upon motion the direction of the resistance R of the
driven upon the driving tooth must lie on the other side
of the normal APQ, being inclined to it at an angle APN
equal to the limiting angle of resistance. Thus the inclination
of R to the normal APQ is in both cases the same, but its
position in respect to that line is in the one case the *reverse*
of its position in the other case.*

The same construction being made as before,

$$m_1 = \mathrm{BM} = \mathrm{B}t + t\mathrm{M} = \mathrm{B}t + \mathrm{A}n = \overline{\mathrm{BA}}.\sin.\mathrm{BA}t + \overline{\mathrm{AP}}.\sin.\mathrm{APO}.$$

Also $\mathrm{BA}t = \mathrm{BOR} = \mathrm{BAP} - \mathrm{APO} = \theta - \varphi$; †

$$\therefore\ m_1 = r_1 \sin.(\theta - \varphi) + \lambda \sin. \varphi,$$

$$m_2 = \mathrm{CN} = \mathrm{C}s - s\mathrm{N} = \mathrm{C}s - \mathrm{A}n = \overline{\mathrm{CA}}.\sin.\mathrm{CA}s - \overline{\mathrm{AP}}.\sin.\mathrm{APO}.$$

But As is parallel to PN,

$$\therefore\ \mathrm{CA}s = \mathrm{BOR} = \mathrm{BAP} - \mathrm{APO} = \theta - \varphi ;$$

$$\therefore\ m_2 = r_2 \sin.(\theta - \varphi) - \lambda \sin. \varphi.$$

Substituting these values of m_1 and m_2 in equation (243),

$$P_1 = \left(\frac{a_2}{a_1}\right) \left\{ \frac{r_1 \sin.(\theta - \phi) + \lambda \sin. \varphi + \left(\dfrac{\rho_1 \mathrm{L}_1}{a_1}\right)\sin. \varphi_1}{r_2 \sin.(\theta - \varphi) - \lambda \sin. \phi - \left(\dfrac{\rho_2 \mathrm{L}_2}{a_2}\right)\sin. \varphi_2} \right\} P_2 . \ (247).$$

This expression differs from the preceding (equation 246)
only in the substitution of $(\theta - \varphi)$ for $(\theta + \varphi)$ in the first terms
of the numerator and denominator.

---

* Hence it follows, that when the point of contact is in the act of cross-
ing the line of centres, the direction of the resultant pressure R is passing
from one side to the other of the perpendicular APQ; and therefore that
when the point of contact is *in* the line of centres, the resultant pressure
is perpendicular to that line, and the angle BOR a right angle; a condition which cannot however be assumed to obtain *approximately* in respect
to positions of any point of contact exceedingly near to the line of centres.

† The angle θ being here taken as before to represent the inclination
BAP of the line AP, joining the point of contact of the pitch circles with
the point of contact of the teeth, to the line of centres.

Dividing numerator and denominator of the fraction in the second member of that equation by sin. $(\theta + \phi)$, and throwing out the factors r_1 and r_2, we have

$$P_1 = \left(\frac{r_1 a_2}{r_2 a_1}\right) \left\{ \frac{1 + \dfrac{\lambda \sin. \phi + \left(\dfrac{\rho_1 L_1}{a_1}\right) \sin. \phi_1}{r_1 \sin. (\theta + \phi)}}{1 - \dfrac{\lambda \sin. \phi + \left(\dfrac{\rho_2 L_2}{a_2}\right) \sin. \phi_2}{r_2 \sin. (\theta + \phi)}} \right\} P_2.$$

Now it is evident, that if in this fractional expression $\theta - \phi$ be substituted for $\theta + \phi$ the numerator will be increased and the denominator diminished, so that the value of P_1 corresponding to any given value of P_2 will be increased. Whence it follows, that the resistance to the motion of the wheels by the friction of the common surfaces of contact of their teeth and of the bearings of their axes is greater when the contact of their teeth takes place *before* than when it takes place, at an equal angular distance, *behind* the line of centres — a principle confirmed by the experience of all practical mechanists.

220. To DETERMINE THE RELATION OF THE STATE BORDER-ING UPON MOTION BETWEEN THE PRESSURE P_1 APPLIED TO THE DRIVING WHEEL AND THE RESISTANCE P_2 OPPOSED TO THE MOTION OF THE DRIVEN WHEEL, THE WEIGHTS OF THE WHEELS BEING TAKEN INTO THE ACCOUNT.

Now let the influence of the weights W_1 and W_2 of the two wheels be taken into the account. The pressures applied to each wheel being now three in number instead of two, the relations between P_1 and R, and P_2 and R are determined by equation (163) instead of equation (158). Substituting W_1 and W_2 for P_3 in the two cases, we obtain, instead of equations (241) and (242), the following,

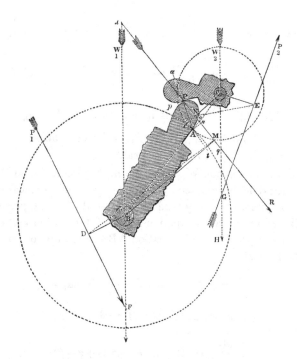

$$P_1 = \frac{1}{a_1}\left\{m_1 + \left(\frac{L_1\rho_1}{a_1}\right)\sin.\,\phi_1\right\}R + \frac{M_1 W_1}{L_1 a_1{}^2}\rho_1\sin.\,\phi_1$$

$$P_2 = \frac{1}{a_2}\left\{m_2 - \left(\frac{L_2\rho_2}{a_2}\right)\sin.\,\phi_2\right\}R - \frac{M_2 W_2}{L_2 a_2{}^2}\rho_2\sin.\,\phi_2 \qquad \cdots\cdots (248);$$

in which equations M_1 and M_2 represent certain functions determined (Art. 166.) by the inclinations of the pressures P_1 and P_2 to the vertical.

Eliminating R between the above equations, neglecting terms above the first dimension in sin. ϕ, sin. ϕ_1 and sin. ϕ_2, and multiplying by $a_1 a_2$,

$$P_1 a_1\left\{m_2 - \frac{L_2\rho_2}{a_2}\sin.\,\phi_2\right\} - P_2 a_2\left\{m_1 + \frac{L_1\rho_1}{a_1}\sin.\,\phi_1\right\} = \frac{M_1 W_1}{L_1 a_1}m_2\rho_1\sin.\,\phi_1 +$$

$$\frac{M_2 W_2}{L_2 a_2}m_1\rho_2\sin.\,\phi_2 \cdots\cdots (249).$$

Substituting the values of m_1 and m_2 from equations (244) and (245),

$$P_1 a_1 \left\{ r_2 \sin. (\theta+\phi) - \lambda \sin. \phi - \frac{L_2 \rho_2}{a_2} \sin. \phi_2 \right\} - P_2 a_2 \left\{ r_1 \sin. (\theta+\phi) + \lambda \sin. \phi + \frac{I_1 \rho_1}{a_1} \sin. \phi_1 \right\}$$

$$= \left\{ \frac{M_1 W_1}{L_1 a_1} r_2 \rho_1 \sin. \phi_1 + \frac{M_2 W_2}{L_2 a_2} r_1 \rho_2 \sin. \phi_2 \right\} \sin. (\theta+\phi) \ \ldots \ (250).$$

Now (Art. 166.) $\dfrac{M_1}{a_1} = m_1 \cos. \iota_{13} + a_1 \cos. \iota_{23}$, where ι_{13} represents the inclination $W_1 F P_1$ of P_1 to the vertical, and ι_{23} the inclination $R r F$ of R to the vertical.*

Let the inclination $W_1 BD$ of the perpendicular upon P_1 to the vertical be represented by α_1, that angle being so measured that the pressure P_1 may tend to increase it ; let α_2 represent, in like manner, the inclination ECG of CE to the vertical; and let β represent the inclination $AB r$ of the line of centres to the vertical,

$$\therefore \ \iota_{13} = W_1 F P_1 = W_1 BD - BDF = \alpha_1 - \frac{\pi}{2},$$

$$\iota_{23} = R r F = BOR - OB r = \theta + \phi - \beta \ ;$$

$$\therefore \ \frac{M_1}{a_1} = m_1 \sin. \alpha_1 + a_1 \cos. (\theta + \phi - \beta).$$

Similarly $\dfrac{M_2}{a_2} = m_2 \cos. P_2 GH + a_2 \cos. R q W_2.$† Now

$P_2 GH = ECG + GEC = \alpha_2 + \dfrac{\pi}{2};$ and $R q W_2 = \pi - R r F,$ and $R r F$ was before shown to be equal to $(\theta + \phi - \beta),$

$$\therefore \ \frac{M_2}{a_2} = - m_2 \sin. \alpha_2 - a_2 \cos. (\theta + \phi - \beta).$$

Substituting the values of m_1 and m_2, from equations (244) and (245),

---

* See note, p. 190.

† It is to be observed that the direction of the arrow in the figure represents that of the resistance opposed by the driven wheel to the motion of the driving wheel, so that the direction of the pressure of the driving upon the driven wheel is opposite to that of the arrow.

$$\frac{M_1}{a_1} = r_1 \sin.(\theta+\phi)\sin.\alpha_1 + \lambda \sin.\alpha_1 \sin.\phi + a_1 \cos.(\theta+\phi-\beta)$$

$$\frac{M_2}{a_2} = -r_2 \sin.(\theta+\phi)\sin.\alpha_2 + \lambda \sin.\alpha_2 \sin.\phi - a_2 \cos.(\theta+\phi-\beta)$$

$$\left.\begin{array}{c} \\ \\ \end{array}\right\} \ldots (251).$$

Let it be supposed that the distances DM and EN, represented by L_1 and L_2, are of finite dimensions, the directions of neither of the pressures P_1 and P_2 approaching to coincidence with the direction of R, — a supposition which has been virtually made in deducing equation (163) from equation (161), on the former of which equations, equations (248) depend. And let it be observed that the terms involving sin. ϕ in the above expressions (equations 251) will be of two dimensions in ϕ_1, ϕ_2, and ϕ, when substituted in equation (250), and may therefore be *neglected*. Moreover, that in all cases the direction of RP is so nearly perpendicular to the line of centres BC, that in those terms of equation (250), which are multiplied by sin. ϕ_1 and sin. ϕ_2, the angle $\theta + \phi$, or BOR, may be assumed $= \frac{\pi}{2}$; any error which that supposition involves, exceedingly small in itself, being rendered exceedingly less by that multiplication. Equations (251) will then become

$$\frac{M_1}{a_1} = r_1 \sin.\alpha_1 + a_1 \sin.\beta, \quad \frac{M_2}{a_2} = -r_2 \sin.\alpha_2 - a_2 \sin.\beta.$$

Substituting these values in the first factor of the second member of equation (250), and representing that factor by $N r_1 r_2$, we have

$$N r_1 r_2 = \frac{W_1}{L_1} r_2 \rho_1 (r_1 \sin.\alpha_1 + a_1 \sin.\beta) \sin.\phi_1 - \frac{W_2}{L_2} r_1 \rho_2 (r_2 \sin.\alpha_2 + a_2 \sin.\beta) \sin.\phi_2;$$

and dividing by $r_1 r_2$,

$$N* = \frac{W_1 \rho_1}{L_1}(\sin.\alpha_1 + \frac{a_1}{r_1}\sin.\beta)\sin.\phi_1 - \frac{W_2 \rho_2}{L_2}(\sin.\alpha_2 + \frac{a_2}{r_2}\sin.\beta)\sin.\phi_2 \ . \ . \ (252).$$

* If the direction of P be that of a tangent at the point of contact A of the wheels, a case of frequent occurrence, the value of L_1 vanishing, that of N would appear to become infinite in this expression. The difficulty will however be removed, if we consider that when a_1 becomes, as in

Substituting Nr_1r_2 for the factor, which it represents in equation (250), we have

$$P_1a_1\{r_2\sin.(\theta+\phi)-\lambda\sin.\phi-\frac{L_2\rho_2}{a_2}\sin.\phi_2\}-P_2a_2\{r_1\sin.(\theta+\phi)+\lambda\sin.\phi+$$

$$\frac{L_1\rho_1}{a_1}\sin.\phi_1\}=Nr_1r_2\sin.(\theta+\phi)\ \ .\ .\ .\ .\ .\ (253).$$

Solving this equation in respect to P_1,

$$P_1=\frac{a_2r_1}{a_1r_2}\left\{\frac{1+\dfrac{\lambda\sin.\phi+\dfrac{L_1\rho_1}{a_1}\sin.\phi_1}{r_1\sin.(\theta+\phi)}}{1-\dfrac{\lambda\sin.\phi+\dfrac{L_2\rho_2}{a_2}\sin.\phi_2}{r_2\sin.(\theta+\phi)}}\right\}P_2+\frac{\dfrac{Nr_1}{a_1}}{1-\dfrac{\lambda\sin.\phi+\dfrac{L_2\rho_2}{a_2}\sin.\phi_2}{r_2\sin.(\theta+\phi)}}.$$

Whence, performing actual division by the denominators of the fractions in the second member of the equation, and omitting terms of two dimensions in sin. ϕ_1, sin. ϕ_2, sin. ϕ (observing that N is already of *one* dimension in those variables), we have

$$P_1=\frac{a_2r_1}{a_1r_2}\left\{1+\{\lambda\left(\frac{1}{r_1}+\frac{1}{r_2}\right)\sin.\phi+\frac{L_1\rho_1}{a_1r_1}\text{ sn. }\phi_1+\frac{L_2\rho_2}{a_2r_2}\sin.\phi_2\}\text{cosec.}(\theta+\phi)\right\}P_2+\frac{Nr_1}{a_1}\ .\ .\ (254).$$

In this expression it is assumed that the contact of the teeth is behind the line of centres.

---

this case, equal to r_1, and the point M is supposed to coincide with A, L_1 becomes a chord of the pitch circle, and is therefore represented by $2r_1\sin.\frac{1}{2}$DBA, or $2r_1\sin.\frac{1}{2}(a_1+\beta)$; so that $\dfrac{\sin.a_1+\dfrac{a_1}{r_1}\sin.\beta}{L_1}=$

$$\frac{\sin.a_1+\sin.\beta}{2r_1\sin.\frac{1}{2}(a_1+\beta)}=\frac{2\sin.\frac{1}{2}(a_1+\beta)\cos.\frac{1}{2}(a_1+\beta)}{2r_1\sin.\frac{1}{2}(a_1+\beta)}=\frac{1}{r_1}\cos.\frac{1}{2}(a_1+\beta).$$

If, therefore, we take the angle $a_1=-\beta$, so as to give to P_1 the direction of a tangent at A, this expression will assume the value, $\dfrac{1}{r_1}\cos.0$, or $\dfrac{1}{r_1}$; so that in this case

$$N=\frac{W_1\rho_1}{r_1}\sin.\phi_1-\frac{W_2\rho_2}{L_2}\left(\sin.a_2+\frac{a_2}{r_2}\sin.\beta\right)\sin.\phi_2.$$

221. THE MODULUS OF A SYSTEM OF TWO TOOTHED WHEELS.

Let n_1 and n_2 represent the numbers of teeth in the driving and driven wheels respectively, and let it be observed that these numbers are one to another as the radii of the pitch circles of the wheels; then, multiplying both sides of equation (254) by $a_1\dfrac{r_2}{r_1}$, we shall obtain

$$P_1 a_1 \frac{r_2}{r_1} = P_2 a_2 \left\{ 1 + \{\lambda\left(\frac{1}{r_1} + \frac{1}{r_2}\right) \sin.\phi + \frac{L_1\rho_1}{a_1 r_1} \sin.\phi_1 + \frac{L_2\rho_2}{a_2 r_2} \sin.\phi_2\} \operatorname{cosec.}(\theta+\phi) \right\} + N r_2.$$

Now let $\Delta\psi$ represent an exceedingly small increment of the angle ψ, through which the driven wheel is supposed to have revolved, after the point of contact P has passed the line of centres; and let it be observed that the first member of the above equation is equal to $P_1 a_1 \dfrac{r_2}{r_1} \dfrac{\Delta\psi}{\Delta\psi}$, and that $\dfrac{r_2}{r_1}\Delta\psi$ represents the angle described by the driving wheel (Art. 206), whilst the driven wheel describes the angle $\Delta\psi$; whence it follows (Art. 50.) that $P_1 a_1 \left(\dfrac{r_2}{r_1}\Delta\psi\right)$ represents the *work* ΔU_1 done by the driving pressure P_1, whilst this angle $\Delta\psi$ is described by the driven wheel,

$$\therefore \frac{\Delta U}{\Delta\psi} = P_2 a_2 \left\{ 1 + \{\lambda\left(\frac{1}{r_1} + \frac{1}{r_2}\right) \sin.\phi + \frac{L_1\rho_1}{a_1 r_1} \sin.\phi_1 + \frac{L_2\rho_2}{a_2 r_2} \sin.\phi_2\} \operatorname{cosec.}(\theta+\phi) \right\} + N r_2.$$

Let now $\Delta\psi$ be conceived infinitely small, so that the first member of the above equation may become the differential co-efficient of U_1, in respect to ψ. Let the equation, then, be integrated between the limits 0 and ψ; P_2, L_1, and L_2, and therefore N (equation 252) being conceived to remain constant, whilst the angle ψ is described; we shall then obtain the equation

$$U_1 = P_2 a_2 \int_0^{\psi} \left\{ 1 + \{\lambda\left(\frac{1}{r_1} + \frac{1}{r_2}\right) \sin.\phi + \frac{L_1\xi_1}{a_1 r_1} \sin.\phi_1 + \frac{L_2\xi_2}{a_2 r_2} \sin.\phi_2\} \operatorname{cosec.}(\theta+\phi) \right\} d\psi + N \cdot S \ldots (255),$$

where S is taken to represent the arc $r_2\psi$ described by
the pitch circle of the driven wheel, and therefore by that of
the driving wheel also, whilst the former revolves through the
angle ψ.

222. THE MODULUS OF A SYSTEM OF TWO TOOTHED WHEELS,
THE NUMBER OF TEETH ON THE DRIVEN WEEEL BEING
CONSIDERABLE, AND THE WEIGHTS OF THE WHEELS BEING
TAKEN INTO ACCOUNT.

It is evident that the space traversed by the point of contact of two teeth on the face of either of them is, in this case,
small as compared with the radius of its pitch circle, and that
the direction of the resultant pressure R (see *fig.* p. 282.)
upon the teeth is very nearly perpendicular to the line of
centres BC, whatever may be the particular forms of the teeth;
provided only that they be of such forms as will cause them
to act truly with one another. In this case, therefore, the
angle BOR represented by $\theta + \varphi$ is very nearly equal to $\frac{\pi}{2}$,
and cosec. $(\theta + \varphi) = 1$.

Since, moreover, RP is very nearly perpendicular to the
line of centres at A, and that the point of contact P of the
teeth deviates but little from that line, it is evident that the
line AP represented by λ differs but little from an arc of
the pitch circle of the driven wheel, and that it differs the
less as the supposition made at the head of this article more
nearly obtains. Let us suppose ψ to represent the angle
subtended by this arc at the centre C of the pitch circle of
the driven wheel, then will the arc itself be represented by
$r_2\psi$, and therefore $\lambda = r_2\psi$ very nearly. Substituting this
value of λ in equation (255), observing that cosec. $(\theta + \varphi) = 1$,
and that $\frac{r_2}{r_1} = \frac{n_2}{n_1}$ (equation 232), and integrating,

$$U_1 = \{1 + \tfrac{1}{2}\psi\left(1 + \frac{n_2}{n_1}\right) \sin. \varphi + \frac{L_1\rho_1}{a_1r_1}\sin. \phi_1 + \frac{L_2\rho_2}{a_2r_2}\sin. \phi_2\} P_2 a_2\psi + Nr_2\psi \, .. (256).$$

But the driven or working pressure P_2 being supposed to

remain constant, whilst any two given teeth are in action, $P_2 a_2 \psi$ represents the work U_2 yielded by that pressure whilst those teeth are in contact: also $r_2 \psi$ represents the space S, described by the circumference of the pitch circle of either wheel whilst this angle is described.   Now let ψ be conceived to represent the angle subtended by the pitch of one of the teeth of the driven wheel, these teeth being supposed to act only *behind* the line of centres, then $\psi = \dfrac{2\pi}{n_2}$, n_2 representing the number of teeth on the driven wheel,

and $\tfrac{1}{2}\psi\left(1 + \dfrac{n_2}{n_2}\right) = \dfrac{\pi}{n_2}\left(1 + \dfrac{n_2}{n_1}\right) = \pi\left(\dfrac{1}{n_1} + \dfrac{1}{n_2}\right)$;

$\therefore\ U_1 = \left\{ 1 + \pi\left(\dfrac{1}{n_1} + \dfrac{1}{n_2}\right) \sin. \phi' + \dfrac{L_1 \rho_1}{a_1 r_1} \sin. \phi_1 + \dfrac{L_2 n_2}{a_2 r_2} \sin. \phi_2 \right\} U_2 + N . S\ \ldots$ (257),

which relation between the work done at the moving and working points, whilst any two given teeth are in contact, is evidently also the relation between the work similarly done, whilst *any given number* of teeth are in contact.   It is therefore the MODULUS of any system of two toothed wheels, the numbers of whose teeth are considerable.

223. THE MODULUS OF A SYSTEM OF TWO WHEELS WITH IN-VOLUTE TEETH OF ANY NUMBERS AND DIMENSIONS.

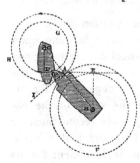

The locus of the points of contact of the teeth has been shown (Art. 203.) to be in this case a straight line DE, which passes through the point of contact A of the pitch circles, and touches the circles (EF and DG) from which the involutes are struck.   Let P represent any position of this point of contact, then is AP measured along the given line DE the distance represented by λ in Art. 218., and the angle CAD, which is in this case constant, is that represented by θ. Since, moreover, the point of contact of the teeth moves precisely as a point P upon a flexible cord DE, unwinding from the circle EF and winding upon DG, would (see note, p. 254.), it is evident that the distance AP, being that which such a point would traverse whilst the pitch circle AH revolved through a certain angle ψ, measured from the line of centres is precisely equal to the length of string which would wind upon DG whilst this angle is described by it; or to the arc of that circle which subtends the angle ψ. If, therefore, we represent the angle ACD by η, so that $CD = \overline{CA} \cos. ACD = r_2 \cos. \eta$, then $\lambda = r_2 \psi \cos. \eta$. Substituting this value for λ in equation (254), and observing that $\theta + \varphi = \frac{\pi}{2} - \eta + \varphi = \frac{\pi}{2} - (\eta - \varphi)$, and that $\frac{r_2}{r_1} = \frac{n_2}{n_1}$, we have

$$P_1 = \frac{a_2 r_1}{a_1 r_2}\left\{ 1 + \left\{\psi\left(1 + \frac{n_2}{n_1}\right) \cos. \eta \sin. \varphi + \frac{L_1 \rho_1}{a_1 r_1} \sin. \varphi_1 + \frac{L_2 \rho_2}{a_2 r_2} \sin. \varphi_2\right\} \sec. (\eta - \varphi)\right\} P_2 + \frac{Nr_1}{a_1} \ldots (258);$$

from which equation we obtain by the same steps as in Art. 221, observing that η is constant,

$$U_1 = \left\{ 1 + \left\{\pi\left(\frac{1}{n_1} + \frac{1}{n_2}\right) \cos. \eta \sin. \varphi + \frac{L_1 \rho_1}{a_1 r_1} \sin. \varphi_1 + \frac{L_2 \rho_2}{a_2 r_2} \sin. \varphi_2\right\} \sec. (\eta - \varphi)\right\} U_2 + NS \ldots (259),$$

which is the modulus of a system of two wheels having any given numbers of involute teeth.

224. THE INVOLUTE TOOTH OF LEAST RESISTANCE.

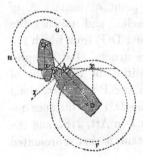

It is evident that the value of U_1 in equation (259), or of the *work* which must be done upon the driving wheel to cause a given amount U_2 to be yielded by the driven wheel is dependent for its amount upon the value of the coefficient of U_2 in the second member of that equation; and that this coefficient, again, is dependent for its value (other things being the same) upon the value of η representing the angle ACD, or its equal the angle DAI, which the tangent DE to the circles from which the involutes are struck makes with a perpendicular AI to the line of centres. Moreover, that the co-efficient N not involving this factor η (equation 252), the variation of the value of U_1, so far as this angle is concerned, is wholly involved in the corresponding variation of the co-efficient of U_2 and becomes a minimum with it; so that the value of η which gives to the function of η represented by this co-efficient, its minimum value, is the value of it which satisfies the condition of the *greatest economy of power*, and determines that inclination DAI of the tangent DE to the perpendicular to the line of centres, and those values, therefore, of the radii CD and BE of the circles whence the involutes are struck, which correspond to the tooth of least resistance.

To determine the value of η which corresponds to a minimum value of this co-efficient, let the latter be represented by u; then, for the required value of η,

$$\frac{du}{d\eta} = 0, \text{ and } \frac{d^2u}{d\eta^2} > 0.$$

Let $\pi\left(\frac{1}{n_1} + \frac{1}{n_2}\right) = A$, $\dfrac{L_1\rho_1}{a_1 r_1}\sin.\ \varphi_1 + \dfrac{L_2\rho_2}{a_2 r_2}\sin.\ \varphi_2 = B$;

$\therefore\ u = 1 + (A\cos.\ \eta\sin.\ \varphi + B)\sec.\ (\eta - \phi)$;

$\therefore\ u = 1 + B\sec.\ (\eta - \phi) + A\sin.\ \varphi\cos.\ \eta\sec.\ (\eta - \phi)$;

$\therefore\ \dfrac{du}{d\eta} = B\sec.\ (\eta - \phi)\tan.\ (\eta - \phi) - A\sin.\ \phi\{\sin.\ \eta\sec.\ (\eta - \phi) - \cos.\ \eta\tan.\ (\eta - \phi)\sec.\ (\eta - \phi)\}$;

$\therefore\ \dfrac{du}{d\eta} = B\sec.\ ^2(\eta - \phi)\sin.\ (\eta - \phi) - A\sin.\ \phi\sec.\ ^2(\eta - \phi)\{\sin.\ \eta\cos.\ (\eta - \phi) - \cos.\ \eta\sin.\ (\eta - \phi)\}$;

$\therefore\ \dfrac{du}{d\eta} = \sec.\ ^2(\eta - \phi)\{B\sin.\ (\eta - \phi) - A\sin.\ ^2\phi\}$ . . . . . (260).

In order, therefore, that $\dfrac{du}{d\eta}$ may vanish for any value of η, one of the factors which compose the second member of the above equation must vanish for that value of η; but this can never be the case in respect to the first factor, for the least value of the square of the secant of an arc is the square of the radius. If, therefore, the function u admit of a minimum value, the second factor of the above equation vanishes when it attains that value; and the corresponding value of η is determined by the equation,

$$\text{B sin. } (\eta - \varphi) - \text{A sin.}^2\varphi = 0 \quad \ldots \ldots \text{ (261).}$$

or by $\sin (\eta - \varphi) = \dfrac{A}{B}\sin.^2\varphi$, or by $\eta = \varphi + \sin.^{-1}\left(\dfrac{A}{B}\sin.^2\varphi\right)$;

or substituting the values of A and B,

$$\eta = \varphi + \sin.^{-1}\left\{\dfrac{\left(\dfrac{1}{n_1}+\dfrac{1}{n_2}\right)\sin.^2\varphi}{\dfrac{L_1\rho_1}{a_1r_1}\sin.\varphi_1+\dfrac{L_2\rho_2}{a_2r_2}\sin.\varphi_2}\right\} \quad \ldots \ldots \text{ (262)}$$

Now the function u admits of a minimum to which this value of η corresponds, provided that when substituted in $\dfrac{d^2u}{d\eta^2}$ this value of η gives to that second differential co-efficient of u in respect to η a *positive* value.

Differentiating equation (260), we have

$$\frac{d^2u}{d\eta^2}=2\sec.^2(\eta-\varphi)\tan.(\eta-\varphi)\{\text{B sin.}^2(\eta-\varphi)-\text{A sin.}^2\varphi\}+\text{B sec.}^2(\eta-\varphi)\cos.(\eta-\varphi).$$

But the proposed value of η (equation 261) has been shown to be that which, being substituted in the factor $\{\text{B sin.}(\eta-\varphi) - \text{A sin.}^2\varphi\}$, will cause it to vanish, and therefore, with it, the whole of the first term of the value of $\dfrac{d^2u}{d\eta^2}$: it corresponds, therefore, to a minimum, if it gives to the second term $\text{B sec.}^2(\eta-\varphi)\cos.(\eta-\varphi)$ a positive value; or, since $\sec.^2(\eta-\varphi)$ is essentially positive, and B does not involve η, if it gives to $\cos.(\eta-\varphi)$ a positive value, or if $\eta-\varphi<\dfrac{\pi}{2}$, or if $\sin.^{-1}\left(\dfrac{A}{B}\sin.^2\varphi\right)$

$<\dfrac{\pi}{2}$, or if $\dfrac{A}{B}\sin.^2\varphi<1$; or if $\text{A sin.}^2\varphi<\text{B}$; or if

$$\pi\left(\frac{1}{n_1}+\frac{1}{n_2}\right)\sin.^2\varphi<\frac{L_1\rho_1}{a_1r_1}\sin.\varphi_1+\frac{L_2\rho_2}{a_2r_2}\sin.\varphi_2 \quad \ldots \ldots \text{ (263).}$$

This condition being satisfied, the value of η, determined by equation (262), corresponds to a minimum, and determines the INVOLUTE TOOTH OF LEAST RESISTANCE.*

225. TO DETERMINE IN WHAT PROPORTION THE ANGLE OF CONTACT OF EACH TOOTH SHOULD BE DIVIDED BY THE LINE OF CENTRES ; OR THROUGH HOW MUCH OF ITS PITCH EACH TOOTH SHOULD DRIVE BEFORE AND BEHIND THE LINE OF CENTRES, THAT THE WORK EXPENDED UPON FRICTION MAY BE THE LEAST POSSIBLE.

Let the proportion in which the angle of contact of each tooth is divided by the line of centres be represented by x, so that $x\dfrac{2\pi}{n_2}$ may represent the angular distance from the line of centres of a line drawn from the centre of the driven wheel to the point of contact of the teeth when they first come into action before the line of centres, and $(1-x)\dfrac{2\pi}{n_2}$ the corresponding angular distance behind the line of centres when they pass out of contact ; and let it be observed that, on this supposition, if U_2 represent as before the work yielded by the driven wheel during the contact of any two teeth, xU_2 will represent the portion of that work done before, and $(1-x)U_2$ that done behind, the line of centres. Then proceeding in respect to equation 258 by the same method as was used in deducing from that equation the modulus (Equation 259), but integrating first between the limits 0 and $x\dfrac{2\pi}{n_2}$, in order to determine the work u_1 done by the driving pressure before the point of contact passes the

---

* It may easily be shown by eliminating η between equations (259) and (261) that the modulus corresponding to this condition of the greatest economy of power, where involute teeth are used, is represented by the formula

$$U_1 = \left\{ 1 + \tfrac{1}{2}A \sin. 2\phi + (B^2 - A^2 \sin.^4\phi) \right\} U_2 + NS.$$

line of centres, and then between the limits 0 and $(1-x)\dfrac{2\pi}{n_2}$
to determine the work u_2 done after the point of contact has
passed the line of centres; observing, moreover, that in the
former case $-\phi$ is to be substituted in sec. $(\eta-\phi)$ for ϕ
(Art. 219.), we have

$$u_1 = \left\{1 + \left\{x\pi\left(\frac{1}{n_1}+\frac{1}{n_2}\right)\cos.\,\eta\,\sin.\,\phi + \frac{L_1\rho_1}{a_1r_1}\sin.\,\phi_1 + \frac{L_2\rho_2}{a_2r_2}\sin.\,\phi_2\right\}\sec.\,(\eta+\phi)\right\}xU_2 + Ns_1;$$

Or assuming

$$\pi\left(\frac{1}{n_1}+\frac{1}{n_2}\right)\cos.\,\eta\,\sin.\,\phi = a,\ \text{and}\ \frac{L_1\rho_1}{a_1r_1}\sin.\,\phi_0 + \frac{L_2\rho_2}{a_2r_2}\sin.\,\phi_2 = b$$

$$u_1 = \{1 + (ax+b)\sec.(\eta+\phi)\}xU_2 + Ns_1.$$

s_1 representing the space described by the pitch circle of
either wheel before the line of centres is passed; similarly,

$$u_2 = \left\{1 + \{a(1-x)+b\}\sec.(\eta-\phi)\right\}(1-x)U_2 + Ns_2.$$

Adding these equations together, and representing by U_1 the
whole work $u_1 + u_2$ done by the driving pressure during the
contact of the teeth, and by S the whole space described by
the circumference of either pitch circle, we have

$$U_1 = \{1 + (ax^2+bx)\sec.\,(\eta+\phi) + \{a(1-x)^2+b(1-x)\}\sec.(\eta-\phi)\}U_2 + NS \ ..\ (264)$$

by which equation is determined the modulus of two wheels
driven by involute teeth, when the contact takes place partly
before and partly behind the line of centres.

Let the portion of the work U_1, which is expended upon
the *friction* of the teeth be represented by u.   Then

$$u = \left\{(ax^2+bx)\sec.\,(\eta+\phi) + \{a(1-x)^2+b(1-x)\}\sec.\,(\eta-\phi)\right\}U_2 + NS.$$

Now the value of x, which gives to this function its minimum, and which therefore determines that division of the
driving arc which corresponds to the greatest economy of
power, is evidently the value which satisfies the conditions

$$\frac{du}{dx} = 0 \qquad\qquad \frac{d^2u}{dx_2} > 0$$

But differentiating and reducing

$$\frac{du}{dx} = \left\{2ax\{\sec.\,(\eta+\phi)+\sec.(\eta-\phi)\} + b\{\sec.\,(\eta+\phi) - \sec.(\eta-\phi)\} - 2a\sec.(\eta-\phi)\right\}U_2;$$

$$\frac{d^2u}{dx^2} = 2a\{\sec.(\eta+\phi) + \sec.(\eta-\phi)\}U_2 :$$

Whence it appears that the second condition is always satisfied, and that the first condition is satisfied by that value of x, which is determined by the equation

$$2ax\{\sec.(\eta+\phi) + \sec.(\eta-\phi)\} + b\{\sec.(\eta+\phi) - \sec.(\eta-\phi)\} - 2a\sec.(\eta-\phi) = 0 ;$$

Whence we obtain by transposition and reduction

$$x = \frac{1}{2}\left\{1 - \left(1+\frac{b}{a}\right)\tan.\eta\tan.\phi\right\}.$$

So that the condition of the greatest economy of power is satisfied in respect to involute teeth, when the teeth first come into contact before the line of centres at a point whose angular distance from it is less than one half the angle subtended by the pitch by that fractional part of the last-mentioned angle, which is represented by the formula $\frac{1}{2}\left(1+\frac{b}{a}\right)$ $\tan.\eta\tan.\phi$, or substituting for b and a their values by the formula

$$\frac{1}{2}\left[1 + \frac{\dfrac{L_1\rho_1}{a_1r_1}\sin.\phi + \dfrac{L_2\rho_2}{a_2r_2}\sin.\phi_2}{\pi\left(\dfrac{1}{n_1}+\dfrac{1}{n_2}\right)\cos.\eta\sin.\phi}\right]\tan.\eta\tan.\phi \quad \ldots \quad (265).$$

That division of the angle of contact of any two teeth by the line of centres, which is consistent with the greatest economy of power, is always, therefore, an unequal division, the less portion being that which lies *before* the line of centres; and its fractional defect from one half the angle of contact, as also the fractional excess of the greater portion above one half that angle, is in every case represented by the above formula, and is therefore dependent upon the dimensions of the wheels, the forms and numbers of the teeth, and the circumstances under which the driving and working pressures are applied to them.*

* The division of the arc of contact which corresponds to the greatest economy of power in epicycloidal teeth, may be determined by precisely the same steps.

226. THE MODULUS OF A SYSTEM OF TWO WHEELS DRIVEN
BY EPICYCLOIDAL TEETH.

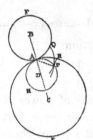

The locus of the point of contact P of any two such teeth
is evidently the generating circle APH of
the epicycloidal face of one of the teeth,
and the hypocycloidal flank of the other
(Art. 204.); for it has been shown (Art.
201.), that if the pitch circles of the wheel
and the generating circle APH of the
teeth be conceived to revolve about fixed
centres B, C, D by their mutual contact
at A, then will a point P in the circum-
ference of the last mentioned circle move at the same time
upon the surfaces of both the teeth which are in contact,
and therefore always coincide with their point of contact,
so that the distance AP of the point of contact P of the
teeth from A, which distance is represented in equation
(255) by λ, is in this case the chord of the arc AP, which
the generating circle, if it revolved by its contact with
the pitch circles, would have described, whilst each of the
pitch circles revolved through a certain angle measured
from the line of centres. Let the angle which the driven
wheel (whose centre is C) describes between the period
when the point of contact P of the teeth passes the line of
centres, and that when it reaches the position shown in the
figure be represented as before by ψ, the arc of the pitch
circle of that wheel which passes over the point A during that
period will then be represented by $r_2\psi$. Now the generating
circle APH having revolved in contact with this pitch circle,
an equal arc of that circle will have passed over the point A ;
whence it follows that the arc $AP = r_2\psi$; and that if the radius
of the generating circle be represented by r, then the angle

ADP subtended by the arc AP is represented by $\dfrac{r_2}{r}\,\psi$, or

by $2e\psi$, if $2e$ be taken to represent the ratio $\dfrac{r_2}{r}$ of the radius of the pitch circle of the driven wheel to the radius of the *gene-rating* circle. Now the chord $AP = 2\overline{AD}$ sin. $\frac{1}{2}$ ADP; therefore $\lambda = 2r$ sin. $e\psi = \frac{r_2}{e}$ sin. $e\psi$. Substituting this value of λ in equation (255); observing, moreover, that the angle PAD represented by θ in that equation is equal to $\frac{\pi}{2} - \frac{1}{2}$ ADP, or to $\frac{\pi}{2} - e\psi$, and that the *whole* angle ψ through which the driven wheel is made to revolve by the contact of each of its teeth is represented by $\frac{2\pi}{n_2}$, we have

$$U_1 = P_2 a_2 \int_0^{\frac{2\pi}{n_2}} 1 \left\{ + \{ \frac{r_2}{e}\left(\frac{1}{r_1}+\frac{1}{r_2}\right) \text{sin.} \ \phi \ \text{sin.} \ e\psi + \frac{L_1\rho_1}{a_1 r_1}\text{sin.} \ \phi_1 + \frac{L_2\rho_2}{a_2 r_2}\text{sin.} \ \phi_2 \} \text{sec.} \ (e\psi - \phi) \right\} d\psi + NS;$$

or, assuming L_1 and L_2 to remain constant during the contact of any two teeth representing the constant $1 + \frac{L_1\rho_1}{a_1 r_1}$ sin. ϕ_1 $+ \frac{L_2\rho_2}{a_2 r_2}$ sin. ϕ_2 by A, and observing that $\frac{r_2}{r_1} = \frac{n_2}{n_1}$,

$$U_1 = P_2 a_2 \left\{ A\int_0^{\frac{2\pi}{n_2}} \text{sec.} \ (e\psi - \phi)d\psi + \frac{1}{e}\left(1+\frac{n_2}{n_1}\right) \text{sin.} \ \phi \int_0^{\frac{2\pi}{n_2}} \text{sin.} \ e\psi \ \text{sec.}(e\psi - \phi)d\psi \right\} + NS.$$

Now the *general* integral, $\int$ sec. $(e\psi - \phi)d\psi$, or $\frac{1}{e}\int$ sec. $(e\psi - \phi) \ d \ (e\psi - \phi)$ being represented * by the function $\frac{1}{e}$ log.$_\varepsilon$ tan. $\left\{ \frac{\pi}{4} + \frac{1}{2}(e\psi - \phi) \right\}$, its *definite* integral between the limits 0 and $\frac{2\pi}{n_2}$ has for its expression,

$$\frac{1}{e} \ \text{log.}_\varepsilon \frac{\text{tan.}\left\{ \frac{\pi}{4}+\frac{1}{2}\left(e\frac{2\pi}{n_2}-\phi\right)\right\}}{\text{tan.} \ \left(\frac{\pi}{4}-\frac{\phi}{2}\right)}.$$

* Hymer's Integ. Cal. Art. 52.

Also $\int_0^{\frac{2\pi}{n_2}} \sec.(e\psi - \phi)\sin. e\psi d\psi = \int_0^{\frac{2\pi}{n_2}} \sec.(e\psi - \phi)\sin.\{(e\psi - \phi) + \phi\}d\psi$

$= \int_0^{\frac{2\pi}{n}} \sec. (e\psi - \phi)\{\sin. (e\psi - \phi) \cos. \phi + \cos. (e\psi - \phi) \sin. \phi\}d\psi$

$= \int^{\frac{2\pi}{n}} \{\cos. \phi \tan. (e\psi - \phi) + \sin. \phi\}d\psi$

$= \frac{1}{e} \cos. \phi \int_0^{\frac{2\pi}{n}} \tan.(e\psi - \phi)d (e\psi - \phi) + \frac{2\pi}{n^2} \sin. \phi.$

Now the *general* integral $\int \tan.(e\psi - \phi)d(e\psi - \phi)$ has for its expression $- \log._\varepsilon \cos. (e\psi - \phi)$.* Taking its *definite* integral between the limits 0 and $\frac{2\pi}{n_2}$, we have, therefore,

$\int_0^{\frac{2\pi}{n_2}} \sec. (e\psi - \phi) \sin. e\psi d\psi = - \frac{1}{e} \cos. \phi \log._\varepsilon \frac{\cos.\left(\frac{2e\pi}{n_2} - \phi\right)}{\cos. \phi.} + \frac{2\pi}{n_2}\sin. \phi.$

Substituting these expressions in the modulus, representing $\frac{\pi}{4} - \frac{\phi}{2}$ by ϕ', and observing that if U_2 represent the work *yielded* by the driven wheel during the action of each tooth, then $P_2a_2 . \frac{2\pi}{n_2} = U_2$, so that $P_2a_2 = \frac{n_2 U_2}{2\pi}$, we have

$U_1 = \frac{n_2}{2e\pi}\left\{ A \log._\varepsilon \frac{\tan.\left(\frac{e\pi}{n_2} + \phi'\right)}{\tan. \phi'} - \sin. \phi \left(1 + \frac{n_2}{n_1}\right)\left\{ \frac{\cos. \phi}{e}\log._\varepsilon \frac{\cos.\left(\frac{2e\pi}{n_2} - \phi\right)}{\cos. \phi} - \frac{2\pi}{n_2}\sin. \phi \right\} \right\}U_2 + NS \cdot \cdot (266).$

Now $\log._\varepsilon \frac{\cos.\left(\frac{2e\pi}{n_2} - \phi\right)}{\cos. \phi} = \log._\varepsilon \left\{ 1 + \tan. \frac{2e\pi}{n_2} \tan. \phi \right\} \cos. \frac{2e\pi}{n_2}$

---

* Hymer's Integ. Cal. Art. 52.

$$=\log._{\varepsilon}\cos.\frac{2e\pi}{n_2}+\log._{\varepsilon}\left\{1+\tan.\frac{2e\pi}{n_2}\tan.\phi\right\}=\log._{\varepsilon}\cos.\frac{2e\pi}{n_2}+$$

$$\tan.\frac{2e\pi}{n_2}\tan.\phi-\tfrac{1}{2}\tan.^2\frac{2e\pi}{n_2}.\tan.^2\phi+\&c.$$ Substituting this expression in the preceding equation, and neglecting terms above the first dimension in tan. ϕ and sin. ϕ,

$$U_1=\frac{n_2}{2e\pi}\left\{A\log._{\varepsilon}\frac{\tan.\left(\frac{e\pi}{n_2}+\phi'\right)}{\tan.\phi'}-\frac{1}{2e}\left(1+\frac{n_2}{n_1}\right)\sin.2\phi\log._{\varepsilon}\cos.\frac{2e\pi}{n_2}\right\}U_2+NS\ \dots\ (267).$$

227. If the radius r of the generating circle be equal to one half the radius r_2 of the pitch circle of the driven wheel, according to the method generally adopted by mechanics (Art. 205.), then $e=\tfrac{1}{2}\frac{r_2}{r}=\tfrac{1}{2}\frac{2r}{r}=1.$

In this case therefore — that is, where the flanks of the driven wheel are straight (Art. 212.) — the modulus becomes

$$U_1=\frac{n_2}{2\pi}\left\{A\log._{\varepsilon}\frac{\tan.\left(\frac{\pi}{n_2}+\phi'\right)}{\tan.\phi'}-\tfrac{1}{2}\left(1+\frac{n_2}{n_1}\right)\sin.2\phi\log._{\varepsilon}\cos.\frac{2\pi}{n_2}\right\}U_2+NS\ \dots\ (268).$$

228. Substituting (in equation 267.) for ϕ' its value $\frac{\pi}{4}-\frac{\phi}{2}$,

$$\log._{\varepsilon}\frac{\tan.\left(\frac{e\pi}{n_2}+\phi'\right)}{\tan.\phi'}=\log._{\varepsilon}\frac{\tan.\left\{\frac{\pi}{4}+\left(\frac{e\pi}{n_2}-\frac{\phi}{2}\right)\right\}}{\tan.\left(\frac{\pi}{4}-\frac{\phi}{2}\right)}=\log._{\varepsilon}\frac{1+\tan.\left(\frac{e\pi}{n_2}-\frac{\phi}{2}\right)}{1-\tan.\left(\frac{e\pi}{n_2}-\frac{\phi}{2}\right)}-$$

$$\log._{\varepsilon}\frac{1-\tan.\frac{\phi}{2}}{1+\tan.\frac{\phi}{2}}=2\tan.\left(\frac{e\pi}{n_2}-\frac{\phi}{2}\right)+2\tan.\frac{\phi}{2}+\tfrac{2}{3}\tan.^3\left(\frac{e\pi}{n}-\frac{\phi}{2}\right)+\tfrac{2}{3}\tan.^3\frac{\phi}{2}+\&c.$$

If, therefore, we assume the teeth in the driven wheel to be so numerous, or n_2 to be so great a number, that the third power and all higher powers of tan. $\left(\frac{e\pi}{n_2}-\frac{\phi}{2}\right)$ may be neglected as compared with its first power, and if we neglect powers of tan. $\frac{\phi}{2}$ above the second,

$$\log_{\epsilon} \frac{\tan. \left(\frac{e\pi}{n_2}-\phi'\right)}{\tan. \phi'} = 2\left[\tan. \left(\frac{e\pi}{n_2}-\frac{\phi}{2}\right) + \tan. \frac{\phi}{2}\right];$$

which expression becomes $\frac{2e\pi}{n_2}$ if we suppose the two arcs which enter into it to be so small as to equal their respective tangents.

Again, $\log_{\epsilon} \cos. \frac{2e\pi}{n_2} = -\frac{1}{2}\left(\frac{2e\pi}{n_2}\right)^2$ very nearly.[*]

Substituting these values in equation (267), and performing actual multiplication by the factor $\frac{n_2}{2e\pi}$, we have

$$U_1 = \left\{A + \frac{1}{2}\pi\left(\frac{1}{n_1}+\frac{1}{n_2}\right) \sin. 2\varphi\right\} U_2 + NS;$$

or substituting for A its value; and assuming $\frac{1}{2} \sin. 2\varphi = \sin. \varphi$, since φ is exceedingly small,

$$U_1 = \left\{\left(1 + \frac{L_1\rho_1}{a_1r_1} \sin. \phi_1 + \frac{L_2\rho_2}{a_2r_2} \sin. \phi_2\right) + \pi\left(\frac{1}{n_1}+\frac{1}{n_2}\right)\sin. \phi\right\} U_2 + NS \ldots \ldots (269),$$

which is the modulus of a wheel and pinion having epicycloidal teeth, the number of teeth n_2 in the driven wheel being considerable (see equation 257.).

It is evident that the value of U_1 in the modulus (equation 266), admits of a *minimum* in respect to the value of e;

---

[*] For assume $\log_{\iota} \cos. x = a_1x^2 + a_2x^4 + a_3x^6 + \ldots$; then differentiating,

$$-\tan. x = 2a_1x + 4a_2x^3 + 6a_3x^5 + \ldots;$$

but (Miller, *Diff. Cal.* p. 95.) $-\tan. x = -x - \frac{1}{3}x^3 - \frac{2}{3.5}x^5 - \ldots$; equating, therefore, the co-efficients of these identical series, we have

$$a_1 = -\frac{1}{2}, \qquad a_1 = -\frac{1}{3.4}, \qquad a_3 = -\frac{2}{3.5.6}, \&c. = \&c.;$$

$$\therefore \log_{\iota} \cos. x = -\frac{x^2}{2} - \frac{x^4}{3.4} - \frac{2x^6}{3.5.6} \ldots$$

there is, therefore, a given relation of the radius of the generating circle of the driving, to that of the driven wheel, which relation being observed in striking the epicycloidal faces and the hypocycloidal flanks of the teeth of two wheels destined to work with one another, those wheels will work with a greater economy of power than they would under any other epicycloidal forms of their teeth.   This value of e may be determined by assuming the differential co-efficient of the co-efficient of U_2 in equation (266) equal to zero, and solving the resulting transcendental equation by the method of *approximation*.

229. THE MODULUS OF THE RACK AND PINION.

If the radius r_2 of the pitch circle of the driven wheel be supposed *infinite* (Art. 215.), that wheel becomes a *rack*, and the radius r_1 of the driving wheel remaining of finite dimensions, the two constitute a rack and pinion.   To determine the modulus of the rack and pinion in the case of teeth of any form, the number upon the pinion being great, or in the case of involute teeth and epicycloidal teeth of any number and dimensions, we have only to give to r_2 an infinite value in the moduli already determined in respect to these several conditions.   But it is to be observed in respect to epicycloidal teeth, that n_2 becomes infinite with r_2, whilst the ratio $\dfrac{r_2}{n_2}$ remains finite, and retains its equality to the ratio $\dfrac{r_1}{n_1}$ (equation 232), so that $\dfrac{e}{n_2} = \dfrac{1}{2}\dfrac{r_2}{n_2 r} = \dfrac{1}{2}\dfrac{r_1}{n_1 r} = \dfrac{e_1}{n_1}$; if we represent the ratio $\dfrac{r_1}{r}$ by $2e_1$.   Making n_2 and r_2 infinite in each of the equations (257), (259), and (266), and substituting $\dfrac{e_1}{n_1}$ for $\dfrac{e}{n_2}$ in equation (266); we have

1. For the modulus of the rack and pinion when the teeth are very *small*, whatever may be their forms, provided that they work truly.

$$U_1 = \left\{ 1 + \frac{L_1\rho_1}{a_1 r_1}\sin. \ \phi_1 + \frac{\pi}{n_1}\sin. \ \phi \right\} U_2 + NS \ \cdots \cdots (270).$$

2. For the modulus of a rack and pinion, with involute teeth of any dimensions (see *fig.* 1. p. 277.),

$$U_1 = \left\{ 1 + \left(\frac{\pi}{n_1}\cos. \ \eta \ \sin. \ \phi + \frac{L_1 r_1}{a_1 r_1}\sin. \ \phi_1 \right) \sec. \ (\eta - \phi) \right\} U_2 + NS \ \cdots (271).$$

3. For the modulus of the rack and pinion, with cycloidal and epicycloidal teeth respectively (Art. 216.),

$$U_1 = \frac{n_1}{2e_1\pi}\left\{ \left(1 + \frac{L_1\rho_1}{a_1 r_1}\sin. \ \phi_1 \right)\log._\varepsilon \frac{\tan. \left(\frac{e_1\pi}{n_1} + \phi' \right)}{\tan. \ \phi'} - \frac{\sin. \ 2\phi^*}{e_1}\log._\varepsilon \frac{\cos. \left(\frac{2e_1\pi}{n_1} \right)}{\cos. \ \phi} \right\} U_2 + NS.$$

In each of which cases the value of N is determined by making r_2 infinite in equation (252).

CONICAL OR BEVIL WHEELS.

230. These wheels are used to communicate a motion of rotation to any given axis from another, inclined to the first at any angle.

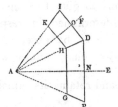

Let AF be an axis to which a motion of rotation is to be communicated from another axis AE inclined to the first at any angle EAF, by means of bevil wheels.

Divide the angle EAF by the straight line AD, so that DO and DN, perpendiculars from any point D in AD upon AE and AF respectively, may be to one another as the numbers of teeth which it is required to place upon the two wheels.†

$$* \ \frac{e}{n_2} = \frac{e_1}{n_1}, \ \therefore \ \frac{e}{e_1} = \frac{n_2}{n_1} ;$$

$\therefore$ (equation 267) $\frac{1}{2e}\left(1 + \frac{n_2}{n_1} \right) = \frac{1}{2e}\left(1 + \frac{e}{e_1} \right) = \frac{1}{2}\left(\frac{1}{e} + \frac{1}{e_1} \right) = \frac{1}{2e_1}$, because e is infinite. The friction of the rack upon its guides is not taken into account in the above equations.

† This division of the angle EAF may be made as follows :—Draw ST and UW from any points S and U in the straight lines AE and AF at

Suppose a cone to be generated by the revolution of the line AD about AE, and another by the revolution of the line AD about AF. Then if these cones were made to revolve *in contact* about the fixed axes AE and AF, their surfaces would *roll* upon one another along their whole line of contact DA, so that no part of the surface of one would slide upon that of the other, and thus the whole surface of the one cone, which passes in a given time over the line of contact AD, be equal to the whole surface of the other, which passes over that line in the same time. For it is evident that if n_1 times the circumference of the circle DP be equal to n_2 times that of the circle DI and these circles be conceived to revolve in contact carrying the cones with them, whilst the cone DAP makes n_1 revolutions, the cone DAI will make n_2 revolutions; so that whilst any other circle GH of the one cone makes n_1 revolutions, the corresponding circle HK of the other cone will make n_2 revolutions; but n_1 times the circumference of the circle GH is equal to n_2 times that of the circle HK, for the diameters of these circles, and therefore their circumferences, are to one another (by similar triangles) in the same proportion as the diameters and the circumferences of the circles DP and DI. Since, then, whilst the cones make n_1 and n_2 revolutions respectively, the circles HG and HK are carried through n_1

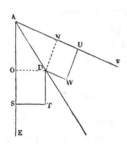

right angles to those lines respectively, and having their lengths in the ratio of the numbers of teeth which it is required to place upon the two wheels; and through the extremities T and W of these lines draw TD and WD parallel to AE and AF respectively, and meeting in D. A straight line drawn from A through D will then make the required division of the angle; for if DO and DN be drawn perpendicular to AE and AF, they will evidently be equal to UW and ST, and therefore in the required proportion of the numbers of the teeth; moreover, any other two lines drawn perpendicular to AE and AF from any other point in AD will manifestly be in the same proportion as DO and DN.

x 2

and n_2 revolutions respectively, and that n_1 times the circumference of HG is equal to n_2 times that of HK, therefore the
circles HG and HK *roll* in contact through the whole of
that space, nowhere sliding upon one another. And the
same is true of any other corresponding circles on the cones;
whence it follows that their whole surfaces are made to *roll*
upon one another by their mutual contact, no two parts being
made to slide upon one another by the rolling of the rest.

The rotation of the one axis might therefore be communicated to the other by the rolling of two such cones in contact, the surface of the one cone carrying with it the surface
of the other, along their line of contact AD, by reason of the
mutual friction of their surfaces, supposing that they could be
so pressed upon one another as to produce a friction equal to
the pressure under which the motion is communicated, or the
work transferred. In such a case the angular velocities of
the two axes would evidently be to one another (equation
232) inversely, as the circumferences of any two corresponding circles DP and DI upon the cones, or inversely as their
radii ND and OD, that is (by construction) inversely as the
numbers and teeth which it is proposed to cut upon the
wheels.

When, however, any considerable pressure accompanies
the motion to be communicated, the friction of two such
cones becomes insufficient, and it becomes necessary to transfer
it by the intervention of bevil teeth. It is the characteristic
property of these teeth that they cause the motion to be
transferred by their successive contact, precisely as it would
by the continued contact of the surfaces of the cones.

231. *To describe the teeth of bevil wheels.**

From D let FDE be drawn at right angles to AD, intersecting the axes AE and AF of the two cones in E and F;

---

* The method here given appears first to have been published by
Mr. Tredgold in his edition of *Buchanan's Essay on Millwork*, 1823,
p. 103.

suppose conical surfaces to be generated by the revolution of the lines DE and DF about AE and AF respectively;

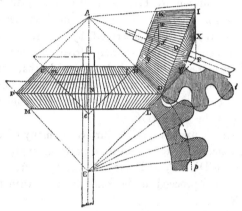

and let these conical surfaces be *truncated* by planes LM and XY respectively perpendicular to their axes AE and AF, leaving the distances DL and DY about equal to the *depths* which it is proposed to assign to the teeth. Let now the conical surface LDPM be conceived to be developed upon a plane perpendicular to AD, and passing through the point D, and let the conical surface XIDY be in like manner developed, and upon the same plane. When thus developed, these conical surfaces will have become the plane surfaces of two segmental annuli MP*pm* and IX*xi*[*], whose centres are in the points E and F of the axes AE and AF, and which touch one another in the point D of the line of contact AD of the cones.

[*] The lines MP and *pm* in the development, coincided upon the cone, as also the lines IX and *ix*, the other letters upon the development in the above figure represent points which are identical with those shown by the same letters in the preceding figure. In that figure the conical surfaces are shown developed, not in a plane perpendicular to AD, but in the plane which contains that line and the lines AE and AF, and which is perpendicular to the last-mentioned plane. It is evidently unnecessary, in the construction of the pattern teeth, actually to develop the conical extremities of the wheels as above described; we have only to determine the lengths of the radii DE and DF by construction, and with them to describe

Let now plane or *spur* teeth be struck upon the circles Pp and Ii, such as would cause them to drive one another as they would be driven by their mutual contact; that is, let these circles Pp and Ii be taken as the pitch circles of such teeth, and let the teeth be described, by any of the methods before explained, so that they may drive one another correctly. Let, moreover, their pitches be such, that there may be placed as many such teeth on the circumference Pp as there are to be teeth upon the bevil wheel HP, and as many on Ii as upon the wheel HI.

Having struck upon a flexible surface as many of the first teeth as are necessary to constitute a pattern, apply it to the conical surface DLMP, and trace off the teeth from it upon that surface, and proceed in the same manner with the surface DIXY.

Take DH equal to the proposed lengths of the teeth, draw *ef* through H perpendicular to AD, and terminate the wheels at their lesser extremities by concave surfaces HG*ml* and HK*xy* described in the same way as the convex surfaces which form their greater extremities. Proceed, moreover, in the construction of pattern teeth precisely in the same way in respect to those surfaces as the others; and trace out the teeth from these patterns on the lesser extremities as on the greater, taking care that any two similar points in the teeth traced upon the greater and lesser extremities shall lie in the same straight line passing through A. The pattern teeth thus traced upon the two extremities of the wheels are the extreme boundaries or edges of the teeth to be placed upon them, and are a sufficient guide to the workman in cutting them.

---

two arcs, Pp, Ii, for the pitch circles of the teeth, and to set off the pitches upon them of the same lengths as the pitches upon the circles DP and DI, which last are determined by the numbers of teeth required to be cut upon the wheels respectively.

232. *To prove that teeth thus constructed will work truly with one another.*

It is evident that if two exceedingly thin wheels had been taken in a plane perpendicular to AD (*fig.* p. 309.) passing through the point D, and having their centres in E and F; and if teeth had been cut upon these wheels according to the pattern above described, then would these wheels have worked truly with one another, and the ratio of their angular velocities have been inversely that of ED to FD, or (by similar triangles) inversely that of ND to OD; which is the ratio required to be given to the angular velocities of the bevil wheels.

Now it is evident that that portion of each of the conical surfaces DPML and DIXY which is at any instant passing through the line LY is at that instant revolving in the plane perpendicular to AD which passes through the point D, the one surface revolving in that plane about the centre E and the other about the centre F; those portions of the teeth of the bevil wheels which lie in these two conical surfaces will therefore drive one another truly, at the instant *when they are passing through the line* LY, if they be cut of the forms which they must have had to drive one another truly (and with the required ratio of their angular velocities) had they acted entirely in the above-mentioned plane perpendicular to AD and round the centres E and F. Now this is precisely the form in which they have been cut. Those portions of the bevil teeth which lie in the conical surfaces DPML and DIXY will therefore drive one another truly at the instant when they pass through the line LY; and therefore they will drive one another truly through an exceedingly small distance on either side of that line. Now it is only through an exceedingly small distance on either side of that line that any two given teeth remain in contact with one another. Thus then it follows, that those portions of the teeth which lie in the conical surfaces DM and DX work truly with one another.

x 4

Now conceive the faces of the teeth to be intersected by an infinity of conical surfaces parallel and similar to DM and DX; precisely in the same way it may be shown that those portions of the teeth which lie in each of this infinite number of conical surfaces work truly with one another; whence it follows that the *whole* surfaces of the teeth, constructed as above, work truly together.

233. THE MODULUS OF A SYSTEM OF TWO CONICAL OR BEVIL WHEELS.

Let the pressures P_1 and P_2 be applied to the conical wheels represented in the accompanying figure at perpendicular distances a_1 and a_2 from their axes CB and CG; let the length AF of their teeth be represented by b; let the distance of any point in this line from F be represented by x, and conceive it to be divided into an exceedingly great number of equal parts, each represented by Δx. Through each of these points of division imagine planes to be drawn

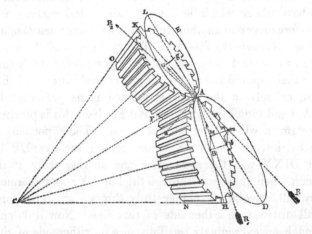

perpendicular to the axes CB and CG of the wheels, dividing the whole of each wheel into elements or laminæ of equal thickness; and let the pressures P_1 and P_2 be conceived to be equally distributed to these laminæ. The pressure thus distributed to each will then be represented by $\dfrac{P_1}{b}\Delta x$ on the one wheel, and $\dfrac{P_2}{b}\Delta x$ on the other. Let p_1 and p_2 represent the two pressures thus applied to the extreme laminæ AH and AK of the wheels, and let them be in equilibrium when thus applied to those sections separately and independently of the rest ; then if R represent the pressure sustained along that narrow portion of the surface of contact of the teeth of the wheels which is included within these laminæ, and if R_1 and R_2 represent the resolved parts of the pressure R in the directions of the planes AH and AK of these laminæ, the pressures p_1 and R_1 applied to the circle AH are pressures in equilibrium, as also the pressures p_2 and R_2 applied to the circle AK. If, therefore, we represent as before (Art. 218.) by m_1 and m_2, the perpendiculars from B and G upon the directions of R_1 and R_2, and by L_1 and L_2, the distances between the feet of the perpendiculars a_1, m_1 and a_2, m_2, we have (equation 241, 242.), neglecting the weights of the wheels,

$$\left.\begin{aligned} p_1 &= \frac{1}{a_1}\left\{ m_1 + \left(\frac{\rho_1 L_1}{a_1}\right)\sin.\,\phi_1 \right\}R_1 \\[2mm] p_2 &= \frac{1}{a_2}\left\{ m_2 - \left(\frac{\rho_2 L_2}{a_2}\right)\sin.\,\phi_2 \right\}R_2 \end{aligned}\right\} \quad \dots \dots (272).$$

ρ_1 and ρ_2 representing the radii of the axes of the two wheels, and ϕ_1 and ϕ_2 the corresponding limiting angles of resistance. Let γ_1 and γ_2 represent the inclinations of the direction of R to the planes of AH and AK respectively; then

$$R_1 = R \cos.\,\gamma_1, \quad R_2 = R \cos.\,\gamma_2.$$

Now it has been shown in the preceding article, that the action of that part of the surface of contact of the teeth which is included in each of the laminæ AH, AK, is identical with the action of teeth of the same form and pitch upon two cylindrical wheels AD and AL of the same small thickness, situated in a plane EAD perpendicular to AC, and having their centres in the intersections, b and g, with that plane of the axes CB and CG produced. The reciprocal pressure R of the teeth of the element has therefore its direction in the plane EAD; and if its direction coincided with the line of centres DL of the two circles EA and AD, then would its inclinations to the planes of AH and AK be represented by DAH and LAK, or by ACB and ACG.

The direction of R is however, in every case, inclined to the line of centres at a certain angle, which has been shown (Art. 218.) to be represented in every position of the teeth, after the point of contact has passed the line of centres by $(\theta + \varphi)$; where θ represents the inclination to AL of the line λ, which is drawn from the point of contact A of the pitch circles to the point of contact of the teeth, and where φ represents the limiting angle of resistance between the surfaces of the teeth. To determine the inclination γ_1 of RA to the plane of the circle AH, its inclination RAD to the line of centres being thus represented by $(\theta + \varphi)$, and the inclination of the plane AD, in which it acts, to the plane AH being DAH, which is equal ACB, let this last angle be represented by ι_1; and let Aa in the accompanying figure represent the intersection of the planes AD and AH; Aard representing a portion of the former plane and Aach of the latter. Let moreover Ar represent the direction of the pressure R in the former plane, and let Ad and Ah be portions of the lines AD and AH of the preceding figure. Draw rc perpendicular to the Aach, and rd and ch parallel to Aa, and join dh; then rAc represents the inclination γ_1 of the direction of R to the plane AD, dAr represents the inclination $(\varphi + \theta)$ of AR to AD, and dAh represents the inclination ι of the planes AD and AH to one another. Also, since Aa is perpendicular to the plane Ahd, therefore dr is perpendicular to that plane,

$$\therefore rc = \overline{Ar}\sin.\gamma_1 = \overline{Ad}\sec.(\theta + \varphi)\sin.\gamma_1.$$

Also $hd = \overline{Ad}\sin.\iota_1$, but $rc = hd$,

$$\therefore \overline{Ad}\sec.(\theta + \varphi)\sin.\gamma_1 = \overline{Ad}\sin.\iota_1;$$
$$\therefore \sin.\gamma_1 = \cos.(\theta + \varphi)\sin.\iota_1.$$

In like manner it may be shown that sin. $\gamma_2 = \cos. (\theta + \varphi) \sin. \iota_2$, ι_2 being taken to represent the inclination KAL of the planes AE and AK, which angle is also equal to the angle ACG.

From the above equations it follows that

$$R_1 = R \cos. \gamma_1 = R\sqrt{1 - \cos.^2(\theta + \varphi) \sin.^2\iota_1} \left.\begin{array}{c} \\ \\ \end{array}\right\} \quad \ldots \ldots (273).$$
$$R_2 = R \cos. \gamma_2 = R\sqrt{1 - \cos.^2(\theta + \varphi) \sin.^2\iota_2}$$

From the centre b of the circle AD draw bm per-

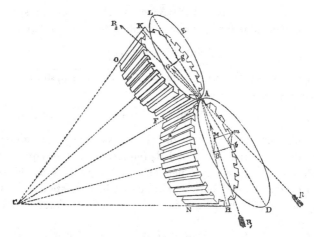

pendicular to RA, then is BM (the perpendicular let fall from the centre of the circle AH upon the direction of R_1) the *projection* of bm upon the plane of the circle AH. To determine the inclination of bm to the plane AH, draw An parallel to bm; the sine of the inclination of An to the plane AH is then determined to be cos. DAn. sin. ι_1, precisely as the sine of the inclination of Am to the same plane was before determined to be cos. DAm . sin.ι_1.

Now $\text{DA}n = \text{A}bm = \dfrac{\pi}{2} - \text{DAR} = \dfrac{\pi}{2} - (\theta + \varphi)$; therefore the sine of the inclination of An, and therefore of bm, to the plane AH is represented by the formula sin. $(\theta + \varphi) \sin.\iota_1$, and the cosine of its inclination by $\sqrt{1 - \sin.^2(\theta + \varphi) \sin. \iota_1}$;

$$\therefore \quad m_1 = \text{BM} = \overline{bm} \sqrt{1 - \sin.^2(\theta + \varphi) \sin.^2\iota_1}.$$

Now it has been shown (Art. 218.) that the perpendicular *bm* let fall from the centre of a spur wheel upon the direction of the pressure upon its teeth is in any position of their point of contact represented (equation 244) by the formula

$$r_1 \sin. (\theta + \varphi) + \lambda \sin. \varphi,$$

where θ, φ, λ represent the same quantities which they have been taken to represent in this article ; but r_1 represents the radius bA of the circle AD, instead of the radius BA of the circle AH; now $bA = \overline{BA} \sec. DAH = r_1 \sec. \iota_1$; substituting this value for r_1 in the preceding formula we have

$$bm = r_1 \sin. (\theta + \varphi) \sec. \iota_1 + \lambda \sin. \varphi \, ;$$

$$\left. \begin{array}{l} \therefore m_1 = \{ r_1 \sin. (\theta + \varphi) \sec. \iota_1 + \lambda \sin. \varphi \} \sqrt{1 - \sin.^2(\theta + \varphi) \sin.^2 \iota_1.} \\ \text{Similarly it may be shown that} \\ m_2 = \{ r_2 \sin. (\theta + \varphi) \sec. \iota_2 - \lambda \sin. \varphi \} \sqrt{1 - \sin.^2(\theta + \varphi) \sin.^2 \iota_2.} \end{array} \right\} \, .. \, (274).$$

Substituting the values of m_1 and m_2 above determined, and also the values of R_1 and R_2 (equations 273) in equations (272), and eliminating R between those equations, a relation will be determined between p_1 and p_2 which is applicable to any distance of the point of contact of the teeth from the line of centres.

Let it now be assumed that the number of the teeth of the driven wheel is considerable, so that the angle $\dfrac{2\pi}{n_2}$ traversed by the point of contact of each tooth may be small, and the greatest value of the line λ, the chord of an exceedingly small arc of the pitch circle of the driven wheel. In this case $\theta + \varphi$ will very nearly equal $\dfrac{\pi}{2}$ (Art. 222.); so that cos. $^2(\theta + \varphi)$ will be an exceedingly small quantity and may be neglected, and sin. $(\theta + \varphi)$ very nearly equal *unity*. Substituting these values in equations (273) and (274) we have

$$R_1 = R, \quad R_2 = R,$$

$$m_1 = r_1 + \lambda \sin. \varphi \cos. \iota_1, \quad m_2 = r_2 - \lambda \sin. \varphi \cos. \iota_2.$$

Substituting these values in equations (272) and dividing those equations by one another so as to eliminate R,

$$\frac{p_1}{p_2} = \frac{a_2}{a_1} \cdot \frac{r_1 + \lambda \sin. \phi \cos. \iota_1 + \left(\frac{\rho_1 L_1}{a_1}\right)\sin. \phi_1}{r_2 - \lambda \sin. \phi \cos. \iota_2 - \left(\frac{\rho_2 L_2}{a_2}\right)\sin. \phi_2};$$

$$\therefore \frac{p_1}{p_2} = \frac{a_2 r_1}{a_1 r_2} \cdot \frac{1 + \frac{\lambda}{r_1}\sin. \phi \cos. \iota_1 + \left(\frac{\rho_1 L_1}{a_1 r_1}\right)\sin. \phi_1}{1 - \frac{\lambda}{r_2}\sin. \phi \cos. \iota_2 - \left(\frac{\rho_2 L_2}{a_2 r_2}\right)\sin. \phi_2}.$$

Whence performing actual division by the denominator of the fraction, and neglecting terms involving dimensions above the first in sin. ϕ, sin. ϕ_1, sin. ϕ_2,

$$\frac{p_1}{p_2} = \frac{a_2 r_1}{a_1 r_2}\left\{1 + \lambda\left(\frac{\cos. \iota_1}{r_1} + \frac{\cos. \iota_2}{r_2}\right)\sin. \phi + \left(\frac{\rho_1 L_1}{a_1 r_1}\right)\sin. \phi_1 + \left(\frac{\rho_2 L_2}{a_2 r_2}\right)\sin. \phi_2\right\}.$$

Now if ψ represent the angle described by the driven wheel or circle ELA, whilst any two teeth are in contact, since λ is very nearly a chord of that circle subtending this small angle ψ (Art. 222.), $\therefore \lambda = r_2\psi$. Let Ψ represent the angle described by the conical wheel FK, whilst the circle ELA describes the angle ψ; then, since the pitch circle of the thin wheel AK and the circle ELA revolve in contact at A, they describe equal arcs whilst they thus revolve, respectively, through the unequal angles ψ and Ψ. Moreover, the radius Ag of the circle AL = $\overline{AG}$ sec. GAg = r_2 sec. ι_2, therefore ψr_2 sec. $\iota_2 = \Psi r_2$;

$$\therefore \psi = \Psi \cos. \iota_2 \quad \ldots \quad (275).$$

Substituting the above values of ψ and λ, and observing that $\frac{r_2}{r_1} = \frac{n_2}{n_1}$,

$$\frac{p_1}{p_2} = \frac{a_2 n_1}{a_1 n_2}\left\{1 + n_2\left(\frac{\cos. \iota_1}{n_1} + \frac{\cos. \iota_2}{n_2}\right)\Psi \cos. \iota_2 \sin. \phi + \left(\frac{\rho_1 L_1}{a_1 r_1}\right)\sin. \phi_1 + \left(\frac{\rho_2 L_2}{a_2 r_2}\right)\sin. \phi_2\right\} \ldots (276).$$

Multiplying both sides of this equation by $p_2\frac{a_1 n_2}{n_1}$, and observing that $p_1 a_1 \dfrac{n_2}{n_1} = p_1 a_1 \dfrac{n_2}{n_1} \dfrac{\Delta\Psi}{\Delta\Psi}$, and that $\dfrac{n_2}{n_1}\Delta\Psi$ is the exceedingly small angle described by the driving wheel AN, whilst the driven wheel describes the angle $\Delta\Psi$, so that if Δu_1 represent the work done by the pressure p_1 upon the lamina AH, whilst the angle $\Delta\Psi$ is described by the driven wheel, then $p_1 a_1 \dfrac{n_2}{n_1}\Delta\Psi = \Delta u_1$, we have

$$\frac{\Delta u_1}{\Delta\Psi} = p_2 a_2 \left\{ 1 + n_2 \left(\frac{\cos. \iota_1}{n_1} + \frac{\cos. \iota_2}{n_2} \right) \Psi \cos. \iota_2 \sin. \phi + \left(\frac{\rho_1 L_1}{a_1 r_1} \right)\sin. \phi_1 + \left(\frac{\rho_2 L_2}{a_2 r_2} \right)\sin. \phi_2 \right\} ;$$

or assuming $\Delta\Psi$ infinitely small, and integrating between the limits 0 and $\dfrac{2\pi}{n_2}$ (Art. 222.),

$$u_1 = \frac{2\pi p_2 a_2}{n_2}\left\{ 1 + \pi \left(\frac{\cos. \iota_1}{n_1} + \frac{\cos. \iota_2}{n_2} \right)\cos. \iota_2 \sin. \phi + \left(\frac{\rho_1 L_1}{a_1 r_1} \right)\sin. \phi_1 + \left(\frac{\rho_2 L_2}{a_2 r_2} \right)\sin. \phi_2 \right\}$$

Now the above relation between the work u_1 done by the pressure p_1 upon the extreme element AH of the driving wheel whilst any two teeth are in contact, and the pressure p_2 opposed to the motion of the corresponding element of the driven wheel, is evidently applicable to any other two corresponding elements; the values of p_2, r_1, r_2, L_1 and L_2 proper to those elements being substituted in the formula. If therefore we represent by ΔU_1 that increment of the whole work U_1 done upon the driving wheel, which is due to any one of the elements into which we have imagined that wheel to be divided, and if we substitute for p_2 its value $\dfrac{P_2}{b}\Delta x$, assign to L_1, L_2, r_1, r_2 their values proper to that element, and represent those values by L, L′, r, $r′$,

$$\Delta U_1 = \frac{2\pi P_2 a_2}{n_2 b}\left\{ 1 + \pi \left(\frac{\cos. \iota_1}{n_1} + \frac{\cos. \iota_2}{n_2} \right)\cos. \iota_2 \sin. \phi + \left(\frac{L \rho_1}{a_1 r} \right)\sin. \phi_1 + \left(\frac{L' \rho_2}{a_2 r'} \right)\sin \phi_2 \right\}\Delta x ;$$

or assuming Δx infinitely small, and integrating between the limits 0 and b, and observing that $P_2 a_2 \dfrac{2\pi}{n_2}$ represents the whole work U_2 done upon the driven wheel under the constant pressure P_2 during the contact of any two teeth,

$$U_1 = U_2\{1 + \pi\left(\frac{\cos. \iota_1}{n_1} + \frac{\cos. \iota_2}{n_2}\right) \cos. \iota_2 \sin. \phi + \frac{\rho_1 \sin. \phi_1}{ba_1}\int_0^b \frac{L}{r} dx + \frac{\rho_2 \sin. \phi_2}{ba_2}\int_0^b \frac{L'}{r'} dx\}. \quad (277).$$

Now $a + x$ being taken to represent the distance of the point of contact of any two such elements from C, and a to represent the distance CF, the radii r and r' of these elements are evidently (by similar triangles) represented by $\frac{a+x}{a} r_1$ or $\left(1 + \frac{x}{a}\right) r_1$, and $\frac{a+x}{a} r_2$ or $\left(1 + \frac{x}{a}\right) r_2$, r_1 and r_2 representing the radii of the extreme elements NF and OF, or of the pitch circles of the lesser extremities of the wheels.

Also assuming, as we have done, the pressures R_1 and R_2 to be perpendicular to the lines BA, GA joining the centre

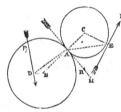

of each element with their point of contact A, so that the points M and N (see *fig.* p. 315.) coincide with the point A (see accompanying figure) * ; and representing the angles ABD and ACE made by the perpendiculars DB and CE with the line of centres by θ_1 and θ_2 respectively ; observing also that $\overline{AD^2} = \overline{BA^2} - \overline{2BA}$.

$\overline{BD} \cos. ABD + \overline{BD^2}$, so that $\left(\frac{AD}{BA}\right)^2 = 1 - 2\left(\frac{BD}{BA}\right) \cos. ABD + \left(\frac{BD}{BA}\right)^2$, we have, substituting, in the second member of this equation, for BA or r its value $r_1\left(1 + \frac{x}{a}\right)$

$$\left(\frac{L}{r}\right)^2 = 1 - 2\left(\frac{a_1}{r_1}\right)\left(1 + \frac{x}{a}\right)^{-1} \cos. \theta_1 + \left(\frac{a_1}{r_1}\right)^2\left(1 + \frac{x}{a}\right)^{-2} ;$$

or expanding the binomials in this expression, observing that $\frac{x}{a}$ is an exceedingly small quantity, neglecting terms involving powers of that quantity above the first, and reducing,

* The circles in this figure represent two of the corresponding laminæ into which wheels have been imagined to be divided ; they are not, therefore, in the same plane. Their planes intersect in AH.

$$\left(\frac{L}{r}\right)^2 = 1 - 2\left(\frac{a_1}{r_1}\right)\cos.\ \theta_1 + \left(\frac{a_1}{r_1}\right)^2 + 2\left(\frac{a_1}{r_1}\right)\left(\cos.\ \theta_1 - \frac{a_1}{r_1}\right)\frac{x}{a}\ .\ (278)$$

Now L_1 representing the value of L when $x=0$, and θ remaining constant,

$$\left(\frac{L_1}{r_1}\right)^2 = 1 - 2\left(\frac{a_1}{r_1}\right)\cos.\ \theta_1 + \left(\frac{a_1}{r_1}\right)^2 ;$$

$$\therefore\ 2\left(\frac{a_1}{r_1}\right)\left(\cos.\ \theta_1 - \frac{a_1}{r_1}\right) = 1 - \left(\frac{a_1}{r_1}\right)^2 - \left(\frac{L_1}{r_1}\right)^2.$$

Let now the angle ADB, made in respect to the first element of the driving wheel between the perpendicular BD or a_1 and the chord AD or L, be represented by η_1, and let η_2 represent the corresponding angle in the driven wheel, then

$$L_1{}^2 - 2L_1 a_1 \cos.\ \eta_1 + a_1{}^2 = r_1{}^2,\ \therefore\ \left(\frac{L_1}{r_1}\right)^2 - 2\frac{L_1 a_1}{r_1{}^2}\cos.\ \eta_1 + \left(\frac{a_1}{r_1}\right)^2 = 1;$$

$$\therefore\ -2\frac{L_1 a_1}{r_1{}^2}\cos.\ \eta_1 = 1 - \left(\frac{a_1}{r_1}\right)^2 - \left(\frac{L_1}{r_1}\right)^2 = 2\left(\frac{a_1}{r_1}\right)\left(\cos.\ \theta_1 - \frac{a_1}{r_1}\right).$$

Substituting these values of $\left(\frac{L_1}{r_1}\right)^2$ and $2\left(\frac{a_1}{r_1}\right)\left(\cos.\ \theta - \frac{a_1}{r_1}\right)$

in equation (278) ;

$$\left(\frac{L}{r}\right)^2 = \left(\frac{L_1}{r_1}\right)^2 - 2\left(\frac{L_1 a_1}{r_1{}^2}\right)\left(\frac{x}{a}\right)\cos.\ \eta_1 = \left(\frac{L_1}{r_1}\right)^2\left\{1 - 2\left(\frac{a_1}{L_1}\right)\left(\frac{x}{a}\right)\cos.\ \eta_1\right\};$$

Extracting the square root of the binomial, and neglecting terms involving powers of $\frac{x}{a}$ above the first,

$$\frac{L}{r} = \frac{L_1}{r_1} - \left(\frac{a_1}{r_1}\right)\left(\frac{x}{a}\right)\cos.\ \eta_1 = \frac{a_1}{r_1}\left\{\frac{L_1}{a_1} - \frac{x}{a}\cos.\ \eta_1\right\};$$

$$\therefore\ \text{(Equation 277)}\ \frac{\rho_1\sin.\ \phi_1}{ba_1}\int_0^b\frac{L}{r}dx = \frac{\rho_1\sin.\ \phi_1}{r_1}\left\{\frac{L_1}{a_1} - \tfrac{1}{2}\frac{b}{a}\cos.\ \eta_1\right\}.$$

Similarly $\dfrac{\rho_2\sin.\ \phi_2}{ba_2}\displaystyle\int_0^b\frac{L'}{r'}dx = \dfrac{\rho_2\sin.\ \phi_2}{r_2}\left\{\dfrac{L_2}{a_2} - \tfrac{1}{2}\dfrac{b}{a}\cos.\ \eta_2\right\}.$

Substituting these values in the modulus (equation 277),

$$U_1 = U_2 \left\{ 1 + \pi \left(\frac{\cos. \iota_1}{n_1} + \frac{\cos. \iota_2}{n_2} \right) \cos. \iota_2 \sin. \phi + \frac{\rho_1 \sin. \phi_1}{r_1} \left(\frac{L_1}{a_1} - \frac{1}{2} \frac{b}{a} \cos. \eta_1 \right) + \right.$$

$$\left. \frac{\rho_2 \sin. \phi_2}{r_2} \left(\frac{L_2}{a_2} - \frac{1}{2} \frac{b}{a} \cos. \eta_2 \right). \right\}$$

Now let the angle BCG, or the inclination of the axes, from one to the other of which motion is transferred by the wheels, be represented by 2ι; therefore $\iota_1 + \iota_2 = 2\iota$. Also $a \sin. \iota_1 = r_1$ and $a \sin. \iota_2 = r_2$,

$$\therefore \frac{\sin. \iota_1}{\sin. \iota_2} = \frac{r_1}{r_2} = \frac{n_1}{n_2};$$

$$\therefore \frac{\sin.^2 \iota_1}{n_1^2} = \frac{\sin.^2 \iota_2}{n_2^2}; \quad \therefore \frac{1}{n_1^2} - \frac{\cos.^2 \iota_1}{n_1^2} = \frac{1}{n_2^2} - \frac{\cos.^2 \iota_2}{n_2^2};$$

$$\therefore \frac{1}{n_1^2} - \frac{1}{n_2^2} = \frac{\cos.^2 \iota_1}{n_1^2} - \frac{\cos.^2 \iota_2}{n_2^2} = \left(\frac{\cos. \iota_1}{n_1} + \frac{\cos. \iota_2}{n_2} \right) \left(\frac{\cos. \iota_1}{n_1} - \frac{\cos. \iota_2}{n_2} \right) =$$

$$\left(\frac{\cos. \iota_1}{n_1} + \frac{\cos. \iota_2}{n_2} \right) \left(\frac{1}{n_1} \frac{\cos. \iota_1}{\cos. \iota_2} - \frac{1}{n_2} \right) \cos. \iota_2 ;$$

$$\therefore \left(\frac{\cos. \iota_1}{n_1} + \frac{\cos. \iota_2}{n_2} \right) \cos. \iota_2 = \frac{\dfrac{1}{n_1^2} - \dfrac{1}{n_2^2}}{\dfrac{1}{n_1} \dfrac{\cos. \iota_1}{\cos. \iota_2} - \dfrac{1}{n_2}}.$$

Now $\dfrac{\cos. \iota_1}{\cos. \iota_2} = \dfrac{\cos. \{\iota + \frac{1}{2}(\iota_1 - \iota_2)\}}{\cos. \{\iota - \frac{1}{2}(\iota_1 - \iota_2)\}} = \dfrac{1 - \tan. \frac{1}{2}(\iota_1 - \iota_2) \tan. \iota}{1 + \tan. \frac{1}{2}(\iota_1 - \iota_2) \tan. \iota}$;

also $\dfrac{n_1}{n_2} = \dfrac{\sin. \iota_1}{\sin. \iota_2} = \dfrac{\sin. \{\iota + \frac{1}{2}(\iota_1 - \iota_2)\}}{\sin. \{\iota - \frac{1}{2}(\iota_1 - \iota_2)\}} = \dfrac{\tan. \iota + \tan. \frac{1}{2}(\iota_1 - \iota_2)}{\tan. \iota - \tan. \frac{1}{2}(\iota_1 - \iota_2)}$

$$\therefore \tan. \frac{1}{2}(\iota_1 - \iota_2) = \frac{n_1 - n_2}{n_1 + n_2} \tan. \iota ;$$

$$\therefore \frac{\cos. \iota_1}{\cos. \iota_2} = \frac{1 - \dfrac{n_1 - n_2}{n_1 + n_2} \tan.^2 \iota}{1 + \dfrac{n_1 - n_2}{n_1 + n_2} \tan.^2 \iota} = \frac{(n_1 + n_2) - (n_1 - n_2) \tan.^2 \iota}{(n_1 + n_2) + (n_1 - n_2) \tan.^2 \iota} ;$$

$$\therefore \frac{1}{n_1} \frac{\cos. \iota_1}{\cos. \iota_2} - \frac{1}{n_2} = \frac{1}{n_1 n_2} \left(n_2 \frac{\cos. \iota_1}{\cos. \iota_2} - n_1 \right) = \frac{-1}{n_1 n_2} \frac{(n_1^2 - n_2^2) + (n_1^2 - n_2^2) \tan.^2 \iota}{(n_1 + n_2) + (n_1 - n_2) \tan.^2 \iota}$$

Y

$$= \frac{\left(\frac{1}{n_1^2}-\frac{1}{n_2^2}\right)\sec.^2\iota}{\left(\frac{1}{n_1}+\frac{1}{n_2}\right)-\left(\frac{1}{n_1}-\frac{1}{n_2}\right)\tan.^2\iota} = \frac{\left(\frac{1}{n_1^2}-\frac{1}{n_2^2}\right)}{\left(\frac{1}{n_1}+\frac{1}{n_2}\right)\cos.^2\iota-\left(\frac{1}{n_1}-\frac{1}{n_2}\right)\sin.^2\iota} ;$$

$$\therefore \left(\frac{\cos.\iota_1}{n_1}+\frac{\cos.\iota_2}{n_2}\right)\cos.\iota_2 = \left(\frac{1}{n_1}+\frac{1}{n_2}\right)\cos.\iota^2-\left(\frac{1}{n_1}-\frac{1}{n_2}\right)\sin.\iota^2 = \left(\frac{1}{n_1}+\frac{1}{n_2}\right)-\frac{2\sin.^2\iota}{n_1}.$$

Substituting in the preceding relation, between U_1 and U_2,

$$U_1 = \left\{1+\pi\left\{\left(\frac{1}{n_1}+\frac{1}{n_2}\right)-\frac{2\sin.^2\iota}{n_1}\right\}\sin.\phi+\frac{\rho_1\sin.\phi_1}{r_1}\left(\frac{L_1}{a_1}-\tfrac{1}{2}\frac{b}{a}\cos.\eta_1\right)+\right.$$
$$\left. \frac{\rho_2\sin.\phi_2}{r_2}\left(\frac{L_2}{a_2}-\tfrac{1}{2}\frac{b}{a}\cos.\eta_2\right)\right\}U_2 \dots\dots (279),$$

which is the modulus of the conical or bevil wheel, neglecting the influence of the weight of the wheel.

If for cos. η_1 and cos. η_2 we substitute their values (see p. 320.), we shall obtain by reduction

$$U_1 = \left\{1+\pi\left\{\left(\frac{1}{n_1}+\frac{1}{n_2}\right)-\frac{2\sin.^2\iota}{n_1}\right\}\sin.\phi+\frac{\rho_1\sin.\phi}{a_1 r_1}\left\{L\left(1-\tfrac{1}{4}\frac{b}{a}\right)+\frac{b(r_1^2-a_1^2)}{4a\,L_1}\right\}\right.$$
$$\left. +\frac{\rho_2\sin.\phi_2}{a_2 r_2}\left\{L_2\left(1-\tfrac{1}{4}\frac{b}{a}\right)+\frac{b(r_2^2-a_2^2)}{4a L_2}\right\}\right\}U_2 \dots\dots (280).$$

from which equation it is manifest that the most favourable directions of the driving and working pressures are those determined by the equations

$$L_1^2 = b\left(\frac{a_1^2-r_1^2}{4a-b}\right), \qquad L_2^2 = b\left(\frac{a_2^2-r_2^2}{4a-b}\right).$$

234. It is evident, that if the plane of the revolution of such a wheel be vertical, the influence of its weight must be very nearly the same as that of a cylindrical or spur wheel of the same weight, having a radius equal to the mean radius of the conical wheel, and revolving also in a vertical plane. If the axis of the wheel be not horizontal, its weight must be resolved into two pressures, one acting in the plane of the wheel, and the other at right angles to it; the latter is ef-

fective only on the extremity of the axis, where it is borne as
by a pivot, so that the work expended by reason of it may be
determined by Art. 177., and will be found to present itself
under the form of $N_2 . S$, where N_2 is a constant and S the
space described by the pitch circle of the wheel, whilst
the work U_1 is done. The resolved weight in the plane
of the wheel must be substituted for the weight of the
wheel in equation (252), which determines the value of N.
Assuming the value of N, this substitution being made, to
be represented by N_1, the whole of the second term of the
modulus will thus present itself under the form $(N_1 + N_2)S$.

$$\therefore U_1 = \left\{ 1 + \pi \left\{ \left(\frac{1}{n_1} + \frac{1}{n_2} \right) - \frac{2 \sin.^2 \iota}{n_1} \right\} \sin. \phi + \frac{\rho_1 \sin. \phi_1}{r_1} \left(\frac{L_1}{a_1} - \tfrac{1}{2} \frac{b}{a} \cos. \eta_1 \right) \right.$$

$$\left. + \frac{\rho_2 \sin. \phi_2}{r_2} \left(\frac{L_2}{a_2} - \tfrac{1}{2} \frac{b}{a} \cos. \eta_2 \right) \right\} U_2 + (N_1 + N_2)S \dots (281).$$

235. Comparing the modulus of a system of two conical
wheels with that of a system of two cylindrical wheels
(equation 257), it will be seen that the fractional excess
of the work U_2 lost by the friction of the latter over
that lost by the friction of the former is represented by the
formula

$$ + \frac{2\pi \sin.^2 \iota \sin. \phi}{n_1} + \tfrac{1}{2} \frac{b}{a} \left(\frac{\rho_1}{r_1} \cos. \eta_1 \sin. \phi_1 + \frac{\rho_2}{r_2} \cos. \eta_2 \sin. \phi_2 \right) \dots (281).$$

The first term of this expression is due to the friction of
the teeth of the wheels alone, as distinguished from the friction of their axes ; the latter is due exclusively to the friction
of the axes. Both terms are essentially positive, since η_1
and η_2 are in every case less than $\frac{\pi}{2}$.

Thus, then, it appears that the loss of power due to the
friction of bevil wheels is (other things being the same)
essentially less than that due to the friction of spur wheels,
so that there is an economy of power in the substitution of a
bevil for a spur wheel wherever such substitution is practicable. This result is entirely consistent with the experience

of engineers, to whom it is well known that bevil wheels *run lighter* than spur wheels.

236. THE MODULUS OF A TRAIN OF WHEELS.

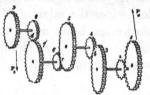

In a train of wheels such as that shown in the accompanying figure, let the radii of their pitch circles be represented in order by $r_1, r_2, r_3 \ldots r_4$, beginning from the driving wheel; and let a_1 represent the perpendicular distance of the driving pressure from the centre of that wheel, and a_2 that of the driven pressure or resistance from the centre of the last wheel of the train; U the work done upon the first wheel, u_2 the work yielded by the second wheel to the third, u_3 that yielded by the fourth to the fifth, &c., and U_2 the work yielded by the last or n^{th} wheel upon the resistance, then is the relation between U_1 and u_2 determined by the modulus (equation 257), it being observed that the point of application of the resistance on the second wheel is its point of contact b with the third wheel, so that in this case $a_2 = r_3$.

These substitutions being made, and L_2 being taken to represent the distance between the point b and the *projection* of the point a upon the third wheel, we have

$$U_1 = \left\{ 1 + \pi \left(\frac{1}{n_1} + \frac{1}{n_2} \right) \sin. \varphi + \frac{L_1 \rho_1}{a_1 r_1} \sin. \varphi_1 + \frac{L_2 \rho_2}{r_2 r_3} \sin. \varphi_2 \right\} u_2 + N_1{}^*. S.$$

To determine, in like manner, the relation between u_2 and u_3, or the modulus of the third and fourth wheels, let it be observed that the work u_2 which drives the third wheel has been considered to be done upon it at its point of contact b with the fourth; so that in this case the distance between the point of contact of the driving and driven wheels and the foot of the perpendicular let fall upon the driving pres-

* See note, p. 289.

sure from the centre of the driving wheel *vanishes*, and the term which involves the value of L representing that line disappears from the modulus, whilst the perpendicular upon the driving pressure from the centre of the driving wheel becomes r_3. Let it also be observed, that the work of the fourth wheel is done at the point of contact c of the fifth and sixth wheels, so that the perpendicular upon the direction of that work from the axis of the driven wheel is r_5. We shall thus obtain for the modulus of the third and fourth wheels,

$$u_2 = \left\{ 1 + \pi\left(\frac{1}{n_3} + \frac{1}{n_4}\right)\sin.\varphi + \frac{L_3\rho_3}{r_4 r_5}\sin.\varphi_3 \right\} u_3 + N_2 S_2.$$

In which expression L_3 represents the distance between the point c and the projection of the point b upon the fifth wheel.

In like manner it may be shown, that the modulus of the fifth and sixth wheels, or the relation between u_3 and u_4, is

$$u_3 = \left\{ 1 + \pi\left(\frac{1}{n_5} + \frac{1}{n_6}\right)\sin.\varphi + \frac{L_4\rho_4}{r_6 r_7}\sin.\varphi_4 \right\} u_4 + N_3 . S_3;$$

and that of the seventh and eighth wheels, or the relation between u_4 and u_5,

$$u_4 = \left\{ 1 + \pi\left(\frac{1}{n_7} + \frac{1}{n_8}\right)\sin.\varphi + \frac{L_5\rho_5}{r_8 r_9}\sin.\varphi_5 \right\} u_5 + N_4 . S_4;$$

and that, if the whole number of wheels be represented by $2p$, or the number of pairs of wheels in the train by p, then is the modulus of the last pair,

$$u_p = \left\{ 1 + \pi\left(\frac{1}{n_{2p-1}} + \frac{1}{n_{2p}}\right)\sin.\varphi + \frac{L_{p+1}\,\rho_{p+1}}{r_{2p}a_2}\sin.\varphi_{p+1} \right\} U_2 + N_p . S_p;$$

In which expressions the symbols N_1, N_2, N_3 ... N_p, are taken to represent, in respect to the successive pairs of wheels of the train, the values of that function (equation 252), which determines the friction due to the *weights* of those wheels; and each of the symbols L_2, L_3, L_4 ... L_p, the distance between the point of contact of a corresponding pair of wheels and the projection upon its plane of the point of contact of the next preceding pair in the train; whilst the symbols n_1, n_2, n_3 ... n_{2p}, represent the numbers of teeth in the wheels; r_1, r_2, r_3, ... r_{2p}, the radii of their pitch circles; and S_1, S_2, S_3 ... S_p, the spaces described by their points of contact a, b, c, &c. whilst the work U_1 is done upon the first wheel of the train.

Let us suppose the coefficients of u_2, u_3, u_4 ... U_2, in these moduli to be represented by $(1+\mu_1)$, $(1+\mu_2)$, $(1+\mu_3)$ .... $(1+\mu_p)$; they will then become

$$U_1 = (1+\mu_1)\, u_2 + N_1 . S_1,$$
$$u_2 = (1+\mu_2)\, u_3 + N_2 . S_2,$$
$$u_3 = (1+\mu_3)\, u_4 + N_3 . S_3,$$
$$\&c. = \&c.$$
$$u_p = (1+\mu_p)\, U_2 + N_p . S_p.$$

Eliminating u_2, u_3, u_4 ... u_p, between these equations, we shall obtain an equation of the form

$$U_1 = (1+\mu_1)(1+\mu_2)(1+\mu_3) \ldots (1+\mu_p)\, U_2 + N . S \ldots (282),$$

where

$$NS = N_1 S_1 + (1+\mu_1) N_2 S_2 + (1+\mu_1)(1+\mu_2) N_3 S_3 + \ldots$$
$$+ (1+\mu_1)(1+\mu_2) \ldots (1+\mu_p) N_p S_p \ldots \ldots (283).$$

Now let it be observed, that the space described by the first wheel, at distance unity from its centre, whilst the space S_1 is described by its circumference, is represented by $\dfrac{S_1}{r_1}$, and that this same space is represented by $\dfrac{S}{a_1}$ if S represent the space described in the same time by the foot of the perpendicular a_1, or the space through which the moving pressure may be conceived to work during that time; so that $\dfrac{S_1}{r_1} = \dfrac{S}{a_1}$. Also let it be observed that the space described by the third wheel, at distance unity from its centre,

is the same with that described at the same distance from
its centre by the second wheel, so that $\dfrac{S_2}{r_3} = \dfrac{S_1}{r_2}$; in like
manner that the spaces described at distances unity from
their centres by the fourth and fifth wheels are the same, so
that $\dfrac{S_3}{r_5} = \dfrac{S_2}{r_4}$; and similarly, that $\dfrac{S_4}{r_7} = \dfrac{S_3}{r_6}$, &c. = &c.; and
finally, $\dfrac{S_p}{r_{2p-1}} = \dfrac{S_{p-1}}{r_{2p-2}}$.

Multiplying the *two* first of these equations together, then
the *three* first, the *four* first, &c., and transposing, we have

$$S_1 = \frac{r_1}{a_1} S, \quad S_2 = \frac{r_1 \cdot r_3}{a_1 \cdot r_2} S = \left(\frac{r_1}{a_1}\right)\left(\frac{n_3}{n_2}\right) S,$$

$$S_3 = \frac{r_1 \cdot r_3 \cdot r_5}{a_1 \cdot r_2 \cdot r_4} S = \left(\frac{r_1}{a_1}\right)\left(\frac{n_3 \cdot n_5}{n_2 \cdot n_4}\right) S,$$

$$S_4 = \frac{r_1 \cdot r_3 \cdot r_5 \cdot r_7}{a_1 \cdot r_2 \cdot r_4 \cdot r_6} S = \left(\frac{r_1}{a_1}\right)\left(\frac{n_3 \cdot n_5 \cdot n_7}{n_2 \cdot n_4 \cdot n_6}\right) S,$$

&c. = &c.

$$S_p = \frac{r_1 \cdot r_3 \cdot r_5 \cdots r_{2p-1}}{a_1 \cdot r_2 \cdot r_4 \cdots r_{2p-2}} S = \left(\frac{r_1}{a_1}\right)\left(\frac{n_3 \cdot n_5 \cdots n_{2p-1}}{n_2 \cdot n_4 \cdots n_{2p-2}}\right) S.$$

Substituting these values of S_1, S_2, &c. in equation (283),
and dividing by S, we have

$$N = \left(\frac{r_1}{a_1}\right)\left\{ N_1 + (1+\mu_1)\left(\frac{n_3}{n_2}\right) N_2 + (1+\mu_1)(1+\mu_2)\left(\frac{n_3 \cdot n_5}{n_2 \cdot n_4}\right) N_3 + \dots \right\};$$

or if we observe that the quantities μ_1, μ_2, μ_3, are composed
of terms all of which are of *one* dimension in sin. φ, sin. φ_1,
sin. ϕ_2, &c. and that the quantities N_1, N_2, N_3, &c. (equation 252) are all likewise of one dimension in those exceedingly small quantities; and if we neglect terms above the first
dimension in those quantities, then

$$N = \left(\frac{r_1}{a_1}\right)\left\{ N_1 + \left(\frac{n_3}{n_2}\right) N_2 + \left(\frac{n_3 n_5}{n_2 n_4}\right) N_3 + \left(\frac{n_3 n_5 n_7}{n_2 n_4 n_6}\right) N_4 + \dots \right\} \dots (284).$$

If in like manner we neglect in equation (282) terms of more than one dimension in μ_1, μ_2, μ_3, &c. we have

$$U_1 = \{1 + \mu_1 + \mu_2 + \mu_3 + \ldots + \mu_p\}U_2 + N . S.$$

Now $\mu_1 = \pi\left(\dfrac{1}{n_1} + \dfrac{1}{n_2}\right)\sin.\phi + \dfrac{L_1\rho_1}{1^r}\sin.\phi_1 + \dfrac{L_2\rho_2}{r_2 r_3}\sin.\phi_2$,

$\mu_2 = \pi\left(\dfrac{1}{n_3} + \dfrac{1}{n_4}\right)\sin.\phi + \dfrac{L_3\rho_3}{r_4 r_5}\sin.\phi_3$,

$\mu_3 = \pi\left(\dfrac{1}{n_5} + \dfrac{1}{n_6}\right)\sin.\phi + \dfrac{L_4\rho_4}{r_6 r_7}\sin.\phi_4$,

&c. = &c.

$\mu_p = \pi\left(\dfrac{1}{n_{2p-1}} + \dfrac{1}{n_{2p}}\right)\sin.\phi + \dfrac{L_p\rho_p}{r_{9p}a_2}\sin.\phi_p$.

Substituting these values of μ_1, μ_2, &c. in the preceding equation,

$$\therefore U_1 = \left\{1 + \pi\left(\frac{1}{n_1} + \frac{1}{n_2} + \frac{1}{n} \cdots \frac{1}{n_{2p}}\right)\sin.\phi + \frac{L_1\rho_1}{a_1 r_2}\sin.\phi_1 + \frac{L_2\rho_2}{r_2 r_3}\sin.\phi_2 + \frac{L_3\rho_3}{r_4 r_5}\sin.\phi_3 \right.$$
$$\left. + \ldots \ldots \frac{L_p\rho_p}{r_{2p}a_2}\sin.\phi_p\right\}U_2 + N . S \ldots (285),$$

which is a general expression for the modulus of a train of any number of wheels.

237. The work U_1 which must be done upon the first wheel of a train to yield a given amount U_2 at the last wheel, exceeds the work U_2, or, in other words, the work done upon the driving point exceeds that yielded at the *working* point, by a quantity which is represented by the expression

$$\pi\left(\frac{1}{n_1} + \frac{1}{n_2} + \cdots + \frac{1}{n_{2p}}\right)\sin.\phi . U_2 + \left(\frac{L_1\rho_1}{a_1 r_1}\sin.\phi_1 + \frac{L_2\rho_2}{r_{2'}3}\sin.\phi_2 + \cdots + \frac{L_p\rho_p}{r_{2p}a_2}\sin.\phi_p\right)U_2 + NS \ldots (286).$$

In which expression the *first* term represents the expenditure of work due to the friction of the teeth, and varies directly as the work U_2, which is done by the machine. The *second* term represents the expenditure of work due to the friction of the axes of the wheels, and varies in like manner directly as the work done. Whilst the *third* term represents the expenditure of work due to the weights of the wheels of the train, and is wholly independent of the work done, but only upon the space S, through which that work is done at the point where the driving pressure is applied to the train.

238. *The expenditure of work due to the friction of the teeth.*

The work expended upon the friction of the teeth is represented by the formula

$$\pi \left(\frac{1}{n_1} + \frac{1}{n_2} + \frac{1}{n_3} + \cdots + \frac{1}{n_{2p}} \right) \sin. \varphi \quad \ldots \ldots (287),$$

whose value is evidently less as the factor sin. φ is less, or as the coefficient of friction between the common surfaces of the teeth is less; and as the numbers of the teeth in the different wheels which compose the train are greater. The number of teeth in any one wheel of the train may, in fact, be taken so small, as to give this formula a considerable value as compared with U_2, or to cause the expenditure of work upon the friction of the teeth to amount to a considerable fraction of the work yielded by the train : and the numbers of teeth of *two* or *more* wheels of such a train might even be taken so small as to cause the work expended upon their friction to *equal* or to *surpass* by any number of times the work yielded by the train at its working point. This will become the more apparent if we consider that the surfaces of contact of the teeth of wheels are for the most part free from unguent after they have remained any considerable time in action, so that the limiting angle of resistance assumes in most cases a much greater value at the surfaces of the teeth of the wheels than at their axes. From this consideration the importance of assigning the greatest possible number of teeth to the wheels of a train individually and collectively is apparent.

239. *The expenditure of work due to the friction of the axes.*

This expenditure is represented by the formula

$$\left(\frac{L_1\rho_1}{a_1 r_1}\sin.\ \varphi_1 + \frac{L_2\rho_2}{r_2 r_3}\sin.\ \varphi_2 + \ldots + \frac{L_p\rho_p}{r_{2p} a_2}\sin.\ \phi_p\right) U_2 \ldots (288),$$

forming the second term of formula 285. Now, evidently, the value of this formula is less as the quantities sin. φ_1, sin. φ_2, &c. are less, or as the limiting angles of resistance between the surfaces of the axes and their bearings are less, or the lubrication of the axes more perfect; and it is less as the fractions $\dfrac{L_1\rho_1}{a_1 r_1}$, $\dfrac{L_2\rho_2}{r_2 r_3}$, $\dfrac{L_3\rho_3}{r_4 r_5}$, &c. are less.

Now, L_2 being the distance between the point of contact b

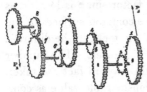

of the third and fourth wheels and the projection of the point of contact a of the first and second upon the plane of those wheels, it follows that, generally, L_2 is least when the projection of a falls on *the same side of the axis* as the point b [*]; and that it is *least of all* when this line falls on that side and in the line joining the axis with the point b; whilst it is *greatest of all* when it falls in this line produced to the opposite side of the axis. In the former case its value is represented by $r_3 - r_2$, and in the latter by $r_3 + r_2$; so that, generally, the maximum and minimum values of L_2 are represented by the expression $r_3 \pm r_2$, and the maximum and minimum values of $\dfrac{L_2\rho_2}{r_2 r_3}$ by $\left(\dfrac{1}{r_2} \pm \dfrac{1}{r_3}\right)\rho_2$. And similarly it appears that the maximum and minimum values of $\dfrac{L_3\rho_3}{r_4 r_5}$ are represented by $\left(\dfrac{1}{r_4} \pm \dfrac{1}{r_5}\right)\rho_3$; and so of the rest. So that the maximum and minimum values of

---

[*] This important condition is but a particular case of the general principle established in Art. 168.; from which principle it follows, that the driving pressure on each wheel should be applied on the same side of the axis as the driven pressure.

the work lost by the friction of the axes are represented
by the expression

$$\left\{ \left(\frac{1}{a_1} \pm \frac{1}{r_1}\right) \rho_1 \sin. \varphi_1 + \left(\frac{1}{r_2} \pm \frac{1}{r_3}\right) \rho_2 \sin. \varphi_2 + \left(\frac{1}{r_4} \pm \frac{1}{r_5}\right) \rho_3 \sin. \varphi + \dots \right\} U_2;$$

from which expression it is manifest, that in every case the
expenditure of work due to the friction of the axes is less as the
radii of the axes are less when compared with the radii of
the wheels; being wholly independent of actual dimensions of
these radii, but only upon the ratio or proportion of the
radius of each axis to that of its corresponding wheel: moreover, that this expenditure of work is the least when the
wheels of the train are so arranged, that the projection of the
point of contact of any pair upon the plane of the next
following pair shall lie in the line of centres of this last pair,
between their point of contact and the axis of the driving wheel
of the pair; whilst the expenditure is greatest when this projection falls in that line but on the other side of the axis.
The difference of the expenditures of work on the friction of
the axes under these two different arrangements of the train
is represented by the formula

$$2 \left\{ \frac{\rho_1}{r_1} \sin. \varphi_1 + \frac{\rho_2}{r_3} \sin. \varphi_2 + \frac{\rho_3}{r_5} \sin. \varphi_3 + \frac{\rho_4}{r_7} \sin. \varphi_4 + \dots \right\} U_2;$$

which, in a train of a great number of wheels, may amount
to a considerable fraction of U_2; that fraction of U_2 representing the amount of power which may be sacrificed by a
false arrangement of the points of contact of the wheels.

240. *The expenditure of work due to the weights of the several
wheels of the train.*

The third and last term $N . S$ of the expression (285) represents the expenditure of work due to the *weights* of the
several wheels of the train; of this term the factor N is
represented by an expression (equation 284), each of the
terms of which, involves as a factor one of the quantities N_1,
N_2, N_3, &c. whose general type or form is that given in

equation (252), it being observed that the direction of the
driving pressure on any pair of the wheels being supposed
that of a tangent to their point of contact; the case is that
discussed in the note to page 289. The other factor of each
term of the expression (equation 284) for N, is a fraction
having the product $n_3 \, n_5 \ldots$ of the numbers of teeth in all
the preceding drivers of the train, except the first, for its
numerator, and the product $n_2 . n_4 . n_6 \ldots$ of the numbers of
teeth in the preceding followers of the train for its denominator; so that if the train be one by which the motion is to
be accelerated, the numbers of teeth in the followers being
small as compared with those in the drivers, or if the multiplying power of the train be great, and if the quantities
$N_1, N_2, N_3,$ &c. be all *positive;* then is the expenditure of work
by reason of the weights of the wheels considerable, as compared with the whole expenditure. Since, moreover, the co-
efficients of $N_1, N_2, N_3,$ &c. in the expression for N (equation 284) increase rapidly in value, this expenditure of
work is the greatest in respect to those wheels of the train
which are farthest removed from its first driving wheel: for
which reason, especially, it is advisable to diminish the weights
of the wheels as they recede from the driving point of the
train, which may readily be done, since the strain upon each
successive wheel is less, as the work is transferred to it under
a more rapid motion.

241. *The modulus of a train in which all the drivers are equal
to one another and all the followers, and in which the points
of contact of the drivers and followers are all similarly
situated.*

The numbers of teeth in the drivers of the train being in
this case supposed equal, and also the radii of these wheels,
$n_1 = n_3 = n_5 = n_7 =$ &c., $r_1 = r_3 = r_5 = r_7 =$ &c. The numbers of
teeth in the followers being also equal, and also the radii of
the followers $n_2 = n_4 = n_6 =$ &c., $r_2 = r_4 = r_6 =$ &c.
If moreover, to simplify the investigation, the *driving* work

U_1 be supposed to be done upon the first wheel of the train

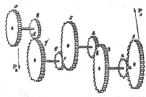

at a point situated in respect to the point of contact a of that wheel with its pinion precisely as that point of contact is in respect to the point of contact b of the next pair of wheels of the train; and if a similar supposition be made in respect to the point at which the *driven* work U_2 is done upon the last pinion of the train, then, evidently, $L_1 = L_2 = L_3 = \ldots = L_p$, and (see equation 252) $N_1 = N_2 = \ldots = N_p$.

The modulus (equation 285) will become, these substitutions being made in it, the axes being, moreover, supposed all to be of the same dimensions and material, and equally lubricated, and it being observed that the drivers and the followers are each p in number,

$$U_1 = \left\{ 1 + \pi p \left(\frac{1}{n_1} + \frac{1}{n_2} \right) \sin. \, \phi + p \, \frac{L_1 \rho_1}{r_1 r_2} \sin. \, \varphi_1 \right\} U_2 + NS \ldots (289),$$

which is the modulus required.

Moreover, the value of N (equation 284) will become by the like substitutions,

$$N = N_1 \left(\frac{r_1}{r_2} \right) \left\{ 1 + \left(\frac{n_1}{n_2} \right) + \left(\frac{n_1}{n_2} \right)^2 + \left(\frac{n_1}{n_2} \right)^{2'} + \ldots + \left(\frac{n_1}{n_2} \right)^{p-1} \right\},$$

or

$$N = N_1 \left(\frac{n_1}{n_2} \right) \left\{ \frac{\left(\frac{n_1}{n_2} \right)^p - 1}{\left(\frac{n_1}{n_2} \right) - 1} \right\} \ldots \ldots (290).$$

THE TRAIN OF LEAST RESISTANCE.

242. *A train of equal driving wheels and equal followers being required to yield at the last wheel of the train a given amount of work U_2, under a velocity m times greater or less than that under which the work U_1 which drives the train is done by the moving power upon the first wheel; it is required to determine what should be the number p of*

pairs of wheels in the train, so that the work U_1 *expended through a given space* S, *in driving it, may be a minimum.*

Since the number of revolutions made by the last wheel of the train is required to be a given multiple or part of the number of revolutions made by the first wheel, which multiple or part is represented by m, therefore (equation 236),

$$m = \left(\frac{n_1}{n_2}\right)^p ; \therefore \frac{n_1}{n_2} = m^{\frac{1}{p}} = \frac{r_1}{r_2};$$

$$\therefore \frac{1}{n_2} = \frac{m^{\frac{1}{p}}}{n_1}, \text{ and } \frac{1}{r_2} = \frac{m^{\frac{1}{p}}}{r_1};$$

$$\therefore \left(\frac{1}{n_1} + \frac{1}{n_2}\right) = \frac{m^{\frac{1}{p}} + 1}{n_1}, \text{ and } \frac{1}{r_1 r_2} = \frac{m^{\frac{1}{p}}}{r_1^2}.$$

Substituting these values in the modulus (equation 289); substituting, moreover, for N its value from equation (290), we have

$$U_1 = \left\{ 1 + \frac{\pi}{n_1} p (m^{\frac{1}{p}} + 1) \sin. \phi + \left(\frac{L_1 \rho_1}{r_1^2}\right) p m^{\frac{1}{p}} \sin. \phi_1 \right\} U_2 + N_1 (m-1) \left(\frac{m^{\frac{1}{p}}}{m^{\frac{1}{p}} - 1}\right) S \ldots (291).$$

It is evident that the question is solved by that value of p which renders this function a minimum, or which satisfies the conditions $\dfrac{dU_1}{dp} = 0$ and $\dfrac{d^2U_1}{dp^2} > 0$. The first condition gives by the differentiation of equation (291),

$$\left\{ m^{\frac{1}{p}} \left(1 - \frac{\log. \iota\, m}{p} \right) \left(\frac{\pi}{n_1} \sin. \phi + \frac{L_1 \rho_1}{r_1^2} \sin. \phi_1 \right) + \frac{\pi}{n_1} \sin. \phi \right\} U_2 + \frac{m^{\frac{1}{p}} (m-1) \log. \iota\, m}{p^2 (m^{\frac{1}{p}} - 1)^2} N_1 S = 0 \ldots (290).$$

This equation may be solved in respect to p, for any given values of the other quantities which enter into it, *by approximation.* If, being differentiated a second time, the above expression represents a positive quantity when the value of p (before determined) is substituted in it, then does that value satisfy both the conditions of a minimum, and supplies, therefore, its solution to the problem.

If we suppose $\phi_1 = 0$ and $N_1 = 0$, or in other words, if we neglect the influence of the *friction of the axes* and of the *weights of the wheels* of the train upon the conditions of the question, we shall obtain

$$m^{\frac{1}{p}}\left(1 - \frac{\log_{\cdot\varepsilon} m}{p}\right)\frac{\pi}{n_1}\sin.\ \phi + \frac{\pi}{n_1}\sin.\ \phi = 0;$$

whence by reduction,

$$p = \frac{\log_{\cdot\varepsilon} m}{1 + m^{-\frac{1}{p}}} * \quad \ldots \ldots \ldots \ (293).$$

* This formula was given by the late Mr. Davis Gilbert in his paper on the " Progressive improvements made in the efficiency of steam engines in Cornwall," published in the Transactions of the Royal Society for 1830. Towards the conclusion of that paper Mr. Gilbert has treated of the methods best adapted for imparting great angular velocities, and, in connection with that subject, of the friction of toothed wheels ; having reference to the friction of the surfaces of their teeth alone, and neglecting all consideration of the influence due to the weights of the wheels and to the friction of their axes.   The author has in vain endeavoured to follow out the condensed reasoning by which Mr. Gilbert has arrived at this remarkable result ; it supplies another example of that rare sagacity which he was accustomed to bring to the discussion of questions of practical science.   Mr. Gilbert has given the following examples of the solution of the formula by the method of approximation : — If $m = 120$, or if the velocity is to be increased by the train 120 times, then the value of p given by the above formula, or the number of pairs of wheels which should compose the train, so that it may work with a minimum resistance, reference being had only to the friction of the surfaces of the teeth, is 3·745; and the value of the factor $p(m^{\frac{1}{p}}+1)$ (equation 291), which being multiplied by $\frac{\pi}{n_1}\sin.\ \phi\ U_2$ represents the work expended on the friction of the surfaces of the teeth ; is in this case 17·9; whereas its value would, according to Mr. Gilbert, be 121 if the velocity were got up by a single pair of wheels.   So that the work lost by the friction of the teeth in the one case would only be one seventh part of that in the other.   In like manner Mr. Gilbert found, that if $m = 100$, then p should equal 3·6 ; in which case the loss by friction of the teeth would amount to the sixth part only of the loss that would result from that cause if $p = 1$, or if the required velocity were got up by one pair of wheels.

If $m=40$, then $p=2·88$, with a gain of one third over a single pair.

If $m=3·59$, then $p=1$.

If $m=12·85$, then $p=2$.

If $m=46·3$, then $p=3$.

If $m=166·4$, then $p=4$.

It is evident that when p in any of the above examples appears under the form of a fraction, the nearest whole number to it, must be taken in practice. The influence of the weights of the wheels of the train, and that of the friction of the axes, so greatly however modify these results, that although they are fully sufficient to show the existence in every case of a certain number of wheels, which being assigned to a train destined to produce a given acceleration of motion shall cause that train to produce the required effect with the least expenditure of power, yet they do not in any case determine correctly what that number of wheels should be.

THE INCLINED PLANE.

243. Let AB represent the surface of an inclined plane on which is supported a body whose centre of gravity is C, and its weight W, by means of a pressure acting in any direction, and which may be supposed to be supplied by the tension of a cord passing over a pulley and carrying at its extremity a weight.

Let OR represent the direction of the resultant of P and W. If the direction of this line be inclined to the perpendicular ST to the surface of the plane, at an angle OST equal to the limiting angle of resistance, on that side of ST which is farthest from the summit B of the plane (as in *fig.* 1.), the body will be upon the point of slipping *upwards;* and if it be inclined to the perpendicular at an angle OST,

(1.) (2.)

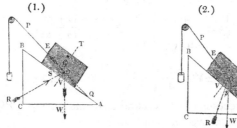

equal to the limiting angle of resistance, but on the side of ST nearest to the summit B (as in *fig.* 2.), then the body will be upon the point of slipping downwards (·Art. 138.); the former condition corresponds to the superior and the latter to the inferior state bordering upon motion (Art. 140.).

Now the resistance of the plane is equal and opposite to the resultant of P and W; let it be represented by R.

There are then three pressures P, W, and R in equilibrium.

$$\therefore \text{ (Art. 14.)} \quad \frac{P}{W} = \frac{\sin. \text{WOR}}{\sin. \text{POR}}.$$

Let $\angle \text{BAC} = \iota$, $\angle \text{OST} = \lim^{s}$. $\angle$ of resistance $= \varphi$, let θ

z

represent the inclination PQB of the direction of P to the surface of the plane, and draw OV perpendicular to AB; then,

in *fig.* 1., $\text{WOR} = \text{WOV} + \text{SOV} = \text{BAC} + \text{OST} = \iota + \varphi$,

and $\text{POR} = \text{PQB} + \text{OSQ} = \text{PQB} + \dfrac{\pi}{2} - \text{OST} = \dfrac{\pi}{2} + \theta - \varphi$;

in *fig.* 2., $\text{WOR} = \text{WOV} - \text{SOV} = \text{BAC} - \text{OST} = \iota - \varphi$,

and $\text{POR} = \text{PQB} + \text{OSQ} = \text{PQB} + \dfrac{\pi}{2} + \text{OST} = \dfrac{\pi}{2} + \theta + \varphi$;

$$\therefore \text{WOR} = \iota \pm \varphi; \quad \text{and} \quad \text{POR} = \dfrac{\pi}{2} + (\theta \mp \varphi);$$

the upper or lower sign being taken according as the body is upon the point of sliding up the plane, as in *fig.* 1., or down the plane, as in *fig.* 2. Or if we suppose the angle φ to be taken positively or negatively according as the body is on the point of slipping upwards or downwards; then generally $\text{WOR} = \iota + \varphi \qquad \text{POR} = \dfrac{\pi}{2} + (\theta - \varphi);$

$$\therefore \frac{P}{W} = \frac{\sin. (\iota + \varphi)}{\sin. \left(\dfrac{\pi}{2} + \theta - \varphi\right)} = \frac{\sin. (\iota + \varphi)}{\cos. (\theta - \varphi)};$$

$$\therefore P = W \cdot \frac{\sin. (\iota + \varphi)}{\cos. (\theta - \varphi)} \cdot \cdot \cdot \cdot \cdot (294).$$

If the direction of P be parallel to the plane, $\angle$ PQB or $\theta = 0$; and the above relation becomes

$$P = W \cdot \frac{\sin. (\iota + \varphi)}{\cos. \varphi} \cdot \cdot \cdot \cdot \cdot (295).$$

(3.)   (4.)

If $\iota = 0$ the plane becomes horizontal (fig. 3.), and the relation between P and W assumes the form

$$P = W \cdot \frac{\sin. \varphi}{\cos. (\theta - \varphi)} \cdot \cdot \cdot \cdot \cdot (296).$$

If $\theta = 0$, $P = W \cdot \tan. \varphi$, as it ought (see Art. 138.).

If the angle PQB or θ (fig. 1.) be increased so as to become $\pi - \theta$, PQ will assume the direction shown in fig. 4., and the relation (equation 294) between P and W will become

$$P = -W \frac{\sin. (\iota + \varphi)}{\cos. (\theta + \varphi)} \quad \cdots \cdots (297).$$

The negative sign showing that the direction of P must, in order that the body may slip up the plane, be opposite to that assumed in fig. 1.; or that it must be a pushing pressure in the direction PO instead of a pulling pressure in the direction OP.

If, however, the body be upon the point of slipping down the plane, so that φ must be taken negatively; and if, moreover, φ be greater than ι, then sin. $(\iota + \varphi)$, will become sin. $(\iota - \varphi) = -\sin. (\varphi - \iota)$, so that P will in this case assume the positive value

$$P = W \cdot \frac{\sin. (\varphi - \iota)}{\cos. (\theta - \varphi)} \quad \cdots \cdots (298),$$

which determines the force just necessary under these circumstances to pull the body down the plane.

If $\iota = \varphi$, $P = 0$, the body will therefore, in this case, be upon the point of slipping down the plane without the application of any pressure whatever to cause it to do so, other than its own weight. The plane is, under these circumstances, said to be inclined at the angle of repose, which angle is therefore equal to the limiting angle of resistance.

244. *The direction of least traction.*

Of the infinite number of different directions in which the pressure P may be applied, each requiring a different amount to be given to that pressure, so as to cause the body to slide up the plane, that direction will require the least value to be assigned to P for this purpose, or will be the direction of least traction, which gives to the denominator of the fraction

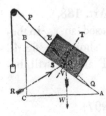

in equation (294) its greatest value, or which makes $\theta - \phi = 0$ or $\theta = \phi$. The direction of P is therefore that of least traction when the angle PQB is equal to the limiting angle, a relation which obtains in respect to each of the cases discussed in the preceding article.

245. THE MOVEABLE INCLINED PLANE.

Let ABC represent an inclined plane, to the back AC of

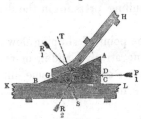

which is applied a given pressure P_1, and which is moveable between the two resisting surfaces GH and KL, of which either remains fixed, and the other is upon the point of yielding to the pressure of the plane.

If we suppose the resultants of the resistances upon the different points of the two surfaces AB and BC of the plane to be represented by R_1 and R_2 respectively, it is evident that the directions of these resistances and of the pressure P_1 will meet, when produced, in the same point O*; and that, since the plane is upon the point of slipping upon each of the surfaces, the direction of each of these resistances is inclined to the perpendicular to the surface of the plane, at the point where it intersects it, at an angle equal to the corresponding limiting angle of resistance.

So that if ET and FS represent perpendiculars to the surfaces AB and BC of the plane at the points E and F and ϕ_1, ϕ_2, the limiting angles of resistance between these surfaces of the plane and the resisting surfaces GH and KL respectively, then $R_1ET = \phi_1$, $R_2FS = \phi_2$.

Now the pressures P_1, R_1, R_2 being in equilibrium (Art. 14.),

$$\frac{P_1}{R_1} = \frac{\sin. EOF}{\sin. DOF}, \text{ and } \frac{P_1}{R_2} = \frac{\sin. EOF}{\sin. DOE}.$$

* Since either is equal and opposite to the resultant of the other two.

But the four angles of the quadrilateral figure BEOF being equal to four right angles (Euc. 1·32.), $EOF = 2\pi - EBF - OEB - OFB$; but $EBF = \iota$, $OEB = \frac{\pi}{2} + \phi_1$, $OFB = \frac{\pi}{2} + \phi_2$. $\therefore EOF = \pi - \iota - \phi_1 - \phi_2$.

Similarly, $DOE = 2\pi - ADO - AEO - DAE$; but $ADO = \frac{\pi}{2}$, $AEO = \frac{\pi}{2} - \phi_1$, $BAC = \frac{\pi}{2} - \iota$: $\therefore DOE = \frac{\pi}{2} + \iota + \phi_1$.

Since, moreover, DO is parallel to BC, both being perpendicular to AC, $\therefore DOF = \pi - OFC$; but $OFC = \frac{\pi}{2} - \phi_2$:

$\therefore DOF = \frac{\pi}{2} + \phi_2$.

$$\therefore \frac{P_1}{R_2} = \frac{\sin.\{\pi - (\iota + \phi_1 + \phi_2)\}}{\sin.\left(\frac{\pi}{2} + \phi_2\right)} = \frac{\sin.(\iota + \phi_1 + \phi_2)}{\cos.\phi_2};$$

$$\therefore P_1 = R_1 \frac{\sin.(\iota + \phi_1 + \phi_2)}{\cos.\phi_2}. \quad \ldots (299.)$$

$$\frac{P_1}{R_2} = \frac{\sin.\{\pi - (\iota + \phi_1 + \phi_2)\}}{\sin.\left(\frac{\pi}{2} + \iota + \phi_1\right)} = \frac{\sin.(\iota + \phi_1 + \phi_2)}{\cos.(\iota + \phi_1)};$$

$$\therefore P_1 = R_2 . \frac{\sin.(\iota + \phi_1 + \phi_2)}{\cos.(\iota + \phi_1)}. \quad \ldots (300.)$$

In the case in which the surface GH yields to the pressure of the plane, KL remaining fixed, we obtain (equation 121.) for the *modulus* (see Art. 148.), observing that $P_1^{(0)} = R_1 \sin.\iota$ (equation 229),

$$U_1 = U_2 \frac{\sin.(\iota + \phi_1 + \phi_2)}{\sin.\iota . \cos.\phi_2}. \quad \ldots (301).$$

In the case in which the surface KL yields, CH remaining fixed, observing that $P_1^{(0)} = R_2 \tan.\iota$ (equation 300), we have,

$$U_1 = U_2 \frac{\sin.(\iota + \phi_1 + \phi_2)}{\cos.(\iota + \phi_1)\tan.\iota}. \quad \ldots (302).$$

Equations 301 and 302 may be placed respectively under the forms

$$U_1 = U_2 \frac{\sin.(\phi_1 + \phi_2)}{\cos.\phi_2}\{\cot.(\phi_1 + \phi_2) + \cot. \iota\}$$

and

$$U_1 = U_2 \frac{\cos.(\phi_1 + \phi_2)}{\sin.\phi_1}\left\{\frac{\tan.\iota + \tan.(\phi_1 + \phi_2)}{(\cot.\phi_1 - \tan.\iota)\tan.\iota}\right\}.$$

The value of U_1 corresponding to a given value of U_2 is in the former equation a *minimum* when $\iota = \frac{\pi}{2}$, and in the latter when

$$\tan.\iota = \left\{\sqrt{\frac{\cos.\phi_2}{\sin.\phi_1 \sin.(\phi_1 + \phi_2)} - 1}\right\}\tan.(\phi_1 + \phi_2). \ldots (303).$$

From the former of these equations it follows, that the work lost by friction (when the driving surface of the plane is its hypothenuse) is less as the inclination of the plane is greater, or as its mechanical advantage is less.

246. *A system of two moveable inclined planes.*

Let A and B represent two inclined planes, of which A

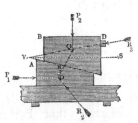

rests upon a horizontal surface, and receives a horizontal motion from the action of the pressure P_1; communicating to B a motion which is restricted to a vertical direction by the resistance of the obstacle D, which vertical motion of the plane is opposed by the pressure P_2 applied to its superior surface. It is required to determine a relation between the pressures P_1 and P_2, in their state bordering upon motion; and the *modulus* of the machine.

Let R_1 represent the pressure of the plane A upon the plane B, or the resistance of the latter plane upon the former, and R_3 the resistance of the obstacle D upon the back of the plane B; then is the relation between R_1 and P_1 determined by equation (299). And since R_1, R_3, P_2 are pressures in equilibrium, the relation between R_1 and P_2 is expressed (Art. 14.) by the relation $\dfrac{R_1}{P_2} = \dfrac{\sin. P_2 Q R_3}{\sin. R_1 Q R_3}$. Now $R_3 Q$ is inclined to a perpendicular to the back of the plane B, at an angle equal to the limiting angle of resistance between the surface of that plane and the obstacle D on which it is upon the point of sliding. Let this angle be represented by φ_3, then is the inclination of R_3 to the back of the plane or $P_2 Q$ represented by $\dfrac{\pi}{2} - \varphi_3$; so that $P_2 Q R_3 = \dfrac{\pi}{2} - \varphi_3$.

And if $R_3 Q$ be produced so as to meet the surface of the plane A in V, and VS be drawn horizontally, $R_1 Q R_3 =$

$$QVR_1 + VR_1Q = R_3 VS + SVA + VR_1Q = \varphi_3 + \iota + \frac{\pi}{2} + \varphi_1,$$

where represents the inclination of the superior surface of the plane A or the inferior surface of the plane B to the horizon. Substituting these values of $P_2 Q R_3$ and $R_1 Q R_3$ we obtain

$$\frac{R_1}{P_2} = \frac{\sin.\left(\dfrac{\pi}{2} - \varphi_3\right)}{\sin.\left(\dfrac{\pi}{2} + \iota + \varphi_3 + \varphi_1\right)} = \frac{\cos. \varphi_3}{\cos.(\iota + \varphi_3 + \varphi_1)}.$$

Multiplying this equation by equation 299, and solving in respect to P_1,

$$P_1 = P_2 \frac{\sin.(\iota + \varphi_1 + \varphi_2)\cos.\varphi_3}{\cos.(\iota + \varphi_1 + \varphi_3)\cos.\varphi_2} \quad \cdots \cdots (304).$$

$\therefore$ (Art. 152.) $U_1 = U_2 \dfrac{\sin.(\iota + \varphi_1 + \varphi_2)\cos.\varphi_3}{\cos.(\iota + \varphi_1 + \varphi_3)\tan.\iota\cos.\varphi_2} \cdot \cdot$ (305).

A system of three inclined planes, two of which are moveable, and the third fixed.

247. The inclined plane A, in the accompanying figure, is fixed in position, the plane B is moveable upon A, having its upper surface inclined to the horizon at a less angle than the lower; and C is an inclined plane resting upon B, which is

z 4

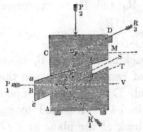

prevented from moving horizontally by the obstacle D, but may be made to slide along this obstacle vertically. It is required to determine a relation between P_1 and P_2, applied, as shown in the figure, when the system is in the state bordering upon motion.

Let R_1, R_2, R_3 represent the resistances of the surfaces on which motion takes place, φ_1 φ_2 φ_3 their limiting angles of resistance respectively, and ι_1 ι_2 the inclinations of the two surfaces of contact of B to the horizon. Since P_1, R_1, R_2 are pressures in equilibrium, as also P_2, R_2, R_3

$$\therefore \frac{P_1}{R_2} = \frac{\sin. R_2OR_1}{\sin. P_1OR_1}, \quad \frac{R_2}{P_2} = \frac{\sin. P_2QR_3}{\sin. R_2QR_3}.$$

Multiplying these equations together,

$$\frac{P_1}{P_2} = \frac{\sin. R_2OR_1 . \sin. R_2QR_3}{\sin. P_1OR_1 . \sin. R_2QR_3}.$$

Draw OS and OT parallel to the faces of the plane B; then

$R_2OR_1 = R_1OS + QOT - TOS$; but $R_1OS = \dfrac{\pi}{2} - \varphi_1$, since OS

is parallel to the inferior face of the plane B, also $QOT = \dfrac{\pi}{2} - \varphi_2$, since OT is parallel to the superior face of the plane B; and TOS = the inclination of the faces of the plane B to one another $= \iota_1 - \iota_2$.

$$\therefore R_2OR_1 = \left(\frac{\pi}{2} - \varphi_1\right) + \left(\frac{\pi}{2} - \varphi_2\right) - (\iota_1 - \iota_2) = \pi - (\varphi_1 + \varphi_2) - (\iota_1 - \iota_2).$$

Also $P_2QR_3 = \dfrac{\pi}{2} - R_3QM = \dfrac{\pi}{2} - \varphi_3$.

Let P_1O be produced to V; therefore $P_1OR_1 = \pi - R_1OV =$

$\pi - (R_1OS - SOV) = \pi - \left\{ \left(\dfrac{\pi}{2} - \varphi_1\right) - \iota_1 \right\} = \dfrac{\pi}{2} + \iota_1 + \varphi_1$. Lastly

$R_2QR_3 = OQM + MQR_3$. Now, $MQR_3 = \varphi_3$; also, $OQM =$

$\pi - QOV = \pi - (QOT + TOV) = \pi - \left\{ \left(\dfrac{\pi}{2} - \varphi_2\right) + \iota_2 \right\} = \dfrac{\pi}{2} - \iota_2$

$+\varphi_2. \quad \therefore R_2QR_3 = \dfrac{\pi}{2} - \iota_2 + \varphi_2 + \varphi_3 = \dfrac{\pi}{2} - (\iota_2 - \varphi_2 - \varphi_3).$

$$\therefore \frac{P_1}{P_2} = \frac{\sin. \left\{\pi - (\varphi_1 + \varphi_2) - (\iota_1 - \iota_2)\right\} \sin. \left(\dfrac{\pi}{2} - \varphi_3\right)}{\sin. \left(\dfrac{\pi}{2} + \iota_1 + \varphi_1\right). \sin. \left\{\dfrac{\pi}{2} - (\iota_2 - \varphi_2 - \varphi_3)\right\}}.$$

$$\therefore P_1 = P_2 . \frac{\sin. \left\{(\varphi_1 + \varphi_2) + (\iota_1 - \iota_2)\right\} \cos. \varphi_3}{\cos. (\iota_1 + \varphi_1) \cos. \left\{\iota_2 - (\varphi_2 + \varphi_3)\right\}} \quad \dots \quad (306).$$

Whence we obtain for the modulus (Art. 152.), observing

that $\Phi_1^{(0)} = \dfrac{\sin. (\iota_1 - \iota_2)}{\cos. \iota_1 \cos. \iota_2}.$

$$U_1 = U_2 \frac{\sin. (\varphi_1 + \varphi_3 + \iota_1 - \iota_2) \cos. \iota_1 \cos. \iota_2 \cos. \varphi_3}{\cos. (\iota_2 - \varphi_2 - \varphi_3) \cos. (\iota_1 + \varphi_1) \sin. (\iota_1 - \iota_2)} \quad \dots \quad (307).$$

THE WEDGE DRIVEN BY PRESSURE.

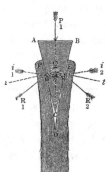

248. Let ACB represent an isosceles wedge, whose angle ACB is represented by 2ι, and which is driven between the two resisting surfaces DE and DF, by the pressure P_1. Let R_1 and R_2 represent the resistances of these surfaces upon the acting surfaces CA and CB of the wedge when it is upon the point of moving forwards. Then are the directions of R_1 and R_2 inclined respectively to the perpendiculars Gs and Rt to the faces CA and CB of the wedge, at angles each equal to the limiting angle of resistance φ. The pressures R_1 and R_2 are therefore equally inclined to the axis of the wedge, and to the direction of P_1, whence it follows that $R_1 = R_2$, and therefore (Art. 13.) that $P_1 = 2R_1 \cos. \frac{1}{2}GOR$. Now, since CGOR is a quadrilateral figure, its four angles are equal to two right angles; therefore $GOR = 2\pi - GCR - OGC - ORC$. But $GCR = 2\iota$; $OGC = ORC = \dfrac{\pi}{2} + \varphi$; $\therefore GOR = \pi - (2\iota + 2\varphi) \quad \therefore \frac{1}{2}GOR = \dfrac{\pi}{2} - (\iota + \varphi).$

$$\therefore \ P_1 = 2R_1 \sin. (\imath + \varphi) \ \ . \ . \ . \ . \ (307).$$

Whence it follows (equation 121) that the modulus of the wedge is

$$U_1 = U_2 \frac{\sin. (\imath + \varphi)}{\sin. \imath} \ \ . \ . \ . \ . \ (309).$$

This equation may be placed under the form

$$U_1 = U_2 \{\cot. \varphi + \cot. \imath\} \sin. \varphi.$$

The work lost by reason of the friction of the wedge is greater, therefore, as the angle of the wedge is less; and infinite for a finite value of φ, and an infinitely small value of $\imath$.

The angle of the wedge.

249. Let the pressure P_1, instead of being that just sufficient to drive the wedge, be now supposed to be that which is only just sufficient to keep it in its place when driven. The

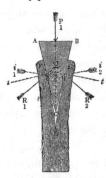

two surfaces of the wedge being, under these circumstances, upon the point of sliding backwards upon those between which the wedge is driven, at their points of contact G and R, it is evident that the directions of the resistances $\imath_1$G and $\imath_2$R upon those points, must be inclined to the normals sG and tR at angles, each equal to the limiting angle of resistance, but measured on the sides of those normals opposite to those on which the resistances R_1G and R_2R are applied.*

In order to adapt equation 307 to this case, we have only then to give to φ a negative value in that equation. It will then become

$$P_1 = 2R_1 \sin. (\imath - \varphi) \ \ . \ . \ . \ . \ (310).$$

---

* This will at once be apparent, if we consider that the direction of the resultant pressure upon the wedge at G must, in the one case, be such, that, if it acted *alone*, it would cause the surface of the wedge to slip downwards on the surface of the mass at that point, and in the other case upwards ; and that the resistance of the mass is in each case opposite to this resultant pressure.

So long as ι is greater than φ, or the angle C of the wedge greater than twice the limiting angle of resistance, P_1 is positive ; whence it follows that a certain pressure acting in the direction in which the wedge is driven, and represented in amount by the above formula, is, in this case, necessary to keep the wedge from receding from any position into which it has been driven.   So that if, in this case, the pressure P_1 be wholly removed, or if its value become less than that represented by the above formula, then the wedge will recede from any position into which it has been driven, or it will be *started.*   If ι be less than φ, or the angle C of the wedge less than twice the limiting angle of resistance, P_1 will become negative ; so that, in this case, a pressure, opposite in direction to that by which the wedge has been driven, will have become necessary to cause it to recede from the position into which it has been driven ; whence it follows, that if the pressure P be now wholly removed, the wedge will remain fixed in that position ; and moreover that it will still remain fixed, although a certain pressure be applied to cause it to recede, provided that pressure do not exceed the negative value of P_1, determined by the formula.

It is this property of remaining fixed in any position into which it is driven when the force which drives it is removed, that characterises the wedge, and renders it superior to every other implement driven by impact.

It is evidently, therefore, a principle in the formation of a wedge to be thus used, that its angle should be less than twice the limiting angle of resistance between the material which forms its surface, and that of the mass into which it is to be driven.

THE WEDGE DRIVEN BY IMPACT.

250. The wedge is usually driven by the impinging of a heavy body with a greater or less velocity upon its back, in the direction of its axis.   Let W represent the weight of such a body, and V its velocity, every element of it being conceived to move with the same velocity.   The work ac-

cumulated in this body, when it strikes the wedge, will then be represented (Art. 66.) by $\frac{1}{2}\frac{W}{g}V^2$. Now the whole of this work is done by it upon the wedge, and by the wedge upon the resistances of the surfaces opposed to its motion; if the bodies are supposed to come to rest after the impact, and if the influence of the elasticity and mutual compression of the surfaces of the striking body and of the wedge are neglected, and if no permanent compression of their surfaces follows the impact.* $\therefore U_1 = \frac{1}{2}\frac{WV^2}{g}$.

Substituting this value of U_1 in equation 308, and solving in respect to U_2, we have

$$U_2 = \frac{1}{2}\frac{WV^2}{g}\frac{\sin. \iota}{\sin. (\iota + \phi)} \quad \cdots \quad (311);$$

by which equation the work U_2 yielded upon the resistances opposed to the motion of the wedge by the impact of a given

---

* The influence of these elements on the result may be deduced from the principles about to be laid down in the chapter upon impact. It results from these, that if the surfaces of the impinging body and the back of the wedge, by which the impact is given and received, be exceedingly *hard*, as compared with the surfaces between which the wedge is driven, then the mutual pressure of the impinging surfaces will be exceedingly great as compared with the resistance opposed to the motion of the wedge. Now, this latter being neglected, as compared with the former, the work received or gained by the wedge from the impact of the hammer will be shown in the chapter upon impact to be represented by $\frac{(1+e)^2 W_1^2 W_2 V^2}{2g(W_1+W_2)^2}$, where W_1 represents the weight of the hammer, W_2 the weight of the wedge, and e that measure of the elasticity whose value is unity when the elasticity is perfect. Equating this expression with the value of U_1 (equation 309), and neglecting the effect of the elasticity and compression of the surfaces G and R, between which the wedge is driven, we shall obtain the approximation

$$U_2 = \frac{(1+e)^2 W_1^2 W_2 V^2}{2g(W_1+W_2)^2}\cdot\frac{\sin. \iota}{\sin. (\iota + \phi)}.$$

From this expression it follows, that the useful work is the greatest, other things being the same, when the weight of the wedge is equal to the weight of the hammer, and when the striking surfaces are hard metals, so that the value of e may approach the nearest possible to unity.

weight W with a given velocity V is determined; or the
weight W necessary to yield a given amount of work when
moving with a given velocity; or, lastly, the velocity V with
which a body of given weight must impinge to yield a given
amount of work.

If the wedge, instead of being isosceles, be of the form of

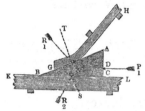

a right angled triangle, as shown
in the accompanying figure, the
relation between the work U_1
done upon its back, and that
yielded upon the resistances opposed to its motion at either of
its faces, is represented by equa-
tions 301 and 302. Supposing therefore this wedge, like the
former, to be driven by impact, substituting as before for U_1
its value $\frac{1}{2}\frac{W}{g}V^2$, and solving in respect to U_2', we have in
the case in which the face AB of the wedge is its driving
surface

$$U_2 = \frac{1}{2}\frac{WV^2}{g} \cdot \frac{\sin. \imath \cos. \varphi_2}{\sin. (\imath + \varphi_1 + \varphi_2)} \cdots \cdots (312);$$

when the base BC of the wedge is its driving surface,

$$U_2 = \frac{1}{2}\frac{WV^2}{g} \cdot \frac{\tan. \imath \cos. (\imath + \varphi_1)}{\sin.(\imath + {}_1 + \varphi_2)} \cdots \cdots (313).$$

251. If the power of the wedge be applied by the interven-

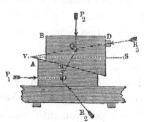

tion of an inclined plane moveable
in a direction at right angles to the
direction of the impact *, as shown
in the accompanying figure, then
substituting for U_1 in equation 305
half the *vis viva* of the impinging
body, and solving, as before, in
respect to U_2, we have

$$U_2 = \frac{1}{2}\frac{WV^2}{g} \cdot \frac{\cos. (\imath + \varphi_1 + \varphi_3) \tan. \imath \cos. \varphi_2}{\sin. (\imath + \varphi_1 + \varphi_2) \cos. \varphi_3} \cdots \cdots (314).$$

* This is the form under which the power of the wedge is applied for
the expressing of oil.

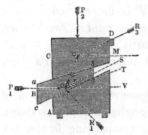

If instead of the base of the plane being parallel to the direction of impact, it be inclined to it, as shown in the accompanying figure, then substituting as above in equation 307, we have

$$U_2 = \frac{1}{2}\frac{WV^2}{g} \cdot \frac{\cos.(\iota_2 - \varphi_2 - \varphi_3)\cos.(\iota_1 + \varphi_1)\sin.(\iota_1 - \iota_2)}{\sin.(\varphi_1 + \varphi_3 + \iota_1 - \iota_2)\cos.\iota_1 \cos.\iota_2 \cos.\varphi_3} . (315).$$

THE MEAN PRESSURE OF IMPACT.

252. It is evident from equations 311, 312, 313, that, since, whatever may be the weight of the impinging body or the velocity of the impact, a certain finite amount of *work* U_2 is yielded upon the resistances opposed to the motion of the wedge; there is in every such case a certain mean resistance R overcome through a certain space S, in the direction in which that resistance acts, which resistance and space are such, that

$$RS = U_2, \text{ and therefore } R = \frac{U_2}{S}.$$

If therefore the space S be exceedingly small as compared with U_2, there will be an exceedingly great resistance R overcome by the impact through that small space, however slight the impact. From this fact the enormous amount of the resistances which the wedge, when struck by the hammer, is made to overcome, is accounted for. The power of thus subduing enormous resistances by impact is not however peculiar to the wedge, it is common to all implements of impact, and belongs to its nature; its effects are rendered *permanent* in the wedge by the property possessed by that implement of retaining permanently any position into which it is driven between two resisting surfaces, and thereby opposing itself effectually to the tendency of those surfaces, by reason of their elasticity, to recover their original form and position. It is equally true of any the slightest *direct* impact of the hammer as of its impact applied through the wedge, that it is sufficient to cause any finite resistance opposed to it to yield through a certain finite space, however great that resistance may be. The difference lies in this, that the surface yielding through this exceedingly small but finite space under the blow, of the hammer, immediately recovers itself after the blow if the limits of elasticity be not passed; whereas the space which the wedge is, by such an impact made to traverse, in the direction of its length, becomes a permanent separation.

THE SCREW.

253. Let the system of two moveable inclined planes 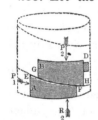 represented in fig. p. 342. be formed of exceedingly thin and pliable laminæ, and conceive one of them, A for instance, to be wound upon a *convex* cylindrical surface, as shown in the accompanying figure, and the other, B, upon a *concave* cylindrical surface having an equal diameter, and the same axis with the other; then will the surfaces EF and GH of these planes represent truly the threads or helices of two screws, one of them of the form called the male screw, and the other the female screw. Let the helix EF be continued, so as to form more than one spire or convolution of the thread; if, then, the cylinder which carries this helix be made to revolve upon its axis by the action of a pressure P_1 applied to its circumference, and the cylinder which carries the helix GH be prevented from revolving upon its axis by the opposition of an obstacle D, which leaves that cylinder nevertheless free to move in a direction parallel to its axis, it is evident that the helix EF will be made to slide beneath GH, and the cylinder which carries the latter helix to traverse longitudinally; moreover, that the conditions of this mutual action of the helical surfaces EF and GH will be precisely analogous to those of the surfaces of contact of the two moveable inclined planes discussed in Art. 246. So that the

conditions of the equilibrium of the pressures P_1 and P_2 in
the state bordering upon motion, and the modulus of the
system, will be the same in the one case as in the other ; with
this single exception, that the resistance R_2 of the mass on
which the plane A rests (see fig. p. 342.) is not, in the case
of the screw, applied only to the thin edge of the base of the
lamina A, but to the whole extremity of the solid cylinder on
which it is fixed, or to a circular projection from that extremity serving it as a pivot.   Now if, in equation 304,
we assume $\varphi_2 = 0$, we shall obtain that relation of the
pressures P_1 and P_2 in their state bordering upon motion,
which would obtain if there were no friction of the extremity
of the cylinder on the mass on which it rests ; and observing
that the pressure P_2 is precisely that by which the pivot at
the extremity of the cylinder is pressed upon this mass, and
therefore the moment (see Art. 177, equation 188) of the
resistance to the rotation of the cylinder produced by the
friction of this pivot by $\frac{2}{3}P_2\rho \tan.\varphi_2$, where ρ represents the
radius of the pivot ; observing, moreover, that the pressure
which must be applied at the circumference of the cylinder
to overcome this resistance, above that which would be required to give motion to the screw if there were no such
friction, is represented by $\frac{2}{3}P_2\frac{\rho}{r} \tan.\varphi_2$, r being taken to represent the radius of the cylinder, we obtain for the entire
value of the pressure P_1 in the state bordering upon motion,

$$P_1 = P_2\frac{\sin.(\iota+\varphi_1)\cos.\varphi_3}{\cos.(\iota+\varphi_1+\varphi_3)} + \frac{2}{3}P_2\frac{\rho}{r} \tan.\varphi_2.$$

The pressure P_1 has here been supposed to be applied to
turn the screw at its *circumference ;* it is customary, however,
to apply it at some distance from its circumference by the
intervention of an arm.   If a represent the length of such an
arm, measuring from the axis of the cylinder, it is evident that
the pressure P_1 applied to the extremity of that arm, would
produce at the circumference of the cylinder a pressure
represented by $P_1\frac{a}{r}$, which expression being substituted for

P_1 in the preceding equation, and that equation solved in respect to P_1, we obtain finally for the relation between P_1 and P_2 in their state bordering upon motion,

$$P_1 = P_2 \left(\frac{r}{a}\right) \left\{ \frac{\sin.(\iota + \varphi_1)\cos.\varphi_3}{\cos.(\iota + \varphi_1 + \varphi_3)} + \frac{2}{3}\left(\frac{\rho}{r}\right)\tan.\varphi_2 \right\} \dots (316).$$

If in like manner we assume in the modulus (equation 305) $\varphi_2 = 0$, and thus determine a relation between the work done at the driving point and that yielded at the working point, on the supposition that no work is expended on the friction of the pivot; and if to the value of U_1 thus obtained we add the work expended upon the resistance of the pivot which is shown (equation 189) to be represented at each revolution by $\frac{4}{3}\pi\rho P_2 \tan.\varphi_1$, and therefore during n revolutions by $\frac{4}{3}\pi n\rho P_2$, we shall obtain the following general expression for the modulus; the whole expenditure of work due to the prejudicial resistances being taken into account.

$$U_1 = U_2 . \frac{\sin.(\iota + \varphi_1)\cos.\varphi_3}{\cos.(\iota + \varphi_1 + \varphi_3)\tan.\iota} + \frac{4}{3}\pi n\rho P_2 \tan.\varphi_2.$$

Representing by λ the common distance between the threads of the screw, i. e. the space which the nut B is made to traverse at each revolution of the screw; and observing that $n\lambda P_2 = U_2$, so that $\frac{4}{3}\pi n\rho P_2 \tan.\varphi_2 = \frac{4}{3}\pi \frac{U_2}{\lambda}\rho \tan.\varphi_2 = \frac{2}{3}\frac{2\pi r}{\lambda} \cdot \frac{\rho}{r} \cdot U_2 \tan.\varphi_2$, in which expression $\frac{2\pi\rho}{\lambda} = \cot.\iota$, we obtain finally for the modulus of the screw

$$U_1 = U_2 \left\{ \frac{\sin.(\iota + \varphi_1)\cos.\varphi_3}{\cos.(\iota + \varphi_1 + \varphi_3)} + \frac{2}{3}\frac{\rho}{r}\tan.\varphi_2 \right\} \cot.\iota \dots (317).$$

It is evidently immaterial to the result at what distance from the axis, the obstacle D is opposed to the revolution of that cylinder which carries the lamina B; since the amount of that resistance does not enter into the result as expressed in the above formula, but only its direction determined by the angle φ_3, which angle depends upon the nature of the resisting surfaces, and not upon the position of the resisting point.

APPLICATIONS OF THE SCREW.

254. The accompanying figure represents an application of the screw under the circumstances described in the preceding article, to the well known machine called the VICE.

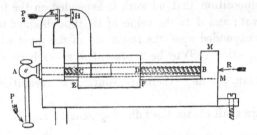

AB is a solid cylinder carrying on its surface the thread of a male screw, and within the piece CD is a hollow cylindrical surface, carrying the corresponding thread of a female screw; this female screw is prevented from revolving with the male screw by a groove in the piece CD, which carries it, and which is received into a corresponding projection EF of the solid frame of the machine, serving it as a guide; which guide nevertheless allows a longitudinal motion to the piece CD. A projection from the frame of the instrument at B, met by a pivot at the extremity of the male screw, opposes itself to the tendency of that screw to traverse in the direction of its length. The pressure P_2 to be overcome is applied between the jaws H and K of the vice, and the driving pressure P_1 to an arm which carries round with it the screw AB.

It is evident that, in the state bordering upon motion, the resistance R upon the pivot at the extremity B of the screw AB, resolved in a direction parallel to the length of that screw, must be equal to the pressure P_2 (see Art. 16.); so that if we imagine the piece CD to become fixed, and the piece BM to become moveable, being prevented from revolving, as CD was, by the intervention of a groove and guide, then might the instrument be applied to overcome any given resistance R opposed to the motion of this piece CD by the constant pressure of its pivot upon that piece.

The screw is applied under these circumstances in the

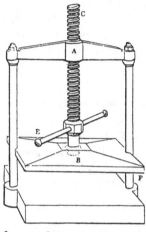

common screw press. The piece A, fixed to the solid frame of the machine, contains a female screw whose thread corresponds to that of the male screw ; this screw, when made to turn by means of a handle fixed across it, presses by the intervention of a pivot B, at its extremity, upon the surface of a solid piece EF moveable vertically, but prevented from turning with the screw by grooves receiving two vertical pieces, which serve it as guides, and form parts of the frame of the machine.

The formulæ determined in Art. 253. for the preceding case of the application of the screw, obtain also in this case, if we assume $\varphi_3 = 0$. The loss of power due to the friction of the piece EF upon its guides will, however, in this calculation, be neglected ; that expenditure is in all cases exceedingly small, the pressure upon the guides, whence their friction results, being itself but the result of the friction of the pivot B upon its bearings ; and the former friction being therefore, in all cases, a quantity of two dimensions in respect to the coefficient of friction.

If, instead of the lamina A (p. 251.) being fixed upon the convex surface of a solid cylinder, and B upon the concave surface of a hollow cylinder, the order be reversed, A being fixed upon the hollow and B on the solid cylinder, it is evident that the conditions of the equilibrium will remain the same, the male instead of the female screw being in this case made to progress in the direction of its length. If, however, the longitudinal motion of the male screw B

(p. 251.) be, under these circumstances, arrested, and that
screw thus become fixed, whilst the obstacle opposed to
the longitudinal motion of the female screw A is removed,
and that screw thus becomes free to revolve upon the
male screw, and also to traverse it longitudinally, except
in as far as the latter motion is opposed by a certain resist-

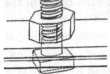

ance R, which the screw is intended,
under these circumstances, to overcome; then will the combination assume the well known form of the screw
and nut.

To adapt the formulæ of Art. 253. to this case, φ_3 must be
made $=0$, and instead of assuming the friction upon the
extremity of the screw (equation 316) to be that of a
solid pivot, we must consider it as that of a hollow pivot,
applying to it (by exactly the same process as in Art. 252.),
the formulæ of Art. 178. instead of Art. 177.

THE DIFFERENTIAL SCREW.

255. In the combination of three inclined planes discussed
in Art. 247., let the plane B be conceived of much greater
width than is given to it in the figure (p. 344.), and.let it
then be conceived to be wrapped upon a convex cylindrical
surface. Its two edges ab and cd will thus become the helices
of two screws, having their threads of different inclinations
wound round different portions of the same cylinder, as

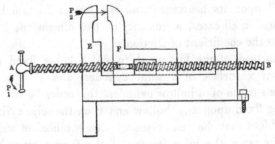

represented in the accompanying figure, where the thread
of one screw is seen winding upon the surface of a solid

cylinder from A to C, and the thread of another, having a different inclination, from D to B.

Let, moreover, the planes A and C (p. 344.) be imagined to be wrapped round two hollow cylindrical surfaces, of equal diameters with the above-mentioned solid cylinder, and contained within the solid pieces E and F, through which hollow cylinders AB passes. Two female screws will thus be generated within the pieces E and F, the helix of the one adapting itself to that of the male screw extending from A to C, and the helix of the other to that upon the male screw extending from D to B. If, then, the piece E be conceived to be fixed, and the piece F moveable in the direction of the length of the screw, but prevented from turning with it by the intervention of a guide, and if a pressure P_1 be applied at A to turn the screw AB, the action of this combination will be precisely analogous to that of the system of inclined planes discussed in Art. 247., and the conditions of the equilibrium precisely the same ; so that the relation between the pressure P_1 applied to turn the screw (when estimated at the circumference of the thread) and that P_2, which it may be made to overcome, are determined by equation (306), and its modulus by equation (307).

The invention of the differential screw has been claimed by M. Prony, and by Mr. White of Manchester. A comparatively small pressure may be made by means of it to yield a pressure enormously greater in magnitude.* It admits of numerous applications, and, among the rest, of that suggested in the preceding engraving.

---

* It will be seen by reference to equation (306), that the working pressure P_2 depends for its amount, not upon the actual inclinations ι_1 ι_2 of the threads, but on the difference of their inclinations ; so that its amount may be enormously increased by making the threads nearly of the same inclination. Thus, neglecting friction, we have, by equation (306),

$P_2 = P_1 \dfrac{\cos. \iota_1 \cos. \iota_2}{\sin. (\iota_1 - \iota_2)}$; which expression becomes exceedingly great when ι_1 nearly equals ι_2.

HUNTER'S SCREW.

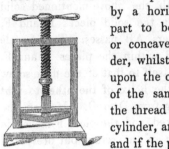

256. If we conceive the plane B (p. 344.) to be divided by a horizontal line, and the upper part to be wrapped upon the inner or concave surface of a hollow cylinder, whilst the lower part is wrapped upon the outer or convex circumference of the same cylinder, thus generating the thread of a female screw within the cylinder, and a male screw without it; and if the plane C be then wrapped upon the convex surface of a solid cylinder just fitting the inside or concave surface of the above-mentioned hollow cylinder, and the plane A upon a concave cylindrical surface just capable of receiving and adapting itself to the outside or convex surface of that cylinder, the *male* screw thus generated adapting itself to the thread of the screw within the hollow cylinder, and the female screw to the thread of that without it; if, moreover, the female screw last mentioned be fixed, and the solid male screw be free to traverse in the direction of its length, but be prevented turning upon its axis by the intervention of a guide; if, lastly, a moving pressure or power be applied to turn the hollow screw, and a resistance be opposed to the longitudinal motion of the solid screw which is received into it; then the combination will be obtained, which is represented in the preceding engraving, and which is well known as Mr. Hunter's screw, having been first described by that gentleman in the seventeenth volume of the *Philosophical Transactions.*

The theory of this screw is identical with that of the preceding, the relation of its driving and working pressures is determined by equation (306), and its modulus by equation (307).

THE THEORY OF THE SCREW WITH A SQUARE THREAD
IN REFERENCE TO THE VARIABLE INCLINATION OF THE
THREAD AT DIFFERENT DISTANCES FROM THE AXIS.

257. In the preceding investigation, the inclined plane
which, being wound upon the cylinder, generates the thread
of the screw, has been imagined to be an exceedingly *thin
sheet*, on which hypothesis every point in the thread may be
conceived to be situated at the same distance from the axis of
the screw; and it is on this supposition that the relation
between the driving and working pressure in the screw and
its modulus have been determined.

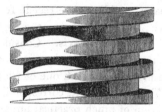
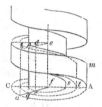

Let us now consider the actual case in which the thread
of the screw is of finite thickness, and different elements of it
situated at different distances from its axis.

Let *mb* represent a portion of the square thread of a screw,
in which form of thread a line *be*, drawn from any point *b* on
the outer edge of the thread perpendicular to the axis *ef*,
touches the thread throughout its whole depth *bd*. Let AC
represent a plane perpendicular to its axis, and *af* the projection of *be* upon this plane. Take *p* any point in *bd*, and
let *q* be the projection of *p*. Let $ep = r$, mean radius of
thread $= R$, inclination of that helix of the thread whose
radius is R* $= I$, inclination of the helix passing through
$p = \iota$, whole depth of thread $= 2T$, distance between threads
(or *pitch*) of screw $= L$. Now, since the helix passing through

* This may be called the *mean* helix of the thread. The term helix
is here taken to represent any spiral line drawn upon the surface of the
thread; the distance of every point in which, from the axis of the screw, is
the same.

p may be considered to be generated by the enwrapping of
an inclined plane whose inclination is ι upon a cylinder whose
radius is r, the base of which inclined plane will then become
the arc tq, we have $pq = tq.$ tan. ι. But, if the angle Afa be
increased to 2π, pq will become equal to the common distance
L between the threads of the screw, and tq will become a
complete circle, whose radius is r; therefore $L = 2\pi r$ tan. ι,
and this being true for all values of r, therefore $L = 2\pi R$
tan. I. Equating the second members of these equations,
and solving in respect to tan. ι,

$$\tan. \iota = \frac{R \tan. I}{r} \quad \ldots \ldots \quad (318).$$

From which expression it appears, that the inclination of the
thread of a square screw increases rapidly as we recede from
its edge and approach its axis, and would become a right
angle if the thread penetrated as far as the axis. Considering, therefore, the thread of the screw as made up of an
infinite number of helices, the modulus of each one of which
is determined by equation (317), in terms of its corresponding
inclination ι, it becomes a question of much practical importance to determine, if the screw act upon the resistance at one
point only of its thread, at what distance from its axis that
point should be situated, and if its pressure be applied at all
the different points of the depth of its thread, as is commonly
the case, to determine how far the conditions of its action
are influenced by the different inclinations of the thread at
these different depths.

We shall omit the discussion of the former case, and proceed to the latter.

Let P_2 represent the pressure parallel to its axis which is
to be overcome by the action of the screw. Now it is evident that the pressure thus produced upon the thread of the
screw is the same as though the whole central portion of it
within the thread were removed, or as though the whole
pressure P_2 were applied to a ring whose thickness is As or
2T. Now the area of this ring is represented by $\pi\{(R + T)^2$
$-(R - T)^2\}$, or by $4\pi RT$. So that the pressure of P_2, upon

every square unit of it, is represented by $\dfrac{P_2}{4\pi RT}$. Let Δr represent the exceedingly small thickness of such a ring whose radius is r, and which may therefore be conceived to represent the termination of the exceedingly thin cylindrical surface passing through the point p; the area of this ring is then represented by $2\pi r\Delta r$, and therefore the pressure upon it by $\dfrac{P_2 . 2\pi r\Delta r}{4\pi RT}$, or by $\dfrac{P_2 r\Delta r}{2RT}$. Now this is evidently the pressure sustained by that elementary portion of the thread which passes through p, whose thickness is Δr, and which may be conceived to be generated by the enwrapping of a thin plane, whose inclination is ι, upon a cylinder whose radius is r; whence it follows (by equation 316) that the elementary pressure ΔP_1, which must be applied to the arm of the screw to overcome this portion of the resistance P_2, thus applied parallel to the axis upon an element of the thread, is represented by

$$\Delta P_1 = \left(\frac{P_2 r\Delta r}{2RT}\right)\left(\frac{r}{a}\right)\left\{\frac{\sin.(\iota+\phi_1)\cos.\phi_3}{\cos.(\iota+\phi_1+\phi_3)} + \frac{2}{3}\frac{\rho}{r}\tan.\phi_2\right\};$$

whence, passing to the limit and integrating, we have

$$P_1 = \frac{P_2}{2RDa}\int_{R-D}^{R+D}\left\{\frac{\sin.(\iota+\phi_1)\cos.\phi_3}{\cos.(\iota+\phi_1+\phi_3)}r^2 + \frac{2}{3}\rho r\tan.\phi_2\right\}dr.$$

Now

$$\frac{\sin.(\iota+\phi_1)\cos.\phi_3}{\cos.(\iota+\phi_1+\phi_3)} = \frac{\tan.\iota+\tan.\phi_1}{1-\tan.\phi_1\tan.\phi_3-\tan.\iota(\tan.\phi_1+\tan.\phi_3)}$$

$$= \frac{\tan.\iota+\tan.\phi_1}{(1-\tan.\phi_1\tan.\phi_3)\{1-\tan.\iota\tan.(\phi_1+\phi_3)\}} = \tan.\phi_1+\tan.\iota$$

$+\tan.(\phi_1+\phi_3)\tan.^2\iota$. Neglecting dimensions of $\tan.\phi_1$ and $\tan.\phi_3$ above the first*,

---

* The integration is readily effected without this omission; and it might be desirable so to effect it where the theory of wooden screws is under discussion, the limiting angle of resistance being, in respect to such screws, considerable. The length and complication of the resulting expression have caused the omission of it in the text.

$$\therefore \; P_1 = \frac{P_2}{2RDa} \int_{R-D}^{R+D} \{(\tan.\,\phi_1 + \tan.\,\iota + \tan.(\phi_1+\phi_3)\tan.^2\iota)r^2 + \tfrac{2}{3}\rho r \tan.\,\phi_2\}dr \ldots (319).$$

Substituting in this expression for tan. its value (equation 318), it becomes

$$P_1 = \frac{P_2}{2RDa} \int_{R-D}^{R+D} \{r^2 \tan.\,\phi_1 + Rr \tan. I + R^2 \tan.^2 I \tan.(\phi_1+\phi_3) + \tfrac{2}{3}\rho r \tan.\,\phi_2\}dr.$$

Integrating and reducing,

$$P_1 = \frac{P_2 R}{a}\left\{\tan. I + \left(1+\tfrac{1}{3}\frac{D^2}{R^2}\right)\tan.\,\phi_1 + \tfrac{2}{3}\left(\frac{\rho}{R}\right)\tan.\,\phi_2 + \tan.^2 I \tan.(\phi_1+\phi_3)\right\} \ldots (320) \; ;$$

whence we obtain (by equation 121) for the modulus,

$$U_1 = U_2\left\{1 + \left\{\left(1+\tfrac{1}{3}\frac{D^2}{R^2}\right)\tan.\,\phi_1 + \tfrac{2}{3}\left(\frac{\rho}{R}\right)\tan.\,\phi_2 + \tan.^2 I \tan.(\phi_1+\phi_3)\right\}\cot. I\right\} \ldots (321)\,\text{;}$$

whence it follows that the best inclination of the thread, in respect to the economy of power in the use of the square screw, is that which satisfies the equation

$$\tan. I = \left\{\frac{\left(1+\tfrac{1}{3}\frac{D^2}{R^2}\right)\tan.\,\phi_1 + \tfrac{2}{3}\left(\frac{\rho}{R}\right)\tan.\,\phi_2}{\tan.(\phi_1+\phi_3)}\right\}^{\frac{1}{2}}$$

258. The inclination of thread of a square screw rarely exceeds 7°, so that the term $\tan.^2 I \tan.(\phi_1+\phi_3)$ rarely exceeds ·015 $\tan.(\phi_1+\phi_3)$, and may therefore be neglected, as compared with the other terms of the expression; as also may the term $\tfrac{1}{3}\left(\frac{D}{R}\right)^2 \tan.\,\phi_2$, since the depth 2D of a square screw being usually made equal to about $\tfrac{1}{8}$th of the diameter, this term does not commonly exceed $\tfrac{1}{192}\tan.\,\phi_1$.

Omitting these terms, observing that $L = 2\pi R \tan. I$, and eliminating tan. I,

$$P_1 = P_2 \cdot \frac{I}{a}\left\{\frac{L}{2\pi} + R \tan.\,\phi_1 + \tfrac{2}{3}\rho \tan.\,\phi_2\right\} \cdots \cdots (322).$$

$$U_1 = U_2\left\{1 + \frac{2\pi}{L}(R \tan.\,\phi_1 + \tfrac{2}{3}\rho \tan.\,\phi_2)\right\} \cdots \cdots (323).$$

THE BEAM OF THE STEAM ENGINE.

259. Let P_1, P_2, P_3, P_4 represent the pressures applied by
the piston rod, the crank rod, the air pump rod, and the cold

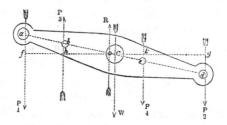

water pump rod, to the beam of a steam engine ; and suppose
the directions of all these pressures to be vertical.*

Let the rods, by which the pressures P_1, P_2, P_3, P_4 are applied to the beam, be moveable upon solid axes or gudgeons,
whose centres are a, d, b, e, situated in the same straight line
passing through the centre C of the solid axis of the beam.

Let ρ_1, ρ_2, ρ_3, ρ_4 represent the radii of these gudgeons, ρ
the radius of the axis of the beam, and φ_1, φ_2, φ_3, φ_4, φ the
limiting angles of resistance of these axes respectively. Then,
if the beam be supposed in the state bordering upon motion
by the preponderance of P_1, each gudgeon or axis being
upon the point of turning on its bearings, the directions of
the pressures P_1, P_2, P_3, P_4, R, will not be through the centres
of their corresponding axes, but separated from them by
perpendicular distances severally represented by $\rho_1 \sin. \varphi_1$,
$\rho_2 \sin. \varphi_2$, $\rho_3 \sin. \varphi_3$, $\rho_4 \sin. \varphi_4$, and $\rho \sin. \varphi$, which distances,
being perpendicular to the directions of the pressures, are all
measured horizontally.

Moreover, it is evident that the direction of the pressure P_1

* A supposition which in no case deviates greatly from the truth, and
any error in which may be neglected, inasmuch as it can only influence the
results about to be obtained in as far as they have reference to the friction
of the beam ; so that any error in the result must be of two dimensions at
least in respect to the coefficient of friction and the small angle by which
any pressure deviates from a vertical direction.

is on that side of the centre a of its axis which is nearest to
the centre of the beam, since the influence of the friction of
the axis a is to *diminish* the effect of that pressure to turn
the beam. And for a like reason it is evident that the directions of the pressures P_2, P_3, P_4 are farther from the
centre of the beam than the centres of their several axes,
since the effect of the friction is, in respect to each of these
pressures, to increase the resistance which it opposes to the
rotation of the beam; moreover, that the resistance R upon
the axis of the beam has its direction upon the same side of
the centre C as P_1, since it is equal and opposite to the resultant pressure upon the beam, and that resultant would,
by itself, turn the beam in the same direction as P_1 turns it.
Let now $a_1 = Ca$, $a_2 = Cd$, $a_3 = Cb$, $a_4 = Ce$. Draw the horizontal line of fCg, and let the angle $aCf = \theta$. Let, moreover,
W be taken to represent the weight of the beam, supposed
to act through the centre of its axis. Then since P_1, P_2, P_3,
P_4, W, R are pressures in equilibrium, we have, by the principle of the equality of moments, taking o as the point from
which the moments are measured, $P_1 . \overline{of} = P_2 . \overline{og} + P_3 . \overline{oh} +$
$P_4 . \overline{ok} + W . \overline{oC}.$

Now $of = Cf - Co = a_1 \cos. \ \theta - \rho_1 \sin. \ \phi_1 - \rho \sin. \ \phi$, $og = Cg$
$+ Co = a_2 \cos. \ \theta + \rho_2 \sin. \ \phi_2 + \rho \sin. \ \phi$, $oh = Ch - Co = a_3 \cos. \ \theta$
$+ \rho_3 \sin. \ \phi_3 - \rho \sin. \ \phi$, $ok = Ck + Co = a_4 \cos. \ \theta + \rho_4 \sin. \ \phi_4 + \rho \sin. \ \phi.$

$$\therefore \ P_1 \left\{ a_1 \cos. \ \theta - (\rho_1 \sin. \ \phi_1 + \rho \sin. \ \phi) \right\} = \left\{ \begin{array}{c} P_2 \left\{ a_2 \cos. \ \theta + (\rho_2 \sin. \ \phi_2 + \rho \sin. \ \phi) \right\} + P_3 \left\{ a_3 \cos. \ \theta + (\rho_3 \sin. \ \phi_3 - \rho \sin. \ \phi) \right\} \\ + P_4 \left\{ a_4 \cos. \ \theta + (\rho_4 \sin. \ \phi_4 + \rho \sin. \ \phi) \right\} + W \rho \sin. \ \phi \end{array} \right\} \ .. (324).$$

Multiplying this equation by θ, observing that $a_1\theta$ represents the space described by the point of application of P_1, so
that $P_1 a_1 \theta$ represents the work U_1 of P_1; and similarly that
$P_2 a_2 \theta$ represents the work U_2 of P_2, $P_3 a_3 \theta$, that U_3 of P_3 and
$P_4 a_4 \theta$, that U_4 of P_4, also that $a_1 \theta$ represents the space S_1 described by the extremity of the piston rod very nearly; we
have

$$U_1 \left\{ \cos. \ \theta - \left(\frac{\rho_1 \sin. \ \phi_1 + \rho \sin. \ \phi}{a_1} \right) \right\} = \left\{ \begin{array}{c} U_2 \left\{ \cos. \ \theta + \left(\frac{\rho_2 \sin. \ \phi_2 + \rho \sin. \ \phi}{a_2} \right) \right\} + U_3 \left\{ \cos. \ \theta + \frac{\rho_3 \sin. \ \phi_3 - \rho \sin. \ \phi}{a_3} \right\} \\ + U_4 \left\{ \cos. \ \theta + \left(\frac{\rho_4 \sin. \ \phi_4 + \rho \sin. \ \phi}{a_4} \right) \right\} + W S_1 \left(\frac{\rho}{a_1} \right) \sin. \ \phi \end{array} \right\} \ .. (325)..$$

which is the modulus of the beam.

Its form will be greatly simplified if we assume cos. $\theta = 1$, since θ is small*, suppose the coefficient of friction at each axis to be the same, so that $\varphi = \varphi_1 = \varphi_2 = \varphi_3 = \varphi_4$, and divide by the coefficient of U_1, omitting terms above the first dimension in $\frac{\rho_1}{a_1}$ sin. φ, &c. ; whence we obtain by reduction

$$U_1 = \left\{ \begin{array}{l} U_2\left\{1 + \left(\frac{\rho + \rho_1}{a_1} + \frac{\rho + \rho_2}{a_2}\right) \sin.\,\phi\right\} + U_3\left\{1 + \left(\frac{\rho + \rho_1}{a_1} - \frac{\rho - \rho_3}{a_3}\right) \sin.\,\phi\right\} \\ \quad + U_4\left\{1 + \left(\frac{\rho + \rho_1}{a_1} + \frac{\rho + \rho_4}{a_4}\right) \sin.\,\phi\right\} + W S_1\left(\frac{\rho}{a_1}\right) \sin.\,\phi \end{array} \right\} \quad \dots (326).$$

260. *The best position of the axis of the beam.*

Let a be taken to represent the length of the beam, and x the distance aC of the centre of its axis from the extremity to which the driving pressure is applied.

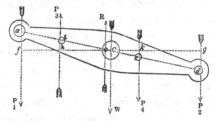

Let the influence of the position of the axis on the economy of the work necessary to open the valves, to work the air-pump, and to overcome the friction produced by the weight of the axis, be neglected; and let it be assumed to be that, by which a given amount of work U_2 may be yielded per stroke upon the crank rod, by the least possible amount U_1 of work done upon the piston rod. If then, in equation (326), we assume the three last terms of the second member to be represented by A, and observe that a_1 in that equation is represented by x, and a_2 by $a - x$, we shall obtain

* In practice the angle θ never exceeds 20°, so that cos. θ never differs from unity by more than 060307. The error, resulting from which difference, in the friction, estimated as above, must in all cases be inconsiderable.

$$U_1 = \left\{ 1 + \left(\frac{\rho + \rho_1}{x} + \frac{\rho + \rho_2}{a - x} \right) \sin. \phi \right\} U_2 + A.$$

The best position of the axis is determined by that value of x which renders this function a minimum; which value of x is represented by the equation

$$x = \frac{a}{1 + \left(\dfrac{\rho + \rho_2}{\rho + \rho_1} \right)^{\frac{1}{3}}} \quad \cdots \cdots \quad (327).$$

If $\rho_2 > \rho_1$, then $\left(\dfrac{\rho + \rho_2}{\rho + \rho_1} \right)^{\frac{1}{3}} > 1$ and $x < \frac{1}{2}a$; in this case, therefore, the axis is to be placed nearer to the driving than to the working end of the beam. If $\rho_2 < \rho_1$, the axis is to be fixed nearer to the working than to the driving end of the beam.

261. It has already been shown (Art.168.), that a machine working, like the beam of a steam engine, under two given pressures about a fixed axis, is worked with the greatest economy of power when both these pressures are applied on the same side of the axis. This principle is manifestly violated in the beam engine; it is observed in the engine worked by Crowther's parallel motion *, and in the marine engines recently introduced by Messrs. Seaward, and known as the Gorgon engines. It is difficult indeed to defend the use of the beam on any other legitimate ground than this; that in some degree it aids the fly-wheel to equalise the revolution of the crank arm †, an explanation which does not extend to its use in pumping engines, where nevertheless it retains its place; adding to the expense of construction, and, by its weight, greatly increasing the prejudicial resistances opposed to the motion of the engine.

* As used in the mining districts of the north of England.

† The reader is referred to an admirable discussion of the equalising power of the beam, by M. Coriolis, contained in the thirteenth volume of the *Journal de l'E'cole Polytechnique*.

THE CRANK.

262. *The modulus of the crank, the direction of the resistance being parallel to that of the driving pressures.*

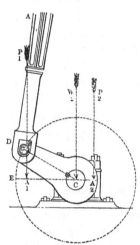

Let CD represent the arm of the crank, and AD the connecting rod. And to simplify the investigation, let the connecting rod be supposed always to retain its vertical position.* Suppose the weight of the crank arm CD, acting through its centre of gravity, to be resolved into two other weights (Art. 16.) one of which W_2 is applied at the centre C of its axis, and the other at the centre c of the axis which unites it with the connecting rod. Let this latter weight, *when added to the weight of the connecting rod,* be represented by W_1. Let P_2 represent a pressure opposed to the revolution of the crank, which would at any instant be just sufficient to balance the driving pressure P_1 transmitted through the connecting rod; and to simplify the investigation, let us suppose the direction of the pressure P_2 to be vertical and downwards.

Let $Cc = a$, $CA_1 = a_1$, $CA_2 = a_2$, $cCW_2 = \theta$, radii of axes C and $c = \rho_1, \rho_2$, lim. $\angle$ s of resistance $= \varphi_1, \varphi_2$, $W =$ whole weight of crank arm and connecting rod $= W_1 + W_2$.

Since the crank arm is in the state bordering upon motion, the perpendicular distance of the direction of the resistance upon its axis C from the centre of that axis, is represented by $\rho_1 \sin. \varphi_1$ (Art. 153.). This resistance is also equal to

---

* Any error resulting from this hypothesis affecting the conditions of the question only in as far as the friction is concerned, and being of two dimensions at least in terms of the coefficient of friction and the small angular deviation of the connecting rod from the vertical.

$P_1 \pm (P_2 + W)$; P_1 being supposed greater than $P_2 + W$, and the sign $\pm$ being taken according as the direction of P_1 is

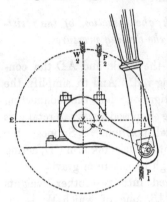

downwards or upwards, or according as the crank arm is describing its descending or ascending arc. Whence it follows, that the moment of the resistance of the axis about its centre is represented by $\{P_1 \pm (P_2 + W)\}\rho_1 \sin.\phi_1$. Now the pressures P_1, P_2, and the resistance of the axis, are pressures in equilibrium. Therefore, by the principle of the equality of moments, observing that the driving pressure is represented by $P_1 \pm W_1$, according as the arm is descending or ascending,

$$(P_1 \pm W_1)a_1 = P_2 a_2 + \{P_1 \pm (P_2 + W)\}\rho_1 \sin. \phi_1.$$

Since moreover the axis c, which unites the connecting rod and the crank arm, is upon the point of turning upon its bearings, the direction of the pressure P_1 is not through the centre of that axis, but distant from it by a quantity represented by $\rho_2 \sin. \phi_2$, which distance is to be measured on that side of the centre c which is nearest to C, since the friction diminishes the effect of P_1 to turn the crank arm.

$$\therefore \quad a_1 = a \sin. \theta - \rho_2 \sin. \phi_2 \quad . \quad . \quad . \quad . \quad (328).$$

Substituting this value of a_1 in the preceding equation,

$$(P_1 \pm W_1)(a \sin. \theta - \rho_2 \sin. \phi_2) = P_2 a_2 + \{P_1 \pm (P_2 + W)\}\rho_1 \sin. \phi_1 \ldots (329).$$

Transposing and reducing

$$P_1\{a \sin. \theta - \rho_2 \sin. \phi_2 - \rho_1 \sin. \phi_1\} = P_2\{a_2 \pm \rho_1 \sin. \phi_1\} \pm W \rho_1 \sin. \phi_1 \mp W_1 (a \sin. \theta - \rho_2 \sin. \phi_2);$$

which is the relation between P_1 and P_2 in their state bordering upon motion. Now if $\Delta\theta$ represent an exceedingly small angle described by the crank arm, $a_2 \Delta\theta$ will represent the space through which the resistance P_2 is overcome whilst that angle is described, and $P_2 a_2 \Delta\theta$ will represent the increment ΔU_2 of the work yielded by the crank whilst that small angle is described. Multiplying the above equation by $a_2\Delta\theta$, we have

$$P_1 a_2 \{a \sin.\theta - \varrho_2 \sin.\phi_2 - \varrho_1 \sin.\phi_1\}\Delta\theta = \{a_2 \pm \varrho_1 \sin.\phi_1\}\Delta U_2 \pm W a_2 \varrho_1 \sin.\phi_1 \Delta\theta \mp W_1 a_2(a \sin.\theta - \varrho_2 \sin.\phi_2)\Delta\theta \ldots \text{(330)}.$$

whence passing to the limit, integrating from $\theta=\Theta$ to $\theta = \pi - \Theta$, and dividing by a_2

$$P_1\{2a\cos.\Theta - (\pi - 2\Theta)(\varrho_2 \sin.\phi_2 + \varrho_1 \sin.\phi_1)\} = \{1 \pm \tfrac{\varrho_1}{a_2}\sin.\phi_1\}U_2 \pm W(\pi - 2\Theta)\varrho_1 \sin.\phi_1 \mp W_1\{2a\cos.\Theta - \varrho_2(\pi - 2\Theta)\sin.\phi_2\}.\text{(331)}.$$

Now, let it be observed that $2a\cos.\Theta$ represents the projection of the path of the point c upon the vertical direction of P_1, whilst the arm revolves between the positions Θ and $\pi - \Theta$; so that $P_1 2a\cos.\Theta$ represents (Art. 52.) the work U_1 done by P_1 upon the crank whilst the arm passes from one of these positions to the other, or whilst the work U_2 is yielded by the crank. Whence it follows that $P_1 = \frac{U_1}{2a}\sec.\Theta$. Substituting this value of P_1, dividing by a, and reducing, we obtain

$$U_1\{1 - (\tfrac{\pi}{2} - \Theta)\sec.\Theta\left(\tfrac{\rho_2}{a}\sin.\phi_2 + \tfrac{\rho_1}{a}\sin.\phi_1\right)\} = \{1 \pm \tfrac{\rho_1}{a_2}\sin.\phi_1\}U_2,$$
$$\pm W(\pi - 2\Theta)\rho_1 \sin.\phi_1 \mp W_1\{2a\cos.\Theta - \rho_2(\pi - 2\Theta)\sin.\phi_2\} \ldots \text{(332)}.$$

By which equation is determined the modulus of the crank in respect to the ascending or descending stroke, according as we take the upper or lower signs of the ambiguous terms.

Adding these two values of the modulus together, and representing by U_1 the whole work of P_1, and by U_2 the whole work of P_2, whilst the crank arm makes a complete revolution, also by u_1 the work of P_2 in the *down* stroke, and u_2 in the *up* stroke, we obtain

$$U_1\left\{1 - \left(\tfrac{\pi}{2} - \Theta\right)\sec.\Theta\left(\tfrac{\rho_2}{a}\sin.\phi_2 + \tfrac{\rho_1}{a}\sin.\phi_1\right)\right\} = U_2 + (u_1 - u_2)\tfrac{\rho_1}{a_2}\sin.\phi_1\ldots\text{(333)},$$

which is the modulus of the crank in respect to a *vertical* direction of the driving pressure and of the resistance, the arm being supposed in each half revolution, first, to receive the action of the driving pressure when at an inclination of Θ to the vertical, and to yield it when it has again attained the same inclination, so as to revolve under the action of the driving pressure through the angle $\pi-2\Theta$.

In the double-acting engine, $u_1-u_2=0$; in the single-acting engine, $u_1=0$. The work expended by reason of the friction of the crank is therefore less in the latter engine than in the former, when the resistance P_2 is applied, as shown in the figure, on the side of the ascending arc.

If the arm sustain the action of the driving pressure *constantly*, $\Theta=0$, and the modulus becomes, for the *double-acting engine*,

$$U_1\left\{ 1-_2\left(\frac{\rho_2}{a}\sin.\varphi_2+\frac{\rho_1}{a}\sin.\varphi_1\right)\right\}=U_2;$$

or, dividing by the coefficient of U_1, and neglecting dimensions above the first in sin. φ_1, sin. φ_2,

$$U_1=\left\{1+\frac{\pi}{2}\left(\frac{\rho_1}{a}\sin.\varphi_1+\frac{\rho_2}{a}\sin.\varphi_2\right)\right\}U_2 \ . \ . \ . \ . \ (334).$$

The modulus not involving the symbol W which represents the weight of the crank, it is evident that so long as P_1 and P_2 are vertical and P_1 greater than P_2+W, the economy of power in the use of the crank is not at all influenced by its weight and that of the connecting rod, the friction being upon the whole as much diminished by reason of that weight in the ascending stroke as it is increased by it in the descending stroke.

It is evident, moreover, that if the friction produced by the weight of the crank be neglected, the modulus above deduced, for the case in which the directions of the pressures P_1 and P_2 are vertical, applies to every case in which the directions of those pressures are parallel.

The condition $P_1>P_2+W$ evidently obtains in every other position of the crank arm, if it obtain in the horizontal position.

Now, in this position, $P_2=\frac{a}{a_2}P_1$, if we neglect friction. The required condition obtains, therefore, if $P_1>\frac{a}{a_2}P_1+W$. To satisfy this condition, a_2 must be greater than a, or the resistance be applied at a perpendicular distance from the axis greater than the length of crank arm, and so much greater, that $P_1 \left(1 - \dfrac{a}{a_2}\right)$ may exceed W. These conditions commonly obtain in the practical application of the crank.

263. Should it, however, be required to determine the modulus in the case in which P_1 is not, in every position of the arm, greater than $P_2 + W$, let it be observed, that this condition does not affect the determination of the modulus (equation 332) in respect to the descending, but only the ascending stroke; there being a certain position of the arm as it ascends in which the resultant pressure upon the axis represented by the formula $\{P_1 - (P_2 + W)\}$, passing through zero, is afterwards represented by $\{(P_2 + W) - P_1\}$; and when the arm has still further ascended so as to be again inclined to the vertical at the same angle, passes again through zero, and is again represented by the same formula as before. The value of this angle may be determined by substituting P_1 for $P_2 + W$ in equation (329), and solving that equation in respect to θ. Let it be represented by θ_1; let equation (330) be integrated in respect to the ascending stroke from $\theta = 0$ to $\theta = \theta_1$, the work of P_2 through this angle being represented by u_1; let the signs of all the terms involving $\rho_1 \sin. \varphi_1$ then be changed, which is equivalent to changing the formula representing the pressure upon the axis from $\{P_1 - (P_2 + W)\}$ to $\{P + W) - P_1\}$; and let the equation then be integrated from $\theta = \theta_1$ to $\theta = \dfrac{\pi}{2}$, the work of P_2 through this angle being represented by u_2; $2(u_1 + u_2)$ will then represent the whole work U_2 done by P_2 in the ascending arc. To determine this sum, divide the first integral by the coefficient of u_1, and the second by that of u_2, add the resulting equations, and multiply their sum by 2; the modulus in respect to the ascending arc will then be determined; and if it be added to the modulus in respect to the descending arc, the modulus in respect to an entire revolution will be known.

THE DEAD POINTS IN THE CRANK.

264. If equation (329) be solved in respect to P_1 it becomes

$$P_1 = P_2 \left\{ \frac{a_2 \pm \rho_1 \sin. \phi_1}{a \sin. \theta - \rho_2 \sin. \phi_2 - \rho_1 \sin. \phi_1} \right\} \pm \frac{W \rho_1 \sin. \phi_1 - W_1 (a \sin. \theta - \rho_2 \sin. \phi_2)}{a \sin. \theta - \rho_2 \sin. \phi_2 - \rho_1 \sin. \phi_1}.$$

In that position of the arm, therefore, in which

$$\sin. \theta = \frac{\rho_2 \sin. \phi_2 + \rho_1 \sin. \phi_1}{a} \quad \ldots \ldots (335),$$

the driving pressure P_1 necessary to overcome any given resistance P_2 opposed to the revolution of the crank, assumes an infinite value. This position, from which no finite pressure acting in the direction of the length of the connecting rod is sufficient to move the arm, when it is at rest in that position, is called its *dead* point.

Since there are four values of θ, which satisfy equation (335), two in the descending and two in the ascending semi-revolution of the arm, there are, on the whole, four dead points of the crank.* The value of P_1 being, however, in all cases exceedingly great between the two highest and the two lowest of these positions, every position between the two former and the two latter, and for some distance on either side of these limits, is practically a dead point.

THE DOUBLE CRANK.

265. To this crank, when applied to the steam engine, are affixed upon the same solid shaft, two arms at right angles to one another, each of which sustains the pressure of the steam in a separate cylinder of the engine, which pressure is transmitted to it, from the piston rod, by the intervention of a beam and connecting rod as in the marine engine, or a guide and connecting rod as in the locomotive engine.

---

* It has been customary to reckon theoretically only two dead points of the crank, one in its highest and the other in its lowest position. Every practical man is acquainted with the fallacy of this conclusion.

In either case, the connecting rods may be supposed to remain constantly parallel to themselves, and the pressures

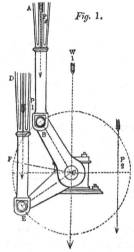

Fig. 1.

applied to them in different planes to act in the same plane*, without materially affecting the results about to be deduced.†

Let the two arms of the crank be supposed to be of the same length a; let the same driving pressure P_1 be supposed to be applied to each; and let the same notation be adopted in other respects as was used in the case of the crank with a single arm; and, first, let us consider the case represented in *fig.* 1., in which both arms of the crank are upon the same side of the centre C.

Let the angle $W_1CB = \theta$; therefore $W_1CE = \dfrac{\pi}{2} + \theta$: whence it follows, by precisely the same reasoning as in Art. 262., that the perpendicular upon the direction of the driving pressure applied by the connecting rod AB is represented (see equation 328) by a sin. $\theta - \rho_2$ sin. φ_2, and the perpendicular upon the pressure applied by the rod CD by a sin. $\left(\dfrac{\pi}{2} + \theta\right) - \rho_2$ sin. φ_2, or a cos. $\theta - \rho_2$ sin. φ_2. Let now a_1 be taken to represent the perpendicular distance from the axis C, at which a single pressure, equal to $2P_1$, must be applied, so as to produce the same effect to turn the crank as is produced by the two pressures actually applied to it by the two connecting rods; then, by the principle of the equality of moments,

$$2P_1 a_1 = P_1(a \text{ sin. } \theta - \rho_2 \text{ sin. } \varphi_2) + P_1(a \text{ cos. } \theta - \rho_2 \text{ sin. } \varphi_2);$$

* This principle will be more fully discussed by a reference to the theory of statical couples. (See Pritchard on Statical Couples.)

† The relative dimensions of the crank arm and the connecting rod are here supposed to be those usually given to these parts of the engine; the supposition does not obtain in the case of a short connecting rod.

$$\therefore\ a_1 = \tfrac{1}{2}a(\sin.\ \theta + \cos.\ \theta) - \rho_2 \sin.\ \varphi_2\ ;$$

$$\therefore\ a_1 = \frac{a}{\sqrt{2}}\left(\sin.\ \theta \cos.\frac{\pi}{4} + \cos.\ \theta \sin.\frac{\pi}{4}\right)^{*} - \rho_2 \sin.\ \varphi_2 = \frac{a}{\sqrt{2}}\sin.\ \left(\theta + \frac{\pi}{4}\right) - \rho_2 \sin.\ \varphi_2\ ;$$

which expression becomes identical with the value of a_1, determined by equation (328), if in the latter equation a be replaced by $\frac{a}{\sqrt{2}}$, and θ by $\theta + \frac{\pi}{4}$. Whence it follows that the conditions of the equilibrium of the double crank in the state bordering upon motion, and therefore the form of the modulus, are, whilst both arms are on the same side of the centre, precisely the same as those of the single crank, the direction of whose arm bisects the right angle BCE, and the length of whose arm equals the length of either arm of the double crank divided by $\sqrt{2}$.

Now, if θ_1 be taken to represent the inclination W_1CF of this imaginary arm to W_1C, both arms will be found on the same side of the centre, from that position in which $\theta_1 = \frac{\pi}{4}$ to that in which it equals $(\pi - \frac{\pi}{4})$. If, therefore, we substitute $\frac{\pi}{4}$ for Θ, in equations (331), and for a, $\frac{a}{\sqrt{2}}$, and add these equations together, the symbol $2U_2$ in the resulting equation will represent the whole work yielded by the working pressure, whilst both arms remain on the same side of the centre, in the ascending and the descending arcs. We thus obtain, representing the sum of the driving pressures upon the two arms by P_1,

$$2P_1\{a - \frac{\pi}{2}(\rho_2 \sin.\ \varphi_2 + \rho_1 \sin._1\varphi)\} = 2U_2\ \ .\ .\ .\ .\ .\ (336).$$

Throughout the remaining two quadrants of the revolution of the crank, the directions of the two equal and parallel pressures applied to it through the connecting rods being opposite, the resultant pressure upon the axis is represented by $(P_2 + W)$, instead of $\{P_1 \pm (P_2 + W)\}$; whilst, in other respects, the conditions of the equilibrium of the state bordering

* Whewell's Mechanics, p. 25.

upon motion remain the same as before; that is, the same

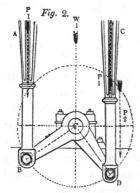

as though the pressure P_1 were applied to an imaginary arm, whose length is $\dfrac{a}{\sqrt{2}}$, and whose position coincides with CF. Now, referring to equation (329), it is apparent that this condition will be satisfied if, in that equation, the ambiguous sign of $(P_2 + W)$ be suppressed, and the value of P_1 in the second member, which is multiplied by $\rho_1 \sin. \varphi_1$, be assumed $= 0$; by which assumption the term $-\rho_1 \sin. \varphi_1$ will be made to disappear from the left-hand member of equation (330), and the ambiguous signs which affect the first and second terms of the right-hand member will become positive. Now, these substitutions being made, and the equation being then integrated, first, between the limits 0 and $\dfrac{\pi}{4}$, and then between the limits $\dfrac{3\pi}{4}$ and π, the symbol U_2 in it will evidently represent the work done during each of those portions of a semi-revolution of the imaginary arm in which the two real arms of the crank are not on the same side of the centre. Moreover, the integral of that equation between the limits 0 and $\dfrac{\pi}{2}$, is evidently the same with its integral between the limits $\dfrac{3\pi}{4}$ and π. Taking, therefore, twice the former integral, we have

$$2P_1 a_2 \left\{ \frac{a}{\sqrt{2}}\left(1 - \cos. \frac{\pi}{4}\right) - \frac{\pi}{4}\rho_2 \sin. \varphi_2 \right\} = \left\{ a_2 + \rho_1 \sin. \varphi_1 \right\} 2U_2$$

$$+ \frac{\pi}{2}W a_2 \rho_1 \sin. \varphi_1 \mp 2 W_1 a_2 \left\{ \frac{a}{\sqrt{2}}\left(1 - \cos. \frac{\pi}{4}\right) - \frac{\pi}{4}\rho_2 \sin. \varphi_2 \right\}.$$

Dividing this equation by $(a_2 + \rho_1 \sin. \varphi_1)$, or by $a_2 \left(1 + \dfrac{\rho_1}{a_2} \sin. \varphi_1\right)$, and neglecting terms above the first dimension in $\sin. \varphi_1$ and $\sin. \varphi_2$,

$$2P_1\left\{\frac{a}{\sqrt{2}}\left(1-\cos.\frac{\pi}{4}\right)\left(1-\frac{\rho_1}{a_2}\sin.\phi_1\right)-\frac{\pi}{4}\rho_2\sin.\phi_2\right\}=2U_2$$

$$+\frac{\pi}{2}W\rho_1\sin.\phi_1\mp2W_1\left\{\frac{\sqrt{2}}{a}\left(1-\cos.\frac{\pi}{4}\right)\left(1-\frac{\rho_1}{a_2}\sin.\phi_1\right)-\frac{\pi}{4}\rho_2\sin.\phi_2\right\};$$

in which equation $2U_2$ represents the work done in the descending or ascending arcs of the imaginary arm, according as the ambiguous sign is taken positively or negatively. Taking, therefore, the sum of the two values of the equation given by the ambiguous sign, and representing by $4U_2$ the whole work done in the descending and ascending arcs, during those portions of each complete revolution when both of the arms are not on the same side of the centre, we have

$$4P_1\left\{\frac{a}{\sqrt{2}}\left(1-\cos.\frac{\pi}{4}\right)\left(1-\frac{\rho_1}{a_2}\sin.\phi_1\right)-\frac{\pi}{4}\rho_2\sin.\phi_2\right\}=4U_2+W\pi\rho_1\sin.\phi_1;$$

or, observing that $\cos.\frac{\pi}{4}=\frac{1}{\sqrt{2}}$,

$$2P_1\left\{a(\sqrt{2}-1)-a(\sqrt{2}-1)\frac{\rho_1}{a_2}\sin.\phi_1-\frac{\pi}{2}\rho_2\sin.\phi_2\right\}=4U_2+W\pi\rho_1\sin.\phi_1.$$

Adding this equation to equation (336), and representing by U_2 the entire work yielded during a complete revolution of the imaginary arm,

$$2P_1\left\{a\sqrt{2}-a(\sqrt{2}-1)\frac{\rho_1}{a_2}\sin.\phi_1-\frac{\pi}{2}(2\rho_2\sin.\phi_2+\rho_1\sin.\phi_1)\right\}=U_2+W\pi\rho_1\sin.\phi_1.$$

But if U_1 represent the whole work done by the driving pressures at each revolution of the imaginary arm, then $4\frac{a}{\sqrt{2}}P_1=U_1$. Since $2\frac{a}{\sqrt{2}}$ is the projection of the space described by the extremity of the arm during the ascending and descending strokes respectively, therefore $2P_1=\frac{U_1}{a\sqrt{2}}$.

Substituting this value for $2P_1$,

$$U_1\left\{1-\frac{\sqrt{2}-1}{\sqrt{2}}\frac{\rho_1}{a_2}\sin.\phi_1-\frac{\pi}{2\sqrt{2}}\left(\frac{2\rho_2}{a}\sin.\phi_2+\frac{\rho_1}{a}\sin.\phi_2\right)\right\}=U_2+W\pi\rho_1\sin.\phi_1\ldots\ (337),$$

which is the modulus of the double crank, the directions of the driving pressure and the resistance being both supposed vertical; or if the friction resulting from the weight of the crank be neglected, and W be therefore assumed $=0$, then does the above equation represent the modulus of the double crank, whatever may be the direction of the driving pressure, provided that the direction of the resistance be parallel to it. Dividing by the coefficient of U_1, and neglecting terms of more than one dimension in sin. φ_1 and sin. φ_2,

$$U_1 = U_2\left\{1 + \frac{\sqrt{2}-1}{\sqrt{2}}\frac{\rho_1}{a_2}\sin. \phi_1 + \frac{\pi}{2\sqrt{2}}\left(\frac{2\rho_2}{a}\sin. \phi_2 + \frac{\rho_1}{a}\sin. \phi_1\right)\right\} + W\pi\rho_1\sin. \phi_1 \dots (338).$$

THE CRANK GUIDE.

266. In some of the most important applications of the steam engine, the crank is made to receive its continuous rotatory motion, from the alternating rectilinear motion of the piston rod, directly through the connecting rod of the crank, without the intervention of the beam or parallel motion; the connecting rod being in this case jointed at one extremity, to the extremity of the piston rod, and the oblique pressure upon it which results from this connexion being sustained by the intervention of a cross piece fixed upon it, and moving between lateral guides.*

Let the length CD of the connecting rod be represented by b, and that BD of the crank arm by a, and let P_1 and P_2 in the following figure be taken, respectively, to represent the pressure upon the piston rod of the engine and the connecting rod of the crank, and RS to represent the direction of the resistance of the guide in the state bordering upon motion by the excess of the driving pressure P_1. Then is

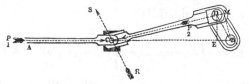

RS inclined to a perpendicular to the direction of the guides,

* The contrivance is that well known as applied to the locomotive carriage.

or of the motion of the piston rod, at an angle equal to the
limiting angle of resistance (Art. 141.) of the surfaces of contact of the guides.

Since, moreover, P_1, P_2, R are pressures in equilibrium,

$$\therefore \frac{P_2}{P_1} = \frac{\sin. P_1 CS}{\sin. P_2 CS}.$$

Let $\angle$ BCD $= \theta$; limiting angle of resistance of guide

$= \varphi$; therefore, $P_1 CS = \dfrac{\pi}{2} - \varphi$, $P_2 CS = \dfrac{\pi}{2} + \varphi - \theta$;

$$\therefore \frac{P_2}{P_1} = \frac{\sin. \left(\dfrac{\pi}{2} - \varphi\right)}{\sin. \left\{\dfrac{\pi}{2} - (\theta - \varphi)\right\}} = \frac{\cos. \varphi}{\cos. (\theta - \varphi)}. \quad \ldots \ldots (339).$$

Let BD $= a$, CD $= b$, and DBC $= \theta_1$, and assume P_2 to remain constant, P being made to vary according to the conditions of the state bordering upon motion;

$\therefore \Delta U_1 = P_1 . \Delta \overline{AC} = -P_1 . \Delta \overline{BC} = -P_1 . \Delta(a \cos. \theta_1 + b \cos. \theta) = P_2 \sec. \varphi \cos.(\theta - \varphi) (a \sin. \theta_1 \Delta \theta_1 + b \sin. \theta \Delta \theta)$;

$$\Delta U_2 = -P_2 (\Delta \overline{AC}) \cos. \theta = P_2 (a \sin. \theta_1 \Delta \theta_1 + b \sin. \theta \Delta \theta) \cos. \theta;$$

$$\therefore U_1 = P_2 \sec. \varphi \{a \int_0^{\frac{\pi}{2}} \sin. \theta_1 \cos.(\theta - \varphi) d\theta_1 + b \int_0^{0} \sin. \theta \cos.(\theta - \varphi) d\theta\},$$

$$U_2 = P_2 \{a \int_0^{\frac{\pi}{2}} \sin. \theta_1 \cos. \theta d\theta_1 + b \int_0^{0} \sin. \theta \cos. \theta d\theta\}.$$

The second integral in each of these equations vanishes
between the prescribed limits; also $\sin. \theta = \dfrac{a}{b} \sin. \theta_1$; there-

fore $\cos. \theta = (1 - \dfrac{a^2}{b^2} \sin.^2 \theta_1)^{\frac{1}{2}}$;

$$\therefore U_2 = P_2 a \int_0^{\frac{\pi}{2}} \sin. \theta_1 \cos. \theta d\theta_1 = P_2 a \int_0^{\frac{\pi}{2}} (1 - \frac{a^2}{b^2} \sin.^2 \theta)^{\frac{1}{2}} \sin. \theta_1 d\theta_1$$

$$= -P_2 \frac{a^2}{b} \int_0^{\frac{\pi}{2}} \left\{\left(\frac{b^2}{a^2} - 1\right) + \cos.^2 \theta_1\right\}^{\frac{1}{2}} d\cos. \theta_1 = P_2 \frac{a^2}{b} \left\{1 - \frac{1}{2}\left(\frac{b^2}{a^2} - 1\right) \log._\varepsilon \left(\frac{b - a}{b + a}\right)\right\}^*,$$

* Hymer's Int. Cal. art. 17.

$$U_1 = P_2 a \sec. \phi \int_0^\pi \sin. \theta_1 \cos.(\theta-\phi)d\theta_1 = P_2 a \int_0^\pi \sin. \theta_1 \cos. \theta d\theta_1 + P_2 a \tan. \phi \int_0^\pi \sin. \theta \sin. \theta_1 d\theta_1$$

$$= U_2^1 + P_2 \frac{a^2}{b} \tan. \phi \int_0^\pi \sin.^2\theta_1 d\theta_1 = U_2 + P_2 \frac{a^2 \pi}{b\, 2} \tan. \phi \,;$$

whence eliminating P_2 and reducing, we obtain

$$U_1 = U_2 \left\{ 1 + \frac{\pi \tan. \phi}{2 - \left(\frac{b^2}{a^2}-1\right) \log._i \left(\frac{b-a}{b+a}\right)} \right\} \quad \ldots \ldots (340),$$

which is the *modulus* of the crank guide.

THE FLY-WHEEL.

267. The angular velocity of the fly-wheel.

Let P_1 be taken to represent a constant pressure applied

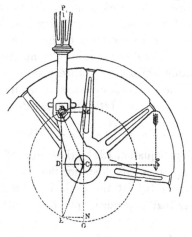

through the connecting rod to the arm of the crank of a
steam engine ; suppose the direction of this pressure to remain
always parallel to itself, and let P_2 represent a *constant* resistance opposed to the revolution of the axis which carries
the fly-wheel, by the useful work done and the prejudicial resistances interposed between the axis of the fly-wheel and the
working points of the machine.

Let the angle $ACB = \theta$, $CB = a$, $CP_2 = a_2$.

Now the projection, upon the direction of P_1, of the path of its point of application B to the crank arm, whilst that arm describes the angle ACB, is AM, therefore (Art. 52.), the work done by P_1 upon the crank, whilst this angle is described, is represented by $P_1 . \overline{AM}$, or by $P_1 \, a$ vers. θ. And whilst the crank arm revolves through the angle θ, the resistance P_2 is overcome through the arc of a circle subtended by the same angle θ, but whose radius is a_2, or through a space represented by $a_2\theta$. So that, neglecting the friction of the crank itself, the work expended upon the resistances opposed to its motion is represented by $P_2 a_2 \theta$, and the excess of the work done upon it through the angle ACB by the moving power, over that expended during the same period upon the resistances, is represented by

$$P_1 a \text{ vers. } \theta - P_2 a_2 \theta \quad \ldots \ldots \text{ (341)}.$$

Now $2aP_1$ represents the work done by the moving pressure P_1 during each effective stroke of the piston, and $2\pi a_2 P_2$ the work expended upon the resistance during each revolution of the fly-wheel; so that if m represent the number of strokes made by the piston whilst the fly-wheel makes one revolution, and if the engine be conceived to have attained its state of uniform or steady action (Art. 146.), then $2maP_1 = 2\pi a_2 P_2$,

$$\therefore \; a_2 P_2 = \frac{m}{\pi} a P_1 \ldots . \text{ (342)}.$$

Eliminating from equation (341) the value of $a_2 P_2$ determined by this equation, we obtain for the excess of the work done by the power (whilst the angle θ is described by the crank arm), over that expended upon the resistance, the expression

$$P_1 a \left\{ \text{vers. } \theta - \frac{m\theta}{\pi} \right\} \quad \ldots \ldots \text{ (343)}.$$

But this excess is equal to the whole work which has been accumulating in the different moving parts of the machine, whilst the angle θ is described by the arm of the crank (Art. 145.). Now, let the whole of this work be conceived to have been accumulated in the fly-wheel, that wheel being proposed

to be constructed of such dimensions as sufficiently to equalise the motion, even if no work accumulated at the same
time in other portions of the machinery (see Art. 150.), or if
the weights of the other moving elements, or their velocities,
were comparatively so small as to cause the work accumulated
in them to be exceedingly small as compared with the work
accumulated during the same period in the fly-wheel. Now,
if I represent the moment of inertia of the fly-wheel, μ the
weight of a cubic foot of its material, α_1 its angular velocity
when the crank arm was in the position CA, and α its angular
velocity when the crank arm has passed into the position
CB; then will $\frac{1}{2}\dfrac{I\mu}{g}(\alpha^2-\alpha_1^2)$ represent the work accumulated
in it (Art. 75.) between these two positions of the crank arm,
so that

$$\frac{1}{2}\frac{I\mu}{g}(\alpha^2-\alpha_1^2)=P_1a\left\{\text{ vers. }\theta-\frac{m\theta}{\pi}\right\};$$

$$\therefore\ \alpha^2-\alpha_1^2=\frac{2P_1ag}{\mu I}\left\{\text{ vers. }\theta-\frac{m\theta}{\pi}\right\}\ \ \cdots\ \ (344).$$

268. *The positions of greatest and least angular velocity of
the fly-wheel.*

If we conceive the engine to have acquired its state of

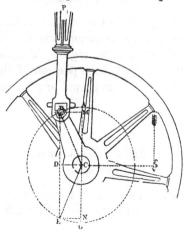

steady or uniform motion, the aggregate work done by the power being equal to that expended upon the resistances, then will the angular velocity of the fly-wheel return to the same value whenever the wheel returns to the same position ; so that the value of α_1 in equation (344) is a *constant*, and the value of α a function of θ : α assumes, therefore, its minimum and maximum values with this function of θ, or it is a minimum when $\frac{d\alpha^2}{d\theta} = 0$, and $\frac{d^2\alpha^2}{d\theta^2} > 0$, and a maximum when $\frac{d\alpha^2}{d\theta} = 0$, and $\frac{d^2\alpha^2}{d\theta^2} < 0$. But $\frac{d\alpha^2}{d\theta} = \sin.\ \theta - \frac{m}{\pi}$ and $\frac{d^2\alpha^2}{d\theta^2} = \cos.\theta$; therefore $\frac{d\alpha^2}{d\theta} = 0$, when

$$\sin.\ \theta = \frac{m}{\pi} \ . \ . \ . \ . \ . \ (345).$$

Now this equation is evidently satisfied by two values of θ, one of which is the supplement of the other, so that if η represent the one, then will $(\pi - \eta)$ represent the other ; which two values of θ give opposite signs to the value cos. θ of the second differential coefficient of α^2, the one being positive or > 0, and the latter negative or < 0. The one value corresponds, therefore, to a minimum and the other to a maximum value of α. If, then, we take the angle ACB in the preceding figure, such that its sine may equal $\frac{m}{\pi}$ (equation 345), then will the position CB of the crank arm be that which corresponds to the minimum angular velocity of the fly-wheel ; and if we make the angle ACE equal to the supplement of ACB, then is CE the position of the crank arm, which corresponds to the maximum angular velocity of the fly-wheel.

269. *The greatest variation of the angular velocity of the fly-wheel.*

Let α_2 be taken to represent the least angular velocity of the fly-wheel, corresponding to the position CB of the crank arm, and α_3 its greatest angular velocity, corresponding to the position CE ; then does $\frac{\mu I}{2g}(\alpha_3{}^2 - \alpha_2{}^2)$ represent the work accumulated in the fly-wheel between these positions, which accumulated work is equal to the excess of that done by the power over that expended upon the resistances whilst the crank arm revolves from the one position into the other, and is therefore represented by the difference of the values given to the formula (343) when the two values $\pi - \eta$ and η, determined by equation (345), are substituted in it for θ. Now this difference is represented by the formula

$$P_1a\left\{ \text{vers.} (\pi - \eta) - \text{vers.} \eta - \frac{m(\pi - \eta - \eta)}{\pi} \right\}, \text{ or by } P_1a\left\{ 2 \cos. \eta - m\left(1 - \frac{2\eta}{\pi}\right) \right\} ;$$

$$\therefore \frac{\mu I}{2g}(\alpha_3{}^2 - \alpha_2{}^2) = P_1a\left\{ 2 \cos. \eta - m\left(1 - \frac{2\eta}{\pi}\right) \right\} ;$$

$$\therefore \alpha_3{}^2 - \alpha_2{}^2 = \frac{2P_1ga}{\mu I}\left\{ 2 \cos. \eta - m\left(1 - \frac{2\eta}{\pi}\right) \right\} \quad \dots \quad (346) ;$$

in which equation η is taken (equation 345) to represent that angle whose sine is $\frac{m}{\pi}$.

270. *The dimensions of the fly-wheel, such that its angular velocity may at no period of a revolution deviate beyond prescribed limits from the mean.*

Let $\frac{1}{2}N$ be taken to represent the mean number of revolutions made by the fly-wheel per minute; then will $\frac{1}{2}\frac{N}{60}$ represent the mean number of revolutions or parts of a revolution made by it per second, and $\frac{1}{2}\frac{N}{60}2\pi$, or $\frac{N\pi}{60}$, the mean space described per second by a point in the fly-wheel whose distance from the centre is unity, or the mean angular velocity of the fly-wheel. Now, let the dimensions of the fly-wheel be supposed to be such as are sufficient to cause its angular velocity to deviate at no period of its revolution by more than $\frac{1}{n}$th from its mean value; or such that the maximum

value α_3 of its angular velocity may equal $\frac{N\pi}{60}\left(1+\frac{1}{n}\right)$, and

that its minimum value α_2 may equal $\frac{N\pi}{60}\left(1-\frac{1}{n}\right)$; then

$$(\alpha_3{}^2-\alpha_2{}^2)=(\alpha_3+\alpha_2)(\alpha_3-\alpha_2)=\frac{2\pi N}{60}\cdot\frac{2\pi N}{60n}=\frac{\pi^2 N^2}{30^2 n}.$$

Substituting in equation (346),

$$\frac{\pi^2 N^2}{30^2 n}=\frac{2P_1 g a}{\mu I}\left\{2\cos.\,\eta-m\left(1-\frac{2\eta}{\pi}\right)\right\}.$$

Let H be taken to represent the horses' power of the engine, estimated at its driving point or piston; then will 33000H represent the number of units of work done per minute, upon the piston. But this number of units of work is also represented by $\frac{1}{2}Nm\,.\,2P_1 a$; since $\frac{1}{2}Nm$ is the number of strokes made by the piston per minute, and $2P_1 a$ is the work done on the piston per stroke,

$$\therefore\;2P_1 a=66000\frac{H}{m}.$$

Substituting this value for $2P_1 a$ in the above equation, we obtain, by reduction,

$$\mu I=\left\{\frac{66000\,.\,30^2 g}{\pi^2}\right\}\left\{2\cos.\,\eta-m\left(1-\frac{2\eta}{\pi}\right)\right\}\frac{Hn}{N^3 m}\ldots(347).$$

Let k be taken to represent the radius of gyration of the wheel, and M its volume; then (Art. 80.) $Mk^2=I$, therefore $\mu M\,.\,k^2=\mu I$. But μM represents the weight of the wheel in lbs.; let W represent its weight in tons; therefore, $\mu M=2240W$. Substituting this value, and solving in respect to W,

$$W=\left\{\frac{66000\,.\,30^2\,.\,g}{2240\pi^2}\right\}\left\{2\cos.\,\eta-m\left(1-\frac{2\eta}{\pi}\right)\right\}\frac{Hn}{N^3 mk^2}.$$

Substituting their values for π and g, and determining the numerical value of the coefficient,

$$W = 86426 \left\{ \frac{2}{m} \cos.\eta - \left(1 - \frac{2\eta}{\pi}\right) \right\} \frac{Hn}{N^3 k^2} \cdots \cdots (348).$$

If the influence of the work accumulated in the arms of the wheel be given in, for an increase of the equalising power beyond the prescribed limits, that accumulated in the heavy rim or ring which forms its periphery being alone taken into the account *; then (Art. 86.) $Mk^2 = I = 2\pi b c R \,(R^2 + \frac{1}{4}c^2)$, where b represents the thickness, c the depth, and R the mean radius of the rim. But by Guldinus's first property (Art. 38.), $2\pi b c R = M$; therefore $k^2 = (R^2 + \frac{1}{4}c^2)$. Substituting in equation (348)

$$W = 86426 \left\{ \frac{2}{m} \cos.\eta - \left(1 - \frac{2\eta}{\pi}\right) \right\} \frac{Hn}{N^3(R^2 + \frac{1}{4}c^2)} \cdots (349).$$

If the depth c of the rim be (as it usually is) small as compared with the mean radius of the wheel, $\frac{1}{4}c^2$ may be neglected as compared with R^2, the above equation then becomes

$$W = 86426 \left\{ \frac{2}{m} \cos.\eta - \left(1 - \frac{2\eta}{\pi}\right) \right\} \frac{Hn}{N^3 R^2} \cdots \cdots (350);$$

by which equation the weight W in tons of a fly-wheel of a given mean radius R is determined, so that being applied to an engine of a given horse power H, making a given number of revolutions per minute $\frac{1}{2}N$, it shall cause the angular velocity of that wheel not to vary by more than $\frac{1}{n}$th from its mean value. It is to be observed that the weight of the wheel varies inversely as the cube of the number of strokes made

---

* If the section of each arm be supposed uniform and represented by κ, and the arms be p in number, it is easily shown from Arts. 79. 81., that the momentum of inertia of each arm about its extremity is very nearly represented by $\frac{1}{3}\kappa(R-\frac{1}{2}c)^3$, where c represents the depth of the rim; so that the whole momentum of inertia of the arms is represented by $\frac{p}{3}\kappa(R-\frac{1}{2}c)^3$, which expression must be added to the momentum of the rim to determine the whole momentum I of the wheel. It appears however expedient to *give* the inertia of the arms to the equalising power of the wheel.

by the engine per minute, so that an engine making twice as many strokes as another of equal horse power, would receive an equal steadiness of motion from a fly-wheel of one eighth the weight; the mean radii of the wheels being the same.

If in equation (347) we substitute for I its value $2\pi bc R^3$, or $2\pi KR^3$ (representing by K the section bc of the rim), and if we suppose the wheel to be formed of cast iron of mean quality, the weight of each cubic foot of which may be assumed to be 450 lb., we shall obtain by reduction

$$R^3 = 68470 \left\{ \frac{2}{m} \cos. \eta - \left(1 - \frac{2\eta}{\pi} \right) \right\} \frac{Hn}{N^3 K} \quad \ldots \quad (351);$$

by which equation is determined the mean radius R of a fly-wheel of cast iron of a given section K, which, being applied to an engine of given horse power H, making a given number of revolutions $\frac{1}{2}$N per minute, shall cause its angular velocity not to deviate more than $\frac{1}{n}$th from the mean; or conversely, the mean radius being given, the section K may be determined according to these conditions.

271. In the above equations, m is taken to represent the number of effective strokes made by the piston of the engine whilst the fly-wheel makes one revolution, and η to represent that angle whose sine is $\frac{m}{\pi}$.

Let now the axis of the fly-wheel be supposed to be a continuation of the axis of the crank, so that both turn with the same angular velocity, as is usually the case; and let its application to the single-acting engine, the double-acting engine, and to the double crank engine, be considered separately.

1. *In the single-acting engine*, but one effective stroke of the piston is made whilst the fly-wheel makes each revolution. In this case, therefore, $m = 1$, and sin. $\eta = \frac{1}{\pi} = 0.3183098$

$= \sin. 18° 33'$; therefore, $\cos. \eta = \cdot9480460$, also $\dfrac{\eta}{\pi} = \dfrac{18° 33'}{180°}$

$= \cdot103055$; therefore, $1 - \dfrac{2\eta}{\pi} = \cdot793888.$

$$\therefore \left\{ \dfrac{2}{m} \cos. \eta - \left(1 - \dfrac{2\eta}{\pi}\right) \right\} = 1\cdot102203.$$

Substituting in equations (350) and (351),

$$W = 95274\cdot35 \dfrac{Hn}{N^3R^2}, \quad R = \dfrac{42\cdot261}{N}\sqrt[3]{\dfrac{Hn}{K}}, \quad K = 75477\dfrac{Hn}{N^3R^3} \ \cdots \ (352);$$

by which equations are determined, according to the proposed conditions, the weight W in tons of a fly-wheel for a *single-acting engine,* its mean radius in feet R being given, and its material being any whatever; and also its mean radius R in feet, its section (in square feet) K being given, and its material being cast iron of mean quality; and lastly, the section K of its rim in square feet, its mean radius R being given, and its material being, as before, cast iron.

2. *In the double-acting engine,* two effective strokes are made per minute by the piston, whilst the fly-wheel makes one revolution. In this case, therefore $m = 2$ and $\sin. \eta = \dfrac{2}{\pi}$

$= 0\cdot636619 = \sin. 39° 32'$; therefore, $\cos. \eta = \cdot7712549$

$\dfrac{\eta}{\pi} = \dfrac{39° 32'}{180} = \cdot21963$; therefore $\left(1 - \dfrac{2\eta}{\pi}\right) = \cdot56074$;

$$\therefore \left\{ \dfrac{2}{m} \cos. \eta - \left(1 - \dfrac{2\eta}{\pi}\right) \right\} = \cdot21051.$$

Substituting in equations (350) and (351),

$$W = 18194\dfrac{Hn}{N^3R^2}, \quad R = \dfrac{24\cdot3365}{N}\sqrt[3]{\dfrac{Hn}{K}}, \quad K = 14414\dfrac{Hn}{N^3R^3} \ \cdots \ (353);$$

by which equations the weight of the fly-wheel in tons, the mean radius in feet, and the section of the rim in square feet, are determined for the double-acting engine.

3. *In the engine working with two cylinders and a double crank,* it has been shown (Art. 265.) that the conditions of the working of the two arms of a double crank are precisely

the same as though the aggregate pressure $2P_1$ upon their extremities, were applied to the axis of the crank by the intervention of a single arm and a single connecting rod; the length of this arm being represented by $\dfrac{a}{\sqrt{2}}$ instead of a, and its direction equally dividing the inclination of the arms of the double crank to one another.

Now, equations (350) and (351) show the proper dimensions of the fly-wheel to be wholly independent of the length of the crank-arm; whence it follows that the dimension of a fly-wheel applicable to the double as well as a single crank, are determined by those equations as applied to the case of a double-acting engine, the pressure upon whose piston rod is represented by $2P_1$. But in assuming $\frac{1}{2}Nm \cdot 2P_1a = 33000H$, we have assumed the pressure upon the piston rod to be represented by P_1; to correct this error, and to adapt equations (350) and (351) to the case of a double crank engine, we must therefore substitute $\frac{1}{2}H$ for H in those equations. We shall thus obtain

$$W = 9097\frac{Hn}{N^3R^2}, \quad R = \frac{19\cdot316}{n}\sqrt[3]{\frac{Hn}{K}}, \quad K = 7207\frac{Hn}{N^3R^3}\ldots(354);$$

by which equations the dimensions of a fly-wheel necessary to give the required steadiness of motion to a double crank engine are determined under the proposed conditions.

THE FRICTION OF THE FLY-WHEEL.

272. W representing the weight of the wheel and φ the limiting angle of resistance between the surface of its axis and that of its bearings, tan. φ will represent its coefficient of friction (Art. 138.), and W tan. φ, the resistance opposed to its revolution by friction at the surface of its axis. Now, whilst the wheel makes one revolution, this resistance is overcome through a space equal to the circumference of the axis, and represented by $2\pi\rho$, if ρ be taken to represent the radius of the axis. The *work* expended upon the friction of the axis, during each complete revolution of the wheel, is therefore represented by

$2\pi\rho W$ tan. φ; and if N represent the number of strokes made by the engine per minute, and therefore $\dfrac{N}{2}$ the number of revolutions made by the fly-wheel per minute, then will the number of units of work expended per minute, upon the friction of the axis be represented by $N\pi\rho W$ tan. φ; and the number of horses' power, or the fractional part of a horse's power thus expended, by

$$\frac{NW\pi\rho \text{ tan. } \varphi}{33000} \quad \ldots \ldots \text{(355)}.$$

If in this equation we substitute for W the weight in lbs. of the fly-wheel necessary to establish a given degree of steadiness in the engine, as determined by equations (352), (353), and (354), we shall obtain for the horse power lost by friction of the fly-wheel, in the *single-acting engine*, the *double-acting engine*, and the *double crank engine*, respectively, the formulæ

$$20317\frac{H n\rho \text{ tan. } \varphi}{N^2R^2}, \; 3879 \cdot 8\frac{H n\rho \text{ tan. } \varphi}{N^2R^2}, \; 1939 \cdot 9\frac{H n\rho \text{ tan. } \varphi}{N^2R^2} \; \ldots \text{(356)}.$$

The Modulus of the Crank and Fly-wheel.

273. If S_1 represent the space traversed by the piston of the engine in any given time, and a the radius of the crank, W the weight of the fly-wheel in lbs., and ρ the radius of its axis, then will $2a$ represent the length of each stroke, $\dfrac{S_1}{2a}$ the number of strokes made in that time, and $2\pi\rho W$ tan. φ . $\dfrac{S_1}{2a}$, or $\pi W S_1 \dfrac{\rho}{a}$ tan. φ the work expended upon the friction of the fly-wheel during that time, which expression being added to the equation (334) representing the work necessary to cause the crank to yield a given amount of work U_2 to the machine driven by it (independently of the work expended on the friction of the fly-wheel), will give the whole amount of work which must be done upon the combination of the crank and fly-wheel, to cause this given amount of work to be

c c 3

yielded by it, on the machine which the crank drives. Let this amount of work be represented by U_1, then in the case in which the directions of the driving pressure and the resistance upon the crank are parallel (equation 334), and the friction of the crank guide is neglected, we obtain for the modulus of the crank and fly-wheel in the double-acting engine

$$U_1 = \left\{ 1 + \frac{\pi}{2}\left(\frac{\rho_1}{a}\sin. \; \varphi_1 + \frac{\rho_2}{a}\sin. \; \varphi_2\right) \right\} U_2 + \pi W S_1 \frac{\rho}{a} \tan. \; \varphi \; . \; . \; . \; (357).$$

The Governor.

274 This instrument is represented in the figure, under that form in which it is most commonly applied to the steam engine. BD and CE are rods jointed at A upon the vertical spindle AF, and at D and E upon the rods DP and EP, which last are again jointed at their extremities to a collar fitted accurately to the surface of the spindle and moveable upon it. At their extremities B and C, the rods DB and EC carry two heavy balls, and being swept round by the spindle — which receives a rapid rotation always proportional to the speed of the machine, whose motion the governor is intended to regulate — these arms by their own centrifugal force, and that of the balls, are made to separate, and thereby to cause the collar at P to descend upon the spindle, carrying with it, by the intervention of the shoulder, the extremity of a lever, whose motion controls the access of the moving power to the driving point of the machine, closing the throttle valve and shutting off the steam from the steam engine, or closing the sluice, and thus diminishing the supply of water to the water-wheel. Let P be taken to represent the pressure of the extremity of the lever upon the collar, Q the strain thereby produced upon each of the rods DP and EP in the direction of its length, W the weight of each of the balls, w the weight

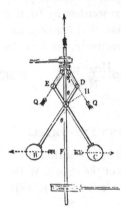

of each of the rods BD and CE, AB$=a$, AD$=b$, DP$=c$, FAB$=\theta$, APD$=\theta_1$. Now upon either of these rods as BD, the following pressures are applied: the weight of the ball and the weight of the rod acting vertically, the centrifugal force of the ball and the centrifugal force of the rod acting horizontally, the strain Q of the rod DP, and the resistance of the axis A. If α represent the angular velocity of the spindle, $\dfrac{W}{g}\alpha^2 . \overline{FB}$, or $\dfrac{W}{g}\alpha^2 a$ sin. θ, will represent the centrifugal force upon the ball (equation 102),

and $\dfrac{W}{g}\alpha^2 a^2$ sin. θ cos. θ its moment about the point A; also the centrifugal force of the rod BD produces the same effect as though its weight were collected in its centre of gravity (Art. 124.), whose distance from A is represented by $\frac{1}{2}(a-b)$, so that the centrifugal force of the rod is represented by

$\frac{1}{2}\dfrac{w}{g}\alpha^2(a-b)$ sin. θ, and its moment about the point A by

$\frac{1}{4}\dfrac{w}{g}\alpha^2(a-b)^2$ sin. θ cos. θ. On the whole, therefore, the sum of the moments of the centrifugal forces of the rod and ball are represented by $\dfrac{\alpha^2}{g}\{Wa^2 + \frac{1}{4}w(a-b)^2\}$ sin. θ cos. θ. Now if μ represent the weight of each unit in the length of the rod, $w=\mu(a+b)$; therefore $Wa^2 + \frac{1}{4}w(a-b)^2 = Wa^2 + \frac{1}{4}\mu(a^2-b^2) (a-b)$. Let this quantity be represented by $W_1 a^2$,

$$\therefore\ W_1 = W + \tfrac{1}{4}\mu\left(1-\frac{b^2}{a^2}\right)(a-b)\ \dots\ (358);$$

then will $\dfrac{\alpha^2}{g}W_1 a^2$sin. θ cos. θ represent the sum of the moments of the centrifugal forces of the rod and ball about A. Moreover, the sum of the moments of the weights of the rod and ball, about the same point, is evidently represented by Wa sin. $\theta + w\frac{1}{2}(a-b)$ sin. θ, or by $\{Wa + \frac{1}{2}\mu(a^2-b^2)\}$sin. θ; let this quantity be represented by $W_2 a$ sin. θ,

$$\therefore\ W_2 = W + \tfrac{1}{2}\mu a\left(1-\frac{b^2}{a^2}\right).\ \dots\dots\ (359).$$

c c 4

Also the moment of Q about $A = Q \cdot \overline{AH} = Qb$ sin. $(\theta + \theta_1)$. Therefore, by the principle of the equality of moments, observing that the centrifugal force of the rod and ball tend to communicate motion in an opposite direction from their weights and the pressure Q,

$$\frac{a^2}{g} W_1 a^2 \sin. \theta \cos. \theta = Qb \sin.(\theta + \theta_1) + W_2 a \sin. \theta.$$

Now P is the resultant of the pressures Q acting in the directions of the rods PD and PE, and inclined to one another at the angle $2\theta_1$; therefore (equation 13),

$$P = 2Q \cos. \theta_1;$$

$$\therefore \quad Q \sin. (\theta + \theta_1) = \tfrac{1}{2} P \frac{\sin.(\theta + \theta_1)}{\cos. \theta_1} = \tfrac{1}{2} P \{ \sin. \theta + \cos. \theta \tan. \theta_1 \}$$

But since the sides b and c of the triangle APD are opposite to the angles θ_1 and θ, therefore $\dfrac{\sin. \theta_1}{\sin. \theta} = \dfrac{b}{c}$; therefore cos. θ_1

$$= \left(1 - \frac{b^2}{c^2} \sin.{}^2\theta \right)^{\frac{1}{2}};$$

$$\therefore \quad Q \sin. (\theta + \theta_1) = \tfrac{1}{2} P \left\{ \sin. \theta + \frac{b}{c} \sin. \theta \cos. \theta \left(1 - \frac{b^2}{c} \sin.{}^2\theta \right)^{-\frac{1}{2}} \right\}.$$

Substituting this value in the preceding equation, dividing by sin. θ, and writing $(1 - \cos.{}^2\theta)$ for sin ${}^2\theta$, we obtain

$$\frac{a^2}{g} W_1 a^2 \cos. \theta = \tfrac{1}{2} Pb \left\{ 1 + \frac{b}{c} \cos. \theta \left(1 - \frac{b^2}{c^2} + \frac{b^2}{c^2} \cos.{}^2\theta \right)^{-\frac{1}{2}} \right\} + W_2 a \; . . \; (360);$$

which equation, of four dimensions in terms of cos. θ, being solved in respect to that variable, determines the inclination of the arms under a given angular velocity of the spindle. It is, however, more commonly the case that the inclination of the arms is given, and that the lengths of the arms, or the weights of the balls, are required to be determined, so that this inclination may, under the proposed conditions, be attained. In this case the values of W_1 and W_2 must be substituted in the above equation from equations (358) and (359), and that equation solved in respect to a or W.

The values of b and c are determined by the position on the spindle, to which it is proposed to make the collar descend, at the given inclination of the arms or value of θ. If the distance AP, of this position of the collar from A, be represented by h, we have $h = b\cos.\theta + c\cos.\theta_1$,

$$\therefore\ h = b\cos.\theta + c\left(1 - \frac{b^2}{c^2}\sin.^2\theta\right)^{\frac{1}{2}}\ \ldots\ (361);$$

from which equation and the preceding, the value of one of the quantities b or c may be determined, according to the proposed conditions, the value of the other being assumed to be any whatever.

If N represent the number of revolutions, or parts of a revolution, made per second by the fly-wheel, and γN the number of revolutions made in the same time by the spindle of the governor, then will $2\pi\gamma$N represent the space α described per second by a point, situated at distance unity from the axis of the spindle. Substituting this value for α in equation (360), and assuming $b = c$, we obtain

$$\frac{4\pi^2\gamma^2 N^2}{g}W_1 a^2\cos.\theta = Pb + W_2 a\ \ldots\ (362):$$

also by equation (361),

$$h = 2b\cos.\theta.\ \ldots\ (363).$$

Eliminating $\cos.\theta$ between these equations, and solving in respect to h,

$$h = \frac{bg(Pb + W_2 a)}{2\pi^2\gamma^2 a^2 W_1}\cdot\frac{1}{N^2}\ \ldots\ (364);$$

Let $P\left(1 + \frac{1}{m}\right)$ and $P\left(1 - \frac{1}{m}\right)$ represent the values of P corresponding to the two states bordering upon motion (Art. 140.), and let $N\left(1 + \frac{1}{n}\right)$ and $N\left(1 - \frac{1}{n}\right)$ be the corresponding values of N ; so that the variation either way of $\frac{1}{n}$th from the mean number N of revolutions, may be upon the point of causing the valve to move. If these values be respectively substituted for P and N in the above formula, it is evident that the corresponding values of h will be equal. Equating those values of h and reducing, we obtain

$$\frac{W_2 a}{P} = \frac{1}{2m}\left(n + \frac{1}{n}\right) - 1.$$

By which equation there is established that relation between the quantities W_2, a, P, m which must obtain, in order that a variation of the number of revolutions, ever so little greater than the $\frac{1}{n}$th part, may cause the valve to move. Neglecting $\frac{1}{n}$ as small when compared with n,

$$n = 2m\left(1 + \frac{W_2 a}{P}\right);$$

which expression, representing that fractional variation in the number of revolutions which is sufficient to give motion to the valve, is the true measure of the SENSIBILITY of the governor.

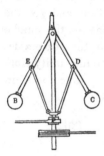

275. The joints E and D are sometimes fixed upon the arms AB and AC as in the accompanying figure, instead of upon the prolongations of those arms as in the preceding figure. All the formulæ of the last Article evidently adapt themselves to this case, if b be assumed $= 0$ (in equations 358, 359). The centrifugal force of the rods EP and DP is neglected in this computation.

THE CARRIAGE-WHEEL.

276. Whatever be the nature of the resistance opposed to the motion of a carriage-wheel, it is evidently equivalent to that of an obstacle, real or imaginary, which the wheel may be supposed, at every instant, to be in the act of surmounting. Indeed it is certain, that, however yielding may be the material of the road, yet by reason of its *compression* before the wheel, such an immoveable obstacle, of exceedingly small height, is continually in the act of being presented to it.

277. *The two-wheeled carriage.*

Let AB represent one of the wheels of a two-wheeled carriage, EF an inclined plane, which it is in the act of ascending, O a solid elevation of the surface of the plane, or an obstacle which it is at any instant in the act of passing over, P the corresponding traction, W the weight of the wheel and of the load which it supports.

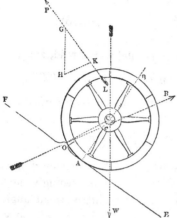

Now the surface of the *box* of the wheel being in the state bordering upon motion on the surface of the axle, the direction of the resistance of the one upon the other is inclined at the limiting angle of resistance, to a radius of the axle at their point of contact (Art. 141.).

This resistance has, moreover, its direction through the point of contact O of the tire of the wheel with the obstacle on which it is in the act of turning. If, therefore, OR be drawn intersecting the circumference of the axis in a point *c*, such that the angle C*c*R may equal the limiting angle of resistance φ, then will its direction be that of the resistance of the obstacle upon the wheel.

Draw the vertical GH representing the weight W, and through H draw HK parallel to OR, then will this line represent (to the same scale) the resistance R, and GK the traction P (Art. 14.);

$$\therefore \frac{P}{W} = \frac{GK}{GH} = \frac{\sin. GHK}{\sin. GKH} = \frac{\sin. GHK}{\sin. (PGH - GHK)} = \frac{\sin. WsO}{\sin. (PLW - WsO)}.$$

Let R = radius of wheel, ρ = radius of axle, ACO = η, ACW = ι = inclination of the road to the horizon, θ = inclination of direction of the traction to the road. Now WsO = WCO + COs, but WCO = ι + η, and $\frac{\sin. COs}{\sin. CcR} = \frac{Cc}{CO}$. Let COs be represented by α, then WsO = ι + η + α, and

$$\sin. \alpha = \frac{\rho}{R} \sin. \phi \;\; . \;\; . \;\; . \;\; . \;\; (365).$$

Also $PLW = \dfrac{\pi}{2} + \iota + \theta$; therefore $PLW - WsO = \dfrac{\pi}{2} - (\eta + \alpha - \theta)$;

$$\therefore \quad P = W \; \frac{\sin.\,(\iota + \eta + \alpha)}{\cos.\,(\eta + \alpha - \theta)} \; \cdots \; (366) ;$$

when the direction of traction is parallel to the road, $\theta = 0$,

$$\therefore \quad P = W \{\sin.\,\iota + \cos.\,\tan.\,(\eta + \alpha)\} \; \cdots \; (367).$$

If the road and the direction of traction be both *horizontal*, $\theta = \iota = 0$, and

$$P = W \tan.\,(\eta + \alpha) \; \cdots \; (368).$$

In all cases of traction with wheels of the common dimensions upon ordinary roads, ACO or η is an exceedingly small angle; α is also, in all cases, an exceedingly small angle (equation 365); therefore $\tan.\,(\eta + \alpha) = \eta + \alpha$ very nearly. Now if A be taken to represent the arc AO, whose length is determined by the height of the obstacle and the radius of the wheel, then

$$\eta = \frac{A}{R} \; \cdots \; (369).$$

Substituting the value of α from equation (365),

$$P = W \cdot \frac{(A + \rho \sin.\,\varphi)}{R} \; \cdots \; (370).$$

278. It remains to determine the value of the arc A intercepted between the lowest point to which the wheel sinks in the road, and the summit O of the obstacle, which it is at every instant surmounting. Now, the experiments of Coulomb, and the more recent experiments of M. Morin *, appear to have fully established the fact, that, on horizontal roads of uniform quality and material, the traction P, when its direction is horizontal, varies directly as the load W, and inversely as the radius R of the wheel; whence it follows (equation 370), that the arc A is constant, or that it is the same for the same quality of road, whatever may be the

---

* Expériences sur le Tirage des Voitures faites en 1837 et 1838. (See APPENDIX.)

weight of the load, or the dimensions of the wheel.* The constant A may therefore be taken as a measure of the resisting quality of the road, and may be called the *modulus of its resistance.*

The mean value of this modulus being determined in respect to a road, whose surface is of any given quality, the value of η will be known from equation (369), and the relation between the traction and the load upon that road, under all circumstances; it being observed, that, since the arc A is *the same* on a horizontal road, whatever be the load, if the traction be parallel, it is also *the same* under the same circumstances upon a sloping road; the effect of the slope being equivalent to a variation of the load. The same substitution may therefore be made for tan. $(\eta + \alpha)$ in equation (367), as was made in equation (368),

$$\therefore \ P = W \left\{ \sin. \iota + \left(\frac{A + \rho \sin. \varphi}{R} \right) \cos. \iota \right\} \ \ \ . \ . \ . \ . \ (371).$$

279. *The best direction of traction in the two-wheeled carriage.*

This best direction of traction is evidently that which gives to the denominator of equation (366) its greatest value ; it is therefore determined by the equation

$$\theta = \eta + \alpha = \frac{A + \rho \sin. \varphi}{R} \ \ . \ . \ . \ (372).$$

280. *The four-wheeled carriage.*

Let $W_1 \ W_2$ represent the loads borne by the fore and hind wheels, together with their own weights, $R_1 \ R_2$ their

---

* In explanation of this fact let it be observed, that although the wheel sinks deeper beneath the surface of the load as the material is softer, yet the obstacle yields, for the same reason, more under the pressure of the wheel, the arc A being by the one cause increased, and by the other diminished. Also, that although by increasing the diameter of the wheel the arc A would be rendered greater if the wheel sank to the same depth as before, yet that it does not sink to the same depth by reason of the corresponding increase of the surface which sustains the pressure.

radii, ρ_1 ρ_2 the radii of their axles, and ϕ_1 ϕ_2 the limiting angles of resistance. Suppose the direction of the traction P parallel to the road, then, since this traction equals the sums of the tractions upon the fore and hind wheels respectively, we have by equation (371)

$$P = W_1 \left\{ \sin. \iota. + \frac{(A + \rho_1 \sin. \phi_1)}{R_1} \cos. \iota \right\} + W_2 \left\{ \sin. \iota + \frac{(A + \rho_2 \sin. \phi_2)}{R_2} \cos. \iota \right\},$$

or,

$$P = (W_1 + W_2) \sin. \iota + A \left(\frac{W_1}{R_1} + \frac{W_2}{R_2} \right) \cos. \iota + \left\{ W_1 \left(\frac{\rho_1}{R_1} \right) \sin. \phi_1 + W_2 \left(\frac{\rho_2}{R_2} \right) \sin. \phi_2 \right\} \cos. \iota \ldots (373).$$

281. *The work accumulated in the carriage-wheel.*

Let I represent the moment of inertia of the wheel about its axis, and M its volume; then will $MR^2 + I$ represent its moment of inertia (Art. 79.) about the point in its circumferences about which it is, at every instant of its motion, in the act of turning. If, therefore, α represent its angular velocity about this point at any instant, U the work at that instant accumulated in it, and μ the weight of each cubical unit of its mass, then (Art. 75.), $U = \frac{1}{2} \alpha^2 \frac{\mu}{g} (MR^2 + I) = \frac{1}{2} \frac{\mu}{g} M (\alpha R)^2 +$

$\frac{1}{2} \alpha^2 \frac{\mu}{g} I.$ Now if V represent the velocity of the axis of the wheel, $\alpha R = V$;

$$\therefore U = \frac{1}{2} \frac{\mu}{g} M V^2 + \frac{1}{2} * \alpha^2 \frac{\mu}{g} I;$$

whence it follows, that the whole work accumulated in the rolling wheel is equal to the sum obtained by adding the work which would have been accumulated in it if it had moved with its motion of translation only, to that which would have been accumulated in it if it had moved with its motion of rotation only. If we represent the radius of gyration (Art. 80.) by K, $I = MK^2$; whence substituting and reducing,

---

* The angular velocity of the wheel would evidently be α, if its centre were fixed, and its circumference made to revolve with the same velocity as now.

$$U = \tfrac{1}{2} \frac{\mu}{g} M \left(1 + \frac{K^2}{R^2} \right) V^2 \ \cdots \ (374).$$

The accumulated work is therefore the same as though the wheel had moved with a motion of translation only, but with a greater velocity, represented by the expression $\left(1 + \frac{K^2}{R^2} \right)^{\frac{1}{2}} V$.

282. On the state of the accelerated or the retarded motion of a machine.

Let the work U_1 done upon the driving point of a machine be conceived to be in excess of that U_2 yielded upon the working points of the machine and that expended upon its prejudicial resistances. Then we have by equation (117)

$$U_1 = AU_2 + BS_1 + \frac{1}{2g}(V^2 - V_1{}^2)\Sigma w\lambda^2 ;$$

where V represents the velocity of the driving point of the machine after the work U_1 has been done upon it, V_1 that when it began to be done, and $\Sigma w\lambda^2$ the coefficient of equable motion. Now let S_1 represent the space through which U_1 is done, and S_2 that through which U_2 is done; and let the above equation be differentiated in respect to S_1,

$$\therefore \ \frac{dU_1}{dS_1} = A\frac{dU_2}{dS_2} \cdot \frac{dS_2}{dS_1} + B + \frac{1}{g}V\frac{dV}{dS_1}\Sigma w\lambda^2 ;$$

but $\dfrac{dU_1}{dS_1} = P_1$ (Art. 51.), if P_1 represent the driving pressure.

Also $\dfrac{dU_2}{dS_2} = P_2$, if P_2 represent the working pressure; also

$$V\frac{dV}{dS_1} = V\frac{dV}{dt} \cdot \frac{dt}{dS_1} = V \cdot \frac{dV}{dt} \cdot \frac{1}{V} = \frac{dV}{dt} = f \text{ (equation 72). If,}$$

therefore, we represent by Λ the relation $\dfrac{dS_2}{dS_1}$, between the spaces described in the same exceedingly small time by the driving and working points, we have

$$P_1 = A\Lambda P_2 + B + \frac{f}{g}\Sigma w\lambda^2 \ \cdots \ (375);$$

$$\therefore f = g \cdot \frac{P_1 - A \Lambda P_2 - B}{\Sigma w \lambda^2} \cdot \cdot \cdot \cdot (376) ;$$

where f (Art. 95.) represents the additional velocity actually acquired per second by the driving point of the machine, if P_1 and P_2 be constant quantities, or, if not, the additional velocity which would be acquired in any given second, if these pressures retained, throughout that second, the values which they had at its commencement.

283. *To determine the coefficient of equable motion.*

$\Sigma w \lambda^2$ represents the sum of the weights of all the moving elements of the machine, each being multiplied by the ratio of its velocity to that of the driving point, which sum has been called (Art. 151.) the *coefficient of equable motion.* If the motion of each element of the machine takes place about a fixed axis, and a_1, a_2, a_3, &c. represent the perpendiculars from their several axes upon the directions in which they receive the driving pressures of the elements which precede them in the series, and b_1, b_2, b_3, &c. the similar perpendiculars upon the tangents to their common surfaces at the points where they drive those that follow them; then, while the first driving point describes the small space ΔS_1, the point of contact of the pth and $p + 1$th elements of the series will be made (Art. 236.) to describe a space represented by

$$\frac{b_1 b_2 \ldots b_p}{a_1 a_2 \ldots a_p} \Delta S_1 ;$$

so that the angular velocity of the pth element will be represented by

$$\frac{b_1 b_2 \ldots b_{p-1}}{a_1 a_2 \ldots a_p} \Delta S_1,$$

and the space described by a particle situated at distance ρ from the axis of that element by

$$\frac{b_1 b_2 \ldots b_{p-1}}{a_1 \cdot a_2 \ldots a_p} \cdot \rho \Delta S,$$

and the ratio λ of this space to that described by the driving point of the machine will be represented by

$$\lambda = \left(\frac{b_1 . b_2 \ldots b_{p-1}}{a_1 . a_2 \ldots a_p}\right)\rho.$$

The sum $\Sigma w\lambda^2$ will therefore be represented in respect to this one element by

$$\left(\frac{b_1 . b_2 \ldots b_{p-1}}{a_1 . a_2 \ldots a_p}\right)\Sigma w\rho^2.$$

Or if I_p represent the moment of inertia of the element, and μ_p the weight of each cubic unit of its mass, that portion of the value of $\Sigma w\lambda^2$ which depends upon this element will be represented by

$$\left(\frac{b_1 b_2 \ldots b_{p-1}}{a_1 a_2 \ldots a_p}\right)^2 I_p \mu_p.$$

And the same being true of every other element of the machine, we have

$$\Sigma w\lambda^2 = \Sigma \left(\frac{b_1 b_2 \ldots b_{p-1}}{a_1 a_2 \ldots a_p}\right)^2 I_p \mu_p \ldots \ldots (377),$$

which is a general expression for the coefficient of equable motion in the case supposed. The value of Λ in equation (366) is evidently represented by

$$\Lambda = \frac{b_1 b_2 b_3 \ldots b_n}{a_1 a_2 a_3 \ldots a_n} \ldots \ldots (378).$$

284. *To determine the pressure upon the point of contact of any two elements of a machine moving with an accelerated or retarded motion.*

Let p_1 be taken to represent the resistance upon the point of contact of the first element with the second, p_2 that upon the point of contact of the second element of the machine with the third, and so on. Then by equation (375), observing that, P_1 and p_1 representing pressures applied to the same element, $\Sigma w\lambda^2$ is to be taken in this case only in respect to that element, so that it is represented by $\mu_1 I_1$, whilst Λ is in this case represented by $\dfrac{b_1}{a_1}$, we have, neglecting friction,

$$P_1 = \frac{b_1}{a_1} p_1 + \frac{f}{g} \mu_1 I_1.$$

Substituting the value of f from equation (376), and solving in respect to p_1,

$$p_1 = \frac{a_1}{b_1} P_1 - \frac{a_1}{b_1}\left(P_1 - \Lambda P_2\right)\frac{\mu_1 I_1}{\Sigma w \lambda^2}. \quad \cdots \cdots \quad (379),$$

where the value of Λ is determined by equation (378), and that of $\Sigma w \lambda^2$ by equation (377). Proceeding similarly in respect to the second element, and observing that the impressed pressures upon that element are p_1 and p_2, we have

$$p_1 = \frac{b_2}{a_2} p + \frac{f_1}{g} \mu_2 I_2,$$

f_1 representing the additional velocity per second of the point of application of p_1, which evidently equals $\dfrac{b_1}{a_1} f$. Substituting, therefore, the value of f from equation (376) as before,

$$p_1 = \frac{b_2}{a_2} p_2 + \frac{b_1}{a_1} \cdot \frac{P_1 - \Lambda P_2}{\Sigma w \lambda^2} \mu_2 I_2.$$

Substituting the value of p_1 from equation (379), and solving in respect to p_2, we have

$$p_2 = \frac{a_1 a_2}{b_1 b_2} P_1 - \frac{a_1 a_2}{b_1 b_2}\left\{ \mu_1 I_1 + \left(\frac{b_1}{a_1}\right)^2 \mu_2 I_2 \right\}\left(\frac{P_1 - \Lambda P_2}{\Sigma w \lambda^2}\right) \quad \cdots \cdots \quad (380).$$

And proceeding similarly in respect to the other points of contact, the pressure upon each may be determined. It is evident, that by assuming values of A and B in equations (375) and (376) to represent the coefficients of the moduli in respect to the several elements of the machine, and to the *whole* machine, the influence of friction might by similar steps have been included in the result.

PART IV.

THE THEORY OF THE STABILITY OF STRUCTURES.

GENERAL CONDITIONS OF THE STABILITY OF A STRUCTURE OF UNCEMENTED STONES.*

A STRUCTURE may yield, under the pressures to which it is subjected, either by the slipping of certain of its surfaces of contact upon one another, or by their turning over upon the edges of one another; and these two conditions involve the whole question of its stability.

THE LINE OF RESISTANCE.

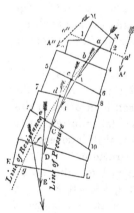

285. Let a structure MNLK, composed of a single row of uncemented stones of any forms, and placed under any given circumstances of pressure, be conceived to be *intersected* by any geometrical surface 1 2, and let the *resultant* aA of all the pressures which act upon one of the parts MN21, into which this intersecting surface divides the structure, be imagined to be taken. Conceive, then, this intersecting surface to change its form and position so as to coincide in succession with all the common surfaces of

---

* Extracted from a Memoir on the Theory of the Arch by the author of this work in the first volume of the " Theoretical and Practical Treatise on Bridges " by Professor Hosking and Mr. Hann of King's College, published by Mr. Weale. These general conditions of the equilibrium of a system of bodies in contact were first discussed by the author in the fifth and sixth volumes of the " Cambridge Philosophical Transactions."

contact 3 4, 5 6, 7 8, 9 10, of the stones which compose the structure; and let bB, cC, dD, eE be the resultants, similarly taken with aA, which correspond to these several planes of intersection.

In each such position of the intersecting surface, the resultant spoken of having its direction produced, will intersect that surface either *within* the mass of the structure, or, when that surface is imagined to be produced, *without* it. If it intersect it *without* the mass of the structure, then the *whole* pressure upon one of the parts, acting in the direction of this resultant, will cause that part to turn over upon the edge of its common surface of contact with the other part; if it intersect it *within* the mass of the structure, it will not.

Thus, for instance, if the direction of the resultant of the forces acting upon the part NM 1 2 had been a'A', not intersecting the surface of contact 1 2 *within* the mass of the structure, but when imagined to be produced beyond it to a'; then the whole pressure upon this part acting in a'A' would have caused it to turn upon the edge 2 of the surface of contact 1 2; and similarly if the resultant had been in a''A'', then it would have caused the mass to revolve upon the edge 1. The resultant having the direction aA, the mass will not be made to revolve on either edge of the surface of contact 1 2.

Thus the condition that no two parts of the mass should be made, by the insistent pressures, to turn over upon the edge of their common surface of contact, is involved in this other, that the direction of the resultant, taken in respect to every position of the intersecting surface, shall intersect that surface actually *within* the mass of the structure.

If the intersecting surface be imagined to take up an *infinite* number of different positions, 1 2, 3 4, 5 6, &c., and the intersections with it, a, b, c, d, &c., of the directions of all the corresponding resultants be found, then the curved line *abcdef*, joining these points of intersection, may with propriety be called the LINE OF RESISTANCE, the resisting points of the resultant pressures upon the contiguous surfaces lying all in that line.

This line can be completely determined by the methods of analysis, in respect to a structure of any given geometrical form, having its parts in contact by surfaces also of given geometrical forms. And, conversely, the form of this line being assumed, and the direction which it shall have through any proposed structure, the geometrical form of that structure may be determined, subject to these conditions; or lastly, certain conditions being assumed, both as it regards the form of the structure and its line of resistance, all that is necessary to the existence of these *assumed* conditions may

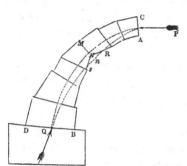

be found. Let the structure ABCD have for its line of resistance the line PQ. Now it is clear that if this line cut the surface MN of any section of the mass in a point *n* without the surface of the mass, then the resultant of the pressures upon the mass CMN will act through *n*, and cause this portion of the mass to revolve about the nearest point N of the intersection of the surface of section MN with the surface of the structure.

Thus, then, it is a condition of the equilibrium that *the line of resistance shall intersect the common surface of contact of each two contiguous portions of the structure actually within the mass of the structure;* or, in other words, that it shall actually go through each joint of the structure, avoiding none: this condition being necessary, that no two portions of the structure may revolve on the edges of their common surface of contact.

THE LINE OF PRESSURE.

286. But besides the condition that no two parts of the structure should turn upon the edges of their common surfaces

of contact, which condition is involved in the determination of the LINE OF RESISTANCE, there is a *second* condition necessary to the stability of the structure. Its surfaces of contact must no where slip upon one another. That this condition may obtain, the resultant corresponding to each surface of contact must have its *direction* within certain limits. These limits are defined by the surface of a right cone (Art. 139.), having the normal to the common surface of contact (at the above-mentioned point of intersection of the resultant) for its axis, and having for its vertical angle twice that whose tangent is the coefficient of friction of the surfaces. If the direction of the resultant be *within* this cone, the surfaces of contact will not slip upon one another; if it be without it, they will.

Thus, then, the *directions* of the consecutive resultants in respect to the normal to the point, where each intersects its corresponding surface of contact, are to be considered as important elements of the theory.

Let then a line ABCDE be taken, which is the locus of

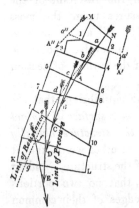

the consecutive intersections of the resultants aA, bB, cC, dD, &c. The direction of the resultant pressure upon every section is a *tangent* to this line; it may therefore with propriety be called the LINE OF PRESSURE. Its geometrical form may be determined under the same circumstances as that of the line of resistance. A straight line cC drawn from the point c, where the LINE OF RESISTANCE abcd intersects any joint 56 of the structure, so as to touch the LINE OF PRESSURE ABCD, will determine the *direction* of the resultant pressure upon that joint: if it lie within the cone spoken of, the structure will not slip upon that joint; if it lie without it, it will.

Thus the whole theory of the equilibrium of any structure is involved in the determination with respect to that struc-

ture of these two lines — the line of resistance, and the line
of pressure : one of these lines, the line of resistance, determining the *point* of application of the resultant of the
pressures upon each of the surfaces of contact of the system ;
and the other, the line of pressure, the *direction* of that
resultant.

The determination of both, under their most general forms,
lies within the resources of analysis ; and general equations
for their determination in that case, in which all the surfaces
of contact, or joints, are planes — the only case which offers
itself as a *practical* case — have been given by the author of
this work in the sixth volume of the " Cambridge Philosophical Transactions."

THE STABILITY OF A SOLID BODY.

287. The stability of a solid body may be considered to be
greater or less, as a greater or less amount of *work* must be
done upon it to overthrow it ; or according as the amount
of work which must be done upon it to bring it into
that position in which it will fall over of its own accord is
greater or less. Thus the stability of the solid represented
in *fig.* 1. resting on a horizontal
plane is greater or less, according as
the work which must be done upon
it, to bring it into the position represented in *fig.* 2., where its centre
of gravity is in the vertical passing
through its point of support, is
greater or less. Now this work is equal (Art. 60.) to that
which would be necessary to raise its whole weight, vertically, through that height by which its centre of gravity is
raised, in passing from the one position into the other.
Whence it follows that the stability of a solid body resting
upon a plane is greater or less, as the product of its weight
by the vertical height through which its centre of gravity is

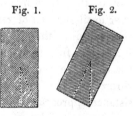

Fig. 1.   Fig. 2.

raised, when the body is brought into a position in which it will fall over of its own accord, is greater or less.

If the base of the body be a plane, and if the vertical height of its centre of gravity when it rests upon a horizontal plane be represented by h, and the distance of the point· or the edge, upon which it is to be overthrown, from the point where its base is intersected by the vertical through its centre of gravity, by k; then is the height through which its centre of gravity is raised, when the body is brought into a position in which it will fall over, evidently represented by $(h^2 + k^2)^{\frac{1}{2}} - h$; so that if W represent its weight, and U the work necessary to overthrow it, then

$$U = W\{(h^2 + k^2)^{\frac{1}{2}} - h\} \quad \ldots \ldots \quad (381).$$

U is a true measure of the stability of the body.

The Stability of a Structure.

288. It is evident that the degree of the stability of a structure, composed of any number of separate but contiguous solid bodies, depends upon the less or greater degree of approach which the line of resistance makes to the extrados or external face of the structure; for the structure cannot be thrown over until the line of resistance is so deflected as to intersect the extrados: the more remote is its direction from that surface, when free from any extraordinary pressure, the less is therefore the probability that any such pressure will overthrow it. The nearest distance to which the line of resistance approaches the extrados will, in the following pages, be represented by m, and will be called the MODULUS OF STABILITY of the structure.

This shortest distance presents itself in the wall and buttress commonly at the lowest section of the structure. It is evidently beneath that point where the line of resistance intersects the lowest section of the structure that the greatest resistance of the foundation should be opposed. If that point be firmly supported, no settlement of the structure can take place under the influence of the pressures to which it is ordinarily subjected.*

THE WALL OR PIER.

289. *The stability of a wall.*

If the pressure upon a wall be uniformly distributed along its length †, and if we conceive it to be intersected by vertical planes, equidistant from one another and perpendicular to its face, dividing it into separate portions, then are the conditions of its stability, in respect to the pressures applied to its entire length, manifestly the same with the conditions of the stability of each of the individual portions into which it is thus divided, in respect to the pressures sustained by that portion of the wall; so that if every such columnar portion or pier into which the wall is thus divided be constructed so as to stand under its insistent pressures with any degree of firmness or stability, then will the whole structure stand with the like degree of firmness or stability; and conversely.

In the following discussion these equal divisions of the length of a wall or pier will be conceived to be made one foot apart; so that in every case the question investigated will be that of the stability of a column of uniform or variable thickness, whose width measured in the direction of the length of the wall is one foot.

290. When a wall is supported by buttresses placed at equal distances apart, the conditions of the stability will be made to resolve themselves into those of a continuous wall,

---

* A practical rule of Vauban, generally adopted in fortification, brings the point where the line of resistance intersects the base of the wall, to a distance from the vertical to its centre of gravity, of $\frac{4}{5}$ths the distance from the latter to the external edge of the base. (See Poncelet, *Mémoire sur la Stabilite des Revêtemens*, note, p. 8.)

† In the wall of a building the pressure of the rafters of the roof is thus uniformly distributed by the intervention of the wall plates.

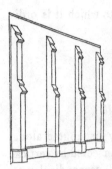

if we conceive each buttress to be extended laterally until it meets the adjacent buttress, its material at the same time so diminishing its specific gravity that its weight when thus spread along the face of the wall may remain the same as before. There will thus be obtained a compound wall whose external and internal portions are of different specific gravities; the conditions of whose equilibrium remain manifestly unchanged by the hypothesis which has been made in respect to it.

THE LINE OF RESISTANCE IN A PIER.

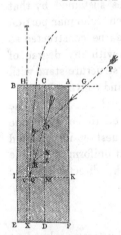

291. Let ABEF be taken to represent a column of uniform dimensions. Let PS be the direction of any pressure P sustained by it, intersecting its axis in O. Draw any horizontal section IK, and take ON to represent the weight of the portion AKIB of the column, and OS on the same scale to represent the pressure P, and complete the parallelogram ONRS; then will OR evidently represent, in magnitude and direction, the resultant of the pressures upon the portion AKIB of the mass (Art. 3.), and its point of intersection Q with IK will represent a point in the line of resistance.

Let PS intersect BA (produced if necessary) in G, and let $GC=k$, $AB=a$, $AK=x$, $MQ=y$, $POC=\alpha$, $\mu=$ weight of each cubic foot of the material of the mass. Draw RL perpendicular to CD; then, by similar triangles,

$$\frac{QM}{OM}=\frac{RL}{OL}$$

But $QM=y$, $OM=CM-CO=x-k$ cot. α, $RL=\overline{RN}$ sin. $RNL=P$ sin. α, $OL=ON+NL=ON+\overline{RN}$ cos. RNL $=\mu ax+P$ cos. α ;

$$\therefore \frac{y}{x-k \text{ cot. } \alpha}=\frac{P \text{ sin. } \alpha}{\mu ax+P \text{ cos. } \alpha} ;$$

$$\therefore y=P \cdot \frac{x \text{ sin. } \alpha-k \text{ cos. } \alpha}{\mu ax+P \text{ cos. } \alpha} \quad \cdots \cdots \quad (382);$$

which is the general equation to the line of resistance of a pier or wall.

292. *The conditions necessary that the stones of the pier may not slip on one another.*

Since in the construction of the parallelogram ONRS, whose diagonal OR determines the direction of the resultant pressure upon any section IK, the side OS, representing the pressure P in magnitude and direction, remains always the same, whatever may be the position of IK; whilst the side ON, representing the weight of AKIB, increases as IK descends : the angle ROM continually diminishes as IK descends. Now, this angle is evidently equal to that made by OR with the perpendicular to IK at Q ; if, therefore, this angle be less than the limiting angle of resistance in the highest position of IK, then will it be less in every subjacent position. But in the highest position of IK, ON=0, so that in this position $ROM=\alpha$. Now, so long as the inclination of OR to the perpendicular to IK is less than the limiting angle of resistance, the two portions of the pier separated by that section cannot slip upon one another (Art. 141.). It is therefore necessary, and sufficient to the condition that no two parts of the structure should slip upon their common surface of contact, that the inclination α of P to the vertical should be less than the limiting angle of resistance of the common surfaces of the stones. All the resultant pressures passing through the point O, it is evident that the *line of pressure* (Art. 286.) resolves itself into that *point*.

293. *The greatest height of the pier.*

At the point where the line of resistance intersects the external face or extrados of the pier, $y = \frac{1}{2}a$; if, therefore, H represents the corresponding value of x, it will manifestly represent the greatest height to which the pier can be built, so as to stand under the given insistent pressure P. Substituting these values for x and y in equation (382), and solving in respect to H,

$$H = \frac{P(\frac{1}{2}a + k)\cos. \alpha}{P \sin. \alpha - \frac{1}{2}\mu a^2} \cdot \cdot \cdot \cdot \cdot (383).$$

If P sin. $\alpha = \frac{1}{2}\mu a^2$, H $= infinity$; whence it follows that in this case the pier will stand under the given pressure P, however great may be the height to which it is raised.

294. *The line of resistance is a rectangular hyperbola.*

Multiplying both sides of equation (382) by the denominator of the fraction in the second member,

$$y(\mu a x + P \cos. \alpha) = P x \sin. \alpha - P k \cos. \alpha \, ;$$

dividing by μa, transposing, and changing the signs of all the terms,

$$\frac{P \sin. \alpha}{\mu a}x - y\left(x + \frac{P \cos. \alpha}{\mu a}\right) = \frac{P \cos. \alpha}{\mu a}k \, ;$$

adding $\dfrac{P^2 \sin. \alpha \cos. \alpha}{\mu^2 a^2}$ to both sides,

$$\frac{P \sin. \alpha}{\mu a}\left(x + \frac{P \cos. \alpha}{\mu a}\right) - y\left(x + \frac{P \cos. \alpha}{\mu a}\right) = \frac{P \cos. \alpha}{\mu a}\left(k + \frac{P \sin. \alpha}{\mu a}\right);$$

$$\therefore \left(\frac{P \sin. \alpha}{\mu a} - y\right)\left(x + \frac{P \cos. \alpha}{\mu a}\right) = \frac{P \cos. \alpha}{\mu a}\left(k + \frac{P \sin. \alpha}{\mu a}\right).$$

Let CH be taken equal to $\dfrac{P \sin. \alpha}{\mu a}$, HT $= \dfrac{P \cos. \alpha}{\mu a}$; and let VQ $= y_1$, TV $= x_1$,

$$\therefore y_1 = VQ = VM - MQ = CH - MQ = \frac{P \sin. \alpha}{\mu a} - y \, ;$$

$$x_1 = TV = HV + TH = x \; \frac{P \cos \alpha}{\mu a};$$

$$\therefore \; x_1 y_1 = \frac{P \cos \alpha}{\mu a} \left(k + \frac{P \sin \alpha}{\mu a} \right) = \text{a constant quantity.}$$

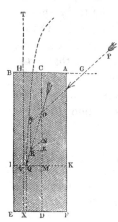

This is the equation to a rectangular hyperbola, whose asymtote is TX. The line of resistance continually approaches TX therefore, but never meets it; whence it follows, that if TX lie (as shown in the figure) within the surface of the mass, or if $CH < CB$ or $\dfrac{P \sin \alpha}{\mu a} < \tfrac{1}{2}a,$ or $2P \sin \alpha < \mu a^2$, then the line of resistance will no where cut the extrados, and the structure will retain its stability under the insistent pressure P, however high it may be built; which agrees with the result obtained in the preceding article.

295. *The thickness of the pier, so that when raised to a given height it may have a given stability.*

Let m be taken to represent the nearest distance to which the line of resistance is intended to approach the extrados of the pier, which distance determines the degree of its stability, and has been called the *modulus* of stability (Art. 288.). It is evident from the last article that this least distance will present itself in the lowest section of the pier. At this lowest section, therefore, $y = \tfrac{1}{2}a - m$. Substituting this value for y in equation (382), and also the height h of the pier for x, and solving the resulting quadratic equation in respect to a, we shall thus obtain

$$a = -\left(\frac{P \cos \alpha}{2 \mu h} - m \right) + \sqrt{ \left(\frac{P \cos \alpha}{2 \mu h} - m \right)^2 + \frac{2P}{\mu} \left\{ \sin \alpha - \left(\frac{k-m}{h} \right) \cos \alpha \right\} } \;\; .. \; (384).$$

296. *To vary the point of application of the pressure* P, *so that any required stability may be given to the pier.*

It is evident, that if in equation (382) we substitute $\frac{1}{2}a-m$

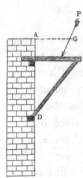

for y and h for x, the modulus of stability m may be made to assume any given value for a given thickness a of the pier, by assigning a corresponding value to k; that is, by moving the point of application G to a certain distance from the axis of the pier, determined by the value of k in that equation. This may be done by various expedients, and among others by that shown in the figure. Solving equation (382) in respect to k, we have

$$k = h \tan. \alpha - (\tfrac{1}{2}a - m)\left(1 + \frac{\mu a h}{\text{P} \cos. \alpha}\right) \ \ldots \ldots (385).$$

It is necessary to the equilibrium of the pier, under these circumstances, that the line of resistance should nowhere intersect its *intrados* below the point D.

The Stability of a Wall supported by Shores.

297. Let the weight of the portion of the wall supported by

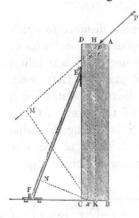

each shore or prop, and the pressure insistent upon it, be imagined to be collected in a single foot of the length of the wall; the conditions of the stability of the wall evidently remain unchanged by this hypothesis. Let ABCD represent one of the columns or piers into which the wall will thus be divided, EF the corresponding shore, P the pressure sustained upon the summit of the wall, Q the thrust upon the shore EF, $2w$ its weight, x the point where the line of resistance intersects the base of the wall, $Cx=m$, $CF=b$, $FEC=\beta$; and let the same notation be taken in other respects as in the preceding articles. Then, since x is a point in the direction of the resultant of the resistances by which the base of the column is sustained, the sum of the moments about that point of the pressure P and half the weight of the shore, supposed to be placed at E *, is equal to the sum of the moments of the thrust Q, and the weight μah of the column; or drawing xM and xN perpendiculars upon the directions of P and Q,

$$P . \overline{xM} + w . \overline{xC} = Q . \overline{xN} + \mu ah . \overline{xK}.$$

Now xM$=\overline{xs}$ sin. xsM$=$(HK$-$Ht) sin.$\alpha=\{h-($H$p+st)$ cot. $\alpha\}$ sin. $\alpha = h$ sin. $\alpha-(k+\frac{1}{2}a-m)$ cos. α, xN$=(b+m)$ cos.β,

$\therefore$ P$\{h$ sin. $a-(k+\frac{1}{2}a-m)$ cos. $a\}+wm=$Q$(b+m)$ cos. $\beta+\mu ah(\frac{1}{2}a-m)$.

Solving this equation in respect to Q, and reducing, we obtain

$$Q=\frac{P\{h \sin. a-(k+\frac{1}{2}a)\cos. a\}-\frac{1}{2}\mu a^2h+m(P\cos. a+\mu ah+w)}{(b+m)\cos. \beta \dagger} \quad . . (386).$$

This expression may be placed under the form

$$=(P\cos. a+\mu ah+w)\sec. \beta-\frac{P\{b\cos. a-h\sin. a+(k+\frac{1}{2}a)\cos. a\}+\mu ah(\frac{1}{2}a+b)+wb}{(b+m)\cos.\beta \dagger}.$$

If the numerator of the fraction in the second member of this equation be a positive quantity (as in all practical cases it will probably be found to be), the value of Q manifestly diminishes with that of m. Now the least value of m, consistent with the stability of the wall, is zero, since the line of resistance no where intersects the extrados; the least value of Q (the shore being supposed *necessary* to the support of the wall) corresponds, therefore, to the value zero of m; moreover this least value of the thrust upon the shore consistent with

* The weight $2w$ of the shore may be conceived to be divided into two equal parts and collected at its extremities.

$\dagger$ The expression $(b+m)$ cos. β may be placed under the form b cot. β sin. $\beta+m$ cos. $\beta=c$ sin. $\beta+m$ cos. β, where c represents the height CE of the point against which the prop rests.

the stability of the wall is manifestly that which it sustains when the wall simply *rests* upon it, the shore not being *driven* so as to increase the thrust sustained by it beyond that just necessary to support the wall. *

This least thrust is represented by the formula

$$Q = \frac{P\{h \sin.\alpha - (k + \tfrac{1}{2}a)\cos.\alpha\} - \tfrac{1}{2}\mu a^2 h}{b \cos.\beta}.$$

The thrust which must be given to the prop in order that there may be given to the wall any required stability, determined by the arbitrary constant m, is determined by equation (386). The stability will diminish as the value of m is increased beyond $\tfrac{1}{2}a$, and the wall will be overthrown inwards when it exceeds a.

298. *The stability of a wall sustained by more than one shore in the same plane.*

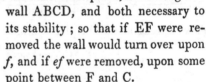

Let EF ef be shores in the same plane, sustaining the wall ABCD, and both necessary to its stability ; so that if EF were removed the wall would turn over upon f, and if ef were removed, upon some point between F and C.

If the thrust of the shore EF be only that just necessary to sustain the tendency of the wall to overturn upon f, it is evident that the line of resistance must pass through that point; but if the thrust exceed that just necessary to the equilibrium, or if the shore be *driven*, then the line of resistance will intersect fg in some point x. Let $fx = m$; then representing the thrust upon EF by Q, the distances fD and fi by h and b, and the angle EFC by β, the value of Q is evidently determined by equation (385).

---

* This case presents an application of the principle of *least resistance*. (See *Theory of the Arch*.)

If z be taken in like manner to represent the point where the line of resistance intersects the base of the wall, and $Cz = m_1$, $CE = b_1$, $Ce = b_2$, $Cfe = \beta_1$, $CD = h_1$, the thrust upon the prop ef by Q_1 and its weight by $2w_1$; then the sum of the moments about the point z of QQ_1, and the weight $\mu a h_1$ of the wall, equals the sum of the moments of P, w, and w_1; or

$$Q_1(b_2 + m_1)\cos.\beta_1 + Q(b_1 + m_1)\cos.\beta + \mu a h_1(\tfrac{1}{2}a - m_1)$$
$$= P\{h_1\sin.\alpha - (k + \tfrac{1}{2}a - m_1)\cos.\alpha\} + (w + w_1)m_1 \ldots . (387).$$

Substituting the value of Q in this equation, from equation (386), and solving in respect to Q_1, the thrust upon the prop ef will be determined, so that the stability of the wall, upon its section fg and upon its base CB, may be m and m_1 respectively.

If $m_1 = m$, the portions of the wall above and below fg are equally stable.

If $m_1 = m = 0$, the thrust upon each shore is only that which is just necessary to support the wall, or which is produced by its actual tendency to overturn. In this case we have

$$Q_1 = \frac{(P\sin.\alpha - \tfrac{1}{2}\mu a^2)(h_1 b - h b_1) + P(b_1 - b)(k + \tfrac{1}{2}a)\cos.\alpha}{b b_2 \cos.\beta_1},$$

the value of Q being determined by equation (385).

299. *The stability of a structure having parallel walls, one of which is supported by means of struts resting on the summit of the other.*

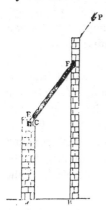

Let AB and CD be taken to represent the walls, and EF one of the struts; the thrust Q upon the strut may be determined precisely as in Art. 297. So that the line of resistance may intersect the base of the wall AB at a given distance m from the extrados (see note, p. 415.).

Let m_1 represent the distance Dx from the extrados at which the line of resistance intersects the base of the wall CD; then taking the moments of the pressures applied to the wall CD about the point x, as in Art. 297., and observing that besides the pressure Q the weight w of one half the strut is applied at E, we have

$$Q\{h_1 \sin. \beta + (k_1 + \tfrac{1}{2}a_1 - m_1) \cos. \beta\} = \mu_1 a_1 h_1 (\tfrac{1}{2}a_1 - m_1) + (k_1 + \tfrac{1}{2}a_1 - m_1)w;$$

in which equation h_1 and a_1 are taken to represent the height and thickness of the wall CD, k_1 the distance of the point E on which the strut rests from the axis of the wall, β the inclination of the strut to the vertical, and μ_1 the weight of a cubic foot of the material of the wall.

Substituting for Q its value from equation (386), and reducing,

$$P\frac{\{h \sin. \alpha - (k+\tfrac{1}{2}a) \cos. \alpha\} - \tfrac{1}{2}\mu a^2 h + m(P \cos. \alpha + \mu a h + w)}{c \sin. \beta + m \cos. \beta} = \frac{\mu_1 a_1 h_1 (\tfrac{1}{2}a_1 - m_1) + (k_1 + \tfrac{1}{2}a_1 - m_1)w}{h_1 \sin. \beta + (k_1 + \tfrac{1}{2}a_1 - m_1) \cos. \beta} \quad .. \text{ (388)}.$$

By this equation is determined that relation between the dimensions of the two walls and the amount of the insistent pressure P, by which any required stability may be assigned to each wall of the structure. If $m = 0$, the pressure upon the strut will be that only which is produced by the tendency of AB to overturn; and the value of m_1 determined from the above equation will give the stability of the external wall on this supposition.

If $m = 0$ and $m_1 = 0$, both walls will be upon the point of overturning, and the above equation will express that relation between the dimensions of the wall and the amount of the insistent pressure, which corresponds to this state of the instability of the structure.

The conditions of the stability, when the wall AB is supported by two struts resting upon the summit of the wall CD, may be determined by a method similar to the above (see Art. 298.).

The general conditions of the stability of the structure discussed in this article evidently include those of a GOTHIC BUILDING having a central nave, whose walls are supported, under the thrust of its roof, by the rafters of the roof of its side aisles. By a reference to the principles of the preceding article, the discussion may readily be made to include the case in which a further support is given to the walls of the nave by *flying buttresses*, which spring from the summits of the walls of the aisles. The influence of the buttresses which support the walls of the aisles upon the conditions of the stability of the structure forms the subject of a subsequent article.

300. *The stability of a wall sustaining the floors of a dwelling.*

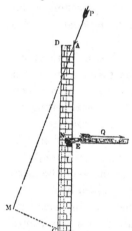

The joists of the floors of a dwelling-house rest at their extremities upon, and are sometimes notched into, pieces of timber called wall-plates, which are imbedded in the masonry of the wall. They serve thus to bind the opposite sides of the house together; and it is upon the support which the thin walls of modern houses receive from these joists, that their stability is sometimes made to depend.*

Representing by w the weight of that portion of the flooring which rests upon the portion ABCD of the wall, and the distance BE by c, taking x, as before, to represent the point where the line of resistance intersects the base of the wall, and measuring the moments from this point, we have

$$\overline{x\mathrm{N}} \cdot \mathrm{Q} + x\mathrm{K} \cdot \mu ah + \overline{x\mathrm{B}} \cdot w = \overline{x\mathrm{M}} \cdot \mathrm{P};$$

whence, taking the same notation as in the preceding articles, and substituting,

$$c\mathrm{Q} + (\tfrac{1}{2}a - m)\mu ah + (a - m)w = \{h\sin.\,\alpha - (k + \tfrac{1}{2}a - m)\cos.\,\alpha\}\mathrm{P};$$

$$\therefore\ \mathrm{Q}c = \{h\sin.\,a - (k + \tfrac{1}{2}a)\cos.\,a\}\mathrm{P} - \tfrac{1}{2}\mu a^2 h - wa + m(\mathrm{P}\cos.\,\alpha + \mu ah + w) \ .\ .\ (389);$$

from which expression it appears that Q is less as m is less. When, therefore, the strain upon the joists is that only which

---

* A house thus constructed evidently becomes unsafe when its wall-plates or the extremities of its joists begin to decay.

is just necessary to preserve the stability of the wall, or which it produces by its tendency to overturn, then $m = 0$. In this case, therefore,

$$Q = \frac{\{h \sin. \alpha - (k + \tfrac{1}{2}a) \cos. \alpha\} P - \tfrac{1}{2}\mu a^2 h - wa}{c} \dots (390).$$

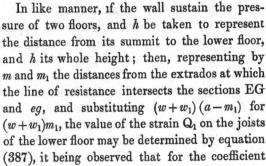

If β be assumed a right angle, and if $(a - m)w$ be substituted for mw, the case discussed in Art. 297. will evidently pass into that which is the subject of the present article, and the preceding equation may thus be deduced from equation (386) (see note, p. 415.).

In like manner, if the wall sustain the pressure of two floors, and h be taken to represent the distance from its summit to the lower floor, and h its whole height; then, representing by m and m_1 the distances from the extrados at which the line of resistance intersects the sections EG and eg, and substituting $(w + w_1)(a - m_1)$ for $(w + w_1)m_1$, the value of the strain Q_1 on the joists of the lower floor may be determined by equation (387), it being observed that for the coefficient of Q_1 in that equation must be substituted (as was shown above) the height $(h_1 - h)$ of the lower floor from the bottom of the wall. If the strain be only that produced by the tendency of the wall to overturn at g and C, then

$$Q_1 c = (h - c)(\tfrac{1}{2}\mu a^2 - P \sin. \alpha) + P(k + \tfrac{1}{2}a) \cos. \alpha + wa - \frac{w_1 ca}{h_1 - h} \dots (391).$$

The value of Q is determined by equation (390), c being taken to represent the distance Ee between the floors. If the joists be not notched into the wall-plates, the friction of their extremities upon them, produced per foot of the length by the weight which they support, must at least equal Q and Q_1 respectively.

301. *The stability of a wall supported by piers or buttresses of uniform thickness.*

Let the piers be imagined to extend along the whole length of the wall, as explained in Art. 290.; and let ABCD represent a section of the compound wall thus produced. Let the weight of each cubic foot of the material of the portion ABFE be represented by μ_1, and that of each cubic foot of GFCD by μ_2, $EA = a_1$, $GD = a_2$, $BC = a$, $AB = h_1$, $CD = h_2$, distance from CD, produced, of the point where P intersects $AE = l$, x the intersection of the line of resistance with CB, $Cx = m$. By the principle of the equality of moments, the moment of P about the point x is equal to the sum of the moments of the weights of GC and AF about that point. But (Art. 297.) moment of $P = P\{h_1 \sin. \ \alpha - (l-m)\cos. \ \alpha\}$; also, moment of weight of $AF = (a - m + \frac{1}{2}a_1)h_1 a_1 \mu_1$; moment of weight of $GC = (\frac{1}{2}a_2 - m)h_2 a_2 \mu_2$.

$\therefore$ $P\{h_1 \sin. a - (l-m)\cos. a\} = (a_2 - m + \frac{1}{2}a_1)h_1 a_1 \mu_1 + (\frac{1}{2}a_2 - m)h_2 a_2 \mu_2$ . . (392).

If the material of the pier be the same with that of the wall ; then, taking b to represent the breadth of each pier, and c the common distance of the piers from centre to centre (Art. 290.), $ca_2\mu_2 = ba_2\mu_1$, therefore $c\mu_2 = b\mu_1$. Representing $\dfrac{c}{b}$ by n, eliminating the value of μ_2 between this equation and equation (392), writing μ for μ_1, and reducing,

$$P(h_1 \sin. a - l \cos. a) = \tfrac{1}{2}\mu\left(a_1^2 h_1 + 2a_1 a_2 h_1 + \frac{1}{n}a_2^2 h_2\right) - m\left\{ P \cos. a + \mu\left(a_1 h_1 + \frac{1}{n}a_2 h_2\right)\right\} \ \ldots (393);$$

by which equation a relation is determined between the dimensions of a wall supported by piers, having a given stability m, and its insistent pressure P. Solving it in respect to a_2, the thickness of the pier necessary to give any required stability to the wall will be determined. (See Appendix.)

If a_2 be assumed to represent that width of the pier by which the wall would just be made to sustain the given pressure P without being overthrown; then taking $m=0$, and solving in respect to a_2,

$$a_2 = -na_1\frac{h_1}{h_2} + \sqrt{\frac{2Pn}{\mu h_2}(h_1 \sin. \ \alpha - l \cos. \ a) + \frac{nh_1}{h_2}\left(\frac{nh_1}{h_2}-1\right)a_1{}^2} \ \text{.. (394)}.$$

302. *The stability of a pier or buttress surmounted by a pinnacle.*

Let W represent the weight of the pinnacle, and e the distance of a vertical through its centre of gravity from the edge C of the pier; then assuming x to be the point where the line of resistance intersects the base of the pier, and taking the same notation as before, equation (392) will evidently become

$$P\{h_1 \sin. \ a - (l-m)\cos. \ a\} = \{a_2-m+\tfrac{1}{2}a_1\}h_1a_1\mu_1 + \{\tfrac{1}{2}a_2-m\}h_2a_2\mu_2 + (e-m)W.$$

Substituting for μ_2 its value $\dfrac{b}{c}\mu_1$ or $\dfrac{\mu_1}{n}$, writing μ for μ_1, and reducing,

$$P(h_1 \sin. \ a - l \cos. \ a) = \tfrac{1}{2}\mu\left(a_1{}^2h_1 + 2a_1a_2h_1 + \frac{1}{n}a_2{}^2h_2\right) + We - m\left\{P\cos. \ a + W + \mu\left(a_1h_1 + \frac{1}{n}a_2h_2\right)\right\} \ \text{.. (395)}$$

If a_2 represent the thickness of that pier by which the wall will just be sustained under the pressure, taking $m=0$, and solving in respect to a_2,

$$a_2 = -na_1\frac{h_1}{h_2} + \sqrt{\frac{2n}{\mu h_2}\{P(h_1 \sin. \ a - l \cos. \ a) - We\} + \frac{nh_1}{h_2}\left(\frac{nh_1}{h_2}-1\right)a_1{}^2} \ \text{.... (396)}.$$

THE GOTHIC BUTTRESS.

303. In Gothic buildings the thickness of a buttress is not unfrequently made to vary at two or three different heights

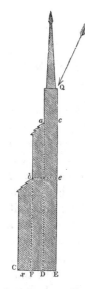

above its base. Such a buttress is represented in the accompanying figure.

The conditions by which any required stability may be assigned to that portion of it whose base is *be* may evidently be determined by equation (295). To determine the conditions of the stability of the whole buttress upon CD, let the heights of the points Q, a, and b above CD be represented by h_1, h_2 and h_3; let $DE=a_1$, $DF=a_2$, $FC=a_3$, $Cx=m_1$; then adopting, in other respects, the same notation as in Arts. 301 and 302.

Since the distances from x of the verticals through the centres of gravity of those portions of the buttress whose bases are DE, DF, and FC respectively, are $(a_3+a_2+\frac{1}{2}a_1-m_1)$, $(a_3+\frac{1}{2}a_2-m_1)$, and $(\frac{1}{2}a_3-m_1)$, we have, by the equality of moments,

$$P\{h_1\sin.\,\alpha-(l-m_1)\cos.\,\alpha\}=(a_3+a_2+\frac{1}{2}a_1-m_1)h_1a_1\mu+(a_3+\frac{1}{2}a_2-m_1)h_2a_2\frac{\mu}{n}+(\frac{1}{2}a_3-m_1)h_3a_3\frac{\mu}{n}+W(e-m_1)\;\cdots\cdots(397).$$

This equation establishes a relation between the dimensions of the buttress and its stability, by which any one of those dimensions which enter into it may be so determined as to give to m_1 any required value, and to the structure any required degree of stability. (See APPENDIX.)

It is evident that, with a view to the greatest economy of the material consistent with a given stability of the buttress, the stability of the portion which rests upon the base *be* should equal that of the whole buttress upon CE; the value of m_1 in the preceding equation should therefore equal that of m in equation (395). If m be eliminated between these two equations, it being observed that h_1 and h_2 in equation (395) are represented by h_1-h_3 and h_2-h_3 in equation (397), a relation will be established between a_1, a_2, a_3, h_1, h_2, h_3, which relation is necessary to the greatest economy of the material; and therefore to the greatest stability of the structure with a given quantity of material.

THE STABILITY OF WALLS SUSTAINING ROOFS.

304. *Thrust upon the feet of the rafters of a roof, the tie-beam not being suspended from the ridge.*

If μ_1 be taken to represent the weight of each square foot of the roofing, $2L$ the span, the inclination BAC of the rafters to the horizon, q the distance between each two principal rafters, and a the inclination to the vertical of the resultant pressure P on the foot of each rafter; then will L sec. represent the length of each rafter, and $\mu_1 Lq$ sec. ι the weight of roofing borne by each rafter. Let the weights thus borne by each of the rafters AB and BC be imagined to be collected in two equal weights at its extremities; the conditions of the equilibrium will remain unchanged, and there will be collected at B the weight supported by one rafter and represented by $\mu_1 Lq$ sec. ι, and at A and C weights, each of which is represented by $\frac{1}{2}\mu_1 Lq$ sec. ι. Now, if Q be taken to represent the thrust produced in the direction of the length of either of the rafters AB and BC, then (Art. 13.) $\mu_1 Lq$ sec. $\iota = 2Q$ cos. $\frac{1}{2}$ABC: but ABC $=\pi-2\iota$.; therefore cos. $\frac{1}{2}$ABC $=$ sin. ι; therefore $2Q$ sin. $\iota = \mu_1 Lq$ sec. ι;

$$\therefore \ Q = \mu_1 Lq \frac{\text{sec.} \ \iota}{2 \ \text{sin.} \ \iota} = \frac{\mu_1 Lq}{2 \ \text{sin.} \ \iota \ \text{cos.} \ \iota} = \frac{\mu_1 Lq}{\text{sin.} \ 2\iota}.$$

The pressures applied to the foot A of the rafter are the thrust Q and the weight $\frac{1}{2}\mu_1 Lq$ sec. ι; and the required pressure P is the resultant of these two pressures. Resolving Q vertically and horizontally, we obtain Q sin. and Q cos. ι, or $\frac{1}{2}\mu_1 Lq$ sec. ι and $\frac{1}{2}\mu_1 Lq$ cosec. ι. The whole pressure applied vertically at A is therefore represented by $\mu_1 Lq$ sec. ι, and the whole horizontal pressure by $\frac{1}{2}\mu_1 Lq$ cosec. ι; whence it follows (Art. 11.) that

$$P = \sqrt{\mu_1{}^2 L^2 q^2 \ \text{sec.}^2 \iota + \tfrac{1}{4}\mu_1{}^2 L^2 q^2 \ \text{cosec.}^2 \iota} = \mu_1 Lq \ \text{sec.} \ \iota \sqrt{1 + \tfrac{1}{4} \cot.^2 \iota} \ . \ . \ (398).$$

$$\tan. \; \alpha = \frac{\frac{1}{2}\mu_1 Lq \; \text{cosec.} \; \imath}{\mu_1 Lq \; \text{sec.} \; \imath} = \tfrac{1}{2} \cot. \imath \quad \dots \dots (399).$$

If the inclination $\imath$ of the roof be made to vary, the span remaining the same, P will attain a minimum value when $\tan. \imath = \dfrac{1}{\sqrt{2}}$, or when

$$\imath = 35° \; 16' \; \dots \dots (400).$$

It is therefore at this inclination of the roof of a given span, whose trusses are of the simple form shown in the figure, that the least pressure will be produced upon the feet of the rafters. If φ represent the limiting angle of resistance between the feet of the rafters and the surface of the tie, the feet of the rafters would not slip even if there were no mortice or notch, provided that α were not greater than φ (Art. 141.), or $\tfrac{1}{2} \cot. \imath$ not greater than $\tan. \varphi$, or

$$\imath \; \text{not greater than} \; \cot.^{-1}(2 \tan. \varphi)^* \; \dots \dots (401).$$

305. *The thrust upon the feet of the rafters of a roof in which the tie-beam is suspended from the ridge by a king-post.*

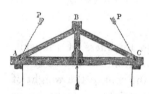

It will be shown in a subsequent portion of this work (see equation 563) that, in this case, the strain upon the king-post BD is equal to $\tfrac{5}{8}$ths of the weight of the tie-beam. Representing, therefore, the weight of each foot in the length of the tie-beam by μ_2, and proceeding exactly as in the last article, we shall obtain for the pressure P upon the feet of the rafters, and its inclination to the vertical, the expressions

* If the surfaces of contact be oak, and thin slips of oak plank be fixed under the feet of the rafters, so that the surfaces of contact may present parallel fibres of the wood to one another (by which arrangement the friction will be greatly increased), $\tan. \varphi = \cdot 48$ (see p. 152.) ; whence it follows that the rafters will not slip, provided that their inclination exceed $\cot.^{-1} \cdot 96$, or $46° \; 10'$.

$$P = \tfrac{1}{2}L\{(2\mu_1 q \sec. \iota + \tfrac{5}{4}\mu_2)^2 + (\mu_1 q \sec.\iota + \tfrac{5}{4}\mu_2)^2 \cot.^2\iota\}^{\frac{1}{2}} \quad .. \ (402).$$

$$\tan. \alpha = \cot. \iota \left(\frac{\mu_1 q \sec. \iota + \tfrac{5}{4}\mu_2}{2\mu_1 q \sec. \iota + \tfrac{5}{4}\mu_2}\right) \quad . . . . . \ (403).$$

306. *The stability of a wall sustaining the thrust of a roof,
having no tie-beam.*

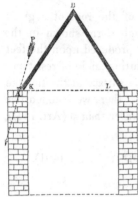

Let it be observed, that in the
equation to the line of resistance
of a wall (equation 382), the terms
P sin. α and P cos. α represent the
horizontal and vertical pressures
on each foot of the length of the
summit of the wall; and that the
former of these pressures is represented in the case of a roof (Art.
304.) by $\tfrac{1}{2}\mu_1 L$ cosec. ι, and the
latter by $\mu_1 L$ sec. ι; whence, substituting these values in equation
(382), we obtain for the equation to the line of resistance in
a wall sustaining the pressure of a roof, without a tie-beam

$$y = L\frac{\tfrac{1}{2}x \cot. \iota - k}{\left(\dfrac{\mu}{\mu_1}\right) ax \cos. \iota + L} \quad . . . . . \ (404);$$

in which expression a represents the thickness of the wall,
k the distance of the feet of the rafters from the centre of the
summit of the wall, L the span of the roof, μ the weight of
a cubic foot of the wall, and μ_1 the weight of each square
foot of the roofing. The thickness a of the wall, so that,
being of a given height h, it may sustain the thrust of a roof
of given dimensions with any given degree of stability, may
be determined precisely, as in Art. 295., by substituting h for x
in the above equation, and $\tfrac{1}{2}a - m$ for y, and solving the
resulting quadratic equation in respect to a.

If, on the other hand, it be required to determine what
must be the inclination ι of the rafters of the roof, so that

being of a given span L it may be supported with a given degree of stability by walls of a given height h and thickness a; then the same substitutions being made as before, the resulting equation must be solved in respect to ι instead of a.

The value of a admits of a minimum in respect to the variable . The value of , which determines such a minimum value of a, is that inclination of the rafters which is consistent with the greatest economy in the material of the wall, its stability being given.

307. *The stability of a wall supported by buttresses, and sustaining the pressure of a roof without a tie-beam.*

The conditions of the stability of such a wall, when supported by buttresses of uniform thickness, will evidently be determined, if in equation (393) we substitute for P cos. α and P sin α their values $\mu_1 L$ sec. ι and $\frac{1}{2}\mu_1 L$ cosec. ι; we shall thus obtain

$$\mu_1 L(\tfrac{1}{2}h_1 \text{ cosec. } \iota - l \text{ sec. } \iota) = \tfrac{1}{2}\mu(a_1{}^2 h_1 + 2a_1 a_2 h_1 + \frac{1}{n}a_2{}^2 h_2) - m\{\mu_1 L \text{ sec. } \iota + \mu(a_1 h_1 + \frac{1}{n}a_2 h_2)\} \ .. \ (405).$$

From which equation the thickness a_2 of the buttresses necessary to give any required stability m to the wall may be determined.

If the thickness of the buttresses be different at different heights, and they be surmounted by pinnacles, the conditions of the stability are similarly determined by substituting for P sin. α and P cos. α the same values in equations (395) and (397).

To determine the conditions of the stability of a Gothic building, whose nave, having a roof without a tie-beam, is supported by the rafters of its two aisles, or by flying buttresses, which rest upon the summits of the walls of its aisles, a similar substitution must be made in equation (388).

If the walls of the aisles be supported by buttresses, equation (388) must be replaced by a similar relation obtained by the methods lad down in Arts. 301. and 303.; the same substitution for P sin. α and P cos. α must then be made.

308. *The conditions of the stability of a wall supporting a shed roof.*

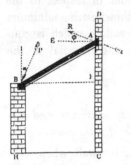

Let AB represent one of the rafters of such a roof, one extremity A resting against the face of the wall of a building contiguous to the shed, and the other B upon the summit of the wall of the shed.

It is evident that when the wall BH is upon the point of being over-thrown, the extremity A will be upon the point of slipping on the face of the wall DC ; so that in this state of the stability of the wall BH, the direction of the resistance R of the wall DC on the extremity A of the rafter will be inclined to the perpendicular AE to its surface at an angle equal to the limiting angle of resistance. Moreover, this direction of the resistance R which corresponds to the state bordering upon motion is common to every other state; for by the principle of least resistance (see *Theory of the Arch*) of all the pressures which might be supplied by the resistance of the wall so as to support the extremity of the rafter, its actual resistance is the least. Now this least resistance is evidently that whose direction is most nearly vertical ; for the pressure upon the rafter is wholly a vertical pressure. But the surface of the wall supplies no resistance whose direction is inclined farther from the horizontal line AE than AR ; AR is therefore the direction of the resistance.

Resolving R vertically and horizontally, it becomes $R \sin.\phi$ and $R \cos.\phi$. Representing the span BF by L, the inclination ABF by ι, the distance of the rafters by q, and the weight of each square foot of roofing by μ_1 (Art. 10.), $R \sin.\phi + P \cos.\alpha = \mu_1 L q \sec. \iota$ and $R \cos.\phi - P \sin.\alpha = 0$; also the perpendiculars let fall from A on P and upon the vertical through the centre of AB, are represented by $L \cos.(\alpha + \iota)$ and $\frac{1}{2} L \cos. \iota$; therefore (Art. 7.) $PL \cos.(\alpha + \iota) = \frac{1}{2} L \cos. \iota . L \mu_1 q \sec. \iota$; therefore $P \cos.(\alpha + \iota) = \frac{1}{2} L \mu_1 q$. Eliminating between these equations we obtain

$$\cot. \alpha = \tan. \phi + 2 \tan. \imath \quad \ldots \ldots \quad (406);$$

$$R = \frac{\frac{1}{2}L\mu_1 q \sec. \imath}{\sin. \phi + \tan. \imath}, \quad P = \frac{1}{2}L\mu_1 q \frac{\{1 + (\tan. \phi + 2 \tan. \imath)^2\}_{\frac{1}{2}}}{\cos. \imath \sec. \phi (\tan. \imath + \sin. \phi)} \cdot (407).$$

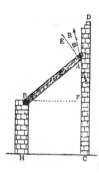 If the rafter, instead of resting at A against the face of the wall, be received into an aperture, as shown in the figure, so that the resistance of the wall may be applied upon its inferior surface instead of at its extremity: then drawing AE perpendicular to the surface of the rafter, the direction AR of the resistance is evidently inclined to that line at the given limiting angle φ. Its inclination to the horizon is therefore represented by $\frac{\pi}{2} - \imath + \phi$. Substituting this angle for φ in equations (406) and (407),

$$\cot. \alpha = \cot. (\imath - \phi) + 2 \tan. \imath \quad \ldots \ldots \quad (408).$$

$$R = \frac{\frac{1}{2}L\mu_1 q \sec. \imath}{\cos. (\imath - \phi) + \tan. \imath}, \quad P = \frac{1}{2}L\mu_1 q \frac{\{1 + (\cot. (\imath - \phi) + 2 \tan. \imath)^2\}^{\frac{1}{2}}}{\cos. \imath \operatorname{cosec}. (\imath - \phi)\{\tan. \imath + \cos. (\imath - \phi)\}} \cdot (409).$$

Substituting in equations (382) and (384) for P sin. α, P cos. α, their values determined above, all the conditions of the stability of a wall supporting such a roof will be determined.

309. THE PLATE BANDE OR STRAIGHT ARCH.

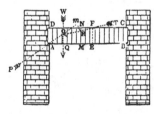 Let MN represent any joint of the plate bande ABCD, whose points of support are A and B ; PA the direction of the resistance at A, WQ a vertical through the centre of gravity of AMND, TR the direction of the resultant pressure upon MN ; the directions of TR, WQ, and PA intersect, therefore, in the same point O.

Let $OAD = a$, $AM = x$, $MR = y$, $AD = H$, $AB = 2L$, weight of cubic foot of material of arch $=\mu_1$. Draw Rm a perpendicular upon PA produced; then by the principle of the equality of moments,

$$\overline{\mathrm{R}m} \cdot \mathrm{P} = \overline{\mathrm{MQ}} \cdot (\text{weight of DM}).$$

But R$m = x \cos. \alpha - y \sin. \alpha$, MQ $= \frac{1}{2}x$, weight of DM $=$ H$\mu_1 x$; also resolving P vertically,

$$\mathrm{P} \cos. \alpha = \mathrm{LH}\mu_1 \ldots \ldots (410).$$

Whence we obtain, by substitution in the preceding equation, and reduction,

$$\mathrm{L}(x - y \tan. \alpha) = \frac{1}{2}x^2 \ldots \ldots (411),$$

which is the equation to the line of resistance, showing it to be a *parabola*. If, in this equation, L be substituted for x, and the corresponding value of y be represented by Y, there will be obtained the equation Y $\tan. \alpha = \frac{1}{2}$L, whence it appears that α is less as Y is greater; but by equation (410), P is less as α is less. P, therefore, is less as Y is greater; but Y can never exceed H, since the line of resistance cannot intersect the extrados. The least value of P, consistent with the stability of the plate bande, is therefore that by which Y is made equal to H, and the line of resistance made to touch the upper surface of the plate bande in F.

Now this least value of P is, by the principle of *least resistance* (see *Theory of the Arch*), the actual value of the resistance at A,

$$\therefore \tan. \alpha = \frac{1}{2}\frac{\mathrm{L}}{\mathrm{H}}. \ldots \ldots (412).$$

Eliminating α between equations (410) and (412),

$$\mathrm{P} = \mathrm{LH}\mu_1 \sqrt{1 + \frac{1}{4}\frac{\mathrm{L}^2}{\mathrm{H}^2}}. \ldots \ldots (413).$$

Multiplying equations (410) and (412) together,

$$\mathrm{P} \sin. \alpha = \frac{1}{2}\mathrm{L}^2\mu_1. \ldots \ldots (414).$$

Now P $\sin. \alpha$ represents the horizontal thrust on the point of support A. From this equation it appears, therefore, that the horizontal thrust upon the abutments of a straight arch is wholly independent of the depth H of the arch, and that it varies as the square of the length L of the arch ; so that the stability of the abutments of such an arch is not at all diminished, but, on the contrary, increased, by increasing the depth of the arch. This increase of the stability of the abutment being the necessary result of an increase of the vertical pressure on the points of support, accompanied by no increase of the horizontal thrust upon them.

310. *The loaded plate bande.*

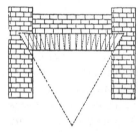

It is evident that the effect of a loading, distributed uniformly over the extrados of the plate bande, upon its stability, is in every respect the same as would be produced if the load were removed, and the weight of the material of the bande increased so as to leave the entire weight of the structure unchanged. Let μ_3 represent the weight of each cubic foot when thus increased, μ_2 the weight of each cubic foot of the load, and H_1 the height of the load ; then $\mu_3 HL = \mu_1 HL + \mu_2 H_1 L,$

$$\therefore \ \mu_3 = \mu_1 + \mu \, \frac{H_1}{H}. \ \ldots \ldots (415).$$

The conditions of the stability of the loaded plate bande are determined by the substitution of this value of μ_3 for μ_1 in the preceding article.

311. *Conditions necessary that the voussoirs of a plate bande may not slip upon one another.*

It is evident that the inclination of every other resultant pressure to the perpendicular to the surface of its corresponding joint, is less than the inclination of the resultant pressure or resistance P, to the perpendicular to the joint AD.

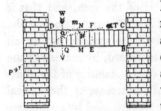

If, therefore, this inclination be not greater than the limiting angle of resistance, then will every other corresponding inclination be less than it, and no voussoir will therefore slip upon the surface of its adjacent voussoir. Now the tangent of the inclination of P to the perpendicular to AD is represented by cot. α or by $\dfrac{2H}{L}$ (equation 412); the required condition is therefore determined by the inequality,

$$\frac{2H}{L} < \tan. \phi. \ \ . \ . \ . \ . \ . \ (416).$$

It is evident that the liability of the arch to failure by the slipping of its voussoirs, is less as its depth is less as compared to its length. In order the more effectually to protect the arch against it, the voussoirs are sometimes cut of the forms shown by the dotted lines in the preceding figure, their joints converging to a point. The pressures upon the points A and B are dependent upon the form of that portion of the arch which lies between those points, and independent of the forms of the voussoirs which compose it: these pressures, and the conditions of the equilibrium of the piers which support the arch, remain therefore unchanged by this change in the forms of the voussoirs.

312. To determine the conditions of the equilibrium of the upright piers or columns of masonry which form the abutments of a straight arch, supposing them to be terminated, as shown in the figure, on a different level from the extrados CD of the arch, let b be taken to represent the elevation of the top of the pier above the point A; then will $b\tan.\alpha$, or $\frac{1}{2}\dfrac{Lb}{H}$ (equation 412), represent the distance AG (p. 410.), or the value of $(k - \frac{1}{2}a)$. Substituting for k in equation (382)

and also the values of $P\sin.\alpha$, $P\cos.\alpha$, from equations (414) and (410), we have

$$y = \tfrac{1}{2}L^2 \frac{x - \left(b + \dfrac{H}{L}a\right)}{\left(\dfrac{\mu}{\mu_1}\right)ax + HL} \quad \ldots\ (417);$$

which is the equation to the line of resistance of the pier, a representing its thickness, b the height of its summit above the springing A of the arch, L the length of the arch, μ the weight of a cubic foot of the material of the arch or abutment (supposed the same).

The conditions of the stability may be determined from this equation as in the preceding articles. If the arch be uniformly loaded, the value of μ_3 given by equation (415) must be substituted for μ_1.

313. The centre of gravity of a buttress whose faces are inclined at any angle to the vertical.

Let the width AB of the buttress at its summit be represented by a, its width CD at the base by b, its vertical height
AF by c, the inclination of its outer face or extrados AC to
the vertical by α_1, that of its intrados AD by α_2.

Let H represent the centre of gravity of
the parallelogram ADEB, and K that of the
triangle BCE, and G that of the buttress;
draw HM, GL, KN, perpendiculars upon
AF. Then representing GL by λ, and observing that the area ADEB is represented
by ac, the area EBC by $\tfrac{1}{2}(b-a)c$, and the
area ADCB by $\tfrac{1}{2}(a+b)c$,

$$\lambda = \frac{ac . \overline{HM} + \tfrac{1}{2}(b-a)c\overline{KN}}{\tfrac{1}{2}(a+b)c} = \frac{2a . \overline{HM} + (b-a)\overline{KN}}{a+b}.$$

Now $HM = Hh + hM = \tfrac{1}{2}a + \tfrac{1}{2}c\tan.\alpha_2 = \tfrac{1}{2}(a + c\tan.\alpha_2)$,

$KN = Kl + lk + kN = \tfrac{2}{3}\tfrac{1}{2}(b-a) + a + \tfrac{2}{3}c\tan.\alpha_2 = \tfrac{1}{3}(b + 2a + 2c\tan.\alpha_2)$;

Substituting these values and reducing,

F F

$$\lambda = \frac{(a^2 + ab + b^2) + (a + 2b)c \tan. \, \alpha_2}{3(a+b)} \quad \ldots \ldots \quad (418).$$

$b = \mathrm{CD} = \mathrm{CF} - \mathrm{DF} = c \tan. \, \alpha_1 + a - c \tan. \, \alpha_2$; also $(a^2 + ab + b^2)$

$= (b-a)^2 + 3ab = c^2 (\tan. \, \alpha_1 - \tan. \, \alpha_2)^2 + 3ac(\tan. \, \alpha_1 - \tan. \, \alpha_2) + 3a^2,$

$(a + 2b)c \, \tan. \, \alpha_2 = \{2c \, (\tan. \, \alpha_1 - \tan. \, \alpha_2) + 3a\}c \, \tan. \, \alpha_2$

$\qquad = 2c^2 (\tan. \, \alpha_1 - \tan. \, \alpha_2) \tan. \, \alpha_2 + 3ac \, \tan. \, \alpha_2;$

$\therefore \; (a^2 + ab + b^2) + (a + 2b)c \, \tan. \, \alpha_2 = c^2 (\tan.^2 \alpha_1 - \tan.^2 \alpha_2)$

$\qquad \qquad + 3ac \, \tan. \, \alpha_1 + 3a^2.$

$$\therefore \; \lambda^* = \frac{\frac{1}{3}c^2 (\tan.^2 \alpha_1 - \tan.^2 \alpha_2) + ac \, \tan. \, \alpha_1 + a^2}{c(\tan. \, \alpha_1 - \tan. \, \alpha_2) + 2a} \quad \ldots \quad (419).$$

314. THE LINE OF RESISTANCE IN A BUTTRESS.

Let LM represent any horizontal section of the buttress TK, a vertical line through the centre of gravity of that portion AMIB of the buttress which rests upon this section. Produce LM to meet the vertical AE in V, and let $\mathrm{KV} = \lambda$ and $\mathrm{AV} = x$; then is the value of λ determined by substituting x for c in equation (419). Let PO be the direction in which a single pressure P is applied to overturn the buttress. Take OS to represent P in magnitude and direction, and ON to represent the weight of the portion AMLB of the buttress; complete the parallelogram SN, and produce its diagonal OR to Q; then will OR evidently be the direction of the resultant pressure upon AMLB, and Q a point in the line of resistance.

Let $\mathrm{VQ} = y$, $\mathrm{AG} = k$, $\angle \mathrm{GOT} = \iota$, $\mu =$ weight of each cubic foot of material; and let the same notation be adopted in other respects as in the last article. By similar triangles,

$$\frac{\mathrm{QK}}{\mathrm{OK}} = \frac{\mathrm{RI}}{\mathrm{OI}},$$

$\mathrm{QK} = \mathrm{QV} - \mathrm{KV} = y - \lambda,$

$\mathrm{OK} = \mathrm{TK} - \mathrm{TO} = \mathrm{TK} - \mathrm{TG} \cot. \, \mathrm{GOT} = x - (\lambda + k) \cot. \, \iota,$

$\mathrm{RI} = \mathrm{RN} \sin. \, \mathrm{RNI} = \mathrm{P} \sin. \, \iota,$

---

* This equation is, of course, to be adapted to the case in which the inclination of AD is on the other side of the vertical, as shown by the dotted line Ad by making α_2, and therefore tan. α_2 negative.

$OI = ON + NI = \frac{1}{2}\mu AV(AB + LM) + RN \cos. RNI = \frac{1}{2}\mu x\{2a + x(\tan. \alpha_1 - \tan. \alpha_2)\} + P \cos. \iota;$

$$\therefore \frac{y - \lambda}{x - (\lambda + k) \cot. \iota} = \frac{P \sin. \iota}{\frac{1}{2}\mu x\{2a + x(\tan. \alpha_1 - \tan. \alpha_2)\} + P \cos. \iota}.$$

Transposing and reducing,

$$y = \frac{\frac{1}{2}\lambda\mu x\{2a + x(\tan. \alpha_1 - \tan. \alpha_2)\} + P(x \sin. \iota - k \cos. \iota)}{\frac{1}{2}\mu x\{2a + x(\tan. \alpha_1 - \tan. \alpha_2)\} + P \cos. \iota};$$

but substituting x for c in equation (419), and multiplying both sides of that equation by the denominator of the fraction in the second member, and by the factor $\frac{1}{2}\mu x$, we have

$$\frac{1}{2}\lambda\mu x\{2a + x(\tan. \alpha_1 - \tan. \alpha_2)\} = \frac{1}{6}\mu x^3(\tan.^2 a - \tan.^2 a_2) + \frac{1}{2}\mu x^2 a \tan. \alpha_1 + \frac{1}{2}\mu x a^2;$$

$$\therefore y = \frac{\frac{1}{3}\mu x^3(\tan.^2 \alpha_1 - \tan.^2 a_2) + \mu x^2 a \tan. \alpha_1 + \mu x a^2 + 2P(x \sin. \iota - k \cos. \iota)}{\mu x\{2a + x(\tan. \alpha_1 - \tan. \alpha_2)\} + 2P \cos. \iota} \quad .. \ (420);$$

which is the equation to the line of resistance in a buttress. If the intrados AD be vertical, tan. α_2 is to be assumed $= 0$. If AD be inclined on the opposite side of the vertical to that shown in the figure, tan. α_2 is to be taken negatively. The line of resistance being of three dimensions in x, it follows that, for *certain values* of y, there are three possible values of x; the curve has therefore a point of contrary flexure. The conditions of the equilibrium of the buttress are determined from its line of resistance precisely as those of the wall.

Thus the thickness a of the buttress at its summit being given, and its height c, and it being observed that the distance CE is represented by $a + c$ tan. α_1, the inclination α_1 of its extrados to the vertical may be determined, so that its line of resistance may intersect its foundation at a given distance m from its extrados, by solving equation (420) in respect to tan. α_1, having first substituted c for x and $a + c$ tan. $\alpha_1 - m$ for y; and any other of the elements determining the conditions of the stability of the buttress may in like manner be determined by solving the equation (the same substitutions being made in it) in respect to that element.

F F 2

315. A WALL OF UNIFORM THICKNESS SUSTAINING THE PRESSURE OF A FLUID.

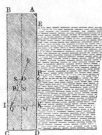

If E be taken to represent the surface of the fluid, IK any section of the wall, and EP two thirds the depth EK; then will P be the centre of pressure* of EK, the tendency of the fluid to overturn the portion AKIB of the wall being the same as would be produced by a single pressure applied perpendicular to its surface at P, and being equal in amount to the weight of a mass of water whose base is equal to EK, and its height to the depth of the centre of gravity of EK, or to $\frac{1}{2}$EK. Let AK$=x$, AE$=e$, weight of each cubic foot of the fluid$=\mu_1$;

$$\therefore \; P = (x-e) \cdot \tfrac{1}{2}(x-e)\mu_1 = \tfrac{1}{2}(x-e)^2\mu_1.$$

Let the direction of P intersect the axis of the wall in O; let it be represented in magnitude by OS; take ON to represent the weight of the portion AKIB of the wall; complete the parallelogram SN, and produce its diagonal to meet IK in Q; then will Q be a point in the line of resistance. Let QM$=y$, AB$=a$, weight of each cubic foot of material of wall$=\mu$. By similar triangles, $\dfrac{QM}{MO}=\dfrac{RN}{NO}$. Now

$$QM=y, \quad MO=PK=\tfrac{1}{3}EK=\tfrac{1}{3}(x-e), \quad RN=OS=P$$
$$=\tfrac{1}{2}\mu_1(x-e)^2, \; NO=\text{weight of ABIK}=\mu ax;$$

$$\therefore \; \frac{y}{\tfrac{1}{3}(x-e)} = \frac{\tfrac{1}{2}\mu_1(x-e)^2}{\mu ax} \; ; \; \therefore \, y = \tfrac{1}{6}\frac{\mu_1(x-e)^3}{\mu ax}.$$

Dividing numerator and denominator of this equation by μ_1, and observing that the fraction $\dfrac{\mu}{\mu_1}$ represents the ratio σ of the specific gravities of the material of the wall and the fluid, we have

* Treatise on " Hydrostatics and Hydrodynamics," by the author of this work, Art. 38. p. 26.

$$y = \tfrac{1}{6}\frac{(x-e)^3}{\sigma a x} \quad \cdots \cdots \quad (421) \; ;$$

which is the equation to the line of resistance in a wall of uniform thickness, sustaining the pressure of a fluid.

316. *To determine the thickness* a *of the wall, so that its height* h *being given, the line of resistance may intersect its foundation at a given distance* m *within the extrados.*

Substituting, in equation (421), h for x, and $\frac{1}{2}a - m$ for y, and solving the resulting equation in respect to a, we obtain

$$a = m + \sqrt{m^2 + \tfrac{1}{3}\frac{(h-e)^3}{\sigma h}} \quad \cdots \cdots \quad (422).$$

Equation (421) may be put under the form $y = \dfrac{1}{6\sigma a}x^\circ\left(1-\dfrac{e}{x}\right)^3$; whence it is apparent that y increases continually with x; so that the nearest approach is made by the line of resistance, to the extrados of the pier, at its lowest section. m therefore represents, in the above expression, the modulus of stability (Art. 288.).

317. *The conditions necessary that the wall should not be overthrown by the slipping of the courses of stones on one another.*

The angle SRO represents the inclination of the resultant pressure upon the section IK to the perpendicular; the proposed condition is therefore satisfied, so long as SRO is less than the limiting angle of resistance φ.

Now, $\tan. \text{SRO} = \dfrac{\text{OS}}{\text{SR}} = \dfrac{\text{RN}}{\text{ON}} = \dfrac{\frac{1}{2}\mu_1(x-e)^2}{\mu a x}$; the proposed condition is therefore satisfied, so long as $\dfrac{(x-e)^2}{2\sigma a x} < \tan. \varphi$; or, reducing this inequality, so long as

$$x < e + \sigma a \tan. \varphi\left\{1 + \left(1 + \frac{2e \cot. \varphi}{\sigma a}\right)^{\frac{1}{2}}\right\} \quad \cdots \cdots \quad (423).$$

318. THE STABILITY OF A WALL OF VARIABLE THICKNESS
SUSTAINING THE PRESSURE OF A FLUID.

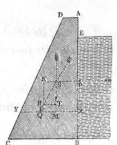

Let us first suppose the internal face AB of the wall to be vertical; let XY be any section of it, P the centre of pressure of AX, and SM a vertical through the centre of gravity of the portion AXYD of the wall. Produce the horizontal direction of the pressure P of the fluid, supposed to be collected in its centre of pressure, to meet MS in S, and let SK be taken to represent it in magnitude, and ST to represent the weight of the portion AXYD of the wall, and complete the parallelogram STRK; then will its diagonal SR represent the direction and amount of the resultant pressure upon the mass AXYD, and if it be produced to intersect XY in Q, Q will be a point in the line of resistance.

Let $AX = x$, $XQ = y$, $MX = \lambda$, $AE = e$, $AD = a$, inclination of DC to vertical $= \alpha$, $\mu =$ weight of cubical foot of wall, $\mu_1 =$ weight of cubical foot of fluid. By similar triangles,

$$\frac{QM}{SM} = \frac{RT}{ST}. \quad \text{Now}$$

$QM = QX - MX = y - \lambda$, $SM = PX = \frac{1}{3}EX* = \frac{1}{3}(x - e)$;

$RT =$ pressure of fluid on $EX = \frac{1}{2}\overline{EX}.\mu_1\overline{EX} = \frac{1}{2}\mu_1(x - e)^2\dagger$;

$ST =$ weight of mass $AY = \frac{1}{2}\{2a + x \tan.\ \alpha\}x\mu.$

$$\therefore \frac{y - \lambda}{\frac{1}{3}(x - e)} = \frac{\frac{1}{2}\mu_1(x - e)^2}{\frac{1}{2}\{2ax + x^2 \tan.\ \alpha\}\mu}.$$

* The centre of pressure of a rectangular plane surface sustaining the pressure of a fluid is situated at two thirds the depth of its immersion. — *Hydrostatics*, p. 26.

† The pressure of a heavy fluid on any plane surface is equal to the weight of a prism of the fluid whose base is equal in area to the surface pressed, and its height to the depth of the centre of gravity of the surface pressed. — *Hydrostatics*, Art. 31.

Let $\dfrac{\mu}{\mu_1} = \sigma$; then, if the fluid be water, σ represents the specific gravity of the material of the wall; and if not, it represents the ratio of the specific gravities of the fluid and wall.

$$\therefore \; y - \lambda = \left(\frac{1}{3\sigma}\right) \frac{(x-e)^3}{2ax + x^2 \tan.\; \alpha}.$$

Now making $\alpha_2 = 0$ in equation (419), and substituting α for α_1, and x for c,

$$\lambda = \frac{\frac{1}{3} x^2 \tan.^2 \alpha + ax \tan.\; \alpha + a^2}{x \tan.\; \alpha + 2a} = \frac{\frac{1}{3} x^3 \tan.^2 \alpha + ax^2 \tan.\; \alpha + a^2 x}{2ax + x^2 \tan.\alpha}$$

Adding this equation to the preceding,

$$y = \frac{\frac{1}{3\sigma}(x-e)^3 + \frac{1}{3} x^3 \tan.^2 \alpha + ax^2 \tan.\; \alpha + a^2 x}{2ax + x^2 \tan.\; \alpha} \quad \ldots \quad (424);$$

which is the equation to the line of resistance to the wall, the conditions of whose stability may be determined from it as before (see Arts. 293. 295.)

319. *The conditions necessary that no course of stones composing the wall may slip upon the subjacent course.*

This condition is satisfied when the inclination of SQ to the perpendicular to the surface of contact at Q is less than the limiting angle of resistance φ; that is, when QSM $< \phi$, or when

$$\tan.\; \varphi > \tan.\; \text{QSM, or} > \frac{RT}{ST}, \text{ or } > \frac{\frac{1}{2}\mu_1(x-e)^2}{\frac{1}{2}(2ax + x^2 \tan.\; \alpha)\mu};$$

$$\text{or } \tan.\; \varphi > \left(\frac{1}{\sigma}\right) \frac{(x-e)^2}{2ax + x \cdot \tan.\; \alpha} \quad \ldots \quad (425).$$

No course of stones will be made by the pressure of the fluid to *slip* upon the subjacent course so long as this condition is satisfied.

It is easily shown that the expression forming the second

member of the above inequality increases continually with x, so that the obliquity of the resultant pressure upon each course, and the probability of its being made to slip upon the next subjacent course, is greater in respect to the lower than the upper courses, increasing with the depth of each course beneath the surface of the fluid.

EARTH WORKS.

320. *The natural slope of earth.*

It has been explained (Art. 243.) that a mass, placed upon an inclined plane and acted upon by no other forces than its weight and the resistance of the plane, will just be supported when the inclination of the plane to the horizon equals the limiting angle of resistance between the surface of the plane and that of the mass which it supports; so that the limiting angle of resistance between the surfaces of the component parts of any mass of earth might be determined by varying continually the slope of its surface until a slope or inclination was attained, at which particular slope small masses of the same earth would only just be supported on its surface, or would just be upon the point of slipping down it. Now this process of experiment is very exactly imitated in the case of embankments, cuttings, and other earth-works, by natural causes. If a slope of earth be artificially constructed at an inclination greater than the particular inclination here spoken of, although, at first, the cohesion of the material may so bind its parts together as to prevent them from sliding upon one another, and its surface from assuming its natural slope, yet by the operation of moisture, penetrating its mass and afterwards drying, or under the influence of frost, congealing, and in the act of congelation *expanding* itself, this cohesion of the particles of the mass is continually in the process of being destroyed; and thus the particles, so long as the slope exceeds the limiting angle of resistance, are continually in the act of sliding down, until, when that angle is at length reached, this descent ceases (except in so far as the particles continue to be *washed* down by the rain), and the surface retains permanently its *natural* slope.

The limiting angle of resistance φ is thus determined by observing what is the natural slope of each description of earth.

The following table contains the results of some such observations * : —

NATURAL SLOPES OF DIFFERENT KINDS OF EARTH.

Nature of Earth.	Natural Slope.	Authority.
Fine dry sand (a single experiment) -	21°	Gadroy.
Ditto - - -	34° 29′	Rondelet.
Ditto - - -	39°	Barlow.
Common earth pulverised and dry -	46° 50′	Rondelet.
Common earth slightly damp -	54°	Rondelet.
Earth the most dense and compact -	55°	Barlow.
Loose shingle perfectly dry -	39°	Pasley.

SPECIFIC GRAVITIES OF DIFFERENT KINDS OF EARTH.

Nature of Earth.	Specific Gravity.
Vegetable earth - - - -	1·4
Sandy earth - - - -	1·6
Marl - - - - -	1·9
Earthy sand - - - -	1·7
Rubble masonry of calcareous earth or siliceous stones -	1·7 to 2·3
Rubble masonry of granite - - -	2·3
Rubble masonry of basaltic stones - -	2·5

321. THE PRESSURE OF EARTH.

Let BD represent the surface of a wall sustaining the pressure of a mass of earth whose surface AE is horizontal.

Let P represent the resultant of the pressures sustained by any portion AX of the wall; and let the cohesion of the

* It is taken from the treatise of M. Navier, entitled *Resumé d'un Cours de Construction*, p. 160.

particles of the earth to one another be neglected, as also their friction on the surface of the wall. It is evident that

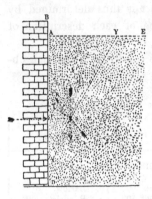

any results deduced in respect to the dimensions of the wall, these elements of the calculation being neglected, will be in *excess*, and err on the safe side.

Now the mass of earth which presses upon AX may yield in the direction of *any* oblique section XY, made from X to the surface AE of the mass. Suppose YX to be the particular direction in which it actually tends to yield ; so that if

AX were removed rupture would first take place along this section, and AXY be the portion of the mass which would first fall. Then is the weight of the mass AYX supported by the resistances of the different elements of the surface AX of the wall, whose resultant is P, and by the resistance of the surface XY on which it tends to slide. Suppose, now, that the mass is upon the point of sliding down the plane XY, the pressure P being that only which is just sufficient to support it ; the resultant SR of the resistances of the different points of XY is therefore inclined (Art. 243.) to the normal ST, at an angle RST equal to the limiting angle of resistance φ between any two contiguous surfaces of the earth.

Now the pressure P, the weight W of the mass AXY, and the resistance R, being pressures in equilibrium, any two of them are to one another inversely as the sines of their inclinations to the third (Art. 14.).

$$\therefore \frac{P}{W}=\frac{\sin. WSR}{\sin. PSR} ; \therefore P = W\frac{\sin. WSR}{\sin. PSR}.$$

But $WSR = WST - RST = AYX - RST = \frac{\pi}{2} - \iota - \varphi$, if $AXY = \iota$;

$$PSR = PST + RST = AXY + RST = \iota + \phi.$$

$$\therefore P = W \cot, (\iota + \phi) \dots (426).$$

Also $W=\frac{1}{2}\mu_1\overline{AX}$. $\overline{AY}=\frac{1}{2}\mu_1 x^2\tan.\iota$; if $\mu=$weight of each cubic foot of earth, and $AX=x$;

$$\therefore\ P=\frac{1}{2}\mu_1 x^2\tan.\iota\cot.(\iota+\phi)\ \ldots\ (427).$$

Now it is evident that this expression, which represents the resistance of the wall necessary to sustain the pressure of the wedge-shaped mass of earth AXY, being dependent for its amount upon the value of (so that different sections, such as XY, being taken, each different mass cut off by such section will require a different resistance of the wall to support it), may admit of a maximum value in respect to that variable.* And if the wall be made strong enough to supply a resistance sufficient to support that wedge-shaped mass of earth whose inclination corresponds to the maximum value of P, and which thus requires the greatest resistance to support it; then will the earth evidently be prevented by it from slipping at any inclination whatever, for it will evidently not slip at *that* angle, the resistance necessary to support it at that angle being supplied; and it will not slip at any other angle, because more than the resistance necessary to prevent it slipping at any other angle is supplied.

If, then, the wall supplies a resistance equal to the maximum value of P in respect to the variable , it will not be overthrown by the pressure of the.earth on AX. Moreover, if it supply any less resistance, it *will* be overthrown; there not being a sufficient resistance supplied by it to prevent the earth from slipping at that inclination which corresponds to the maximum value of P.

To determine the actual pressure of the earth on AX, we have then only to determine the maximum value of P in respect to

* The existence of this maximum will subsequently be shown: it is, however, sufficiently evident, that, as the angle ι is greater, the wedge-shaped mass to be supported is heavier; for which cause, if it operated alone, P would become greater as ι increased. But as ι increases, the plane XY becomes less inclined; for which cause, if it operated alone, P would become less as ι increased. These two causes thus operating to counteract one another, determine a certain inclination in respect to which their neutralising influence is the least, and P therefore the greatest

This maximum value is that which satisfies the conditions

$$\frac{d\mathrm{P}}{d\iota}=0, \text{ and } \frac{d^2\mathrm{P}}{d\iota^2}<0.$$

But differentiating equation (427) in respect to ι, we obtain by reduction

$$\frac{d\mathrm{P}}{d\iota}=\tfrac{1}{4}\mu_1 x^2 \frac{\sin.2(\iota+\phi)-\sin.2\iota}{\cos.^2\iota\sin.^2(\iota+\phi)} \quad \ldots \text{(428)}.$$

Let the numerator and denominator of the fraction in the second member of this equation be represented respectively by p and q; therefore $\frac{d^2\mathrm{P}}{d\iota^2}=\tfrac{1}{4}\mu_1 x^2 \cdot \frac{1}{q^2}\left(\frac{dp}{d\iota}q-\frac{dq}{d\iota}p\right)$; but when $\frac{d\mathrm{P}}{d\iota}=0, p=0$; in this case, therefore, $\frac{d^2\mathrm{P}}{d\iota^2}=\tfrac{1}{4}\mu_1 x^2 \frac{1}{q}\frac{dp}{d\iota}$. Whence it follows, by substitution, that for every value of ι by which the first condition of a maximum is satisfied, the second differential coefficient becomes

$$\frac{d^2\mathrm{P}}{d\iota^2}=\tfrac{1}{2}\mu_1 x^2 \frac{\cos.2(\iota+\phi)-\cos.2\iota}{\cos.^2\iota\sin.^2(\iota+\phi)} \quad \ldots \text{(429)}.$$

Now it is evident from equation (428) that the condition $\frac{d\mathrm{P}}{d\iota}=0$ is satisfied by that value of ι which makes $2(\iota+\phi)=\pi-2\iota$, or

$$\iota=\frac{\pi}{4}-\frac{\phi}{2} \quad \ldots \text{(430)}.$$

And if this value be substituted for ι in equation (429), it. becomes

$$\frac{d^2\mathrm{P}}{d\iota^2}=\mu_1 x^2 \frac{-\sin.\phi}{\cos.^2\left(\frac{\pi}{4}-\frac{\phi}{2}\right)\sin.^2\left(\frac{\pi}{4}+\frac{\phi}{2}\right)} \quad \ldots \text{(431)};$$

which expression is essentially negative, so that the second condition is also satisfied by this value of   It is that, therefore, which corresponds to the maximum value of P; and substituting in equation (427), and reducing, we obtain for this maximum value of P the expression

$$P = \tfrac{1}{2}\mu_1 x^2 \tan.^2\left(\frac{\pi}{4}-\frac{\phi}{2}\right) \ \ldots \ (432);$$

which expression represents the actual pressure of the earth on a surface AX of the wall, whose width is one foot and its depth x.

REVETEMENT WALLS.

322. If, instead of a revetement wall sustaining the pres-

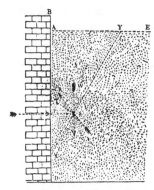

sure of a mass of earth, the weight of each cubic foot of which is represented by μ_1, it had sustained the pressure of a *fluid*, the weight of each cubic foot of which was represented by $\mu_1 \tan.^2\left(\frac{\pi}{4}-\frac{\phi}{2}\right)$, then would the pressure of that fluid upon the surface AX have been represented * by $\tfrac{1}{2}\mu_1 x^2 \tan.^2\left(\frac{\pi}{4}-\frac{\phi}{2}\right)$; so that the pressure of a mass of earth upon a revetement wall (equation 432), when its surface is horizontal (and when its horizontal surface extends, as shown in the figure, to the very surface of the wall), is identical with that of an imaginary fluid whose specific gravity is such as to cause each cubic foot of it to have a weight μ_2, represented in pounds by the formula

$$\mu_2 = \mu_1 \tan.^2\left(\frac{\pi}{4}-\frac{\phi}{2}\right) \ \ldots \ \ldots \ (433).$$

Substituting this value for μ_1 in equations (421) and (424), we determine therefore, at once, the lines of resistance in revetement walls of uniform and variable thickness, under the conditions supposed, to be respectively

$$y = \tfrac{1}{6}\tan.^2\left(\frac{\pi}{4}-\frac{\phi}{2}\right)\cdot\frac{(x-e)^3}{\sigma a x} \ \ldots \ (434);$$

* Hydrostatics, Art. 31.

$$y = \frac{\frac{1}{3\sigma}\tan.^2\left(\frac{\pi}{4}-\frac{\phi}{2}\right)(x-e)^3 + \frac{1}{3}x^3\tan.^2\alpha + ax^2\tan.\alpha + a^2x}{2ax + x^2\tan.\alpha} \dots (435);$$

where σ represents the ratio of the specific gravity of the material of the wall to that of the earth. The conditions of the equilibrium of the revetement wall may be determined from the equation to its line of resistance, as explained in the case of the ordinary wall.

323. *The conditions necessary that a revetement wall may not be overthrown by the slipping of the stones of any course upon those of the subjacent course.*

These are evidently determined from the inequality (425) by substituting μ_2 (equation 433) for μ_1 in that inequality; we thus obtain, representing the limiting angle of resistance of the stones composing the wall by ϕ_1 to distinguish it from that ϕ of the earth,

$$\tan.\phi_1 > \frac{1}{\sigma}\tan.^2\left(\frac{\pi}{4}-\frac{\phi}{2}\right)\frac{(x-e)^2}{2ax + x^2\tan.\alpha} \dots (436);$$

where σ represents the ratio of the specific gravity of the material of the wall to that of the earth.

As before, it may be shown from this expression that the tendency of the courses to slip upon one another is greater in the lower courses than the higher.

324. *The pressure of earth whose surface is inclined to the horizon.*

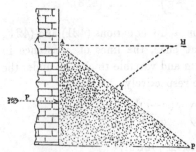

Let AB represent the surface of such a mass of earth, YX the plane along which the rupture of the mass in contact with the surface AX of a revetement wall tends to take place, AX=x, AXY=ι,   XAB=β.

Then if W be taken to represent the weight of the mass AXY, it may be shown, as in Art. 321., equation (426), that $P = W \cot.(\iota + \phi)$.

Now $W = \frac{1}{2}\mu_1 \overline{AX} . \overline{AY} . \sin. \beta$, $AY = \frac{x \sin. \iota}{\sin.(\iota + \beta)}$; there-

fore $W = \frac{1}{2}\mu_1 \dfrac{x \sin. \iota \sin. \beta}{\sin.(\iota + \beta)} = \frac{1}{2}\mu_1 \dfrac{x^2}{\cot. \iota + \cot. \beta}$;

$\therefore P = \frac{1}{2}\mu_1 x^2 . \dfrac{\cot.(\iota + \phi)}{\cot. \iota + \cot. \beta}$ .... (437).

Now the value of ι in this function is that which renders it a maximum (Art. 321.). Expanding $\cot.(\iota + \phi)$, and differentiating in respect to $\tan. \iota$, this value of   is readily determined to be that which satisfies the equation

$$\cot. \iota = \tan. \phi + \sec. \phi \sqrt{1 + \cot. \beta \cot. \phi} \ . \ . \ . \ . \ (438).$$

Substituting in equation (437), and reducing,

$$P = \frac{1}{2}\mu_1 x^2 \left\{ \frac{\cos. \phi}{1 + \sin. \phi \sqrt{1 + \cot. \phi \cot. \beta}} \right\}^2 \ . \ . \ . \ . \ . \ (439).$$

From which equation it is apparent, that the pressure of the earth is, in this case, identical with that of a fluid, of such a density that the weight μ_3, of each cubic foot of it, is represented by the formula

$$\mu_3 = \mu_1 \left\{ \frac{\cos. \phi}{1 + \sin. \phi \sqrt{1 + \cot. \phi \cot. \beta}} \right\}^2 \ . \ . \ . \ . \ (440).$$

The conditions of the equilibrium of a revetement wall sustaining the pressure of such a mass of earth are therefore determined by the same conditions as those of the river wall (Arts. 315. and 318.).

325. THE RESISTANCE OF EARTH.

Let the wall BDEF be supported by the *resistance* of a mass of earth upon its surface AD, a pressure P, applied to its opposite face, tending to overthrow it. Let the surface AH of the earth be horizontal ; and let Q represent the pres-

sure which, being applied to AX, would just be sufficient to
cause the mass of earth in
contact with that portion of
the wall to yield; the prism
AXY slipping backwards upon the surface XY. Adopt-
ing the same notation as in
Art. 321., and proceeding in
the same manner, but ob-
serving that RS is to be measured here on the opposite
side of TS (Art. 243.), since
the mass of earth is supposed to be upon the point of slipping upwards instead of downwards, we shall obtain

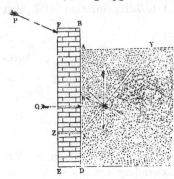

$$Q = \tfrac{1}{2}\mu_1 x^2 \tan \iota \cot (\iota - \phi) \quad \ldots \quad (441).$$

Now it is evident that XY is that plane along which rupture may be made to take place by the *least value* of Q : ι in
the above expression is therefore that angle which gives to that
expression its minimum value. Hence, observing that equation
(441) differs from equation (427) only in the sign of ϕ, and
that the second differential (equation 431) is rendered essentially positive by changing the sign of ϕ, it is apparent
(equation 432) that the value of Q necessary to overcome the
pressure of the earth upon AX is represented by

$$Q = \tfrac{1}{2}\mu_1 x^2 \tan^2\left(\frac{\pi}{4} + \frac{\phi}{2}\right) \quad \ldots \quad (442).$$

326. It is evident that a fluid would oppose the same
resistance to the overthrow of the wall as the resistance of
the earth does, provided that the weight μ_4 of each cubic
foot of the fluid were such that

$$\mu_4 = \mu_1 \tan^2\left(\frac{\pi}{4} + \frac{\phi}{2}\right) \quad \ldots \quad (443);$$

so that the point in AX at which the pressure Q may be
conceived to be applied, is situated at $\tfrac{2}{3}$ds the distance AX.

327. *The stability of a wall of uniform thickness which a given pressure P tends to overthrow, and which is sustained by the resistance of earth.*

Let y be the point in which any section XZ of the wall

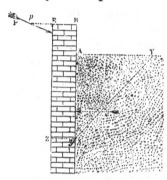

would be intersected by the resultant of the pressures upon the wall above that section, if the whole resistance Q, which the earth in contact with AX is capable of supplying, were called into action. Let BX $=x$, $Xy=y$, BA$=e$, BE$=a$, B$p=k$, weight of cubic feet of material of wall$=\mu$, inclination of P to vertical $=\theta$. Taking the moments about the point y of the pressures applied to BXZE, we have, by the principle of the equality of moments, observing that XQ$=\frac{1}{3}(x-e)$, and that the perpendicular from Y upon P is represented by $x\sin.\theta-(k-y)\cos.\theta$,

$$P\{x\sin.\theta-(k-y)\cos.\theta\}=\tfrac{1}{3}(x-e)Q+(\tfrac{1}{2}a-y)\mu ax ;$$

or substituting for Q its value (equation 442), and solving in respect to y,

$$y=\frac{\tfrac{1}{6}\mu_4(x-e)^3+\tfrac{1}{2}\mu a^2 x-P(x\sin.\theta-k\cos.\theta)}{P\cos.\theta+\mu ax} \quad \ldots \ldots (444).$$

Now it is evident that the wall will not be overthrown upon any section XZ, so long as the greatest resistance Q, which the superincumbent earth is capable of supplying, is sufficient to cause the resultant pressure upon EX to intersect that section, or so long as y in the above equation has a positive value; *moreover, that the stability of the wall is determined by the minimum value of* y *in respect to* x *in that equation, and the greatest height to which the wall can be built, so as to stand, by that value of* x *which makes* y $= 0$.

328. *The stability of a wall which a given pressure tends to overthrow, and which is supported by a mass of earth whose surface is not horizontal.*

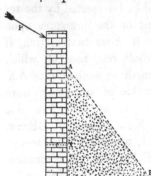

Let β represent the inclination of the surface AB of earth to the horizon. By reasoning similar to that of Art. 324. it is apparent that the resistance Q of the earth in contact with any given portion AX of the wall to displacement, is determined by assigning to φ a negative value in equation (439). Whence it follows, that this resistance is equivalent to that which would be produced by the pressure of a fluid upon the wall, the weight μ_5 of each cubic foot of which was represented by the formula

$$\mu_5 = \mu_1 \left\{ \frac{\cos. \varphi}{1 - \sin. \varphi \sqrt{1 - \cot. \varphi \cot. \beta}} \right\}^2 \ \cdots \ (445).$$

The conditions of the stability of an upright wall subjected to any given pressure P tending to overthrow it, and sustained by the pressure of such a mass of earth, are therefore precisely the same as those discussed in the last article; the symbol μ_4 (equation 444) being replaced by μ_5 (equation 445).

329. *The stability of a revetement wall whose interior face is inclined to the vertical at any angle; taking into account the friction of the earth upon the face of the wall.*

Let α_2 represent the inclination of the face BD of such a wall to the vertical, φ_2 the limiting angle of resistance between the mass of earth and the surface of the wall; and let the same notation be adopted as in the last article in respect to the other elements of the question, and the same construction made. Draw PQ perpendicular to BD;

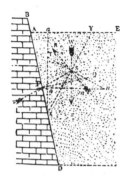

then is the direction PS of the resistance of the wall upon the mass of earth, evidently inclined to QP at an angle QPS equal to the limiting angle of resistance φ_2, in the state bordering upon motion by the overthrow of the wall * (Art. 243.).

Draw Pn horizontally and Xa vertically, produce TS and RS to meet it in m and n, and let $aXY = \iota$,

$$\therefore \frac{P}{W} = \frac{\sin. WSR}{\sin. PSR} = \frac{\sin. (WST - TSR)}{\sin. (RmP + SPm)}.$$

But $WST = AYX = \frac{\pi}{2} - aXY = \frac{\pi}{2} - \iota$, $TSR = \varphi$,

$RmP = TnP + mSn = oXY + RST = \iota + \varphi$,
$SPm = SPQ + QPn = \varphi_2 + \alpha_2$;

$$\therefore \frac{P}{W} = \frac{\sin. \left(\frac{\pi}{2} - \iota - \varphi\right)}{\sin. (\iota + \varphi + \varphi_2 + \alpha_2)} = \frac{\cos. (\iota + \varphi)}{\sin. (\iota + \alpha_2 + \varphi + \varphi_2)};$$

Also $W = \frac{1}{2}\mu_1 \overline{aX} . \overline{AY} = \frac{1}{2}\mu_1 x^2 (\tan. \iota + \tan. \alpha_2)$; if $aX = x$,

$$\therefore P = \frac{1}{2}\mu_1 x^2 \frac{\cos. (\iota + \varphi) (\tan. \iota + \tan. \alpha_2)}{\sin. (\iota + \alpha_2 + \varphi + \varphi_2)} \quad \ldots \ldots (446).$$

Assuming $\alpha_2 + \varphi + \varphi_2 = \beta$, then differentiating in respect to ι,

and assuming $\dfrac{dP}{d\iota} = 0$, we obtain by reduction

$-(\tan. \iota + \tan. \alpha_2) \cos. (\beta - \varphi) + \cos. (\iota + \varphi) \sin. (\iota + \beta) \sec.^2 \iota = 0$; or,

$(\tan. \iota + \tan. \alpha_2)(1 + \tan. \beta \tan. \varphi) + (1 - \tan. \iota \tan.\varphi)(\tan. \iota + \tan. \beta) = 0$;

$\therefore \tan.^2 \iota + 2 \tan. \iota \tan. \beta - \tan. \beta \cot. \varphi + (\cot. \varphi + \tan. \beta) \tan. \alpha_2 = 0.$

* It is not *only* in the state of the wall bordering upon motion that this direction of the resistance obtains, but in every state in which the stability of the wall is maintained. (See *the Principle of Least Resistance*.)

Solving this quadratic in respect to tan. ι, neglecting the negative root, since tan. ι is essentially positive, and reducing,

$$\tan. \iota = (\tan. \beta - \tan. \alpha_2)^{\frac{1}{2}}(\tan. \beta + \cot. \varphi)^{\frac{1}{2}} - \tan. \beta \quad \dots (447).$$

Now the value of determined by this equation, when substituted in the second differential coefficient of P in respect to ι, gives to that coefficient a negative value; it therefore corresponds to a maximum value of P, which maximum determines (Art. 321.) the thrust of the earth upon the portion AX of the wall. To obtain this maximum value of P by substitution in equation (446), let it be observed that

$$\frac{\cos. (\iota + \varphi)}{\sin. (\iota + \beta)} = \frac{1 - \tan. \iota \tan. \varphi}{(\tan. \iota + \tan. \beta)} \left(\frac{\cos. \varphi}{\cos. \beta},\right)$$

$$1 - \tan. \iota \tan. \varphi = 1 + \tan. \beta \tan. \varphi - \tan. \varphi (\tan. \beta - \tan. \alpha_2)^{\frac{1}{2}}(\tan. \beta + \cot. \varphi)^{\frac{1}{2}},$$

$$= \tan. \varphi (\tan. \beta + \cot. \varphi)^{\frac{1}{2}}\{(\tan. \beta + \cot. \varphi)^{\frac{1}{2}} - (\tan. \beta - \tan. \alpha_2)^{\frac{1}{2}}\}$$

$$\tan. \iota + \tan. \beta = (\tan. \beta + \cot. \varphi)^{\frac{1}{2}}(\tan. \beta - \tan. \alpha_2)^{\frac{1}{2}};$$

$$\therefore \frac{\cos. (\iota + \varphi)}{\sin. (\iota + \beta)} = \frac{\sin. \varphi}{\cos. \beta} \left\{ \left(\frac{\tan. \beta + \cot. \varphi}{\tan. \beta - \tan. \alpha_2}\right)^{\frac{1}{2}} - 1 \right\}$$

Also $\tan. \iota + \tan. \alpha_2 = (\tan. \beta + \cot. \varphi)^{\frac{1}{2}}(\tan. \beta - \tan. \alpha_2)^{\frac{1}{2}} - (\tan. \beta - \tan. \alpha_2)$

$$= (\tan. \beta - \tan. \alpha_2)^{\frac{1}{2}}\{(\tan. \beta + \cot. \varphi)^{\frac{1}{2}} - (\tan. \beta - \tan. \alpha_2)^{\frac{1}{2}}\},$$

$$\therefore P = \tfrac{1}{2}\mu_1 x^2 \frac{\sin. \varphi}{\cos. \beta}\{(\tan. \beta + \cot. \varphi)^{\frac{1}{2}} - (\tan. \beta - \tan. \alpha_2)^{\frac{1}{2}}\}^2;$$

which expression may be placed under the following form, better adapted to logarithmic calculation,

$$P = \tfrac{1}{2}\mu_1 x^2 \frac{\sin. \varphi}{\cos.^2 \beta} \left\{ \left(\frac{\cos. (\beta - \varphi)}{\sin. \varphi}\right)^{\frac{1}{2}} - \left(\frac{\sin. (\beta - \alpha_2)}{\cos. \alpha_2}\right)^{\frac{1}{2}} \right\}^2;$$

or substituting for β its value $\alpha_2 + \varphi + \varphi_2$,

$$P = \tfrac{1}{2}\frac{\mu_1 x^2 \sin. \varphi}{\cos.^2 (\alpha_2 + \varphi + \varphi_2)} \left\{ \left(\frac{\cos. (\alpha_2 + \varphi_2)}{\sin. \varphi}\right)^{\frac{1}{2}} - \left(\frac{\sin. (\varphi + \varphi_2)}{\cos. \alpha_2}\right)^{\frac{1}{2}} \right\}^2 \dots (448).$$

By a comparison of this equation with equation (432) it is apparent, that the pressure of a mass of earth upon a revetement wall, under the supposed conditions, is identical with that which it would produce if it were perfectly fluid, provided that the weight of each cubic foot of that fluid had a value represented by the coefficient of $\frac{1}{2}x^2$ in the above equation; so that the conditions of the stability of such a revetement wall are identical (this value being supposed) with the conditions of the stability of a wall sustaining the pressure of a fluid, except that the pressure of the earth is not exerted upon the wall in a direction perpendicular to its surface, as that of a fluid is, but in a direction inclined to the perpendicular at a given angle, namely, the limiting angle of resistance.

330. The pressure of earth which surmounts a revetement wall and slopes to its summit.

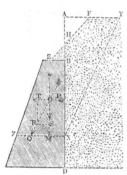

Hitherto we have supposed the surface of the earth whose pressure is sustained by a revetement wall to be horizontal: let us now suppose its surface to be elevated above the summit of the wall, and to descend to it by the natural slope; the wall is then said to be *surcharged*, or to carry a parapet. Let EF represent the natural slope of the earth, FY its horizontal surface, BX any portion of the internal face or intrados of the wall, P the horizontal pressure just necessary to support the mass of earth HXYF, whose weight is W, upon the inclined plane XY. Produce XB and YF to meet in A, and let $AX = x$, $AH = c$, $AXY = \iota$, $\mu_1 =$ weight of each cubic foot of the earth, φ the natural slope of its surface FE. Now it may be shown, precisely by the same reasoning as before, that the actual pressure of the earth upon the portion BX of the wall is represented by that value of P which is a maximum in respect to the variable ι; moreover, that the relation of P and ι is expressed by the function $P = W \cot.(\iota + \varphi)$; where $W = \mu_1$(area HXYF)$= \mu_1$(AXY$-$ AHF)$=\mu_1(\frac{1}{2}x^2 \tan. \iota - \frac{1}{2}c^2 \cot. \varphi)$;

$$\therefore \ \mathrm{P} = \tfrac{1}{2}\mu_1(x^2\tan. \iota - c^2\cot. \varphi)\cot.(\iota + \varphi) \ \cdot \ \cdot \ \cdot \ \cdot \ (449).$$

Expanding $\cot.(\iota + \varphi)$,

$$\mathrm{P} = \tfrac{1}{2}\mu_1 \frac{(x^2\tan. \iota - c^2\cot. \varphi)(1 - \tan. \iota \tan. \varphi)}{\tan. \iota + \tan. \varphi}.$$

To facilitate the differentiation of this function, let $\tan. \iota + \tan. \varphi$ be represented by z, and let it be observed that whatever conditions determine the maximum value of P in respect to z determine also its maximum value in respect to .* Then $\tan. \iota = z - \tan. \varphi$; therefore $1 - \tan. \iota \tan. \varphi = 1 - z\tan. \varphi + \tan.^2\varphi = -z\tan. \varphi + \sec.^2\varphi$. Also, $x^2\tan. \iota - c^2\cot. \varphi = x^2z - (x^2\tan. \varphi + c^2\cot. \varphi)$.

Substituting these values in the preceding expression for P, and reducing,

$$\mathrm{P} = \tfrac{1}{2}\mu_1\left\{ -zx^2\tan. \varphi - \frac{(x^2\tan. \varphi + c^2\cot. \varphi)\sec.^2\varphi}{z} + x^2(\sec.^2\varphi + \tan.^2\varphi) + c^2\right\} \ . \ . \ (450)$$

$$\therefore \ \frac{d\mathrm{P}}{dz} = \tfrac{1}{2}\mu_1\left\{ -x^2\tan. \varphi + \frac{(x^2\tan. \varphi + c^2\cot. \varphi)\sec.^2\varphi}{z^2}\right\},$$

$$\frac{d^2\mathrm{P}}{dz^2} = -\mu_1\frac{(x^2\tan. \varphi + c^2\cot. \varphi)\sec.^2\varphi}{z^3}.$$

The first condition of a maximum is therefore satisfied by the equation

$$-x^2\tan. \varphi + \frac{(x^2\tan. \varphi + c^2\cot. \varphi)\sec.^2\varphi}{z^2} = 0 \ \cdot \ \cdot \ \cdot \ \cdot \ (451);$$

or, solving this equation in respect to z, and reducing, it is satisfied by the equation

$$z = \pm\left(\sec.^2\varphi + \frac{c^2}{x^2}\cosec.^2\varphi\right)^{\frac{1}{2}}.$$

* For $\dfrac{d\mathrm{P}}{d\iota} = \dfrac{d\mathrm{P}}{dz}\dfrac{dz}{d\iota}$, and $\dfrac{d^2\mathrm{P}}{d\iota^2} = \dfrac{d^2\mathrm{P}}{dz^2}\left(\dfrac{dz}{d\iota}\right)^2 + \dfrac{d\mathrm{P}}{dz}\dfrac{d^2z}{d\iota^2}$; now $\dfrac{dz}{d\iota} = \sec.^2\iota$, therefore $\dfrac{d\mathrm{P}}{d\iota} = \dfrac{d\mathrm{P}}{dz}\sec.^2\iota$; and for all values of ι less than $\dfrac{\pi}{2}$, $\sec.^2\iota$ has a finite value, so that $\dfrac{d\mathrm{P}}{d\iota} = 0$ when $\dfrac{d\mathrm{P}}{dz} = 0$.

When, moreover, $\dfrac{d\mathrm{P}}{dz} = 0$, $\dfrac{d^2\mathrm{P}}{d\iota^2} = \dfrac{d^2\mathrm{P}}{dz^2}\left(\dfrac{dz}{d\iota}\right)^2$; so that, when $\dfrac{d^2\mathrm{P}}{dz^2}$ is negative, $\dfrac{d^2\mathrm{P}}{d\iota^2}$ is also negative.

Now the second condition of a maximum is evidently satisfied by any positive value of z, and therefore by the positive root of this equation. Taking, therefore, the positive sign, substituting for z its value, and transposing,

$$\tan.\iota = \left(\sec.^2\phi + \frac{c^2}{x^2}\csc.^2\phi\right)^{\frac{1}{2}} - \tan.\phi \quad \dots \quad (452);$$

which equation determines the tangent of the inclination AXY to the vertical, of the base XY of that wedge-like mass of earth HXYF, whose pressure is borne by the surface BX of the wall. To determine the actual pressure upon the wall, this value of $\tan.\iota$ must be substituted in the expression for P (equation 450). Now the two first terms of the expression within the brackets in the second member of that equation may be placed under the form

$$-z\left\{x^2\tan.\phi + \frac{(x^2\tan.\phi + c^2\cot.\phi)\sec.^2\phi}{z^2}\right\}$$

But it appears by equation (451) that the two terms which compose this expression are *equal*, so that the expression is equivalent to $-2zx^2\tan.\phi$; or, substituting for the value of z, to $-2x^2\tan.\phi(\sec.^2\phi + \frac{c^2}{x^2}\csc.^2\phi)^{\frac{1}{2}}$, or to $-2x\sec.\phi(x^2\tan.^2\phi + c^2)^{\frac{1}{2}}$. Substituting in equation (450),

$$P = \tfrac{1}{2}\mu_1\{-2x\sec.\phi(x^2\tan.^2\phi + c^2)^{\frac{1}{2}} + (x^2\tan.^2\phi + c^2) + x^2\sec.^2\phi\},$$

$$\therefore P = \tfrac{1}{2}\mu_1\{x\sec.\phi - (x^2\tan.^2\phi + c^2)^{\frac{1}{2}}\} \quad \dots \quad (453);$$

by which expression is determined the actual pressure upon a portion of the wall, the distance of whose lowest point from A is represented by x.

331. *The conditions necessary that a revetement wall carrying a parapet may not be overthrown by the slipping of any course of stones on the subjacent course.*

Let ϕ_1 represent the limiting angle of the resistance of the stones of the wall upon one another; and let OQ represent the direction of the resultant pressure on the course XZ.

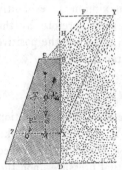

The proposed conditions are then involved (Art. 141.) in the inequality $\varphi_1 > QOM$, or $\tan. \varphi_1 > \tan. QOM$, or

$$\tan. \varphi_1 > \frac{RS}{OS} > \frac{P}{\text{weight of } BZ} * ; \text{ or}$$

substituting for P its value (equation 453), and $\frac{1}{2}\mu(2ax + x^2 \tan.\alpha)$ for the weight of BZ, it appears that the proposed conditions are determined by the inequality

$$\tan. \varphi_1 > \left(\frac{\mu_1}{\mu}\right) \frac{\{x\sec.\varphi - (x^2\tan.^2\varphi + c^2)^{\frac{1}{2}}\}^2}{2ax + x^2\tan.\alpha} \quad \ldots \ldots (454).$$

332. *The line of resistance in a revetement wall carrying a parapet.*

Let OT be taken to represent the pressure P, and OS the weight of BZ. Complete the parallelogram ST, and produce its diagonal OR to Q; then will Q be a point in the line of resistance. Let $AX = x$, $QX = y$, $AB = b$, $AP = X$, $XM = \lambda$, $W = $ weight of BZ [†]. By similar triangles, $\frac{QM}{OM} = \frac{RS}{OS}$; but $QM = (y-\lambda)$, $OM = x - X$, $RS = P$, $OS = W$;

$$\therefore \frac{y-\lambda}{x-X} = \frac{P}{W}, \therefore y = \frac{W\lambda + Px - PX}{W} \quad \ldots \ldots (455).$$

Now the value of λ is determined, from equation (419), by substituting in that equation $(x - b)$ for c; whence we obtain, observing that $\tan. \alpha_2 = 0$, and substituting α for α_1,

$$\lambda = \frac{\frac{1}{3}(x-b)^2 \tan.^2\alpha + a(x-b)\tan. \alpha + a^2}{(x-b)\tan.\alpha + 2a} ;$$

Also $W = \frac{1}{2}\mu(x-b)\{(x-b)\tan.\alpha + 2a\}. \ldots . (456) ;$

* The influence, upon the equilibrium of the wall, of the small portion of earth BHE is neglected in this and the subsequent computation.

† The influence of the weight of the small mass of earth BEH which rests on the summit of the wall is here again neglected.

$$\therefore \ \mathrm{W}\lambda = \tfrac{1}{2}\mu(x-b)\{\tfrac{1}{3}(x-b)^2\tan.^2\alpha + a(x-b)\tan.\alpha + a^2\}.$$

It remains, therefore, only to determine the value of the term P . X. Now it is evident (Art. 16.) that the product P . X is equal to the sum of the moments of the pressures upon the elementary surfaces which compose the whole surface BX. But the pressure upon any such elementary surface, whose distance from A is x, is evidently represented by $\dfrac{d\mathrm{P}}{dx}\Delta x$ *; its moment is therefore represented by $\dfrac{d\mathrm{P}}{dx}x\Delta x$, and the sum of the moments of all such elementary pressures by $\Sigma\dfrac{d\mathrm{P}}{dx}x\Delta x$, or, when Δx is infinitely small, by $\displaystyle\int_b^x \dfrac{d\mathrm{P}}{dx}xdx$; therefore P . X $=\displaystyle\int_b^x \dfrac{d\mathrm{P}}{dx}xdx.$

But differentiating equation (453),

$$\frac{d\mathrm{P}}{dx}=\mu_1\{x\sec.\phi-(x^2\tan.^2\phi+c^2)^{\frac{1}{2}}\}\left\{\sec.\phi-\frac{x\tan.^2\phi}{(x^2\tan.^2\phi+c^2)^{\frac{1}{2}}}\right\}$$

Performing the actual multiplication of the factors in the second member of this equation, observing that $\dfrac{x^2\tan.^2\phi}{(x^2\tan.^2\phi+c^2)^{\frac{1}{2}}}$

$$=\frac{(x^2\tan.^2\phi+c^2)-c^2}{(x^2\tan.^2\phi+c^2)^{\frac{1}{2}}}=(x^2\tan.^2\phi+c^2)^{\frac{1}{2}}-\frac{c^2}{(x^2\tan.^2\phi+c^2)^{\frac{1}{2}}},\ \text{and}$$

reducing, we obtain

$$\frac{d\mathrm{P}}{dx}=\mu_1\left\{x(\sec.^2\phi+\tan.^2\phi)-2\sec.\phi(x^2\tan.^2\phi+c^2)^{\frac{1}{2}}+\frac{c^2\sec.\phi}{(x^2\tan.^2\phi+c^2)^{\frac{1}{2}}}\right\}.$$

Multiplying this equation by x, and integrating between the limits b and x,

* P being a function of x, let it be represented by $f(x)$; then will $f(x)$ represent the pressure upon a portion of the surface BX terminated at the distance x from A, and $f(x+\Delta x)$ that upon a portion terminated at the distance $x+\Delta x$; therefore $f(x+\Delta x)-fx$ will represent the pressure upon the small element Δx of the surface included between these two distances. But by Taylor's theorem, $f(x+\Delta x)-fx=\dfrac{d\mathrm{P}}{dx}\Delta x+\dfrac{d^2\mathrm{P}}{dx^2}\dfrac{(\Delta x)^2}{1\cdot2}+$, &c. ; therefore, neglecting terms involving powers of Δx above the first, pressure on element $=\dfrac{d\mathrm{P}}{dx}\Delta x.$

$$P . X = \mu_1 \left\{ \begin{array}{l} (\tfrac{1}{3}\sec.^2\phi + \tan.^2\phi)(x^3 - b^3) - \tfrac{2}{3}\sec.\phi\cot.^2\phi\{(x^2\tan.^2\phi + c^2)^{\frac{3}{2}} - (b^2\tan.^2\phi + c^2)^{\frac{3}{2}}\} \\ + c^2\sec.\phi\cot.^2\phi\{(x^2\tan.^2\phi + c^2)^{\frac{1}{2}} - (b^2\tan.^2\phi + c^2)^{\frac{1}{2}}\} \end{array} \right\} . (457$$

This value of P . X being substituted in equation (455), and
the values of Wλ, W, P, from equations (453) and (456),
the line of resistance to the revetement wall will be determined, and thence all the conditions of its stability may be
found as before.*

THE ARCH.

333. Each of the structures, the conditions of whose stability (considered as a system of bodies in contact) have hitherto
been discussed, whatever may have been the pressures supposed to be insistent upon it, has been supposed to rest ultimately upon a *single* resisting surface, the resultant of the
resistances on the different elements of which was at once determined in magnitude and direction by the resultant of the
given insistent pressures† being equal and opposite to that
resultant.

The arch is a system of bodies in contact which reposes
ultimately upon *two* resisting surfaces called its abutments.
The resistances of these surfaces are in equilibrium with the
given pressures insistent upon the arch (inclusive of its
weight), but the direction and amount of the resultant pressure upon each surface is dependent upon the unknown resistance of the opposite surface; and thus the general method applicable to the determination of the line of resistance,

* The limits which the author has in this work imposed upon himself
do not leave him space to enter further upon the discussion of this case of
the revetement wall, whose application to the theory of fortification is so
direct and obvious. The reader desirous of further information is referred
to the treatise of M. Poncelet, entitled " Memoire sur la Stabilité des
Revetements et de leurs Fondations." He will there find the subject
developed in all its practical relations, and treated with the accustomed
originality and power of that illustrious author. The above method of
investigation has nothing in common with the method adopted by M. Poncelet except Coulomb's principle of the wedge of maximum pressure.

† The weight of the structure itself is supposed to be included among
these pressures.

and thence of the conditions of stability, in that large class of structures which repose on a single resisting surface, fails in the case of the arch.

334. THE PRINCIPLE OF LEAST RESISTANCE.

*If there be a system of pressures in equilibrium, among which are a given number of resistances, then is each of these a minimum, subject to the conditions imposed by the equilibrium of the whole.**

Let the pressures of the system, which are not resistances, be represented by A, and the resistances by B ; also let any other system of pressures which may be made to replace the pressures B and sustain A, be represented by C.

Suppose the system B to be replaced by C ; then it is apparent that each pressure of the system C is equal to the pressure propagated to its point of application from the pressures of the system A ; or it is equal to that pressure, together with the pressure so propagated to it from the other pressures of the system C.   In the former case it is identical with one of the resistances of the system B ; in the latter case it is greater than it.   Hence, therefore, it appears that each pressure of the system B is a *minimum*, subject to the conditions imposed by the equilibrium of the whole.

If the resultant of the pressures applied to a body, other than the resistances, be taken, it is evident from the above that these resistances are the least possible so as to sustain that resultant ; and therefore that if each resisting point be capable of supplying its resistance in *any* direction, then are all the resistances parallel to one another and to the resultant of the other pressures applied to the body.

* The principle of least resistance was first published by the author of this work in the Philosophical Magazine for October, 1833.

335. *Of all the pressures which can be applied to the highest voussoir of a semi-arch, different in their amounts and points of application, but all consistent with the equilibrium of the semi-arch, that which it would sustain from the pressure of an opposite and equal semi-arch is the least.*

Let EB represent the surface by which an arch rests upon

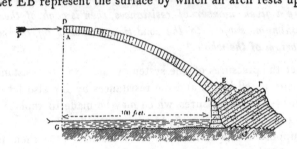

either of its abutments; then are the resistances upon the different points of that surface (Art. 333.) the least pressures, which, being applied to those points, are consistent with the equilibrium of the arch. They are, moreover, parallel to one another: their resultant is therefore the least single pressure, which, being applied to the surface EB, would be sufficient to maintain the equilibrium of the arch, if the abutment were removed.

Now, if this resultant be resolved vertically and horizontally, its component in a vertical direction will evidently be equal to the weight of the semi-arch: it is therefore *given* in amount. In order that the resultant may be a minimum, its vertical component being thus given, it is therefore necessary that its horizontal component should be a minimum; but this horizontal component of the resistance upon the abutment is evidently equal to the pressure P of the opposite semi-arch upon its key-stone: that pressure is therefore a minimum; or, if the semi-arches be equal in every respect, it is the least pressure which, being applied to the side of the key-stone, would be sufficient to support either semi-arch; which was to be proved.

The following proof of this property may be more intelli-

gible to some readers than the preceding. It is independent
of the more general demonstration of the principle of least
resistance. *

The pressure which an opposite semi-arch would produce
upon the side AD of the key-stone, is equal to the tendency
of that semi-arch to revolve forwards upon the inferior edges
of one or more of its voussoirs. Now this tendency to motion
is evidently equal to the least force which would support the
opposite semi-arch. If the arches be equal and equally loaded,
it is therefore equal to the least force which would support
the semi-arch ABED.

336. General conditions of the stability of an arch.†

Suppose the mass ABDC to be acted upon by any num-

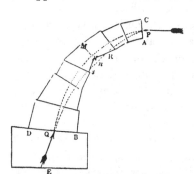

ber of pressures, among
which is the pressure Q,
being the resultant of certain resistances, supplied
by different points in a
surface BD ; common to
the mass and to an im-
moveable obstacle BE.

Now it is clear that under
these circumstances we may
vary the pressure P, both
as to its amount, direction, and point of application in AC,
without disturbing the equilibrium, provided only the form
and direction of the line of resistance continue to satisfy the
conditions imposed by the equilibrium of the system.

These have been shown (Art. 285.) to be the following, —
that it no where *cut* the surface of the mass, except at P,
and within the space BD; and that the resultant pressure
upon no section MN of the mass, or the common surface BD

* See Memoir by the author of this work in Mr. Hann's " Treatise on
the Theory of Bridges," p. 10.

† Theoretical and Practical Treatise on Bridges, vol. i. ; Memoir by the
author of this work, p. 11.

of the mass and obstacle, be inclined to the perpendicular to that surface, at an angle greater than the limiting angle of resistance.

Thus, varying the pressure P, we may destroy the equilibrium, either, first, by causing the resultant pressure to take a direction without the limits prescribed by the resistance of any section MN through which it passes, that is, without the cone of resistance at the point where it intersects that surface; or, secondly, by causing the point Q to fall *without* the surface BD, in which case *no resistance* can be opposed to the resultant force acting in that point; or, thirdly, the point Q lying within the surface BD, we may destroy the equilibrium by causing the line of resistance to cut the surface of the mass somewhere between that point and P.

Let us suppose the limits of the variation of P, within which the first two conditions are satisfied, to be known; and varying it, within those limits, let us consider what may be its *least* and *greatest* values so as to satisfy the third condition.

Let P act at a given point in AC, and in a given direction. It is evident that by diminishing it under these circumstances the line of resistance will be made continually to assume more nearly that direction which it would have if P were entirely removed.

Provided, then, that if P *were* thus removed, the line of resistance would cut the surface, — that is, provided the force P be necessary to the equilibrium, — it follows that by diminishing it we may vary the direction and curvature of the line of resistance, until we at length make it *touch* some point or other in the surface of the mass.

And this is the limit; for if the diminution be carried further, it will *cut* the surface, and the equilibrium will be destroyed. It appears, then, that under the circumstances supposed, when P, acting at a given point and in a given direction, is the least possible, the line of resistance *touches the interior surface or intrados of the mass.*

In the same manner it may be shown that when it is the greatest possible, the line of resistance touches the exterior surface or extrados of the mass.

The direction and point of application of P in AC have here been supposed to be given; but by varying this direction and point of application, the contact of the line of resistance with the intrados of the arch may be made to take place in an infinite variety of different points, and each such variation supplies a new value of P. Among these, therefore, it remains to seek the *absolute* maximum and minimum values of that pressure.

In respect to the direction of the pressure P, or its inclination to AC, it is at once apparent that the least value of that pressure is obtained, whatever be its point of application, when it is *horizontal.*

There remain, then, two conditions to which P is to be subjected, and which involve its condition of a minimum. The first is, *that its amount shall be such as will give to the line of resistance a point of contact with the intrados;* the second, *that its point of application in the key-stone AC shall be such as to give it the least value which it can receive, subject to the first condition.*

337. PRACTICAL CONDITIONS OF THE STABILITY OF AN ARCH OF UNCEMENTED STONES.

The condition, however, that the resultant pressure upon the key-stone is subject, in respect to the *position* of its point of application on the key-stone, to the condition of a minimum, is dependent upon hypothetical qualities of the masonry. It supposes an unyielding material for the arch-stones, and a mathematical adjustment of their surfaces. These have no existence in the uncemented arch. On the striking of the centres the arch invariably sinks at the crown, its voussoirs there slightly opening at their lower edges, and pressing upon one another exclusively by their upper edges. Practically, the line of resistance then, in an arch of *uncemented* stones, *touches the extrados* at the crown; so that only the first of the two conditions of the minimum stated above actually obtains: that, namely, which gives to the line of resistance a contact with the intrados of the arch. This condition being assumed, all consideration of the yielding quality of the material of the arch and its abutments is *eliminated*.

The form of the solid has hitherto been assumed to be given, together with the positions of the different sections made through it ; and the forms of its lines of resistance and pressure, and their directions through its mass have thence been determined.

It is manifest that the converse of this operation is possible.

Having given the form and position of the line of resistance or of pressure, and the positions of the different sections to be made through the mass, it may, for instance, be inquired what form these conditions impose upon the surface which bounds it.

Or the direction of the line of resistance or pressure and the form of the bounding surface may be subjected to certain conditions not absolutely determining either.

If, for instance, the form of the intrados of an *arch* be given, and the direction of the intersecting plane be always perpendicular to it, and if the line of pressure be supposed to intersect this plane always at the same given angle with the perpendicular to it, so that the tendency of the pressure to thrust each from its place may be the same, we may determine what, under these circumstances, must be the extrados of the arch.

If this angle *equal* constantly the limiting angle of resistance, the arch is in a state bordering upon motion, each voussoir being upon the point of slipping downwards or upwards, according as the constant angle is measured above or below the perpendicular to the surface of the voussoir.

The systems of voussoirs which satisfy these two conditions are the greatest and least possible.

If the constant angle be zero, the line of pressure being every where perpendicular to the joints of the voussoirs, the arch would stand even if there were no friction of their surfaces. It is then technically said to be equilibrated ; and the equilibrium of the arch, according to this single condition, constituted the theory of the arch so long in vogue, and so well known from the works of Emerson, Hutton, and Whewell. It is impossible to conceive any arrangement of the parts of an arch by which its stability can be more effectually secured, *so far as the tendency of its voussoirs to slide upon one another is concerned:* there is, however, probably, no practical case in which this tendency really affects the equilibrium. So great is the *limiting angle of resistance* in respect to all the kinds of stone used in the construction of arches, that it would perhaps be *difficult* to construct an arch, the resultant pressure upon any of the joints of which above the springing should lie *without* this angle, or which should yield by the *slipping* of any of its voussoirs.

Traced to the abutment of the arch, the line of resistance ascertains the point where the direction of the resultant pressure intersects it, and the line of pressure determines the inclination to the vertical of that resultant *; these elements determine all the conditions of the equilibrium of the abutments, and therefore of the whole structure; they associate themselves directly with the conditions of the loading of the arch, and enable us so to distribute it as to throw the points of rupture into any given position on the intrados, and give to the line of resistance any direction which shall best conduce to the stability of the structure; from known dimensions, and a known loading of the arch, they determine the dimensions of piers which will support it; or conversely from known dimensions of the piers they ascertain the dimensions and loading of the arch, which may safely be made to span the space between them.

* The inclination of the resultant pressure at the springing to the vertical may be determined independently of the line of pressure, as will hereafter be shown.

338. To DETERMINE THE LINE OF RESISTANCE IN AN ARCH
WHOSE INTRADOS IS A CIRCLE, AND WHOSE LOAD IS COL-
LECTED OVER TWO POINTS OF ITS EXTRADOS SYMMETRI-
CALLY PLACED IN RESPECT TO THE CROWN OF THE ARCH.

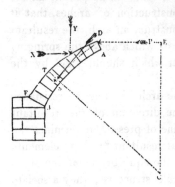

Let ADBF represent any portion of such an arch, P a
pressure applied at its extreme
voussoir, and X and Y the
horizontal and vertical components of any pressure borne
upon the portion DT of its extrados, or of the *resultant* of
any number of such pressures;
let, moreover, the co-ordinates,
from the centre C, of the point
of application of this pressure,
or of this *resultant* pressure, be
x and y.

Let the horizontal force P be applied in AD at a vertical
distance p from C; also let CT represent any plane which,
passing through C, intersects the arch in a direction parallel
to the joints of its voussoirs.

Let this plane be intersected by the *resultant* of the
pressures applied to the mass ASTD in R. These pressures
are the weight of the mass ASTD, the load X and Y, and
the pressure P. Now if pressures equal and parallel to these,
but in opposite directions, were applied at R, they would of
themselves support the mass, and the whole of the subjacent
mass TSB might be removed without affecting the equilibrium.
(Art. 8.) Imagine this to be done; call M the weight of the
mass ASTD, and h the horizontal distance of its centre of
gravity from C, and let CR be represented by ρ, and the
angle ECS by θ, then the perpendicular distances from C of
the pressures M + Y and P − X, imagined to be applied to
R, are $\rho \sin. \theta$ and $\rho \cos. \theta$; therefore, by the condition of the
equality of moments,

$$(M + Y)\rho \sin. \theta + (P - X)\rho \cos. \theta = Mh + Yx - Xy + Pp;$$

$$\therefore\ \rho=\frac{Mh+Yx-Xy+Pp}{(M+Y)\sin.\theta+(P-X)\cos.\theta}\ \cdot\ \cdot\ \cdot\ \cdot\ (458),$$

which is the equation to the line of resistance.

M and h are given functions of θ; as also are X and Y, if the pressure of the load extend *continuously* over the surface of the extrados from D to T.

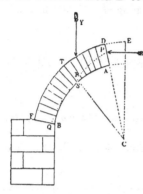

It remains from this equation to determine the pressure P, being that supplied by the opposite semi-arch. As the simplest case, let all the voussoirs of the arch be of the same depth, and let the inclination ECP of the first joint of the semi-arch to the vertical be represented by Θ, and the radii of the extrados and intrados by R and r. Then, by the known principles of statics,

$$Mh=\int_{r}^{R}\int_{\Theta}^{\theta}r^2\sin.\theta d\theta dr=-\tfrac{1}{3}(R^3-r^3)(\cos.\theta-\cos.\Theta);$$

also, $M=\tfrac{1}{2}(R^2-r^2)(\theta-\Theta)$;

$$\therefore\ \rho\{\tfrac{1}{2}(R^2-r^2)(\theta-\Theta)\sin.\theta+Y\sin.\theta-X\cos.\theta+P\cos.\theta\}$$
$$=\tfrac{1}{3}(R^3-r^3)(\cos.\Theta-\cos.\theta)+Yx-Xy+Pp\ \cdot\ \cdot\ \cdot\ \cdot\ \cdot\ (459),$$

which is the general equation to the line of resistance.

The Angle of Rupture.

339. At the points of rupture the line of resistance *meets* the intrados, so that there $\rho=r$: if then Ψ be the corresponding value of θ,

$$r\{\tfrac{1}{2}(R^2-r^2)(\Psi-\Theta)\sin.\Psi+Y\sin.\Psi-X\cos.\Psi+P\cos.\Psi\}$$
$$=\tfrac{1}{3}(R^3-r^3)(\cos.\Theta-\cos.\Psi)+Yx-Xy+Pp\ \cdot\ \cdot\ \cdot\ \cdot\ \cdot\ (460).$$

Also at the points of rupture the line of resistance *touches* the intrados, so that there $\dfrac{d\rho}{d\theta}=\dfrac{dr}{d\theta}=0$; assuming then, to simplify the results, that the pressure of the load is wholly in a vertical direction, so that $X=0$, and that it is collected over a single point of the extrados, so that $\dfrac{dY}{d\theta}=0$, and differentiating equation (459), and assuming $\dfrac{d\rho}{d\theta}=0$, when $\theta=\Psi$ and $\rho=r$, we obtain

$$r\{\tfrac{1}{2}(R^2-r^2)(\Psi-\Theta)\cos.\Psi+\tfrac{1}{2}(R^2-r^2)\sin.\Psi+Y\cos.\Psi-P\sin.\Psi\}=\tfrac{1}{3}(R^3-r^3)\sin.\Psi;$$

hence, assuming $R=r(1+\alpha)$,

$$\left\{\frac{6P}{r^2}+\alpha^2(2\alpha+3)\right\}\tan.\Psi=\left\{\frac{6Y}{r^2}-3\alpha(\alpha+2)\Theta\right\}+3\alpha(\alpha+2)\Psi\ldots\ldots(461).$$

Eliminating $(\Psi-\Theta)$ between equations (460) and (461), we have

$$\left\{\frac{P}{r^2}+\alpha^2(\tfrac{1}{3}\alpha+\tfrac{1}{2})\right\}\sec.^2\Psi-\left\{\frac{Yx+Pp}{r^3}+\alpha(\tfrac{1}{3}\alpha^2+\alpha+1)\cos.\Theta\right\}\sec.\Psi=-\alpha(\tfrac{1}{3}\alpha+1)\ldots\ldots(462).$$

Eliminating P between equations (460) and (461), and reducing,

$$\frac{Y^*}{r^2}\left[\frac{p\cos.\Psi+x\sin.\Psi}{r}-1\right]=(\tfrac{1}{2}\alpha^2+\alpha)(1-\frac{p}{r}\cos.\Psi)(\Psi-\Theta)+\frac{p}{r}(\tfrac{1}{2}\alpha^2+\tfrac{1}{3}\alpha^3)\sin.\Psi$$

$$-\{(\alpha+\alpha^2+\tfrac{1}{3}\alpha^3)\cos.\Theta-(\tfrac{1}{2}\alpha^2+\alpha)\cos.\Psi\}\sin.\Psi\ldots(463).$$

* This equation might have been obtained by differentiating equation (459) in respect to P and θ, and assuming $\dfrac{dP}{d\theta}=0$ when r and Ψ are substituted for ρ and θ; for if that equation be represented by $u=0$, u being a function of P, ρ and θ, $\dfrac{du}{dP}\dfrac{dP}{d\theta}+\dfrac{du}{d\theta}=0$, and $\dfrac{du}{d\rho}\dfrac{d\rho}{d\theta}+\dfrac{du}{d\theta}=0$. The same result $\dfrac{du}{d\theta}=0$ is therefore obtained, whether we assume $\dfrac{dP}{d\theta}=0$, or $\dfrac{d\rho}{d\theta}=0$, which last supposition is that made in equation (461), whence equation (463) has resulted. The hypotheses $\dfrac{dP}{d\theta}=0$, $\rho=r$, determine the minimum of the pressures P, which being applied to a given point of the key-stone will prevent the semi-arch from turning on any of the successive joints of its voussoirs.

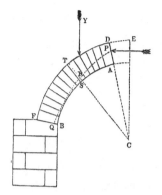

Let $AP = \lambda r$; therefore $\dfrac{p}{r} = (1+\lambda)\cos.\ \Theta.$ Substituting this value of $\dfrac{p}{r}$,

$$\frac{Y}{r^2}\left\{\frac{x}{r}\sin.\ \Psi + (1+\lambda)\cos.\ \Theta\cos.\ \Psi - 1\right\} = (\tfrac{1}{2}a^2 + a)\left\{\{1-(1+\lambda)\cos.\ \Theta\cos.\ \Psi\}\ (\Psi - \Theta)\right.$$

$$\left. + (\cos.\ \Psi - \cos.\ \Theta)\sin.\ \Psi\right\} + \lambda(\tfrac{1}{2}a^2 + \tfrac{1}{3}a^3)\sin.\ \Psi\cos.\ \Theta \ \ldots\ \ldots\ (464),$$

by which equation the angle of rupture Ψ is determined.

If the arch be a continuous segment the joint AD is vertically above the centre, and CD coinciding with CE, $\Theta = 0$; if it be a broken segment, as in the Gothic arch, Θ has a given value determined by the character of the arch. In the pure or equilateral Gothic arch, $\Theta = 30^\circ$. Assuming $\Theta = 0$, and reducing,

$$\frac{Y}{r^2}\left\{\frac{x}{r} - \left(\tan.\frac{\Psi}{2} - \lambda\cot.\ \Psi\right)\right\} = (\tfrac{1}{2}a^2 + a)\left\{\left(\tan.\frac{\Psi}{2} - \lambda\cot.\ \Psi\right)\Psi - \text{vers.}\ \Psi\right\} + \lambda(\tfrac{1}{2}a^2 + \tfrac{1}{3}a^3) \ldots\ (465).$$

It may easily be shown that as Ψ increases in this equation Y increases, and conversely; so that as the load is increased the points of rupture descend. When $Y = 0$, or there is no load upon the extrados,

$$\left(\tan.\frac{\Psi}{2} - \lambda\cot.\Psi\right)\Psi - \text{vers.}\ \Psi + \tfrac{1}{3}\lambda a\frac{3+2a}{2+a} = 0. \ \ldots\ \ldots\ (466).$$

When $x = 0$, or the load is placed on the crown of the arch,

$$\frac{Y}{r^2} = \frac{(\tfrac{1}{2}a^2 + a)\,\text{vers.}\ \Psi - \lambda(\tfrac{1}{2}a^2 + \tfrac{1}{3}a^3)}{\tan.\dfrac{\Psi}{2} - \lambda\cot.\psi} - (\tfrac{1}{2}a^2 + a)\Psi. \ \ldots\ \ldots\ (467).$$

When $\dfrac{x}{r}-\left(\tan.\dfrac{\Psi}{2}-\lambda\cot.\Psi\right)=0$, $\dfrac{Y}{r^2}$ becomes infinite; an

infinite load is therefore required to give that value to the
angle of rupture which is determined by this equation. Solved
in respect to $\tan.\dfrac{\Psi}{2}$, it gives,

$$\tan.\dfrac{\Psi}{2}=\dfrac{\left(\dfrac{x}{r}\right)+\sqrt{\left(\dfrac{x}{r}\right)^2+\lambda(2+\lambda).}}{2+\lambda}\quad\dots\ (468).$$

No loading placed upon the arch can cause the angle of rupture to exceed that determined by this equation.

The Line of Resistance in a circular Arch whose Voussoirs are equal, and whose Load is distributed over different Points of its Extrados.

340. Let it be supposed that the pressure of the load is

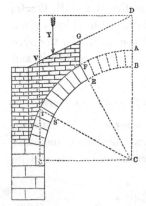

wholly vertical, and such that any
portion FT of the extrados sustains
the weight of a mass GFTV immediately superincumbent to it, and
bounded by the straight line GV
inclined to the horizon at the
angle ι; let, moreover, the weight
of each cubical unit of the load be
equal to that of the same unit of
the material of the arch, multiplied
by the constant factor μ; then, representing AD by Rβ, ACF by Θ,
ACT by θ, and DZ by z, we have,

$$\text{area GFTV}=\int_{\Theta}^{\theta}\overline{\text{TV}}\ .\ dz:$$

but $\text{TV}=\text{MZ}-(\text{MT}+\text{VZ})$, and $\text{MZ}=\text{CD}=\text{R}+\text{R}\beta$, $\text{MT}=$
$\text{R}\cos.\theta$, $\text{VZ}=\overline{\text{DZ}}\tan.\iota=\text{R}\sin.\theta\tan.\iota$. Therefore $\text{MT}+\text{VZ}=$
$\text{R}\cos.\theta+\text{R}\sin.\theta\tan.\iota=\text{R}\{\cos.\theta\cos.\iota+\sin.\theta\sin.\iota\}\sec.\iota=$
$\text{R}\cos.(\theta-\iota)\sec.\iota$;

$$\therefore \ \mathrm{TV} = \mathrm{R}\{1+\beta-\cos.(\theta-\iota)\sec.\iota\}\ ;$$

also, $z=\overline{\mathrm{DZ}}=\mathrm{R}\sin.\theta\ ;$

$$\therefore \text{area } \mathrm{GFTV}=\int_{\Theta}^{\theta}\mathrm{TV}\cdot\frac{dz}{d\theta}d\theta=\mathrm{R}^2\int_{\Theta}^{\theta}\{1+\beta-\cos.(\theta-\iota)\sec.\iota\}\cos.\theta d\theta\ ;$$

$$\therefore \mathrm{Y}=\text{weight of mass } \mathrm{GFTV}=\mu\,\mathrm{R}^2\int_{\Theta}^{\theta}\{1+\beta-\sec.\iota\cos.(\theta-\iota)\}\cos.\theta d\theta=$$

$$\mu\,\mathrm{R}^2\left\{(1+\beta)(\sin.\theta-\sin.\Theta)-\tfrac{1}{4}\sec.\iota\{\sin.(2\theta-\iota)-\sin.(2\Theta-\iota)\}-\tfrac{1}{2}(\theta-\Theta)\right\}..(469).$$

$$\mathrm{Y}x=\text{momentum of } \mathrm{GFTV}=\mu\mathrm{R}^3\int_{\Theta}^{\theta}\{(1+\beta)-\sec.\iota\cos.(\theta-\iota)\}\sin.\theta\cos.\theta d\theta=$$

$$\mu\mathrm{R}^3\{\tfrac{1}{2}(1+\beta)(\cos.^2\Theta-\cos.^2\theta)-\tfrac{1}{3}(\cos.^3\theta-\cos.^3\Theta)-\tfrac{1}{3}\tan.\iota(\sin.^3\theta-\sin.^3\Theta)\}..(470).$$

A SEGMENTAL ARCH WHOSE EXTRADOS IS HORIZONTAL.

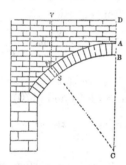

341. As the simplest case, let us first suppose DV horizontal, the material of the loading similar to that of the arch, and the crown of the arch at A, so that $\iota=0$, $\mu=1$, and $\Theta=0$. Substituting the values of Y and Yx (equations 469, 470) which result from these suppositions, in equation (460), solving that equation in respect to $\dfrac{\mathrm{P}}{r^2}$, and reducing, we have,

$$\frac{\mathrm{P}}{r^2}=\frac{\tfrac{1}{2}(1-\alpha)(1+\alpha)^2(1+\beta)\sin.^2\Psi+\tfrac{1}{6}(1+\alpha)^2(1-2\alpha)\cos.^3\Psi+(\tfrac{1}{2}\alpha^2+\tfrac{1}{3}\alpha^3-\tfrac{1}{3})\cos.\Psi-\tfrac{1}{2}\Psi\sin.\Psi+\tfrac{1}{3}}{1+\lambda-\cos.\Psi}..(471).$$

Assuming $\dfrac{d\mathrm{P}}{d\Psi}=0$ (see note, page 468.), and $\lambda=\alpha$, and reducing,

$$\tfrac{2}{3}(1-2\alpha)\cos.^3\Psi-\left\{(1-\alpha)(1+\beta)+(1+\alpha)(1-2\alpha)\right\}\cos.^2\Psi+\left\{\frac{1}{(1+\alpha)^2}+2(1-\alpha^2)(1+\beta)\right\}\cos.\Psi$$

$$+\frac{1}{(1+\alpha)^2}\{1-(1+\alpha)\cos.\Psi\}\frac{\Psi}{\sin.\Psi}-(1-\alpha)(1+\beta)-\frac{2}{3(1+\alpha)^2}-\frac{1+\tfrac{2}{3}\alpha^3}{1+\alpha}=0....(472).$$

In the case in which the line of resistance passes through

H H 4

the bottom of the key-stone, so that $\lambda=0$, equation (471) becomes

$$\frac{P}{r^2}=\tfrac{1}{2}(1+\alpha)^2(1+\beta)(1-\alpha)(1+\cos.\Psi)-\tfrac{1}{2}(1+\alpha)^2(1-2\alpha)(1+\cos.\Psi)\cos.\Psi-\tfrac{1}{2}\Psi\cot.\tfrac{1}{2}\Psi+\tfrac{1}{2}=0\ldots(473);$$

whence assuming $\frac{dP}{d\Psi}=0$, we have

$$\tfrac{2}{3}(1+\alpha)^2(1-2\alpha)\cos.{}^2\Psi+(1+\alpha)^2\{(1-\alpha)\beta+\tfrac{1}{3}(4-5\alpha)\}\cos.\Psi+\frac{\Psi}{\sin.\Psi}$$

$$+\{(1+\alpha)^2(1-\alpha)(1+\beta)+\tfrac{2}{3}+\alpha^2(1+\tfrac{2}{3}\alpha)\}=0\ldots.(474).$$

A GOTHIC ARCH, THE EXTRADOS OF EACH SEMI-ARCH BEING A STRAIGHT LINE INCLINED AT ANY GIVEN ANGLE TO THE HORIZON, AND THE MATERIAL OF THE LOADING DIFFERENT FROM THAT OF THE ARCH.

342. Proceeding in respect to this general case of the stability of the circular arch, by precisely the same steps as in the preceding simpler case, we obtain from equation (460),

$$\frac{P}{r^2}=\frac{(\tfrac{1}{3}\alpha^3+\alpha^2+\alpha)(\cos.\Theta-\cos.\Psi)-(\tfrac{1}{2}\alpha^2+\alpha)(\Psi-\Theta)\sin.\Psi+\frac{Yx}{r^3}-\frac{Y}{r^2}\sin.\Psi}{\{\cos.\Psi-(1+\lambda)\cos.\Theta\}}\ldots(475).$$

in which equation the values of Y and Yx are those determined by substituting Ψ for θ in equations (469) and (470). Differentiating it in respect to Ψ, assuming $\frac{dP}{d\Psi}=0$ (note, p. 468.), and $\lambda=\alpha$, we obtain

$$(\alpha+\tfrac{1}{2}\alpha^2-\tfrac{1}{3}\alpha^3-\tfrac{1}{3}\alpha^4)\cos.\Theta\sin.\Psi-(\tfrac{1}{2}\alpha^2+\alpha)\sin.\Psi\cos.\Psi-(\tfrac{1}{2}\alpha^2+\alpha)\{1-(1+\alpha)\cos.\Theta\cos.\Psi\}(\Psi-\Theta)$$

$$-\{1-(1+\alpha)\cos.\Psi\cos.\Theta\}\frac{Y}{r^2}+\frac{Yx}{r^3}\sin.\Psi+\{\cos.\Psi-(1+\alpha)\cos.\Theta\}\{\frac{1}{r^3}\frac{d(Yx)}{d\Psi}-\frac{\sin.\Psi}{r^2}\frac{dY}{d\Psi}\}=0\ldots(476).$$

Substituting in this equation the values of $\frac{Y}{r^2}$ and $\frac{Yx}{r^3}$, determined by equations (469) and (470), the following equation will be obtained after a laborious reduction: it determines the value of Ψ:

$$A+B\cos.\Psi-C\cos.{}^2\Psi+D\cos.{}^3\Psi+E\sin.\Psi-F\sin.\Psi\cos.\Psi-G\sin.{}^3\Psi-H\cot.\Psi$$

$$+I(1-K\cos.\Psi)\frac{(\Psi-\Theta)}{\sin.\Psi}+\frac{L}{\sin.\Psi}=0\ldots.(477).$$

where

$$A = \mu(1+\alpha)^2 \left\{ \tfrac{2}{3}(1+\alpha)\tan. \imath \sin.^3\Theta - (1+\beta)\{2-(1+\alpha)\cos.^2\Theta\} \right.$$
$$\left. - \tfrac{2}{3}(1+\alpha)\cos.\Theta \right\} + (2\alpha+\alpha^2-\alpha^3-\tfrac{2}{3}\alpha^4)\cos.\Theta.$$

$$B = (1+\alpha)^2\{2\mu(1-\alpha^2)(1+\beta)\cos.\Theta - (1-\mu)\} + 1.$$

$$C = \mu(1+\alpha)^2\{(1-\alpha)(1+\beta) + (1+\alpha)(1-2\alpha)\cos.\Theta\}.$$

$$D = \tfrac{2}{3}\mu(1+\alpha)^2(1-2\alpha).$$

$$E = \mu(1+\alpha)^2(1-2\alpha)\tan.\imath = \tfrac{3}{2}D\tan.\imath.$$

$$F = \mu(1+\alpha)^3(1-2\alpha)\tan.\imath\cos.\Theta = E(1+\alpha)\cos.\Theta.$$

$$G = \tfrac{2}{3}\mu(1+\alpha)^2(1-2\alpha)\tan.\imath = D\tan.\imath.$$

$$H = \tfrac{1}{2}\mu\{2(1+\beta)-\sec.\imath\cos.(\Theta-\imath)\}\sin.2\Theta.$$

$$I = 1-(1-\mu)(1+\alpha)^2.$$

$$K = (1+\alpha)\cos.\Theta.$$

$$L = \mu(1+\alpha)^2\{2(1+\beta)-\sec.\imath\cos.(\Theta-\imath)\}\sin.\Theta.$$

Tables might readily be constructed from this or any of the preceding equations by assuming a series of values of Ψ, and calculating the corresponding values of β for each given value of α, $\imath$, μ, Θ. The tabulated results of such a series of calculations would show the values of Ψ corresponding to given values of α, β, $\imath$, μ, Θ. These values of Ψ being substituted in equation (475), the corresponding values of the horizontal thrust would be determined, and thence the polar equation to the line of resistance (equation 459).

A CIRCULAR ARCH HAVING EQUAL VOUSSOIRS AND SUSTAINING THE PRESSURE OF WATER.

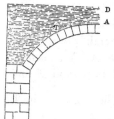

343. Let us next take a case of *oblique* pressure on the extrados, and let us suppose it to be the pressure of *water*, whose surface stands at a height βR above the summit of the key-stone. The pressure of this water being perpendicular to the extrados will every where have its direction through the centre C, so that its moment about that point will vanish, and $Yx - Xy = 0$; more-

over, by the principles of hydrostatics *, the vertical component Y of the pressure of the water, superincumbent to the
portion AT of the extrados, will equal the weight of that
mass of water, and will be represented by the formula (469),
if we assume $\iota = 0$. The horizontal component X of the pressure of this mass of water is represented by the formula

$$X = \mu R^2 \int_{\Theta}^{\theta} \{1 + \beta - \cos.\ \theta\} \sin.\ \theta d\theta = \mu\,(1 + \alpha)^2 r^2 \{(1 + \beta)\,(\cos.\ \Theta - \cos.\ \theta) - \tfrac{1}{2}\,(\cos.\ ^2\Theta - \cos.\ ^2\theta)\} \ .\ .\ (478).$$

Assuming then $\Theta = 0$, we have (equation 469), in respect to that portion of the extrados which lies between the
crown and the points of rupture,

$$\frac{Y}{r^2} = \mu(1 + \alpha)^2 \{(1 + \beta)\sin.\ \Psi - \tfrac{1}{4}\sin.\ 2\Psi - \tfrac{1}{2}\Psi\},$$

and (equation 478) $\dfrac{X}{r^2} = \mu(1 + \alpha)^2\{(1 + \beta)\,\text{vers.}\ \Psi - \tfrac{1}{2}\sin.\ ^2\Psi\}$,

$$\therefore\ \frac{Y}{r^2}\sin.\ \Psi - \frac{X}{r^2}\cos.\ \Psi = \mu(1 + \alpha)^2\{(1 + \beta)\text{vers.}\ \Psi - \tfrac{1}{2}\Psi\sin.\ \Psi\} \ .\ .\ .\ .\ .\ (479).$$

Substituting this value in equation (460), making $Yx - Xy = 0$,
solving that equation in respect to $\dfrac{P}{r^2}$, and making $\dfrac{p}{r} = 1 + \lambda$,
we have

$$\frac{P}{r^2} = \frac{\{\tfrac{1}{2}\alpha^2 + \alpha - \tfrac{1}{2}\mu(1 + \alpha)^2\}\Psi\sin.\ \Psi - \{\alpha + \alpha^2 + \tfrac{1}{3}\alpha^3 - \mu\,(1 + \alpha)^2\,(1 + \beta)\}\text{vers.}\ \Psi}{\lambda + \text{vers.}\ \Psi} \ .\ .\ .\ (480).$$

If, instead of supposing the pressure of the water to be
borne by the extrados, we suppose it to
take effect upon the intrados, tending to
blow up the arch, and if β represent the
height of the water above the crown of
the intrados, we shall obtain precisely the
same expressions for X and Y as before,
except that r must be substituted for $(1 + \alpha)r$, and X and
Y must be taken *negatively;* in this case, therefore, $\dfrac{Y}{r^2}\sin.\ \Psi$

---

* See Hydrostatics and Hydrodynamics, p. 30, 31.

$-\dfrac{X}{r^2}\cos.\Psi=-\mu\{(1+\beta)\text{vers.}\Psi-\tfrac{1}{2}\Psi\sin.\Psi\}$; whence, by substitution in equation (460), and reduction,

$$\dfrac{P}{r^2}=\dfrac{(\tfrac{1}{2}\alpha^2+\alpha+\tfrac{1}{2}\mu)\Psi\sin.\Psi-\{\alpha+\alpha^2+\tfrac{1}{3}\alpha^3+\mu(1+\beta)\}\text{vers.}\Psi}{\lambda+\text{vers.}\,\Psi}\ \ldots\ldots (481).$$

Now by note, page 468., $\dfrac{d\left(\dfrac{P}{r^2}\right)}{d\Psi}=0$; differentiating equations (480) and (481), therefore, and reducing, we have

$$\Psi\left\{\tan.\dfrac{\Psi}{2}-\lambda\cot.\Psi\right\}-\text{vers.}\Psi+A\lambda=0\ \ldots\ldots (482)\ ;$$

which equation applies to both the cases of the pressure of a fluid upon an arch with equal voussoirs; that in which its pressure is borne by the extrados, and that in which it is borne by the intrados; the constant A representing in the first case the quantity $\dfrac{\tfrac{1}{2}\alpha^2+\tfrac{1}{3}\alpha^3-\mu(\tfrac{1}{2}+\beta)(1+\alpha)}{\tfrac{1}{2}\alpha^2+\alpha-\tfrac{1}{2}\mu(1+\alpha)^2}$, and in the second case $\dfrac{\tfrac{1}{2}\alpha^2+\tfrac{1}{3}\alpha^3+\mu(\tfrac{1}{2}+\beta)}{\tfrac{1}{2}\alpha^2+\alpha+\tfrac{1}{2}\mu}$. If the line of resistance pass through the summit of the key-stone, λ must be taken $=\alpha$. If it pass along the inferior edge of the key-stone, $\lambda=0$. In this second case, $\tan.\dfrac{\Psi}{2}\{\Psi-\sin.\Psi\}=0$, therefore $\Psi=0$; so that the point of rupture is at the crown of the arch. For this value of Ψ equations (480) and (481) become vanishing fractions, whose values are determined by known methods of the different calculus to be, when the pressure is on the extrados,

$$\dfrac{P}{r^2}=\alpha-\tfrac{1}{3}\alpha^3+\beta\mu(1+\alpha)^2.\ \ldots\ldots (483)\ ;$$

when the pressure is on the intrados,

$$\dfrac{P}{r^2}=\alpha-\tfrac{1}{3}\alpha^3-\beta\mu.\ \ldots\ldots (484).$$

It is evident that the line of resistance thus passes through the inferior edge of the key-stone, in that state of its equilibrium which precedes its rupture, by the *ascent* of its crown. The corresponding equation to the line of resistance is determined by substituting the above values of $\dfrac{P}{r^2}$ in equation (459). In the case in which the pressure of the water is sustained by the intrados, we thus obtain, observing that

$$\frac{Y}{r^2}\sin.\,\theta - \frac{Y}{r^2}\cos.\,\theta = -\mu\{(1+\beta)\text{ vers. }\theta - \tfrac{1}{2}\theta\sin.\,\theta\}\ ;$$

$$\rho = r\frac{\alpha^2 + 2\alpha - \beta\mu - (\tfrac{1}{3}\alpha^3 + \alpha^2 + \alpha)\cos.\,\theta}{(\tfrac{1}{2}\alpha^2 + \alpha + \tfrac{1}{2}\mu)\theta\sin.\,\theta + (\alpha - \tfrac{1}{3}\alpha^3 + \mu)\cos.\,\theta - \mu(1+\beta)}\ \cdots\cdots\ (485).$$

If for any value of θ in this equation, less than the angle of the semi-arch, the corresponding value of ρ exceed $(1+\alpha)r$, the line of resistance will intersect the extrados, and the arch will *blow up*.

The Equilibrium of an Arch, the Contact of whose Voussoirs is geometrically accurate.

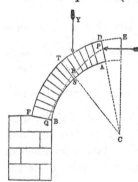

344. The equations (464) and (461) completely determine the value of P, subject to the first of the two conditions stated in Art. 335., viz. that the line of resistance passing through a given point in the key-stone, determined by a given value of λ, shall have a point of geometrical *contact* with the intrados. It remains now to determine it subject to the second condition, viz. that its point of application P on the key-stone shall be such as to give it the least value which it can receive subject to the first condition. It is evident that, subject to this first condition, every different value of λ will give a different value of Ψ; and that of these values of Ψ that which gives the least value of P, and which corresponds to a *positive* value of λ not greater than α, will be the true angle of rupture, on the hypothesis of a mathematical adjustment of the surfaces of the voussoirs to one another. To determine this minimum value of P, in respect to the variation of Ψ dependent on the

variation of λ or of p, let it be observed that λ does not enter into equation (461); let that equation, therefore, be differentiated in respect to P and Ψ, and let $\dfrac{d\mathrm{P}}{d\Psi}$ be assumed $=0$, and Y constant, we shall thence obtain the equation

$$\sec.^2\Psi = \frac{3\alpha(\alpha+2)}{\dfrac{6\mathrm{P}}{r^2} + \alpha^2(2\alpha+3)} \ldots\ldots (486).$$

whence, observing that $\tfrac{1}{2}\sin. 2\Psi = \dfrac{\tan.\Psi}{\sec.^2\Psi} = \left\{\dfrac{\dfrac{6\mathrm{P}}{r^2} + \alpha^2(2a+3)}{3\alpha(\alpha+2)}\right\}\tan.\Psi,$

we obtain by elimination in equation (461)

$$\sin. 2\Psi - 2\Psi = \frac{4\mathrm{Y}}{\alpha(\alpha+2)r^2} - 2\Theta \ldots\ldots (487),$$

from which equation Ψ may be determined.   Also by equation (486),

$$\frac{\mathrm{P}}{r^2} = \tfrac{1}{6}\{3\alpha(\alpha+2)\cos.^2\Psi - \alpha^2(2\alpha+3)\} \ldots\ldots (488);$$

and by eliminating $\sec.\Psi$ between equations (462) and (486), and reducing,

$$\frac{p}{r} = (1+\lambda)\cos.\theta = \frac{r^2}{\mathrm{P}}\left\{\sqrt{a(a+2)\left\{\frac{2\mathrm{P}}{r^2} + a^2(\tfrac{2}{3}a+1)\right\}} - a(\tfrac{1}{3}a^2+a+1)\cos.\theta - \frac{\mathrm{Y}x}{r^3}\right\} \ldots (489).$$

The value of λ given by this equation determines the actual direction of the line of resistance through the key-stone, on the hypothesis made, only in the case in which it is a *positive* quantity, and not greater than α; if it be negative, the line of resistance passes through the bottom of the key-stone, or if it be greater than α, it passes through the top.

Such a mathematical adjustment of the surfaces of contact of the voussoirs as is supposed in this article is, in fact, supplied by the cement of an arch. It may therefore be considered to involve the theory of the cemented arch, the influence on the conditions of its stability of the adhesion of its voussoirs to one another being neglected. In its *settlement*, an arch is liable to disruption in some of those directions in which this adhesion might be necessary to its stability. That old principle, then, which assigns to it such proportions as would cause it to stand firmly did no such adhesion exist, will always retain its authority with the judicious engineer.

APPLICATIONS OF THE THEORY OF THE ARCH.

345. It will be observed that equation (464) or (477) determines the angle Ψ of rupture in terms of the load Y, and the horizontal distance x of its centre of gravity from the centre C of the arch, its radius r, and the depth ar of its voussoirs ; moreover, that this determination is wholly independent of the angle of the arch, and is the same whether its arc be the half or the third of a circle ; also, that if the angle of the semi-arch be less than that given by the above equation as the value of Ψ, there are no points of rupture, such as they have been defined, the line of resistance passing through the springing of the arch and *cutting* the intrados there.

The value of Ψ being known from this equation, P is determined from equation (461), and this value of P being substituted in equation (459), the line of resistance is completely determined ; and assigning to θ the value ACB (p. 467.), the corresponding value of ρ gives us the position of the point Q, where the line of resistance intersects the lowest voussoir of the arch, or the summit of the pier. Moreover, P is evidently equal to the horizontal thrust on the top of the pier, and the vertical pressure upon it is the weight of the arch and load : thus all the elements are known, which determine the conditions of the stability of a pier or buttress (Arts. 295. and 314.) of given dimensions sustaining the proposed arch and its loading.

Every element of the theory of the arch and its abutments is involved, ultimately, in the solution in respect to Ψ of equation (464) or equation (477). Unfortunately this solution presents great analytical difficulties. In the failure of any direct means of solution, there are, however, various methods by which the numerical relation of Ψ and Y may be arrived at indirectly. Among them one of the simplest is this :—

Let it be observed that that equation is readily soluble in respect to Y; instead, then, of determining the value of Ψ for an assumed value of Y, determine conversely the value of Y for a series of assumed values of Ψ. Knowing the distribution of the load Y, the values of x, will be known in respect to these values of Ψ, and thus the values of Y may be numerically determined, and may be tabulated. From such tables may be found, by inspection, values of Ψ corresponding to given values of Y.

The values of Ψ, P, and r are completely determined by equations (487, 488, 489), and all the circumstances of the equilibrium of the circular arch are thence known, on the hypothesis, there made, of a true mathematical adjustment of the surfaces of the voussoirs to one another; and although this adjustment can have no existence in practice when the voussoirs are put together without cement, yet may it obtain in the cemented arch. The cement, by reason of its yielding qualities when fresh, is made to enter into so intimate a contact with the surfaces of the stones between which it is interposed that it takes, when dry, in respect to each joint (abstraction being made of its *adhesive* properties), the character of an exceedingly thin voussoir, having its surfaces mathematically adjusted to those of the adjacent voussoirs; so that if we imagine, not the adhesive properties of the cement of an arch, but only those which tend to the more uniform diffusion of the pressures through its mass, to enter into the conditions of its equilibrium, these equations embrace the entire theory of the cemented arch. The hypothesis here made probably includes all that can be relied upon in the properties of cement as applied to large structures.

An arch may FALL either by the sinking or the rising of its crown. In the former case, the line of resistance passing through the top of the key-stone is made to cut the extrados beneath the points of rupture; in the latter, passing through the bottom of the key-stone, it is made to cut the extrados between the points of rupture and the crown.

In the first case, the values of X, Y, and P, being determined as before, and substituted in equation (459), and p being assumed $=(1+\alpha)r$, the value of θ, which corresponds to $\rho=(1+\alpha)r$, will indicate the point at which the line of resistance cuts the extrados. If this value of θ be less than the angle of the semi-arch, the intersection of the line of resistance with the extrados will take place above the springing, and the arch will fall.

In the second case, in which the crown ascends, let the *maximum* value of ρ be determined from equation (459), p being assumed $=r$; if this value of ρ be greater than R, and the corresponding value of θ less than the angle of rupture, the line of resistance will cut the extrados, the arch will open at the intrados, and it will fall by the descent of the crown.

If the load be collected over a *single* point of the arch, the intersection of the line of resistance with the extrados will take place between this point and the crown; it is that portion only of the line of resistance which lies *between* these points which enters therefore into the discussion. Now if we refer to Art. 338., it will be apparent that in respect to this portion of the line, the values of X and Y in equations (458) and (459) are to be neglected; the only influence of these quantities being found in the value of P.

Example 1. — Let a circular arch of equal voussoirs have

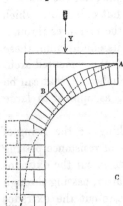

the depth of each voussoir equal to $\frac{1}{10}$th the diameter of its intrados, so that $\alpha=\cdot2$, and let the load rest upon it by three points A, B, D of its extrados, of which A is at the crown and B D are each distant from it 45°; and let it be so distributed that $\frac{3}{8}$ths of it may rest upon each of the points B and D, and the remaining $\frac{1}{4}$ upon A; or let it be so distributed within 60° on either side of the crown as to produce the same effect as though it rested upon these points.

Then assigning one half of the load upon the crown to each semi-arch, and calling x the horizontal distance of the centre of gravity of the load upon either semi-arch from C, it may easily be calculated that $\frac{x}{r}=\frac{3}{4}\sin.45 =\cdot5303301$. Hence it appears from equation (468) that no loading can cause the angle of rupture to exceed 65°. Assume it to equal 60°; the amount of the load necessary to produce this angle of rupture, when distributed as above, will then be determined by assuming in equation (465), $\Psi=60°$, and substituting α for λ, $\cdot2$ for α, and $\cdot5303301$ for $\frac{x}{r}$. We thus obtain $\frac{Y}{r^2}=\cdot0138$.

Substituting this value of $\frac{Y}{r^2}$, and also the given values of α and Ψ in equation (462), and observing that in this equation $\frac{p}{r}$ is to be taken $=1+\alpha$ and $\Theta=0$, we find $\frac{P}{r^2}=\cdot11832$. Substituting this value of $\frac{P}{r^2}$ in the equation (459), we have for the final equation to the line of resistance beneath the point B

$$\rho=r\frac{\cdot2426\,\text{vers.}\,\theta+\cdot1493}{\cdot0138\sin.\theta+\cdot1183\cos.\theta+\cdot22\,\theta\sin.\theta}.$$

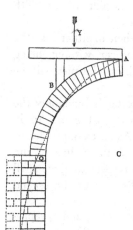

If the arc of the arch be a complete semicircle, the value of ρ in this equation corresponding to $\theta=\frac{\pi}{2}$ will determine the point Q, where the line of resistance intersects the abutment; this value is $\rho=1\cdot09\,r$.

If the arc of the arch be the third of a circle, the value of ρ at the abutment is that corresponding to $\theta=\frac{\pi}{3}$; this will be found to be r, as it manifestly ought to be, since the points of rupture are in this case at the springing.

I I

In the first case the volume of the semi-arch and load is represented by the formula

$$r^2 \left\{ (\tfrac{1}{2}a^2 + a)\frac{\pi}{2} + \frac{Y}{r^2} \right\} = \cdot 3594\, r^2,$$

and in the second case by

$$r^2 \left\{ (\tfrac{1}{2}a^2 + a)\frac{\pi}{3} + \frac{Y}{r^2} \right\} = \cdot 2442\, r^2.$$

Thus, supposing the pier to be of the same material as the arch, the volume of its material, which would have a weight equal to the *vertical* pressure upon its summit, would in the first case be $\cdot 3594\, r^2$, and in the second case $\cdot 2442\, r^2$, whilst the *horizontal* pressures P would in both cases be the same, viz. $\cdot 11832\, r^2$; substituting these values of the vertical and horizontal pressures on the summit of the pier, in equation (382), and for k writing $\tfrac{1}{2}a - (\rho - r)$, we have in the first case

$$H = \frac{\cdot 3594\,(a - \cdot 09\, r)\, r^2}{\cdot 11832\, r^2 - \tfrac{1}{2}a^2};$$

and in the second case,

$$H = \frac{\cdot 2442\, a r^2}{\cdot 11832\, r^2 - \tfrac{1}{2}a_2};$$

where H is the greatest height to which a pier, whose width is a, can be built so as to support the arch.

If $\tfrac{1}{2}a^2 - \cdot 11832\, r^2 = 0$, or $a = \cdot 4864\, r$, then in either case the pier may be built to any height whatever, without being overthrown. In this case the breadth of the pier will be nearly equal to $\frac{1}{4}$th of the span.

The height of the pier being *given* (as is commonly the case), its breadth, so that the arch may júst stand firmly upon it, may readily be determined. As an example, let us suppose the height of the pier to equal the radius of the arch. Solving the above equations in respect to a, we shall then obtain in the first case $a = \cdot 2978\, r$, and in the second $a = \cdot 3\, r$.

If the span of each arch be the same, and r_1 and r_2 represent their radii respectively, then $r_1 = r_2 \sin. 60°$; supposing then the height of the pier in the second arch to be the same as that in the first, viz. r_1, then in the second equation we must write for H, $r_2 \sin. 60°$. We shall thus obtain for a the value $\cdot 28 r_2$.

The piers shown by the dark lines in the preceding figures are of such dimensions as just to be sufficient to sustain the arches which rest upon them, and their loads, both being of a height equal to the radius of the semicircular arch. It will be observed, that in both cases the load $Y = \cdot 0138 r^2$, being that which corresponds to the supposed angle of rupture 60°, is exceedingly small.

Example 2.—Let us next take the example of a Gothic arch, and let us suppose, as in the last examples, that the angle of rupture is 60°, and that $a = \cdot 2$; but let the load in this case be imagined to be collected wholly over the crown of the arch, so that $\dfrac{x}{r} = \sin. 30°$. Substituting in equation (464), 30° for Θ, and 60° for Ψ, and $\cdot 2$ for α, and $\sin. 30°$ for $\dfrac{x}{r}$, we shall obtain the value $1\cdot 3101$ for $\dfrac{Y}{r}$; whence by equation (462) $\dfrac{P}{r^3} = \cdot 67008$, and this value being substituted, equation (459) gives $1\cdot 048$ for the value of ρ when $\theta = \dfrac{\pi}{2}$. We have thus all the data for determining the width of a pier of given height which will just support the arch. Let the height of the pier be supposed, as before, to equal the radius of the intrados; then, since the weight of the semi-arch and its load is $1\cdot 54 r^2$, and the horizontal thrust $\cdot 67008 r^2$, the width a of the pier is found by equation (384) to be $\cdot 422 r$.

The accompanying figure represents this arch; the square, formed by dotted lines over the crown, shows the dimensions of the load of the same materials as the arch which will cause the angle of the rupture to become 60°; the piers are of the required width $\cdot 422 r$, such that when their height is equal to AB, as shown in the figure, and the arch bears this

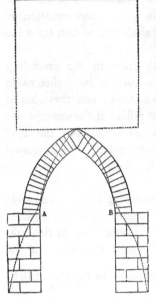

insistent pressure, they may be on the point of overturning. The great amount of the load Y, which in this arch corresponds to an angle of rupture of 60°, shows that angle to be much less than 60° in the great majority of the cases which are offered in the practice of the Gothic arch.

TABLES OF THE THRUST OF ARCHES.

346. It is not possible, within the limits necessarily assigned to a work like this, to enter further upon the discussion of those questions whose solution is involved in the equations which have been given; these can, after all, become accessible to the *general* reader, only when tables shall be formed from them.

Such tables have been calculated with great accuracy by M. Garidel in respect to that case of a segmental arch * whose loading is of the same material as the voussoirs, and the extrados of each semi-arch a straight line inclined at any given angle to the horizon.

Adopting the theory of Coulomb†, M. Garidel has arrived

---

* The term segmental arch is used, here and elsewhere, to distinguish that form of the circular arch in which the intrados is a continuous segment from that in which it is composed of two segments struck from different centres, as in the Gothic arch.

† See Mr. Hann's Theory of Bridges, Art. 16.; also p. 24. of the Memoir on the Arch by the Author of this work, contained in the same volume.

at an equation * which becomes identical with equation (477) in respect to that particular case of the more general conditions embraced by that equation, in which $\mu = 1$ and $\Theta = 0$.

By an ingenious method of approximation, for the details of which the reader is referred to his work, M. Garidel has determined the values of the angle of rupture Ψ, and the quantity $\frac{P}{r^2}$, in respect to a series of different values of α and β. The results are contained in the tables which will be found at the end of this volume.

The value of $\frac{P}{r^2}$ being known from the tables, and the values of Y and Yx from equations 469, 470, the line of resistance is determined by the substitution of these values in equation 459. The line of resistance determines the *point of intersection* of the resultant pressure with the summit of pier; the vertical and horizontal components of this resultant pressure are moreover known, the former being the weight of the semi-arch, and the other the horizontal thrust on the key. All the elements necessary to the determination of the stability of the piers (Arts. 291 and 314) are therefore known.

It will be observed that the amount of the horizontal thrust for each foot of the width of the soffit is determined by multiplying the value of $\frac{P}{r^2}$, shown by the tables, by the square of the radius of the intrados in feet, and by the weight of a cubic foot of the material.

* Tables des Poussées des Voutes, p. 44. Paris, 1837. Bachelier.

PART V.

THE STRENGTH OF MATERIALS.

ELASTICITY.

347. FROM numerous experiments which have been made
upon the elongation, flexure, and torsion of solid bodies
under the action of given pressures, it appears that the
displacement of their particles is subject to the following
laws.

1st. That when this displacement does not extend beyond
a certain distance, each particle tends to return to the place
which it before occupied in the mass, with a force exactly
proportional to the distance through which it has been displaced.

2dly. That if this displacement be carried beyond a certain
distance, the particle remains passively in the new position
which it has been made to take up, or passes finally into some
other position different from that from which it was originally
moved.

The effect of the first of these laws, when exhibited in the
joint tendency of the particles which compose any finite
mass to return to any position in respect to the rest of the
mass, or in respect to one another, from which they have
been displaced, is called *elasticity*. There is every reason to
believe that it exists in all bodies within the limits, more or
less extensive, which are imposed by the second law stated
above.

The force with which each separate particle of a body
tends to return to the position from which it has been displaced varying as the displacement, it follows that the force
with which any *aggregation* of such particles, constituting a
finite portion of the body, when extended or compressed within the limits of elasticity, tends to recover its form, that is,

the force necessary to keep it extended or compressed, is proportional to the amount of the extension or compression ; so that each equal increment of the extending or compressing force produces an equal increment of its extension or compression. This law, which constitutes perfect elasticity, and which obtains in respect to fluid and gaseous bodies as well as solids, appears first to have been established by the direct experiments of S. Gravesande on the elongation of thin wires.[*]

It is, however, by its influence on the conditions of deflexion and the torsion that it is most easily recognised as characterising the elasticity of matter, under all its solid forms[†], within certain limits of the displacement of its particles or elements, called its elastic limits.

[*] For a description of the apparatus of S. Gravesande, see *Illustrations of Mechanics*, by the Author of this work, 2d edition, p. 30. In one of his experiments, Mr. Barlow subjected a bar of wrought iron, one square inch in section, to strains increasing successively from four to nine tons, and found the elongations corresponding to the successive additional strains, each of one ton, to be, in millionths of the whole length of the bar, 120, 110, 120, 120, 120. In a second experiment, made with a bar two square inches in section, under strains increasing from 10 tons to 30 tons, he found the additional elongations, produced by successive additional strains, each of two tons, to be, in millionths of the whole length, 110, 110, 110, 110, 100, 100, 100, 100, 95, 90. From an extensive series of similar results, obtained from iron of different qualities, he deduced the conclusion that a bar of iron of mean quality might be assumed to elongate by 100 millionth parts, or the 10,000th part, of its whole length, under every additional ton strain per square inch of its section. (*Report to Directors of London and Birmingham Railway*. Fellowes. 1835.)

The French engineers of the Pont des Invalides assigned 82 millionth parts to this elongation, their experiments having probably been made upon iron of inferior quality. M. Vicat has assigned 91 millionth parts to the elongation of cables of iron wire (No. 18.) under the same circumstances, MM. Minard and Desormes, 1,176 millionth parts to the elongation of bars of oak. (*Illust. Mech.*, p. 393.)

[†] The experiments of Prof. Robison on torsion show the existence of this property in substances where it might little be expected ; in pipe-clay, for instance.

ELONGATION.

348. To determine the elongation or compression of a bar of a given section under a given strain.

Let K be taken to represent the section of the bar in square inches, L its length in feet, l its elongation or compression in feet under a strain of P pounds, and E the strain or thrust in pounds which would be required to extend a bar of the same material to double its length, or to compress it to one half its length, if the elastic limit of the material were such as to allow it to be so far elongated or compressed, the law of elasticity remaining the same.*

Now, suppose the bar, whose section is K square inches, to be made up of others of the same length L, each one inch in section ; these will evidently be K in number, and the strain or the thrust upon each will be represented by $\frac{P}{K}$. More-over, each bar will be elongated or compressed, by this strain or thrust, by l feet ; so that each foot of the length of it (being elongated or compressed by the same quantity as each other foot of its length) will be elongated or compressed by a quantity represented, in feet, by $\frac{l}{L}$. But to elongate or compress a foot of the length of one of these bars, by one foot, requires (by supposition) E pounds strain or thrust; to elongate or compress it by $\frac{l}{L}$ feet requires, therefore, $E\frac{l}{L}$ pounds. But the strain or thrust which actually produces this elongation is $\frac{P}{K}$ pounds. Therefore, $\frac{P}{K}=E\frac{l}{L}$.

$$\therefore \ l=\frac{PL}{EK} \ \cdots \cdots (490).$$

---

* The value of E in respect to any material is called the *modulus* of its elasticity. The value of the moduli of elasticity of the principal materials of construction have been determined by experiment, and will be found in a table at the end of the volume.

349. *To find the number of units of work expended upon the elongation by a given quantity (l) of a bar whose section is* K *and its length* L.

If x represent any elongation of the bar (x being a part of l), then is the strain P corresponding to that elongation represented (equation 490) by $\frac{KE}{L}x$; therefore the work done in elongating the bar through the small additional space Δx, is represented by $\frac{KE}{L}x\Delta x$ (considering the strain to remain the same through the small space Δx); and the whole work U done is, on this supposition, represented by $\frac{KE}{L}\Sigma x\Delta x$, or (supposing Δx to be infinitely small) by

$$\frac{KE}{L}\int_0^l x\,dx \text{ or by } \tfrac{1}{2}\frac{KE}{L}l^2.$$

$$\therefore \ U = \tfrac{1}{2}\frac{KEl^2}{L} \ \ldots \ldots (491).$$

350. By equation (490) $P = \frac{KE}{L}l$, therefore $U = \tfrac{1}{2}Pl$; whence it follows that the work of elongating the bar is one half that which would have been required to elongate it by the same quantity, if the resistance opposed to its elongation had been, throughout, the same as at its extreme elongation l. If, therefore, the whole strain P corresponding to the elongation l had been put on at once, then, when the elongation l had been attained, *twice* as much work would have been done upon the bar as had been *expended* upon its elasticity. This work would therefore have been *accumulated* in the bar, and in the body producing the strain under which it yields; and if both had been free to move on (as, for instance, when the strain of the bar is produced by a weight suspended freely from its extremity), then would this accumulated work have

been just sufficient yet further to elongate the bar by the same distance $l*$, which whole elongation of $2l$ could not have remained; because the strain upon the bar is only that necessary to keep it elongated by l. The extremity of the bar would therefore, under these circumstances, have oscillated on either side of that point which corresponds to the elongation l.

351. Eliminating l between equations (490) and (491), we obtain

$$U = \tfrac{1}{2}\frac{P^2 L}{KE} \ldots \ldots (492);$$

whence it appears that the work expended upon the elongation of a bar under any strain varies directly as the square of the strain and the length of the bar, and inversely as the area of its section. †

* The mechanical principle involved in this result has numerous applications; one of these is to the effect of a sudden variation of the pressure on a mercurial column. The pressure of such a column varying directly with its elevation or depression, follows the same law as the elasticity of a bar; whence it follows that if any pressure be thrown *at once* or instantaneously upon the surface of the mercury, the variation of the height of the column will be twice that which it would receive from an equal pressure gradually accumulated. Some singular errors appear to have resulted from a neglect of this principle in the discussion of experiments upon the pressure of steam, made with the mercurial column. No such pressure can of course be made to operate in the mathematical sense of the term *instantaneously;* and the term *gradually* has a relative meaning. All that is meant is, that a certain relation must obtain between the rate of the increase of the pressure and the amplitude of the motion, so that when the pressure no longer increases the motion may cease.

† From this formula may be determined the amount of work expended prejudicially upon the elasticity of rods used for transmitting work in machinery, under a reciprocating motion — pump rods, for instance. A *sudden* effort of the pressure transmitted in the nature of an impact may make the loss of work double that represented by the formula; the one limit being the minimum, and the other the maximum, of the possible loss.

THE MODULI OF RESILIENCE AND FRAGILITY.

352. Since $U = \frac{1}{2}E\left(\frac{l}{L}\right)^2 KL$ (equation 491), it is evident that the different amounts of work which must be done upon different bars of the same material to elongate them by equal fractional parts $\left(\frac{l}{L}\right)$, are to one another as the product KL. Let now two such bars be supposed to have sustained that fractional elongation which corresponds to their *elastic limit*; let U_e represent the work which must have been done upon the one to bring it to this elongation, and M_e that upon the other; and let the section of the latter bar be one square inch and its length one foot; then evidently

$$U_e = M_e KL \ . \ . \ . \ . \ . \ (493).$$

M_e is in this case called the *modulus of longitudinal resilience.**

It is evidently a measure of that resistance which the material of the bar opposes to a strain in the nature of an *impact*, tending to elongate it beyond its elastic limits.

If M_f be taken to represent the work which must be similarly done upon a bar one foot long and one square inch in section to produce fracture, it will be a measure of that resistance which the bar opposes to fracture under the like circumstances, and which resistance is opposed to its fragility; it may therefore be distinguished from the last mentioned as the *modulus of fragility*. If U_f represent the work which must be done upon a bar whose section is K square inches and its length L feet to produce fracture; then, as before,

$$U_f = M_f KL \ . \ . \ . \ . \ . \ (494).$$

If P_e and P_f represent respectively the strains which would elongate a bar, whose length is L feet and section K inches,

* The term "modulus of resilience" appears first to have been used by Mr. Tredgold in his work on "the Strength of Cast Iron," Art. 304.

to its elastic limits and to rupture; then, equation (492)

$$U = M_e \cdot KL = \tfrac{1}{2}\frac{P_e^2 L}{KL};$$

$$\therefore \; M_e = \tfrac{1}{2}\frac{P_e^2}{K^2 E}. \quad \text{Similarly } M_f = \tfrac{1}{2}\frac{P_f^2}{K^2 E} \; \ldots \ldots \; (495).$$

These equations serve to determine the values of the moduli M_e and M_f by experiment.*

353. *The elongation of a bar suspended vertically, and sustaining a given strain in the direction of its length, the influence of its own weight being taken into the account.*

Let x represent any length of the bar before its elongation, Δx an element of that length, L the whole length of the bar before elongation, w the weight of each foot of its length, and K its section. Also let the length x have become x_1 when the bar is elongated, under the strain P and its own weight. The length of the bar, below the point whose distance from the point of suspension was x before the elongation, having then been $L - x$, and the weight of that portion of the bar remaining unchanged by its elongation, it is still represented by $(L - x) w$. Now this weight, increased by P, constitutes the strain upon the element Δx; its elongation under this strain is therefore represented (equation 490) by $\frac{P + (L - x) w}{KE} \Delta x$, and the length Δx_1 of the element when thus elongated, by $\Delta x + \frac{P + (L - x) w}{KE} \Delta x$, whence dividing by Δx and passing to the limit, we obtain

$$\frac{dx_1}{dx} = 1 + \frac{P + (L - x) w}{KE} \; \ldots \ldots \; (496).$$

* The experiments required to this determination, in respect to the principal materials of construction, have been made, and are to be found in the published papers of Mr. Hodgkinson and Mr. Barlow. A table of the moduli of resilience and fragility, collected from these valuable data, is a desideratum in practical science.

Integrating between the limits 0 and L, and representing by L_1 the length of the elongated rod,

$$L_1 = \left(1 + \frac{P}{KE}\right) L + \frac{w}{2KE} L^2 * \ldots (497).$$

If the *strain* be converted into a *thrust*, P must be made to assume the negative sign; and if this thrust equal one half the weight of the bar, there will be no elongation at all.

354. THE VERTICAL OSCILLATIONS OF AN ELASTIC ROD OR CORD SUSTAINING A GIVEN WEIGHT SUSPENDED FROM ITS EXTREMITY.

Let A represent the point of suspension of the rod (*fig.* 1. on the next page), L its length AB before its elongation, and $\frac{1}{2}l$ the elongation produced in it by a given weight W suspended from its extremity, and C the corresponding position of the extremity of the rod.

Let the rod be conceived to be elongated through an additional distance CD $=c$ by the application of any other given strain, and then allowed to oscillate freely, carrying with it the weight W; and let P be any position of its extremity during any one of the oscillations which it will thus be made to perform. If, then, CP be represented by x, the corresponding elongation BP of the rod will be represented by $\frac{1}{2}l + x$, and the strain which would retain it permanently at this elongation (equation 490) by $\frac{KE}{L}(\frac{1}{2}l + x)$; the unbalanced pressure or *moving force* (Art. 92) upon the weight W, at the period of this elongation, will therefore be represented by $\frac{KE}{L}(\frac{1}{2}l + x) - W$, or by $\frac{KE}{L}x$; since W, being the strain which would retain the rod at the elongation $\frac{1}{2}l$, is represented by $\frac{KE}{L}\frac{1}{2}l$ (equation 490).

* Whewell's Analytical Statics, p. 113.

The unbalanced pressure, or moving force, upon the mass W varies, therefore, as the distance x of the point P from the given point C; whence it follows by the general principle established in Art. 97., that the oscillations of the point P extend to equal distances on either side of the point C, as a centre, and are performed *isochronously*, the time T of each oscillation being represented by the formula

$$T = \left(\frac{WL}{KE}\right)^{\frac{1}{2}} \pi \ \ldots \ldots (498).$$

The distance from A of the centre C, about which the oscillations of the point P take place, is represented by $L + \frac{1}{2}l$; so that, representing this distance by L_1, and substituting its value for $\frac{1}{2}l$, we have

(1.)   (2.)

$$L_1 = L + \frac{WL}{KE} \ \ldots \ldots (499).$$

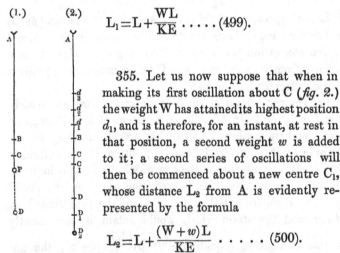

355. Let us now suppose that when in making its first oscillation about C (*fig. 2.*) the weight W has attained its highest position d_1, and is therefore, for an instant, at rest in that position, a second weight w is added to it; a second series of oscillations will then be commenced about a new centre C_1, whose distance L_2 from A is evidently represented by the formula

$$L_2 = L + \frac{(W + w)L}{KE} \ \ldots \ldots (500).$$

So that the distance CC_1 of the two centres is $\frac{wL}{KE}$; and the greatest distance C_1D_1, beneath the centre C_1, attained in the second oscillation, equal to the distance C_1d_1 at which the oscillation commenced above that point. Now $C_1D_1 = C_1d_1 = Cd_1$ $+ CC_1 = CD + CC_1 = c + \frac{wL}{KE}$; the amplitude d_1D_1 of the second oscillation is therefore $2\left(c + \frac{wL}{KE}\right)$.

Let the weight w be conceived to be removed when the lowest point D_1 of the second oscillation is attained, a third series of oscillations will then be commenced, the position of whose centre being determined by equation (499), is identical with that of the centre C, about which the first oscillation was performed. In its third oscillation the extremity of the rod will therefore ascend to a point d_2, as far above the point C as D_1 is below it; so that the amplitude of this third oscillation is represented by $2\overline{CD_1}$, or by $2\overline{C_1D_1 + CC_1}$, or by $2\left(c + \dfrac{2wL}{KE}\right)$. When the highest point d_2 of this third oscillation is attained, let the weight w be again added; a fourth oscillation will then be commenced, the position of whose centre will be determined by equation 500, and will therefore be identical with the centre C_1, about which the second oscillation was performed; so that the greatest distance C_1D_2 beneath that point attained in this fourth oscillation will be equal to C_1d_2, or to $CC_1 + CD_1$; and its amplitude will be represented by $2\left(c + \dfrac{3wL}{KE}\right)$. And if the weight w be thus conceived to be added continually, when the highest point of each oscillation is attained, and taken off at the lowest point, it is evident that the amplitudes of these oscillations will thus continually increase in an arithmetical series; so that the amplitude A_n of the nth oscillation will be represented by the formula

$$A_n = 2\left\{ c + (n-1)\frac{wL}{KE} \right\} \ \ldots \ (501).$$

The ascending oscillations of the series being made about the centre C, and the descending oscillations about C_1, if n be an even number, the centre of the nth oscillation is C_1; the elongation c_n of the rod corresponding to the lowest point of this oscillation is therefore equal to $\overline{BC_1} + \frac{1}{2}A_n$; or substituting for $\overline{BC_1}$ its value given by equation (500), and for A_n its value from equation (501),

$$c_n = c + \frac{(W + nw)L}{KE} \ \ldots \ (502).$$

Thus it is apparent that by the long-continued and periodical addition and subtraction of a weight w, so small as to produce but a slight elongation or contraction of the rod when first added or removed from it, an elongation c_n may eventually be produced, so great as to pass limits of its elasticity, or even to break it. Numerous observations have verified this fact : the chains of suspension bridges have been broken by the measured tread of soldiers * ; and M. Savart has shown, that by fixing an elastic rod at its centre, and drawing the wetted finger along it at measured intervals, it may, by the strain resulting from the slight friction received thus periodically upon its surface, be made with great ease to receive an oscillatory movement of sufficient amplitude to be measured. M. Poncelet has compared the measurement of M. Savart with theoretical deductions analogous to those of the preceding article, and has shown their accordance with it.

DEFLEXION.

356. *The neutral surface of a deflected beam.*

One surface of a beam becoming, when deflected, convex, and the other concave, it is evident that the material forming that side of the beam which is bounded by the one surface is, in the act of flexure, *extended*, and that of the other compressed. The surface which separates these two portions of the material being that where its extension terminates and its compression begins, and which sustains, therefore, neither extension nor compression, is called the NEUTRAL SURFACE.

---

* Such was the fate of the suspension bridge at Broughton near Manchester, the circumstances of which have been ably detailed by Mr. E. Hodgkinson in the fourth volume of the *Manchester Philosophical Transactions*. M. Navier has shown, in his treatise on the theory of suspension bridges (*Sur les Ponts Suspendus*, Paris, 1823), that the duration of the oscillations of the chains of a suspension bridge may in certain cases extend to nearly six seconds ; there might easily, in such cases, arise that isochronism at each interval, or after any number of intervals, between the marching step of the troops and the oscillations of the bridge, whence would result a continually increasing elongation of the suspending chains.

357. THE POSITION OF THE NEUTRAL SURFACE OF A BEAM.

Let ABCD be taken to represent any thin lamina of the
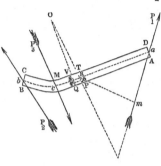 beam contained by planes pa-
rallel to the plane of its de-
flexion, and P_1, P_2, P_3 the
resultants of all the pressures
applied to it; acb that portion
of the neutral surface of the
beam which is contained within
this lamina, and may be called
its neutral line; PT and QV
planes exceedingly near to one
another, and perpendicular to the neutral line at the points
where they intersect it; and O the intersection of PT and
QV when produced.

Now let it be observed that the portion APTD of the
beam is held in equilibrium by the resultant pressure P_1, and
by the elastic forces called into operation upon the surface
PT; of which elastic forces those acting in PR (where the
material of the beam is extended) tend to bring the points to
which they are severally applied nearer to the plane SQ, and
those acting in RT (where the material is compressed), to
carry their several points of application farther from the
plane SV.

Let $aR = x$, $SR = \Delta x$, and imagine the lamina PQVT to
be made up of fibres parallel to SR; then will Δx represent
the length of each of these fibres before the deflexion of the
beam, since the length of the neutral fibre SR has remained
unaltered by the deflexion. Let δx represent the quantity
by which the fibre pq has been elongated by the deflexion of
the beam, then is the actual length of that fibre represented
by $\Delta x + \delta x$. Whence it follows (equation 490), that the
pressure which must have operated to produce this elongation
is represented by $E\dfrac{\delta x}{\Delta x}\Delta k$, Δk being taken to represent the
section of the fibre, or an exceedingly small element of the

K K

section PT of the lamina. Now PT and QV being normals to SR, the point O in which they meet, when produced, is the centre of curvature to the neutral line in R. Let the radius of curvature OR be represented by R, and the distance Rp by ρ. By similar triangles, $\dfrac{Op}{OR}=\dfrac{pq}{SR}$, or $\dfrac{R+\rho}{R}=\dfrac{\Delta x + \delta x}{\Delta x}$,

or $1+\dfrac{\rho}{R}=1+\dfrac{\delta x}{\Delta x}$; therefore, $\dfrac{\rho}{R}=\dfrac{\delta x}{\Delta x}$. Substituting this

value of $\dfrac{\delta x}{\Delta x}$ in the expression for the pressure which must have operated to produce the elongation of the fibre pq, and representing that pressure by ΔP, we have

$$\Delta P = \frac{E\rho}{R}\Delta k \;\;\ldots\; (503).$$

If, therefore, RP be represented by k_1 and RT by k_2, then the sum of the elastic forces developed by the extension of

the fibres in RPQS is represented by $\dfrac{E}{R}\Sigma_0^{k_1}\rho\Delta k$; and, similarly, the sum of those developed by the compression of the

fibres in RTVS is represented by $\dfrac{E}{R}\Sigma_0^{k_2}\rho\Delta k$. Now let it be observed that (since the pressures applied to APTD, and in equilibrium, are the forces of extension and compression acting in RP and RT respectively, and the pressure P_1), if the pressure P_1 be resolved in a direction perpendicular to the plane PT, or parallel to the tangent to the neutral line in R, this resolved pressure will be equal (Art. 16.) to the difference of the sums of the forces of extension and compression applied (in directions perpendicular to that plane, but opposite to one another) to the portions RP and RT of it respectively. Representing, therefore, by θ the inclination ReP$_1$ of the direction of P_1 to the normal to the neutral line in R, we have

$$P_1 \sin.\theta = \frac{E}{R}\Sigma_0^{k_1}\rho\Delta k - \frac{E}{R}\Sigma_0^{k_2}\rho\Delta k.$$

But if k be taken to represent the whole section PT, and h

the distance of the point R from its centre of gravity, then (Art. 18.)

$$kh = \Sigma_0^{\frac{k}{1}}\rho\Delta k - \Sigma_0^{\frac{k}{2}}\rho\Delta k; \quad \therefore \; P_1\sin.\theta = \frac{Ekh}{R};$$

$$\therefore \; h = \frac{RP_1}{Ek}\sin.\theta \; . \; . \; . \; . \; (504);$$

which expression represents the distance of the neutral line from the centre of gravity of any section PT of the lamina, that distance being measured towards the extended or the compressed side of the lamina according as θ is positive or negative; so that the neutral line passes from one side to the other of the line joining the centres of gravity of the cross sections of the lamina, at the point where $\theta = 0$, or at the point where the normal to the neutral line is parallel to the direction of P_1.

358. Case of a rectangular beam.

If the form of the beam be such that it may be divided into laminæ parallel to ABCD of similar forms and equal dimensions, and if the pressure P_1 applied to each lamina may be conceived to be the same ; or if its section be a rectangle, and the pressures applied to it be applied (as they usually are) uniformly across its width, then will the distance h of the neutral line of each lamina from the centre of gravity of any cross section of that lamina, such as PT, be the same, in respect to corresponding points of all the laminæ, whatever may be the deflexion of the beam; so that in this case the neutral surface is always a cylindrical surface.

359. Case in which the deflecting pressure P_1 is nearly perpendicular to the length of the beam.

In this case θ, and therefore $\sin.\theta$, is exceedingly small, so long as the deflexion is small at every point R of the neutral line; so that h is exceedingly small, and the neutral line of the lamina passes very nearly, or accurately, through the centre of gravity of its section PT.

360. The radius of curvature of the neutral surface of a beam.

Since the pressures applied to the portion APTD of the

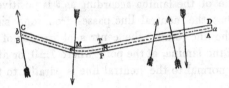

lamina ABCD are in equilibrium, the principle of the equality of moments must obtain in respect to them; taking, therefore, the point R, where the neutral axis of the lamina intersects PT, as the point from which the moments are measured, and observing that the elastic pressures developed by the extension of the material in RP and its compression in RT both tend to turn the mass APTD in the same direction about the point R, and that each such pressure upon an element Δk of the section PT is represented (equation 498) by $\frac{E}{R}\rho\Delta k$, and therefore the moment of that pressure about the point R by $\frac{E}{R}\rho^2\Delta k$, it follows that the sum of the moments about the point R of all these elastic pressures upon PT is represented by $\frac{E}{R}\Sigma\rho^2\Delta k$, or by $\frac{EI}{R}$, if I be taken to represent the moment of inertia of PT about R. Observing, moreover, that if p represent the length of the perpendicular let fall from R upon the direction of any pressure P applied to the portion APTD of the beam, Pp will represent its moment, and ΣPp will represent the sum of the moments of all the similar pressures applied to that portion of the beam;

we have, by the principle of the equality of moments,

$$\frac{EI}{R} = \Sigma P p \;;$$

$$\therefore \; \frac{1}{R} = \frac{\Sigma P p}{EI} \; \ldots \ldots (505).$$

361. The neutral surface of the beam is a cylindrical surface, whatever may be its deflexion or the direction of its deflecting pressure, provided that its section is a rectangle (Art. 355); or whatever may be its section, provided that its deflexion be small, the direction of the deflecting pressure nearly perpendicular to its length, and its form before deflexion symmetrical in respect to a plane perpendicular to the plane of deflexion. In every such case, therefore, the neutral lines of all the laminæ similar to ABCD, into which the beam may be divided, will have equal radii of curvature at points similar to R lying in the same right line perpendicular to the plane of deflexion; taking, therefore, equations similar to the above in respect to all the laminæ, multiplying both sides of each by I, adding them together, and observing that R and E are the same in all, we have $\frac{\Sigma I}{R} = \frac{\Sigma P p}{E}$. In this case, therefore, I may be taken in equation (505) to represent the moment of inertia of the *whole section* of the beam, and P the pressure applied across its whole width.

362. *The radius of curvature of a beam whose deflexion is small, and the direction of the deflecting pressures nearly perpendicular to the length of the beam.*

In this case the neutral line is very nearly a straight line, perpendicular to the directions of the deflecting pressures; so that, representing its length by x, we have, in this case, $p = x$; and equation 505 becomes

$$\frac{1}{R} = \frac{\Sigma P x}{EI} \; \ldots \ldots (506) \;;$$

which relation obtains, whatever may be the form of the

transverse section of the beam, I representing its moment of inertia in respect to an axis passing through its centre of gravity and perpendicular to the plane of deflexion.

363. *The moment of inertia* I *of the transverse section of a beam about the centre of gravity of the section.*

In treating of the moments of inertia of bodies of different geometrical forms in a preceding part of this work (Art. 82, &c.), we have considered them as solids ; whereas the moment of inertia I of the section of a beam which enters into equation 505 and determines the curvature of the beam when deflected, is that of the geometrical *area* of the section. Knowing, however, the moment of inertia of a solid about any axis, whose section perpendicular to that axis is of a given geometrical form, we can evidently determine the moment of the area of that section about the same axis, by supposing the solid in the first place to become an exceedingly thin lamina (*i. e.* by making that dimension of the solid which is parallel to the axis exceedingly small in the expression for the moment of inertia), and then dividing the resulting expression by the exceedingly small thickness of this lamina. We shall thus obtain the following values of I : —

364. For a beam with a *rectangular section,* whose breadth is represented by b and its depth by c (equation 61), $\quad I = \tfrac{1}{12}bc^3.$

365. For a beam with a triangular section, whose base is b and its height c (equation 63), $\quad I = \tfrac{1}{12}bc(\tfrac{1}{4}b^2 + \tfrac{1}{3}c^2).$

366. For a beam or column with a circular section, whose radius is c (equation 66), $\quad I = \tfrac{1}{4}\pi c^4.$

367. To determine the moment of inertia I in respect to a beam whose transverse section is 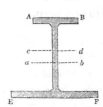 of the form represented in the accompanying figure, about an axis ab passing through its centre of gravity; let the breadth of the rectangle AB be represented by b_1 and its depth by d_1, and let b_2 and d_2 be similarly taken in respect to the rectangle EF, and b_3 and d_3 in respect to CD; also let I_1 represent the moment of inertia of the section about the axis cd passing through the centre of CD, A_1, A_2, A_3, the areas of the rectangles respectively, and A the area of the whole section.

Now the moments of inertia of the several rectangles, about axes passing through their centres of gravity, are represented by $\frac{1}{12}b_1d_1^3$, $\frac{1}{12}b_2d_2^3$, $\frac{1}{12}b_3d_3^3$, and the distances of these axes from the axis cd are respectively $\frac{1}{2}(d_1+d_3)$, $\frac{1}{2}(d_2+d_3)$, 0. Therefore (equation 58),

$$I_1 = \tfrac{1}{12}b_1d_1^3 + \tfrac{1}{4}(d_1+d_3)^2A_1 + \tfrac{1}{12}b_2d_2^3 + \tfrac{1}{4}(d_2+d_3)^2A_2 + \tfrac{1}{12}b_3d_3^3 ;$$

but $A_1 = b_1d_1$, $A_2 = b_2d_2$, $A_3 = b_3d_3$;

$$\therefore I_1 = \tfrac{1}{12}(A_1d_1^2 + A_2d_2^2 + A_3d_3^2) + \tfrac{1}{4}(d_1+d_3)^2A_1 + \tfrac{1}{4}(d_2+d_3)^2A_2.$$

Also if h represent the distance between the axes ab and cd, then (Art. 18.) $hA = \frac{1}{2}(d_2+d_3)A_3 - \frac{1}{2}(d_1+d_3)A_1$, and (equation 58) $I = I_1 - h^2A$.

$$\therefore I = \tfrac{1}{12}(A_1d_1^2 + A_2d_2^2 + A_3d_3^2) + \tfrac{1}{4}\{(d_1+d_3)^2A_1 + (d_2+d_3)^2A_2\} - \tfrac{1}{4}\frac{\{(d_2+d_3)A_2 - (d_1+d_3)A_1\}}{A} \quad ..(507).$$

If d_1 and d_2 be exceedingly small as compared with d_3, neglecting their values in the two last terms of the equation and reducing, we obtain

$$I = \tfrac{1}{12}(A_1d_1^2 + A_2d_2^2 + A_3d_3^2) + \tfrac{1}{4}\left(\frac{4A_1A_2 + A_1A_3 + A_2A_3}{A}\right)d_3^2 \dots (508).$$

If the areas AB and EF be equal in every respect,

$$I = \tfrac{1}{6}\{d_1^2 + 3(d_1+d_3)^2\}A_1 + \tfrac{1}{12}A_3d_3^2 \dots \dots (509).$$

K K 4

368. THE WORK EXPENDED UPON THE DEFLEXION OF A
BEAM TO WHICH GIVEN PRESSURES ARE APPLIED.

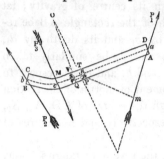

If ΔP represent the pressure which must have operated to
produce the elongation or com-
pression which the elementary
fibre pq receives, by reason of
the deflexion of the beam, Δx
the length of that fibre before
the deflexion of the beam, and
Δk its section ; then the work
which must have been done
upon it, thus to elongate or
compress it, is represented
(equation 492) by $\frac{1}{2}\frac{(\Delta P)^2 . \Delta x}{E . \Delta k}$. But (equation 503) $\Delta P =$
$\frac{E\rho}{R}\Delta k$. The work expended upon the extension or com-
pression of pq is therefore represented by

$$\frac{1}{2}\frac{E . \Delta x}{R^2}(\rho^2\Delta k).$$

And the same being true of the work expended on the compression or extension of every other fibre composing the
elementary solid VTPQ, it follows that the whole work expended upon the deflexion of that element of the beam is
represented by $\frac{1}{2}\frac{E\Delta x}{R^2}\Sigma\rho^2\Delta k$, or by $\frac{1}{2}\frac{EI}{R^2}\Delta x$; for $\Sigma\rho^2\Delta k$ represents the moment of inertia I of the section PT, about an
axis perpendicular to the plane of ABCD, and passing through
the point R. If, therefore, a_1 be taken to represent the
length of that portion of the beam which lies between D and
M before its deflexion, and therefore the length of the portion ac of its neutral line after deflexion, then *the whole work
expended upon the deflexion of the part* AM *of the beam is
represented by* $\frac{1}{2}E\Sigma_0^{a_1}\frac{I}{R^2}\Delta x$. But (equation 505) $\frac{I}{R^2} =$

$\dfrac{P_1{}^2 p_1{}^2}{E^2 I^2}$; whence, by substitution, the above expression becomes

$\frac{1}{2}\dfrac{P_1{}^2}{E}\sum_0^{a_1}\dfrac{p_1{}^2}{I}\Delta x$. Passing to the limit, and representing the work expended upon the deflexion of the part AM of the beam by u_1,

$$u_1 = \frac{P_1{}^2}{2E}\int_0^{a_1}\frac{p_1{}^2}{I}dx \ \ .\ .\ .\ .\ (510).$$

369. *The work expended upon the deflexion of a beam of uniform dimensions, when the deflecting pressures are nearly perpendicular to the surface of the beam.*

In this case I is constant, and $p_1 = x$; whence we obtain by integrating (equation 510) between the limits 0 and a_1,

$$u_1 = \tfrac{1}{6}\frac{P_1{}^2 a_1{}^3}{EI} \ \ .\ .\ .\ .\ (511).$$

where u_1 represents the work expended upon the deflexion of the portion AM of the beam. Similarly, if $bc = a_2$, the work expended upon the deflexion of the portion BM of the beam is represented by

$$u_2 = \tfrac{1}{6}\frac{P_2{}^2 a_2{}^3}{EI};$$

so that the whole work U_3 expended upon the deflexion of the beam is represented by

$$U_3 = \frac{P_1{}^2 a_1{}^3 + P_2{}^2 a_2{}^3}{6EI}.$$

But by the principle of the equality of moments, if a represent the whole length of the beam,

$$P_1 a = P_3 a_2, \quad P_2 a = P_3 a_1.$$

Eliminating P_1 and P_2 between these equations and the preceding, we obtain by reduction

$$U_3 = \tfrac{1}{6} \frac{(a_1 a_2)^2 P_3{}^2}{aEI} \ \ldots \ (512).$$

If the pressure P_3 be applied in the centre of the beam, $a_1 = a_2 = \tfrac{1}{2}a;$

$$\therefore \ U_3 = \frac{a^3 P_3{}^2}{96EI} \ \ldots \ (513).$$

370. THE LINEAR DEFLEXION OF A BEAM WHEN THE DI-
RECTION OF THE DEFLECTING PRESSURE IS PERPENDICU-
LAR TO ITS SURFACE.

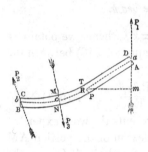

Let the section MN remain fixed, the deflexion taking place on either side of that section; then u_1 representing the work expended upon the deflexion of the portion AM of the beam, and D_1 the deflexion of the point to which P_1 is applied, measured in a direction perpendicular to the surface, we have (equation 40),

$$u_1 = \int P_1 dD_1 \ ; \ \text{therefore} \ P_1 = \frac{du_1}{dD_1}$$

$= \dfrac{du_1}{dP_1} \cdot \dfrac{dP_1{}^*}{dD_1}.$ But by equation (511), $\dfrac{du_1}{dP_1} = \tfrac{1}{3}\dfrac{P_1 a_1{}^3}{EI}\ ;$

therefore $P_1 = \tfrac{1}{3}\dfrac{P_1 a_1{}^3}{EI} \cdot \dfrac{dP_1}{dD_1}\ ;$ therefore $\dfrac{dD_1}{dP_1} = \tfrac{1}{3}\dfrac{a_1{}^3}{EI}\ ;$ whence

we obtain by integration

$$D_1 = \frac{a_1{}^3 P_1}{3EI} \ \ldots \ (514).$$

If the whole work of deflecting the beam be done by the pressure P_3, the points of application of P_1 and P_2 having no motions in the directions of these pressures (Art. 52.), then proceeding in respect to equation (512) precisely as before in respect to equation (511), and representing the deflexion

* Hall's Diff. Cal. p. 30.

perpendicular to the surface of the beam at the point of application of P_3 by D_3, we shall obtain

$$D_3 = \frac{(a_1 a_2)^2 P_3}{3EIa} ^* \quad \dots \quad (515).$$

If the pressure P_3 be applied at the centre of the beam $a_1 = a_2 = \frac{1}{2}a$,

$$\therefore D_3 = \frac{a^3 P_3}{48EI} \quad \dots \quad (516).$$

Eliminating P_1 between equations (511) and (514), and P_3 between equations (512) and (515), we obtain

$$u_1 = \frac{3EID_1^2}{2a_1^3}, \quad U_3 = \frac{3aEID_3^2}{2(a_1 a_2)^2} \quad \dots \quad (517);$$

by which equations the work expended upon the deflexion of a beam is determined in terms of the *deflexion* itself, as by equations (511) and (512) it was determined in terms of the deflecting *pressures*.

371. CONDITIONS OF THE DEFLEXION OF A BEAM TO WHICH ARE APPLIED THREE PRESSURES, WHOSE DIRECTIONS ARE NEARLY PERPENDICULAR TO ITS SURFACE.

Let AB represent any lamina of the beam parallel to its

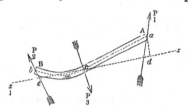

plane of deflexion, and acb the neutral line of that lamina intersected by the direction of P_3 in the point c.

Draw xx_1 parallel to the length of the beam before its deflexion, and take this line as the axis of the abscissæ, and the point c as the origin; then, representing by x and y the co-

---

* This result is identical with that obtained by a different method of investigation by M. Navier (*Resumè de Leçons de Construction*, Art. 359.).

ordinates of any point in *ac;* and by R the radius of curvature of that point, we have *

$$\frac{1}{R}=\frac{d^2y}{dx^2}\left(1+\frac{dy^2}{dx^2}\right)^{-\frac{3}{2}}.$$

Now the deflexion of the beam being supposed exceedingly small, the inclination to *cx* of the tangent to the neutral line is, at all points, exceedingly small, so that $\left(\frac{dy}{dx}\right)^2$ may be neglected as compared with unity; therefore $\frac{1}{R}=\frac{d^2y}{dx^2}$. Substituting this value in equation (506), and observing that in this case p is represented by (a_1-x) instead of x,

$$\frac{d^2y}{dx^2}=\frac{P_1(a_1-x)}{EI}\ \ .\ \ .\ \ .\ \ .\ \ (518);$$

the direction of the pressure P_1 being supposed nearly perpendicular to the surface of the beam, and I constant. Let the above equation be integrated between the limits 0 and x, β being taken to represent the inclination of the tangent at c to cx, so that the value of $\frac{dy}{dx}$ at c may be represented by tan. β,

$$\frac{dy}{dx}-\tan.\ \beta=\frac{P_1}{EI}\{a_1x-\tfrac{1}{2}x^2\}\ \ .\ \ .\ \ .\ \ .\ \ (519).$$

Integrating a second time between the limits 0 and x, and observing that when $x=0$, $y=0$,

$$y=\frac{P_1}{EI}\{\tfrac{1}{2}a_1x^2-\tfrac{1}{6}x^3\}+x\tan.\ \beta\ \ .\ \ .\ \ .\ \ .\ \ (520).$$

Proceeding similarly in respect to the portion *bc* of the neutral line, but observing that in respect to this curve the value of $-\frac{dy}{dx}$ at the point c is represented by tan. β, we have

$$\frac{d^2y}{dx^2}=\frac{P_2(a_2-x)}{EI},$$

* Hall's Diff. Cal., Art. 136.

$$\frac{dy}{dx} + \tan. \beta = \frac{P_2}{EI}\{a_2 x - \tfrac{1}{2}x^2\} \ . \ . \ . \ . \ (521).$$

$$y = \frac{P_2}{EI}\{\tfrac{1}{2}a_2 x^2 - \tfrac{1}{6}x^3\} - x \tan. \beta \ . \ . \ . \ . \ (522).$$

If D_1 and D_2 be taken to represent the deflexions at the points a and b, and ca and cb be assumed respectively equal to cd and ce,

by equation (520), $D_1 = \dfrac{P_1 a_1^3}{3EI} + a_1 \tan. \beta$,

by equation (522), $D_2 = \dfrac{P_2 a_2^3}{3EI} - a_2 \tan. \beta$.

If the pressures P_1 and P_2 be supplied by the resistances of fixed surfaces, then $D_1 = D_2$. Subtracting the above equation we obtain, on this supposition,

$$0 = \frac{P_1 a_1^3 - P_2 a_2^3}{3EI} + (a_1 + a_2) \tan. \beta.$$

Now $P_1 a_1^3 - P_2 a_2^3 = \dfrac{P_3 a_2 a_1^3 - P_3 a_1 a_2^3}{a} = P_3 a_1 a_2 (a_1 - a_2)$;

observing that $P_1 a = P_3 a_2$, $P_2 a = P_3 a_1$, and $a_1 + a_2 = a$,

$$\therefore \ \tan. \beta = \frac{P_3 a_1 a_2 (a_2 - a_1)}{3EIa} \ . \ . \ . \ . \ (523).$$

If β_1, β_2 represent the inclinations of the neutral line to xx_1 at the points a and b, then by equations (519) and (521)

$$\tan. \beta_1 - \tan. \beta = \frac{P_1 a_1^2}{2EI}, \ \tan. \beta_2 + \tan. \beta = \frac{P_2 a_2^2}{2EI}.$$

Substituting for $\tan. \beta$ its value from equation (523), eliminating and reducing,

$$\tan. \beta_1 = \frac{P_3 a_1 a_2 (a_1 + 2a_2)}{6EIa}, \ \tan. \beta = \frac{P_3 a_1 a_2 (a_2 + 2a_1)}{6EIa} \ . \ . \ . \ (524).$$

To determine the point m where the tangent to the neutral line is parallel to cxx_1, or to the undeflected position of the beam, we must assume $\dfrac{dy}{dx} = 0$ in equatio n(521); if we then substitute for $\tan. \beta$ its value from equation (523), substitute for P_2 its value in terms of P_3, and solve the resulting equation in respect to x, we shall obtain for the distance of the point m from c the expression

$$a_2 + \sqrt{\tfrac{1}{3}a_2(a_2 + 2a_1)} \quad \ldots \quad (525).$$

372. The length of the neutral line, the beam being loaded in the centre.

Let the directions of the resistances upon the extremities

of the beam be supposed nearly perpendicular to its surface; then if x and y be the co-ordinates of the neutral line from the point a, we have (equation 506), representing the horizontal distance AB by $2a$, and observing that in this case $\dfrac{1}{R} = -\dfrac{d^2y}{dx^2}$, and that the resistance at A or B $= \tfrac{1}{2}$P,

$$-\mathrm{EI}\frac{d^2y}{dx^2} = \tfrac{1}{2}\mathrm{P}x.$$

Integrating between the limits x and a, and observing that at the latter limit $\dfrac{dy}{dx} = 0$,

$$\mathrm{EI}\frac{dy}{dx} = \tfrac{1}{4}\mathrm{P}(a^2 - x^2).$$

Now if s represent the length of the curve ac,

$$s = \int_0^a \left(1 + \frac{dy^2}{dx^2}\right)^{\tfrac{1}{2}} dx = \int_0^a \left\{1 + \tfrac{1}{2}\left(\frac{dy}{dx}\right)^2\right\} dx \quad \text{nearly; since}$$

the deflexion being small, $\dfrac{dy}{dx}$ is exceedingly small at every point of the neutral line.

$$\therefore s = \int_0^a \left\{1 + \frac{\mathrm{P}_2}{32\mathrm{E}^2\mathrm{I}^2}(a^4 - 2a^2x^2 + x^4)\right\} dx;$$

$$\therefore \quad s = a + \frac{P^2 a^5}{60 E^2 I^2} \quad \cdots \cdots (526).$$

Eliminating P between this equation and equation (516), and representing the deflexion by D,

$$s = a + \frac{3}{5}\frac{D^2}{a} \ast .$$

373. A BEAM, ONE PORTION OF WHICH IS FIRMLY INSERTED IN MASONRY, AND WHICH SUSTAINS A LOAD UNIFORMLY DISTRIBUTED OVER ITS REMAINING PORTION.

Let the co-ordinates of the neutral line be measured from the point B where the beam is inserted in the masonry, and let the length of the portion AD which sustains the load be represented by a, and the load upon each unit of its length by μ; then, representing by x and y the co-ordinates of any point P of the neutral line,

* The following experiments were made by Mr. Hatcher, superintendant of the work-shop at King's College, to verify this result, which is identical with that obtained by M. Navier (*Resumé de Leçons*, Art. 86.). Wrought iron rollers ·7 inch in diameter were placed loosely on wrought iron bars, the surfaces of contact being smoothed with the file and well oiled. The bar to be tested had a square section, whose side was ·7 inch, and was supported on the two rollers, which were adjusted to 10 feet apart (centre to centre) when the deflecting weight had been put on the bar. On removing the weights carefully, the distance to which the rollers receded as the bar recovered its horizontal position was noted.

Deflecting Weight in lbs.	Deflection in Inches.	Distance through which each Roller receded in Inches.	Distance through which each Roller should have receded by Formula.
56	3·7	·1	·13
84	5·45	·2	·29

and observing that the pressures applied to AP, and in equilibrium, are the load $\mu(a-x)$ and the elastic forces developed upon the transverse section at P, we have by the principle of the equality of moments, taking P as the point from which the moments are measured, and observing that since the load μx is uniformly distributed over AP it produces the same effect as though it were collected over the centre of that line, or at distance $\frac{1}{2}(a-x)$ from P; observing, moreover, that the sum of the moments of the elastic forces upon the section at P, about that point, is represented (Art. 360.) by $\frac{EI}{R}$, or by $EI\frac{d^2y}{dx^2}$ (Art. 371.);

$$EI\frac{d^2y}{dx^2}=\tfrac{1}{2}\mu(a-x)^2 \ \ldots \ (527).$$

Integrating twice between the limits 0 and x, and observing that when $x=0$, $\frac{dy}{dx}=0$ and $y=0$, since the portion BC of the beam is rigid, we obtain

$$EI\frac{dy}{dx}=-\tfrac{1}{6}\mu(a-x)^3+\tfrac{1}{6}\mu a^3 \ \ldots \ (528),$$

$$EIy=\tfrac{1}{24}\mu(a-x)^4+\tfrac{1}{6}\mu a^3x-\tfrac{1}{24}\mu a^4 \ \ldots \ (529),$$

which is the equation to the neutral line.

Let, now, a be substituted for x in the above equation; and let it be observed that the corresponding value of y represents the deflexion D at the extremity A of the beam; we shall thus obtain by reduction

$$D=\frac{\mu a^4}{8EI} \ \ldots \ (530).$$

Representing by β the inclination to the horizon of the tangent to the neutral line at A, substituting a for x in equation (528), and observing that when $x=a$, $\frac{dy}{dx}=\tan.\beta$, we obtain

$$\tan.\beta=\frac{\mu a^3}{6EI} \ \ldots \ (531).$$

374. A BEAM SUPPORTED AT ITS EXTREMITIES, AND SUSTAIN-
ING A LOAD UNIFORMLY DISTRIBUTED OVER ITS LENGTH.

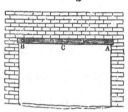

Let the length of the beam be represented by $2a$, the load
upon each unit of length by μ; take
x and y as the co-ordinates of any
point P of the neutral line, from the
origin A; and let it be observed that
the forces applied to AP, and in equilibrium, are the load μx upon that
portion of the beam, which may be
supposed collected over its middle point, the resistance upon
the point A, which is represented by μa, and the elastic forces
developed upon the section at P; then by Art. 362.,

$$\text{EI}\frac{d^2y}{dx^2}=\tfrac{1}{2}\mu x^2-\mu ax \ \ldots \ (532).$$

Integrating this equation between the limits x and a, and
observing that at the latter limit $\frac{dy}{dx}=0$, since y evidently
attains its maximum value at the middle C of the beam,

$$\text{EI}\frac{dy}{dx}=\tfrac{1}{6}\mu(x^3-a^3)-\tfrac{1}{2}\mu a(x^2-a^2) \ \ldots \ (533).$$

Integrating a second time between the limits 0 and x, and
observing that when $x=0$, $y=0$,

$$\text{EI}y=\tfrac{1}{6}\mu(\tfrac{1}{4}x^4-a^3x)-\tfrac{1}{2}\mu a(\tfrac{1}{3}x^3-a^2x) \ \ldots \ (534),$$

which is the equation to the neutral line. Substituting a for
x in this equation, and observing that the corresponding value
of y represents the deflexion D in the centre of the beam, we
have by reduction

$$\text{D}=\frac{5\mu a^4}{24\text{EI}} \ \ldots \ (535).$$

Representing by β the inclination to the horizon of the tangent to the neutral line at A or B, and observing that when

$x=0$ in equation (533), $\dfrac{dy}{dx}=\tan.\,\beta$,

$$\tan.\beta=\frac{\mu a^3}{3EI}\;\cdots\;(536).$$

Let it be observed that the length of the beam, which in equation (516) is represented by a, is here represented by $2a$, and that equation (535) may be placed under the form $D=\frac{5}{8}\cdot\dfrac{(2a\mu)\,(2a)^3}{48EI}$; whence it is apparent that the deflexion of a beam, when uniformly loaded throughout, is the same as though $\frac{5}{8}$ths of that load $(2a\mu)$ were suspended from its middle point.

375. A BEAM IS SUPPORTED BY TWO STRUTS PLACED SYM-
METRICALLY, AND IT IS LOADED UNIFORMLY THROUGHOUT
ITS WHOLE LENGTH; TO DETERMINE ITS DEFLEXION.

Let $CD=2a$, $CA=a_1$, load upon each foot of the length of the beam $=\mu$; then load on each point of support $=\mu a$.

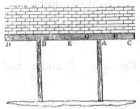

Take C as the origin of the co-ordinates; then, observing that the forces impressed upon any portion CP of the beam, terminating between C and A, are the elastic forces upon the transverse section of the beam at P, and the weight of the load upon CP; and observing that the weight $\mu\overline{CP}$ of the load upon CP, produces the same effect as though it were collected over the centre of that portion of the beam, so that its moment about the point P is represented by $\mu\,.\,\overline{CP}\,.\,\frac{1}{2}CP$, or by $\frac{1}{2}\mu\overline{CP}^2$; we obtain for the equation to the neutral line in respect to the part CA of the beam (Art. 362.)

$$EI\frac{d^2y}{dx^2}=\tfrac{1}{2}\mu.x^2\;\cdots\;(537).$$

Since, moreover, the forces impressed upon any portion CQ of the beam, terminating between A and E, are the elastic forces developed upon the transverse section at Q, the resistance μa of the support at A, and the load upon CQ, whose moment about Q is represented by $\frac{1}{2}\mu\overline{CQ^2}$, we have (equation 506), representing CQ by x,

$$\mathrm{EI}\frac{d^2y}{dx^2}=\tfrac{1}{2}\mu x^2-\mu a(x-a_1) \ldots (538).$$

Representing the inclination to the horizon of the tangent to the neutral line at A by β, dividing equation (537) by μ, integrating it between the limits x and a_1, and observing that at the latter limit $\frac{dy}{dx}=\tan.\beta$, we have, in respect to the portion CA of the beam,

$$\frac{\mathrm{EI}}{\mu}\left(\frac{dy}{dx}-\tan.\beta\right)=\tfrac{1}{6}x^3-\tfrac{1}{6}a_1^3 \ldots (539).$$

Integrating equation (538) between the limits x and a, and observing that at the latter limit $\frac{dy}{dx}=0$, since the neutral line at E is parallel to the horizon,

$$\frac{\mathrm{EI}}{\mu}\frac{dy}{dx}=\tfrac{1}{6}x^3-\tfrac{1}{2}a(x-a_1)^2-\tfrac{1}{6}a^3+\tfrac{1}{2}a(a-a_1)^2 \ldots (540);$$

which equation having reference to the portion AE of the beam, it is evident that when $x=a_1$, $\frac{dy}{dx}=\tan.\beta$.

$$\therefore \frac{\mathrm{EI}}{\mu}\tan.\beta=\tfrac{1}{2}a(a-a_1)^2-\tfrac{1}{6}(a^3-a_1^3)=\tfrac{1}{6}(a-a_1)(2a^2-4aa_1-a_1^2) \ldots (541).$$

Substituting, therefore, for $\tan.\beta$ in equation (539), and reducing, that equation becomes

$$\frac{\mathrm{EI}}{\mu}\frac{dy}{dx}=\tfrac{1}{6}x^3+\tfrac{1}{2}a(a-a_1)^2-\tfrac{1}{6}a^3 \ldots (542).$$

Integrating equation (540) between the limits a_1 and x, and equation (542) between the limits 0 and x, and representing the deflexion at C, and therefore the value of y at A, by D_1,

$$\frac{EI}{\mu}(y-D_1)=\tfrac{1}{24}x^4-\tfrac{1}{6}a(x-a_1)^3-\{\tfrac{1}{6}a^3-\tfrac{1}{2}a(a-a_1)^2\}x-\tfrac{1}{24}a_1{}^4+\tfrac{1}{6}a^3a_1-\tfrac{1}{2}aa_1(a-a)^2$$

$$\frac{EI}{\mu}y=\tfrac{1}{24}x^4+\{\tfrac{1}{2}a(a-a_1)^2-\tfrac{1}{6}a^3\}x \ \ \ldots \ \ (543);$$

the former of which equations determines the neutral line of the portion AE, and the latter that of the portion CA of the beam. Substituting a_1 for x in the latter, and observing that y then becomes D_1; then substituting this value of D_1 in the former equation, and reducing,

$$D_1=\frac{\mu a_1}{24EI}\{12a(a-a_1)^2-(4a^3-a_1{}^3)\} \ \ \ldots \ \ (544);$$

$$\frac{EI}{\mu}y=\tfrac{1}{24}x^4-\tfrac{1}{6}a(x-a_1)^3+\tfrac{1}{6}a\{3(a-a_1)^2-a^2\}x \ \ \ldots \ \ (545).$$

The latter equation being that to the neutral line of the portion AE of the beam, if we substitute a in it for x, and represent the ordinate of the neutral line at E by y_1, we shall obtain by reduction

$$y_1=\frac{\mu a}{24EI}\{4(a_1+2a)(a-a_1)^2-3a^3\} \ \ \ldots \ \ (546).$$

If $a_1=0$, or if the loading commence at the point A of the beam, the *value* of y_1 will be found to be that already determined for the deflexion in this case (equation 535).

Now, representing the deflexion at E by D_2, we have evidently $D_2=D_1-y_1$.

$$D_2=\frac{\mu(a-a_1)^2}{24EI}\{-5a^2+10aa_1+a_1{}^2\} \ \ \ldots \ \ (547).$$

376. The conditions of the deflexion of a beam loaded uniformly throughout its length, and supported at its extremities A and D, and at two points B and C situated at equal distances from them, and in the same horizontal straight line.

Let $AB=a_1$, $AD=2a$.

Let A be taken as the origin of the co-ordinates; let the

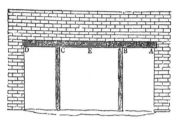

pressure upon that point be represented by P_1, and the pressure upon B by P_2; also the load upon each unit of the length of the beam by μ.

If P be any point in the neutral line to the portion AB of the beam, whose co-ordinates are x and y, the pressures applied to AP, and in equilibrium, are the pressure P_1 at A, the load μx supported by AP, and producing the same effect as though it were collected over the centre of that portion of the beam, and the elastic forces developed upon the transverse section of the beam at P; whence it follows (Art. 362.) by the principle of the equality of moments, taking P as the point from which the moments are measured, that

$$\text{EI}\frac{d^2y}{dx^2} = \tfrac{1}{2}\mu x^2 - P_1 x \ \ \ldots \ (548).$$

Integrating this equation between the limits a_1 and x, and representing the inclination to the horizon of the tangent to the neutral line at B by β_2,

$$\text{EI}\left(\frac{dy}{dx} - \tan.\beta_2\right) = \tfrac{1}{6}\mu(x^3 - a_1{}^3) - \tfrac{1}{2}P_1(x^2 - a_1{}^2) \ \ldots \ (549).$$

Integrating again between the limits 0 and x,

$$\text{EI}(y - x\tan.\beta_2) = \tfrac{1}{6}\mu(\tfrac{1}{4}x^4 - a_1{}^3x) - \tfrac{1}{2}P_1(\tfrac{1}{3}x^3 - a_1{}^2x) \ \ldots \ (550).$$

Whence observing that when $x = a_1$, $y = 0$,

$$\text{EI}\tan.\beta_2 = \tfrac{1}{8}\mu a_1{}^3 - \tfrac{1}{3}P_1 a_1{}^2 \ \ \ldots \ (551).$$

Similarly observing, that if x and y be taken to represent the co-ordinates of a point Q in the beam between B and C, the pressures applied to AQ are the elastic forces upon the section at Q, the pressures P_1 and P_2 and the load μx, we have

$$\text{EI}\frac{d^2y}{dx^2} = \tfrac{1}{2}\mu x^2 - P_1 x - P_2(x - a_1) \ \ \ldots \ (552).$$

Integrating this equation between the limits a_1 and x, and observing that at the former limit the value of $\frac{dy}{dx}$ is represented by $\tan.\beta_2$, we have

$$\mathrm{EI}\left(\frac{dy}{dx} - \tan.\beta_2\right) = \tfrac{1}{6}\mu(x^3 - a_1^3) - \tfrac{1}{2}P_1(x^2 - a_1^2) - \tfrac{1}{2}P_2(x - a_1)^2 \; .. \; (553).$$

Now it is evident that, since the props B and C are placed symmetrically, the lowest point of the beam, and therefore of the neutral line, is in the middle, between B and C; so that $\frac{dy}{dx} = 0$, when $x = a$. Making this substitution in equation (553),

$$\mathrm{EI}\tan.\beta_2 = \tfrac{1}{6}\mu(a^3 - a_1^3) - \tfrac{1}{2}P_1(a^2 - a_1^2) - \tfrac{1}{2}P_2(a - a_1)^2 \; ... \; (554).$$

Since, moreover, the resistances at C and D are equal to those at A and B, and that the whole load upon the beam is sustained by these four resistances, we have

$$P_1 + P_2 = \mu a \; .... \; (555).$$

Assuming $a_1 = na$, and eliminating P_1, P_2, $\tan.\beta_2$, between the equations (551), (554), and (555), we obtain

$$P_1 = \frac{\mu a}{8n}\left\{\frac{n^3 + 12n^2 - 24n + 8}{2n - 3}\right\} \; .... \; (556);$$

$$P_2 = \frac{\mu a}{8n} \cdot \left\{\frac{4n^2 - n^3 - 8}{2n - 3}\right\} = \frac{\mu a(n-2)}{8n}\left\{\frac{-n^2 + 2n + 4}{2n - 3}\right\} \; ..... \; (557);$$

$$\tan.\beta_2 = \frac{\mu a^3 n}{24\mathrm{EI}} \cdot \left\{\frac{5n^3 - 21n^2 + 24n - 8}{2n - 3}\right\} = \frac{\mu a^3 n(n-1)}{24\mathrm{EI}} \cdot \left\{\frac{5n^2 - 16n + 8}{2n - 3}\right\} \; ... \; (558).$$

Making $x = 0$ in equation (549); and observing that the corresponding value of $\frac{dy}{dx}$ is represented by $\tan.\beta_1$, we have

$$\mathrm{EI}(\tan.\beta_1 - \tan.\beta_2) = -\tfrac{1}{6}\mu a_1^3 + \tfrac{1}{2}P_1 a_1^2.$$

Substituting for $\tan.\beta_2$ and P_1' their values from equations (558) and (556), and reducing,

$$\tan.\beta_1 = \frac{\mu a^3 n}{48\mathrm{EI}}\left\{\frac{-3n^3 + 18n^2 - 24n + 8}{2n - 3}\right\} \; .... \; (559).$$

Representing the greatest deflexions of the portions AB and BC of the beam, respectively, by D_1 and D_2, and by x_1 the distance from A at which the deflexion D_1 is attained, we have, by equations (549) and (550),

$$\left.\begin{array}{l} -\text{EI}\tan.\beta_2 = \tfrac{1}{6}\mu(x_1{}^3 - a_1{}^3) - \tfrac{1}{2}\text{P}_1(x_1{}^2 - a_1{}^2) \\ \text{EI}(D_1 - x_1\tan.\beta_2) = \tfrac{1}{6}\mu(\tfrac{1}{4}x^4{}_1 - a_1{}^3x_1) - \tfrac{1}{2}\text{P}_1(\tfrac{1}{3}x_1{}^3 - a_1{}^2x_1) \end{array}\right\} \cdots \cdots (560).$$

The value of D_1 is determined by eliminating x_1 between these equations, and substituting the values of P_1 and $\tan.\beta_2$ from equations (556) and (558).

Integrating equation (553) between the limits a_1 and a, and observing that at the latter limit $y = D_2$, we have

$$\text{EID}_2 = \text{EI}(a-a_1)\tan.\beta_2 + \tfrac{1}{6}\mu\left\{\tfrac{1}{4}(a^4 - a_1{}^4) - a_1{}^3(a-a_1)\right\} - \tfrac{1}{2}\text{P}_1\left\{\tfrac{1}{3}(a^3 - a_1{}^3) - a_1{}^2(a-a_1)\right\} - \tfrac{1}{6}\text{P}_2(a-a_1)^3.$$

Substituting in this equation for the values of $\tan.\beta_2$, P_1, P_2, and reducing, we obtain

$$D_2 = \frac{\mu a^4(1-n)^2}{48\text{EI}(3-2n)}\{n^3 - 2n^2 - 8n + 6\} \cdots \cdots (561).$$

Representing BC by $2a_2$, and observing that $a_2 = \text{AE} - \text{BE} = a - na = (1-n)a$,

$$D_2 = \frac{\mu a_2{}^4}{48\text{EI}} \cdot \frac{n^3 - 2n^2 - 8n + 6}{(3-2n)(1-n)^2} \cdots \cdots (562).$$

377. A BEAM, HAVING A UNIFORM LOAD, SUPPORTED AT EACH EXTREMITY, AND BY A SINGLE STRUT IN THE MIDDLE.

If in the preceding article a_1 be assumed equal to a, or $n = 1$, the two props B and C will coincide in the centre; and the pressure P_2 upon the single prop, resulting from their coincidence, will be represented by twice the corresponding value of P_2 in equation (557); we thus obtain

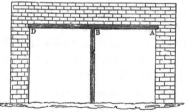

$$P_2 = \tfrac{5}{8}\mu a, \; P_1 = \tfrac{3}{8}\mu a ;$$

$$\left. \tan. \beta_1 = \frac{\mu a^3}{48EI}, \;\; \tan.\beta_2 = 0. \right\} \; \cdots \; (563).$$

The distance x_1 of the point of greatest deflexion of either portion of the beam from its extremities A or D, and the amount D_1 of that greatest deflexion, are determined from equations (560). Making tan. $\beta_2 = 0$ in those equations, substituting for P_1 its value, solving the former in respect to x_1, and the latter in respect to D_1, we obtain

$$x_1 = \frac{1 + \sqrt{33}}{16} a = \cdot 421535 a \; \ldots \; (564).$$

$$D_1 = \frac{\mu x_1 (a - x_1)^2 (2x_1 + a)}{48EI} = \frac{\cdot 25997 \mu a^3}{48EI} \; \cdots \; (565).$$

378. A beam which sustains a uniform load through-out its whole length, and whose extremities are so firmly imbedded in a solid mass of masonry as to become rigid.

Let the ratio of the lengths of the two portions AB and AE of a beam, supported by two props (p. 517.), be assumed to be such as will satisfy the condition $5n^2 - 16n + 8 = 0$; or, solving this equation, let

$$n = \tfrac{2}{5}(4 \pm \sqrt{6}) \; \ldots \ldots \; (566).$$

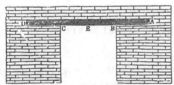

The value of tan. β_2 (equation 558) will then become zero; so that when this relation obtains, the neutral line will, at the point B, be parallel to the axis of the absoissæ; or, in other words, the tangent to the neutral line at the point B will retain, after the deflexion of the beam, the position which it had before; i. e. its position will be that which it would have retained if the beam had been, at that point, *rigid*. Now this condition of rigidity is precisely that which results from the insertion of the beam at its extremities in a mass of masonry, as shown in the accompanying figure; whence it follows that the deflexion in the middle of the beam is the same in the two cases. Taking, therefore, the negative sign in equation (566), and substituting for n its value $\frac{2}{7}(4-\sqrt{6})$ or $\cdot6202041$ in equation (562), and observing that, in that equation, $2a_2$ represents the distance BC in the accompanying figure, we obtain

$$D_2 = \frac{\mu a_2{}^4}{24\text{EI}} \; \cdots \cdots \; (567).$$

By a comparison of this equation with equation (535), it appears that *the deflexion of a beam sustaining a pressure uniformly distributed over its whole length, and having its extremities prolonged and firmly imbedded, is only one fifth of that which it would exhibit if its extremities were free.*[*]

If the masonry which rests upon each inch of the portion AB of the beam be of the same weight as that which rests 'upon each inch of BC, the depth AB of the insertion of each end should equal $\cdot62$ of AE, or about three tenths of the whole length of the beam.

379. *Conditions of the equilibrium of a beam supported at any number of points and deflected by given pressures.*

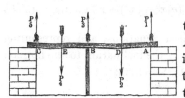

To simplify the investigation, let the points of support ABC be supposed to be three in number, and let the directions of the pressures bisect the distances between them;

[*] The following experiment was made by Mr. Hatcher to verify this result. A strip of deal $\frac{1}{2}$ inch square, was supported with its extremities resting loosely on rollers six feet apart, and was observed to deflect $1\cdot2$ inch in the middle by its own weight. The extremities were then made rigid by confining them between straight edges, and, the distance between the points of support remaining the same, the deflexion was observed to be $\cdot22$ inch. The theory would have given it $\cdot24$.

the same analysis which determines the conditions of the equilibrium in this case will be found applicable in the more
general case. Let P_1, P_3, P_5 be taken to represent the resistances of the several points of support, a_1 and a_2 the distances
between them, P_2 P_4 the deflecting pressures, and x, y the
co-ordinates of any point in the neutral line from the origin
B. Substituting in equation (505) for $\dfrac{1}{R}$ its value $\dfrac{d^2y}{dx^2}$, and
observing that in respect to the portion BD of the beam
$\Sigma Pp = P_2(\frac{1}{2}a_1 - x) - P_1(a_1 - x)$, and that in respect to the
portion DA of the beam, $\Sigma Pp = -P_1(a_1 - x)$, we have
for the differential equation to the neutral line between B
and D

$$EI\frac{d^2y}{dx^2} = P_2(\tfrac{1}{2}a_1 - x) - P_1(a_1 - x) \ \ . \ . \ . \ . \ (568),$$

between D and A

$$EI\frac{d^2y}{dx^2} = -P_1(a_1 - x) \ . \ . \ . \ . \ (569).$$

Representing by β the inclination of the tangent at B to
the axis of the abscissæ, and integrating the former of these
equations twice between the limits 0 and x,

$$EI\frac{dy}{dx} = \tfrac{1}{2}P_2(a_1 x - x^2) - P_1(a_1 x - \tfrac{1}{2}x^2) + EI \tan. \beta \ . \ . \ . \ (570);$$

$$EIy = \tfrac{1}{2}P_2(\tfrac{1}{2}a_1 x^2 - \tfrac{1}{3}x^3) - \tfrac{1}{2}P_1(a_1 x^2 - \tfrac{1}{3}x^3) + EIx \tan. \beta \ . \ . \ . (571).$$

Substituting $\frac{1}{2}a_1$ for x in these equations, and representing
by D_1 the value of y, and by γ the inclination to the horizon
of the tangent at the point D, we obtain

$$EI \tan. \gamma = \tfrac{1}{8}P_2 a_1{}^2 - \tfrac{3}{8}P_1 a_1{}^2 + EI \tan. \beta \ . \ . \ . \ , \ (572),$$

$$EID_1 = \tfrac{1}{24}P_2 a_1{}^3 - \tfrac{5}{48}P_1 a_1{}^3 + \tfrac{1}{2}EIa_1 \tan. \beta \ . \ . \ . \ . \ (573).$$

Integrating equation (569) between the limits $\dfrac{a_1}{2}$ and x,

$$EI\frac{dy}{dx} = -P_1(a_1 x - \tfrac{1}{2}x^2) + EI \tan. \gamma + \tfrac{3}{8}P_1 a^2.$$

Eliminating tan. γ between this equation and equation (572) and reducing,

$$EI\frac{dy}{dx} = -P_1(a_1 x - \tfrac{1}{2}x^2) + EI\tan.\beta + \tfrac{1}{8}P_2 a_1{}^2 \ldots \text{ (574)}.$$

Integrating again between the limits $\frac{a}{2}$ and x, and eliminating the value of D_1 from equation (573),

$$EIy = -\tfrac{1}{2}P_1(a_1 x^2 - \tfrac{1}{3}x^3) + (EI\tan.\beta + \tfrac{1}{8}P_2 a_1{}^2)x - \tfrac{1}{48}P_2 a_1{}^3 \ldots \text{ (575)}.$$

Now it is evident that the equation to the neutral line in respect to the portion CE of the beam, will be determined by writing in the above equation P_5 and P_4 for P_1 and P_2 respectively.

Making this substitution in equation (575), and writing $-\tan.\beta$ for $+\tan.\beta$ in the resulting equation; then assuming $x = a_1$ in equation (575), and $x = a_2$ in the equation thus derived from it, and observing that y then becomes zero in both, we obtain

$$0 = -\tfrac{1}{3}P_1 a_1{}^3 + \tfrac{5}{48}P_2 a_1{}^3 + EI a_1 \tan.\beta,$$
$$0 = -\tfrac{1}{3}P_5 a_2{}^3 + \tfrac{5}{48}P_4 a_2{}^3 - EI a_2 \tan.\beta.$$

Also, by the general conditions of the equilibrium of parallel pressures (Art. 15.),

$$P_1 a_1 + \tfrac{1}{2}P_4 a_2 = P_5 a_2 + \tfrac{1}{2}P_2 a_1,$$
$$P_1 + P_3 + P_5 = P_2 + P_4.$$

Eliminating between these equations and the preceding, assuming $a_1 + a_2 = a$, and reducing, we obtain

$$P_1 = \frac{P_2 a_1(8a_2 + 5a_1) - 3P_4 a_2{}^2}{16 a a_1} \ldots \text{ (576)}.$$

$$P_5 = \frac{P_4 a_2(8a_1 + 5a_2) - 3P_2 a_1{}^2}{16 a a_2} \ldots \text{ (577)}.$$

$$P_3 = \tfrac{1}{2}\left\{ P_2\left(1 + \frac{3a_1}{8a_2}\right) + P_4\left(1 + \frac{3a_2}{8a_1}\right) \right\} \ldots \text{ (578)}.$$

By equation (573),

$$D_1 = \frac{a_1{}^2}{768\text{E}Ia} \left\{ P_2 a_1 (16a_2 + 7a_1) - 9P_4 a_2{}^2 \right\} \ \dots \ (579).$$

Similarly,

$$D_2 = \frac{a_2{}^2}{768\text{E}Ia} \left\{ P_4 a_2 (16a_1 + 7a_2) - 9P_2 a_1{}^2 \right\} \ \dots \ (580);$$

$$\tan.\beta = \frac{a_1}{48\text{E}Ia} \left\{ P_2 (8a_2{}^2 - 5a_1{}^2) - 3P_4 a_2{}^2 \right\} \ \dots \ (581).$$

By equation (572),

$$\tan.\gamma = \frac{a_1}{128\text{E}Ia} \left\{ P_2 a_1{}^2 + P_4 a_2{}^2 \right\} \ \dots \ (582).$$

If a_1 be substituted for x in equation (574), and for P_1 and $\tan.\beta$ their values from equations (576) and (581); and if the inclination of the tangent at A to the axis of x be represented by β_1, we shall obtain by reduction

$$\tan.\beta_1 = \frac{a_1}{32\text{E}Ia} \left\{ P_4 a_2{}^2 - P_2 a_1 (2a_2 + a_1) \right\} \ \dots \ (583).$$

Similarly, if β_2 represent the inclination of the tangent at C to the axis of x,

$$\tan.\beta_2 = \frac{a_2}{32\text{E}Ia} \left\{ P_2 a_1{}^2 - P_4 a_2 (2a_1 + a_2) \right\} \ \dots \ (584).$$

380. If the pressures P_2 and P_4, and also the distances a_1 and a_2, be equal,

$$P_1 = P_5 = \tfrac{5}{16}P_2, \ P_3 = \tfrac{11}{8}P_2, \ \tan.\beta = 0, \ \tan.\beta_1 = \tan.\beta_2 = -\frac{P_2 a_1{}^2}{32\text{E}I}.$$

381. If the distances a_1 and a_2 be equal, and $P_4 = 3P_2$,

$$P_1 = \tfrac{1}{8}P_2, \ P_3 = \tfrac{11}{4}P_2, \ P_5 = \tfrac{9}{8}P_2, \ \tan.\beta = -\frac{P_2 a_1{}^2}{16\text{E}I}, \ \tan.\beta_1 = 0.^*$$

---

* The following experiments were made by Mr. Hatcher to verify this result. The bar ACB, on which the experiment was to be tried, was supported on knife edges of wrought iron at A, C, and B, whose distances AC and CB were each five feet. The angles of the knife edges

382. If $a_1 = a_2$ and $3P_4 = 13P_2$, $P_1 = 0$, $P_3 = \tfrac{11}{3}P_2$, $P_5 = \tfrac{5}{3}P_2$.

383. CURVATURE OF A RECTANGULAR BEAM, THE DIRECTION OF THE DEFLECTING PRESSURE AND THE AMOUNT OF THE DEFLEXION BEING ANY WHATEVER.

The moment of inertia I (Art. 360.) is to be taken, about an axis perpendicular to the plane of deflexion, and passing through the neutral line, the distance h of which neutral line from the centre of gravity of the section is determined by equation (504).

---

were 90°, and the edges were oiled previous to the experiments. The weights were suspended at points D and E intermediate between the

points of support. In measuring the angles of deflexion the instrument (which was a common weighted index-hand turning on a centre in front of a graduated arc) was placed so that the angle c of the parallelogram of wood carrying the arc was just over the knife edge A, the side cd of the parallelogram resting on the deflected bar. This position gave the angle at the point of support. To measure the angle at the centre the parallelogram was placed so that its angle c came as nearly as possible to the centre point of the deflected bar.

1st experiment.—A bar of wrought iron half an inch square, being loaded at D with a weight of 18 lb. 13 oz., and at E with 52 lb. 3 oz., assumed a perfectly horizontal position at A, as shown by the needle. The proportion of these weights is 2·77 : 1.

2d experiment. — A bar ·7 inch square, being loaded at D with a weight of 37·3 lb. and at E with a weight of 112 lb., assumed a perfectly horizontal position at A. The weights were in this experiment accurately in the proportion 3 : 1.

3d experiment. — A round bar ·75 inch in diameter, being loaded at D with 37·3 lb. and at E with 112 lb., showed a deviation from the horizontal position amounting to not more than 20′. The weights were in the proportion of 3 : 1.

The influence of the weight of the bar is not taken into account.

Now $\frac{1}{12}bc^3$ representing (Art. 364.) the moment of inertia of the rectangular section of the beam about an axis passing through its centre of gravity, it follows (Art. 79.) that the moment I about an axis parallel to this passing through a point at distance h from it is represented by

$$I = h^2 bc + \tfrac{1}{12}bc^3.$$

Substituting, therefore, the value of h from equation (504),

$$I = \frac{R^2 P_1^2}{E^2 bc}\sin.^2\theta + \tfrac{1}{12}bc^3 \ \ \ \ . \ . \ . \ . \ (585).$$

Substituting this value in equation (505), and reducing,

$$\frac{1}{R} = \frac{12 P_1 E bc p_1}{12 R^2 P^2 \sin.^2\theta + E^2 b^2 c^4} \ \ \ \ . \ . \ . \ . \ (586).$$

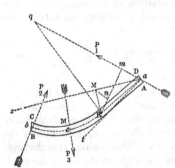

Draw ax parallel to the position of the beam before deflexion; take this line as the axis of the abscissæ and a as the origin; then $p_1 = Rm$ $= Rn + nm = \overline{MR}$ cos. MRm $+ \overline{aM}$ sin. M$am = y$ cos. Mam $+ x$ sin. Mam.

Let, now, the inclination DaP$_1$ of the direction of P$_1$ to the normal at a be represented by θ_1, and the inclination Mat of the tangent to the neutral line at a to ax, by β_1; then

$$\text{M}am = \frac{\pi}{2} - (\theta_1 + \beta_1).$$

$$\therefore \ p_1 = y\sin.(\theta_1 + \beta_1) + x\cos.(\theta_1 + \beta_1).$$

Substituting this value of P$_1$ in the preceding equation,

$$\frac{1}{R} = \frac{12 P_1 E bc\{y\sin.(\theta_1 + \beta_1) + x\cos.(\theta_1 + \beta_1)\}}{12 R^2 P_1^2 \sin.^2\theta + E b^2 c^4} \ \ \ \ . \ . \ . \ . \ . \ (587);$$

where θ represents (Art. 357.) the inclination Rqa of the normal at the point R to the direction of P$_1$.

384. *Case in which the deflexion of the beam is small.*

If the deflexion be small, and the inclination θ_1, of the direction of P_1 to the normal at its point of application, be not greater than $\frac{\pi}{4}$; then $y \sin.(\theta_1 + \beta_1)$ is exceedingly small, and may be neglected as compared with $x \cos.(\theta_1 + \beta_1)$; in this case, moreover, θ is, for all positions of R, very nearly equal to θ_1. Neglecting, therefore, β_1 as exceedingly small, we have

$$\frac{1}{R} = \frac{12P_1 Ebcx \cos.\theta_1}{12R^2 P_1^2 \sin.^2\theta_1 + E^2 b^2 c^4} \quad \cdots \cdots (588).$$

Solving this equation, of two dimensions, in respect to $\frac{1}{R}$, and taking the greater root,

$$\frac{1}{R} = \frac{6P_1}{Ebc^3} \{ x \cos.\theta_1 + \sqrt{x^2 \cos.^2\theta_1 - \tfrac{1}{3}c^2 \sin.^2\theta_1} \} \quad \cdots \cdots (589).$$

385. THE WORK EXPENDED UPON THE DEFLEXION OF A UNIFORM RECTANGULAR BEAM, WHEN THE DEFLECTING PRESSURES ARE INCLINED AT ANY ANGLE GREATER THAN HALF A RIGHT ANGLE TO THE SURFACE OF THE BEAM.

If u_1 represent work expended on the deflexion of the portion AM of the beam, then (equation 510)

$$u_1 = \frac{P_1^2}{2E} \int_0^{a_1} \frac{p_1^2}{I} dx \, ;$$

but by equation (505) $\dfrac{p_1^2}{I} = \dfrac{E}{P_1} \cdot \dfrac{p_1}{R}$,

$$\therefore \; u_1 = \tfrac{1}{2}P_1 \int_0^{a_1} \frac{p_1}{R} dx \quad \cdots \cdots (590).$$

But $\dfrac{p_1}{R} = \dfrac{6P_1}{Ebc^3} \{ x \cos.\theta_1 + \sqrt{x^2 \cos.^2\theta_1 - \tfrac{1}{3}c^2 \sin.\theta_1} \} \; x \cos.\theta_1$;

by equation (589), observing that the deflexion being small,

$p_1 = x\cos.\theta_1$ very nearly. Now the value of $\dfrac{1}{R}$ (equation 589) becomes impossible at the point where $x\cos.\theta_1$ becomes less than $\dfrac{1}{\sqrt{3}}c\sin.\theta_1$; the curvature of the neutral line commences therefore at that point, according to the hypotheses on which that equation is founded. Assuming, then, the corresponding value $\dfrac{1}{\sqrt{3}}c\tan.\theta_1^1$ of x to be represented by x_1, the integral (equation 590) must be taken between the limits x_1 and a_1, instead of 0 and a_1;

$$\therefore\ u_1 = \frac{3P_1{}^2\cos.\theta_1}{Ebc^3}\int_{x_1}^{a_1}\{x^2\cos.\theta_1 + x\sqrt{x^2\cos.^2\theta_1 - \tfrac{1}{3}c^2\sin.^2\theta_1}\}dx;$$

$$\therefore\ u_1 = \frac{P_1{}^2\cos.^2\theta_1}{Ebc^3}\left\{a_1{}^3 - \frac{1}{3\sqrt{3}}c^3\tan.^3\theta_1 + (a_1{}^2 - \tfrac{1}{3}c^2\tan.^2\theta_1)^{\frac{3}{2}}\right\}\ ..\,(591).$$

And a similar expression being evidently obtained for the work expended in the deflexion of the portion BM of the beam, it follows, neglecting the term involving c^3 as exceedingly small when compared with $a_1{}^3$, that the whole work U_1 expended upon the deflexion is represented by the equation

$$U_3 = \frac{1}{Ebc^3}\left\{P_1{}^2\cos.^2\theta_1\{a_1{}^3 + (a_1{}^2 - \tfrac{1}{3}c^2\tan.^2\theta_1)^{\frac{3}{2}}\} + P_2{}^2\cos.^2\theta_2\{a_2{}^3 + (a_2{}^2 - \tfrac{1}{3}c^2\tan.^2\theta_2)^{\frac{3}{2}}\}\right\}.$$

But if θ_3 be taken to represent the inclination of P_3 to the normal to the surface of the beam, as θ_1 and θ_2 represent the similar inclinations of P_1 and P_2, then, the deflexion being small,

$$P_1 a\cos.\theta_1 = P_3 a_2\cos.\theta_3,\quad P_2 a\cos.\theta_2 = P_3 a_1\cos.\theta_3.$$

Eliminating P_1 and P_2 between these equations and the preceding,

$$U_3 = \frac{P_3{}^2\cos.^2\theta_3}{Ea^2bc^3}\left\{a_2{}^2\{a_1{}^3 + (a_1{}^2 - \tfrac{1}{3}c^2\tan.^2\theta_1)^{\frac{3}{2}}\} + a_1{}^2\{a_2{}^3 + (a_2{}^2 - \tfrac{1}{3}c^2\tan.^2\theta_1)^{\frac{3}{2}}\}\right\}\ ..\,(592).$$

If the pressure P_3' be applied perpendicularly in the centre of the beam, and the pressures P_1 and P_2 be applied at its extremities in directions equally inclined to its surface; then

$a_1 = a_2 = \frac{1}{2}a$, $\theta_1 = \theta_2 = \theta$, and $\theta_3 = 0$. Substituting these values in the preceding equations, and reducing,

$$U_3 = \frac{P_3^2\{a^3 + (a^2 - \frac{4}{3}c^2 \tan.^2\theta)_2^3\}}{16 E b c^3} \ \ldots \ (593).$$

386. THE LINEAR DEFLEXION OF A RECTANGULAR BEAM.

D_1 being taken as before (Art. 370.) to represent the deflexion of the extremity A measured in a direction perpendicular to the surface of the beam, we have (Art. 52.)

$$u_1 = \int P_1 \cos.\theta_1 dD_1,$$

$$\therefore \ P_1 \cos.\theta = \frac{du_1}{dD_1} = \frac{du_1}{dP_1} \cdot \frac{dP_1}{dD_1}.$$

But by equation (591),

$$\frac{du_1}{dP_1} = \frac{2P_1}{Ebc^3} \cos.^2\theta_1 \{a_1^3 + (a_1^2 - \frac{1}{3}c^2 \tan.^2\theta_1)^{\frac{3}{2}}\}.$$

$$\therefore \ P_1 \cos.\theta_1 = \frac{dP_1}{dD_1} \cdot \frac{2P_1}{Ebc^3} \cos.^2\theta_1 \{a_1^3 + (a_1^2 - \frac{1}{3}c^2 \tan.^2\theta_1)^{\frac{3}{2}}\}.$$

Dividing both sides by P_1, reducing, and integrating,

$$D_1 = \frac{2P_1}{Ebc^3} \cos.\theta_1 \{a_1^3 + (a_1^2 - \frac{1}{3}c^2 \tan.^2\theta_1)^{\frac{3}{2}} \ \ldots \ (594)$$

Proceeding similarly in respect to the deflexion D_3 perpendicular to the surface of the beam at the point of application of P_3, we obtain from equation (592)

$$D_3 = \frac{2P_3 \cos.\theta_3}{Ea^2bc^3} \left\{ a_2^2\{a_1^3 + (a_1^2 - \frac{1}{3}c^2 \tan.^2\theta_1)^{\frac{3}{2}}\} + a_1^2\{a_2^3 + (a_2^2 - \frac{1}{3}c^2 \tan.^2\theta_2)^{\frac{3}{2}}\} \right\} \ \cdot \cdot \ (595).$$

In the case in which P_1 and P_2 are equally inclined to the extremities of the beam and the direction of P_3 bisects it, this equation becomes

$$D_3 = \frac{P_3\{a^3 + (a^2 - \frac{4}{3}c^2 \tan.^2\theta)^{\frac{3}{2}}\}}{8Ebc^3} \ \ldots \ (596).$$

M M

387. *The work expended upon the deflexion of a beam subjected to the action of pressures applied to its extremities, and to a single intervening point, and also to the action of a system of parallel pressures uniformly distributed over its length.*

Let μ represent the aggregate amount of the parallel

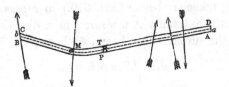

pressures distributed over each unit of the length of the beam, and α their common inclination to the perpendicular to the surface; then will μx represent the aggregate of those distributed uniformly over the surface DT, and these will manifestly produce the same effect as though they were collected in the centre of DT. Their moment about the point R is therefore represented by $\mu x \frac{1}{2} x \cos. \alpha$, or by $\frac{1}{2}\mu x^2 \cos. \alpha$; and the sum of the moments of the pressures applied to AT is represented by $(P_1 x \cos. \theta_1 - \frac{1}{2}\mu x^2 \cos. \alpha)$. Substituting this value of the sum of the moments for $P_1 p_1$ in equation (510), we obtain

$$u_1 = \frac{1}{2E} \int_0^{a_1} \frac{(P_1 x \cos. \theta_1 - \frac{1}{2}\mu x^2 \cos. \alpha)^2}{I} dx.$$

388. *If the pressures be all perpendicular to the surface of the beam,* $\theta_1 = 0$, $\alpha = 0$, *and* I *is constant (equation 504)*; whence we obtain, by integration and reduction,

$$u_1 = \frac{a_1^3}{2EI}\{\tfrac{1}{3}P_1^2 - \tfrac{1}{4}P_1 \mu a_1 + \tfrac{1}{20}\mu^2 a_1^2\} \cdots (597).$$

If the pressure P_3 be applied in the centre of the beam, $P^1 = \frac{1}{2}P_3 + \frac{1}{2}\mu a$, and $a_1 = \frac{1}{2}a$, also the whole work U_3 of deflecting the beam is equal to $2u_1$; whence, substituting and reducing,

$$U_3 = \frac{a^3}{48\mathrm{EI}}\{\tfrac{1}{2}P_3{}^2 + \tfrac{5}{3}P_3\mu a + \tfrac{1}{3}\mu^2 a^2\} \cdots (598).$$

389. A RECTANGULAR BEAM IS SUPPORTED AT ITS EXTRE-
MITIES BY TWO FIXED SURFACES, AND LOADED IN THE
MIDDLE : IT IS REQUIRED TO DETERMINE THE DEFLEXION,
THE FRICTION OF THE SURFACES ON WHICH THE EXTRE-
MITIES REST BEING TAKEN INTO ACCOUNT.

It is evident that the work which produces the deflexion of the beam is done upon it partly by the deflecting

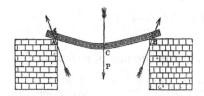

pressure P, and partly by the friction of the surface of the
beam upon the fixed points A and B, over which it moves
whilst in the act of deflecting. Representing by φ the limiting angle of resistance between the surface of the beam and
either of the surfaces upon which its extremity rests, the
friction Q_1 or Q_2 upon either extremity will be represented
by $\tfrac{1}{2}P\tan.\varphi$; and representing by s the length of the curve
ca or cb, and by $2a$ the horizontal distance between the
points of support; the space through which the surface of the
beam would have moved over each of its points of support, if
the point of support had been in the neutral line, is represented by $s-a$, and therefore the whole work done upon the
beam by the friction of each point of support by $\tfrac{1}{2}\tan.\varphi\int Pds$.

Moreover, D representing the deflexion of the beam under
any pressure P, the whole work done by P is represented by
$\int PdD$. Substituting, therefore, for the work expended upon

the elastic forces opposed to the deflexion of the beam its value from equation (593), and observing that the directions of the resistances at A and B are inclined to the normals at those points at angles equal to the limiting angle of resistance, we have

$$\int P dD + \tan.\phi \int P ds = \frac{P^2 \{ a^3 + (a^2 - \frac{4}{3}c \ \tan.^2\phi)^{\frac{3}{2}} \}}{16 E b c^3}.$$

But $\int P dD = \int P \frac{dD}{dP} dP$; and $\int P ds = \int P \frac{ds}{dP} dP =$

$\dfrac{a^5}{30 E^2 I^2} \int P^2 dP$ by equation (526).

Substituting these values in the above equation, and differentiating in respect to P, we have

$$P \frac{dD}{dP} = \frac{P\{ a^3 + (a^2 - \frac{4}{3}c \ \tan.^2\phi)^{\frac{3}{2}} \}}{8 E b c^3} - \frac{P^2 a^5}{30 E^2 I^2} \tan.\phi.$$

Dividing by P, and integrating in respect to P,

$$D = \frac{P\{ a^3 + (a^2 - \frac{4}{3}c^2 \tan.^2\phi)^{\frac{3}{2}} \}}{8 E b c^3} - \frac{P^2 a^5}{60 E^2 I^2} \tan.\phi \dots (599).$$

390. THE SOLID OF THE STRONGEST FORM WITH A GIVEN
QUANTITY OF MATERIAL.

The strongest form which can be given to a solid body in the formation of which a given quantity of material is to be used, and to which the strain is to be applied under given circumstances, is that form which *renders it equally liable to rupture at every point*. So that when, by increasing the strain to its utmost limit, the solid is brought into the state bordering upon rupture at one point, it may be in the state bordering upon rupture at every other point. For let it be supposed to be constructed of any other form, so that its rupture may be about to take place at one point when it is not about to take place at another point, then may a portion of the material evidently be removed from the first point without placing the solid there *in* the state bordering upon rupture, and added at the second point, so as to take it *out* of the state bordering upon rupture at that point; and thus the solid being no longer in the state bordering upon rupture at any point, may be made to bear a strain greater than that which was before upon the point of breaking it, and will have been rendered stronger than it was before.   The first form was not therefore the strongest form of which it could have been constructed with the given quantity of material; nor is any form the strongest which does not satisfy the condition of *an equal liability to rupture at every point.*

The solid, constructed of the strongest form, with a given quantity of a given material, so as to be of a given strength under a given strain, is evidently that which can be constructed, of the same strength, with the least material; so that the *strongest form* is also the form of *the greatest economy of material.*

Rupture.

391. The rupture of a bar of wood or metal may take place either by a *strain* or tension in the direction of its length, to which is opposed its Tenacity; or by a *thrust* or compressing force in the direction of its length, to which is opposed its resistance to Compression; or each of these forces of resistance may oppose themselves to its rupture transversely, the one being called into operation on one side of it, and the other on the other side, as in the case of a Transverse Strain.

Tenacity.

392. The tenacities of different materials as they have been determined by the best authorities, and by the mean results of numerous experiments, will be found stated in a table at the end of this volume.   The unit of tenacity is that opposed to the tearing asunder of a bar one square inch in section, and is estimated in pounds.   It is evident that the

tenacity of a fascile of n such bars placed side by side, or of a single bar n square inches in section, would be equal to n such units, or to n times the tenacity of one bar.

To find, therefore, the tenacity of a bar of any material in pounds, multiply the number of square inches in its section by its tenacity per square inch, as shown by the table.

393. A BAR, CORD, OR CHAIN IS SUSPENDED VERTICALLY, CARRYING A WEIGHT AT ITS EXTREMITY : TO DETERMINE THE CONDITIONS OF ITS RUPTURE.

First. Let the bar be conceived to have a uniform section represented in square inches by K; let its length in inches be L, the weight of each cubic inch μ, the weight suspended from its extremity W, the tenacity of its material per square inch τ; and let it be supposed capable of bearing m times the strain to which it is subjected. The weight of the bar will then be represented by μLK, and the strain upon its highest section by μLK + W. Now the strain on this section is evidently greater than that on any other; it is therefore at this section that the rupture will take place. But the resistance opposed to its rupture is represented by Kτ; whence it follows (since this resistance is m times the strain) that

$$K\tau = m(\mu LK + W),$$

$$\therefore K = \frac{mW}{\tau - \mu m L} \cdot \cdot \cdot \cdot \cdot (600).$$

By which equation is determined the uniform section K of a bar, cord, or chain, so that being of a given length it may be capable of bearing a strain m times greater than that to which it is actually subjected when suspended vertically.

The weight W_1 of the bar is represented by the formula KLμ,

$$\therefore W_1 = \frac{m\mu WL}{\tau - m\mu L} \cdot \cdot \cdot \cdot \cdot (601).$$

394. *Secondly.* Let the section of the rod be variable; and let this variation of the section be such that its strength, *at every point*, may be that which would cause it to bear, without breaking, m times as great a strain as that which it actually bears there. Let K represent this section at a point whose distance from the extremity which carries the weight W is x; then will the weight of the rod beneath that point be represented by $\int \mu K dx$; or, supposing the specific gravity of the material to be every where the same, by $\mu \int K dx$: also the resistance of this section to rupture is Kτ.

$$\therefore \; m\left(W + \mu \int K dx\right) = K\tau$$

Differentiating this expression in respect to x, observing that K is a function of x, and dividing by Kτ, we obtain

$$\frac{1}{K}\frac{dK}{dx} = \frac{m\mu}{\tau};$$

Integrating this expression between the limits 0 and x, and representing by K_0 the area of the lowest section of the rod,

$$\log_\varepsilon \frac{K}{K_0} = \frac{m\mu}{\tau}x; \; \therefore \; K = K_0 \varepsilon^{\frac{m\mu}{\tau}x}.$$

But the strain sustained by the section K_0 is W, therefore $K_0\tau = mW$;

$$\therefore \; K = \frac{mW}{\tau}\varepsilon^{\frac{m\mu}{\tau}x} \quad \ldots \ldots \ldots (602).$$

The whole weight W_2 of the rod, cord, or chain, is represented by the formula

$$W_2 = \mu \int_0^L K dx = \frac{\mu m W}{\tau}\int_0^L \varepsilon^{\frac{m\mu}{\tau}x}, dx = W\left(\varepsilon^{\frac{\mu m L}{\tau}} - 1\right) \ldots \ldots (603).$$

A rope or chain, constructed according to these conditions, is evidently as strong as the rope or chain of uniform section whose weight W_1 is determined by equation (601), the value of m being taken the same in both cases. The saving of material effected by giving to the cord or chain a section varying according to the law determined by equation (603) is represented by $W_1 - W_2$, or by the formula

$$\frac{m\mu WL}{\tau - m\mu L} - W\left(\varepsilon^{\frac{\mu m L}{\tau}} - 1\right) \quad . \quad . \quad . \quad . \quad (604).$$

THE SUSPENSION BRIDGE.

395. *General conditions of the equilibrium of a loaded chain.*

Let AEH represent a chain or cord hanging freely from

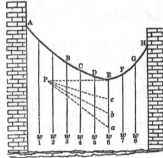

two fixed points A and H, and having certain weights w_1, w_2, w_3, &c., suspended by rods or cords from given points B, C, D, &c., in its length. Through the lowest point E of the chain draw the vertical Ea, containing as many equal parts as there are units in the weight of the chain between E and any point of suspension B, together with the suspending rods attached to it, and the weights which they severally carry; draw aP parallel to the direction of a tangent to the curve at B, and produce the tangent at E to meet aP in P; then will aP and EP contain as many equal parts as there are units in the tensions at B and E respectively; and if Eb and Ec be taken to represent the whole weights sustained by EC and ED, and Pb and Pc be joined, these lines will in like manner represent the tensions upon the points C and D. For the pressures applied to EB, and in equilibrium, being the weight of the chain, the weights of the suspending rods, the weights attached to the rods, and the tensions upon B and E, the principle of the polygon of pressures (Art. 9.) obtains in respect to these pressures. Now the lines drawn to complete this polygon, parallel to the *weights*, form together the vertical line Ea, and the polygon (resolving itself into a triangle) is completed by the lines aP and EP drawn parallel to the *tensions* upon B and E. Each line contains, therefore, as many equal parts (Art. 9.) as there are units in the corresponding tension. Also, the pressures applied to the portion EC of the curve, being the weights whose aggregate is represented by Eb, and the tensions upon E and C, of which the former is represented in direction and amount by EP, it follows (Art. 9.) that the latter is represented also in direction and amount by the line Pb, which completes the triangle aPb; so that bP is parallel to the tangent at C.

In like manner it is evident that the tension upon D is represented in magnitude and direction by cP; so that cP is parallel to the tangent to the curve at D.

The Catenary.

396. *If a chain of uniform section be suspended freely between two fixed points A and B, being acted upon by no other pressures than the weights of its parts, then it will assume the geometrical form of a curve called the catenary.*

Let PT be a tangent to any point P of the curve intersecting the vertical CD passing through its lowest point D in

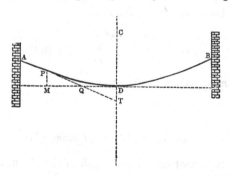

T; draw the horizontal line DM intersecting PT in Q; take this line as the axis of the abscissæ; and let DM $=x$, MP $=y$, DP $=s$, weight of each unit in the length of the chain $=\mu$, tension at D $=c$. Now DT being taken to represent the

weight μs of DP, it has been shown (Art. 395.) that DQ will represent the tension c at D, and TQ that at P.

Also, $\dfrac{dy}{dx} = \tan. \text{PQM} = \tan. \text{DQT} = \dfrac{\text{DT}}{\text{DQ}} = \dfrac{\mu s}{c}$,

$$\therefore \dfrac{dy}{dx} = \dfrac{\mu s}{c} \ \ldots \ (605).$$

Again, $\dfrac{dx}{ds} = \left(1 + \dfrac{dy}{dx}\right)^{-\frac{1}{2}} = \left(1 + \dfrac{\mu^2 s^2}{c^2}\right)^{-\frac{1}{2}}$. Integrating between the limits 0 and s*, and observing that when $s=0$, $x=0$,

$$x = \dfrac{c}{\mu} \log._\varepsilon \left\{ \dfrac{\mu s}{c} + \left(1 + \dfrac{\mu^2 s^2}{c}\right)^{\frac{1}{2}} \right\} \ \ldots \ (606).$$

$$\therefore \dfrac{\mu s}{c} + \left(1 + \dfrac{\mu^2 s^2}{c^2}\right)^{\frac{1}{2}} = \varepsilon^{\frac{\mu x}{c}},$$

$$\therefore \dfrac{\mu s}{c} - \left(1 + \dfrac{\mu^2 s^2}{c}\right)^{\frac{1}{2}} = -\varepsilon^{\frac{-\mu x}{c}}$$

By addition and reduction,

$$s = \tfrac{1}{2} \dfrac{c}{\mu} \left(\varepsilon^{\frac{\mu x}{c}} - \varepsilon^{\frac{-\mu x}{c}} \right) \ \ldots \ (607).$$

Substituting this value for s in equation (605), and integrating between the limits 0 and x,

$$y = \tfrac{1}{2} \dfrac{c}{\mu} \left(\varepsilon^{\frac{\mu x}{c}} + \varepsilon^{\frac{-\mu x}{c}} \right) \ \ldots \ (608);$$

which is the equation to the catenary.

397. The tension (c) on the lowest point of the catenary.

Let 2S represent the whole length of the chain, and $2a$ the horizontal distance between the points of attachment. Now when $x=a$, $s=\text{S}$; therefore (equation 607),

---

* See Hymer's Int. Cal., art. 15.

$$S = \frac{1}{2} \frac{c}{\mu} \left(\varepsilon^{\frac{\mu a}{c}} - \varepsilon^{\frac{-\mu a}{c}} \right) \cdot \cdot \cdot \cdot (609);$$

from which expression the value of c may be determined by approximation.

398. *The tension at any point of the chain.*

The tension T at P is represented by $\overline{TQ} = \sqrt{\overline{DQ^2 + DT^2}};$

$$\therefore T = (c^2 + \mu^2 s^2)^{\frac{1}{2}} \cdot \cdot \cdot \cdot \cdot (610).$$

Now the value of c has been determined in the preceding article; the tension upon any point of the chain whose distance from its lowest point is s is therefore known.

399. *The inclination of the curve to the vertical at any point.*

Let ι represent this inclination, then $\cot \iota = \dfrac{dy}{dx};$

$$\therefore \text{(equation 608)} \quad \cot \iota = \frac{1}{2} \left(\varepsilon^{\frac{\mu x}{c}} - \varepsilon^{\frac{-x\mu}{c}} \right) \cdot \cdot \cdot \cdot (611).$$

The inclination may be determined without having first determined the value of c, by substituting $\cot \iota$ for $\dfrac{\mu s}{c}$ in equation (606); we thus obtain, writing also a and S for x and s,

$$\frac{a}{S} = \tan \iota \cdot \log_\varepsilon (\cot \iota + \operatorname{cosec} \iota) = \tan \iota \cdot \log_\varepsilon \cot \tfrac{1}{2} \iota;$$

$$\therefore -\tan \iota \cdot \log_\varepsilon \tan \tfrac{1}{2} \iota = \frac{a}{S} \cdot \cdot \cdot \cdot \cdot (612).$$

This equation may readily be solved by approximation; and the value of c may then be determined by the equation $c = \mu S \tan \iota.$

400. *A chain of given length being suspended between two given points in the same horizontal line : to determine the depth of the lowest point beneath the points of attachment ; and, conversely, to determine the length of the chain whose lowest point shall hang at a given depth below its points of attachment.*

The same notation being taken as before,

$$\frac{dy}{ds} = \left(1 + \frac{dx^2}{dy^2}\right)^{-\frac{1}{2}} = \left(1 + \frac{c^2}{\mu^2 s^2}\right)^{-\frac{1}{2}} = (c^2 + \mu^2 s^2)^{-\frac{1}{2}} \mu s.$$

Integrating between the limits 0 and s, and observing that $y = 0$ when $s = 0$,

$$y = \frac{1}{\mu}\{(c^2 + \mu^2 s^2)^{\frac{1}{2}} - c\} \quad \ldots \quad (613).$$

Solving this equation in respect to s,

$$s = \sqrt{y\left(y + \frac{2c}{\mu}\right)} \quad \ldots \quad (614).$$

If H represent the depth of the lowest point, or the *versed sine* of the curve, then $y = H$ when $s = S$.

$$H = \frac{1}{\mu}\{(c^2 + \mu^2 S^2)^{\frac{1}{2}} - c\} \quad \ldots \quad (615).$$

$$S = \sqrt{H\left(H + \frac{2c}{\mu}\right)} \quad \ldots \quad (616).$$

401. *The centre of gravity of the catenary.*

If G represent the height of the centre of gravity above the lowest point, we have (Art. 32.)

$$G = \int y\, ds = \int y \frac{ds}{dx} dx.$$

Substituting, therefore, for y and $\frac{ds}{dx}$ their values from equations (607) and (608), we have

$$G = \tfrac{1}{4}\,\frac{c}{\mu'}\int_0^a \left(\varepsilon^{\frac{\mu x}{c}} + \varepsilon^{\frac{-\mu x}{c}}\right)^2 dx.$$

Squaring the binomial, and integrating,

$$G = \tfrac{1}{4}\frac{c}{\mu}\left\{\frac{c}{2\mu}\left(\varepsilon^{\frac{2\mu a}{c}} - \varepsilon^{\frac{-2\mu a}{c}}\right) + 2a\right\} \quad \cdots \cdots (617).$$

Substituting H and S for their values, as given by equations (607) and (608), and reducing,

$$G = \tfrac{1}{2}\left(SH + \frac{ca}{\mu}\right) \quad \cdots \cdots (618).$$

402. The suspension bridge of greatest strength, the weight of the suspending rods being neglected.

Let ADB represent the chain, EF the road-way; and let

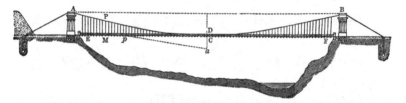

the weight of a bar of the material of the chain, one square inch in section and one foot long, be represented by μ_1, the weight of each foot in the length of the road-way by μ_2, the aggregate section of the chains at any point P (in square inches) by K, the co-ordinates DM and MP of P by x and y, and the length of the portion DP of the chain by s. Then will the weight of DP be represented by $\mu_1\!\int\! Kds$, and the weight of the portion CM of the road-way by $\mu_2 x$; so that the whole load (u) borne by the portion DP of the chain will be represented (neglecting the weight of the suspending rods) by

$$\mu_1\!\int\! Kds + \mu_2 x.$$

$$\therefore \quad u = \mu_1\!\int\! Kds + \mu_2 x. \quad \cdots \cdots (619).$$

Let this load (u), supported by the portion DP of the chain, be represented by the line Da, and draw Dp in the direction of a tangent at D, representing on the same scale the tension c at that point; then will ap be parallel to a tangent to the chain at P (Art. 395).

$$\therefore \frac{dy}{dx} = \frac{u}{c} \quad \ldots \ldots \text{(620)}.$$

Now let it be assumed that the aggregate section of the chains is made so to vary its dimensions, that their strength may at every point be equal to m times the strain which they have there to sustain. But this strain is represented in magnitude by the line ap (Art. 395.), or by $(c^2 + u^2)^{\frac{1}{2}}$; if, therefore, τ be taken to represent the tenacity of the material of the chain, per square inch of the section, then

$$K\tau = m(c^2 + u^2)^{\frac{1}{2}}. \quad \ldots \ldots \text{(621)}.$$

Therefore $K\tau = mc\left(1 + \frac{u^2}{c^2}\right)^{\frac{1}{2}} = mc\left(1 + \frac{dy^2}{dx^2}\right)^{\frac{1}{2}}$ (equation 620)

$= mc\dfrac{ds}{dx}$; therefore $\dfrac{ds}{dx} = \dfrac{K\tau}{mc}$. Also $\displaystyle\int K ds = \int K \frac{ds}{dx} dx = \frac{\tau}{mc}$

$\displaystyle\int K^2 dx = \frac{m}{\tau c} \int (c^2 + u^2) dx$ (equation 621);

$$\therefore \text{ (equation 619)} \quad u = \frac{m\mu_1}{\tau c} \int (c^2 + u^2) dx + \mu_2 x.$$

Differentiating in respect to x, and observing that $\dfrac{du}{dx} =$ $\dfrac{du}{dy}\dfrac{dy}{dx} = \dfrac{u}{c}\dfrac{du}{dy}$ (equation 613), we have

$$\frac{du}{dx} = \frac{u}{c}\frac{du}{dy} = \frac{m\mu_1}{\tau c}(c^2 + u^2) + \mu_2 = \frac{m\mu_1}{\tau c}\left(u^2 + c^2 + \frac{\tau c \mu_2}{m\mu_1}\right);$$

$$\therefore \quad x = \frac{\tau c}{m\mu_1} \int_0^u \frac{du}{u^2 + c^2 + \frac{\tau c \mu_2}{m\mu_1}}, \quad y = \frac{\tau}{m\mu_1} \int_0^u \frac{u\,du}{u^2 + c^2 + \frac{\tau c \mu_2}{m}}.$$

Integrating these expressions *, we obtain

$$x = \frac{\tau c}{m u_1}\left(c^2 + \frac{\tau c \mu_2}{m\mu_1}\right)^{-\frac{1}{2}} \tan.^{-1}\left(c^2 + \frac{\tau c\mu_2}{m\mu_1}\right)^{-\frac{1}{2}} u \quad \ldots \ldots \text{(622)}.$$

$$y = \frac{\tau}{2m\mu_1} \log._\varepsilon \left\{ \frac{u^2 + c^2 + \dfrac{\tau c \mu_2}{m\mu_1}}{c^2 + \dfrac{\tau c\mu_2}{m\mu}} \right\}.$$

Substituting in this equation the value of u given by the preceding equation, and reducing,

$$y = \frac{\tau}{m\mu_1} \log._\varepsilon \sec.\left\{ \frac{m\mu_1}{\tau}\left(1 + \frac{\tau\mu_2}{cm\mu_1}\right)^{\frac{1}{2}} x \right\} \quad \ldots \ldots \text{(623)};$$

which is the equation to the suspension chain of uniform strength, and therefore OF THE GREATEST STRENGTH WITH A GIVEN QUANTITY OF MATERIAL.

403. *To determine the variation of the section* K *of the chain of the suspension bridge of the greatest strength.*

Let the value of u determined by equation (622) be substituted in equation (621); we shall thus obtain by reduction

$$\text{K} = \frac{mc}{\tau}\left\{ 1 + \left(1 + \frac{\tau\mu_2}{mc\mu_1}\right)\tan.^2 \frac{m\mu_1}{\tau}\left(1 + \frac{\tau\mu_2}{cm\mu_1}\right)x \right\}^{\frac{1}{2}} \ . \ . \text{(624)}.\dagger$$

It is evident from this expression that the area of the section of the chains, of the suspension bridge of uniform strength, and therefore of the greatest economy of material, increases from the lowest point towards the points of suspension, where it is greatest.‡

* Hall's Diff. Cal. pp. 280, 283.

† $\dfrac{ds}{dx} = \dfrac{\tau\text{K}}{mc}$; $\therefore s = \dfrac{\tau}{\mu c}\int \text{K} dx$. Now the function K (equation 624) may be integrated in respect to x by known rules of the integral calculus; the value of s may therefore be determined in terms of x, and thence the length in terms of the span. The formula is omitted by reason of its length.

‡ This variation of the section of the chains is exhibited in a suspension bridge recently invented by Mr. Dredge, and appears to constitute the whole merit of that invention.

404. *To determine the weight* W *of the chain of the suspension bridge of the greatest strength.*

Let it be observed that $W = \mu_1 \int K ds = u - \mu_2 x$ (equation 619); substituting the value of u from equation (622), we have

$$W = c\left(1 + \frac{\tau\mu_2}{mc\mu_1}\right)^{\frac{1}{2}} \tan.\left\{\frac{m\mu_1}{\tau}\left(1 + \frac{\tau\mu_2}{cm\mu_1}\right)^{\frac{1}{2}} x\right\} - \mu_2 x \dots (625).$$

405. *To determine the tension* c *upon the lowest point* D *of the chain of uniform strength.*

Let H be taken to represent the depth of the lowest point D, beneath the points of suspension, and $2a$ the horizontal distance of those points; and let it be observed that H and a are corresponding values of y and x (equation 623);

$$\therefore \; H = \frac{\tau}{m\mu_1} \log. \varepsilon \, \sec. \left\{\frac{m\mu_1}{\tau}\left(1 + \frac{\tau\mu_2}{cm\mu_1}\right)^{\frac{1}{2}} a\right\}.$$

Solving this equation in respect to c,

$$c = \frac{\tau\mu_2}{m\mu_1}\left\{\left(\frac{\tau}{m\mu_1 a} \sec.^{-1} \varepsilon^{\frac{m\mu_1}{\tau}H}\right)^2 - 1\right\}^{-1} \dots (626).$$

406. The suspension bridge of greatest strength, the weight of the suspending rods being taken into account.

Conceive the suspending rods to be replaced by a con-

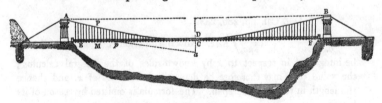

tinuous flexible lamina or plate connecting the roadway with the chain, and of such a uniform thickness that the material contained in it may be precisely equal in weight to the material of the suspending rods. It is evident that the conditions of the equilibrium will, on this hypothesis, be very nearly the same as in the actual case. Let μ_3 represent the weight of each square foot of this plate, then will $\mu_3\int y dx$ represent the weight of that portion of it which is suspended from the portion DP of the chain, and the whole load u upon that portion of the chain will be represented by

$$u = \mu_1 \int K ds + \mu_2 x + \mu_3 \int y dx \ \ldots \ldots \ (627).$$

It may be shown, as before (Art. 402.), that

$$\frac{dy}{dx} = \frac{u}{c}, \ \ K\tau = m(c^2 + u^2)^{\frac{1}{2}} \ \ldots \ldots \ (628),$$

$\int K ds = \frac{m}{\tau c}\int (c^2 + u^2) dx.$ Substituting in equation (627), differentiating in respect to x, and observing that $\frac{du}{dx} = \frac{u}{c}\frac{du}{dy}$,

$$\frac{du}{dx} = \frac{u}{c}\frac{du}{dy} = \frac{m\mu_1}{\tau c}(c^2 + u^2) + \mu_2 + \mu_3 y \ \ldots \ldots \ (629).$$

Transposing, reducing, and assuming,

$$\frac{m\mu_1}{\tau} = \alpha \ \ldots \ldots \ (630);$$

$$\frac{du^2}{dy} - 2\alpha u^2 = 2c(\mu_3 y + \alpha c + \mu_2).$$

A linear equation in u^2, the integration of which by a well-known method (Hall's *Diff. Cal.* p. 397.) gives

$$u^2 \varepsilon^{-2\alpha y} = 2c\int (\mu_3 y + \alpha c + \mu_2)\varepsilon^{-2\alpha y} dy + C.$$

Assuming the length of the shortest connecting rod DC to be represented by b, integrating between the limits b and y, and observing that when $y = b$, $u = 0$,

$$u^2 \varepsilon^{-2\alpha y} = \frac{c}{\alpha}\left\{ \mu_3\left(b\varepsilon^{-2\alpha b} - y\varepsilon^{-2\alpha y}\right) + \left(\frac{\mu_3}{2\alpha} + \alpha c + \mu_2\right)\left(\varepsilon^{-2\alpha} - \varepsilon^{-2\alpha}\right) \right\};$$

$$\therefore \; u^2 = \frac{c}{\alpha} \left\{ \mu_3 \left(b \varepsilon^{2\alpha(y-b)} - y \right) + \left(\frac{\mu_3}{2\alpha} + \alpha c + \mu_2 \right) \left(\varepsilon^{2\alpha(y-b)} - 1 \right) \right\} \; .. (631).$$

Substituting this value of u^2 in equation (628), and reducing,

$$K = \frac{(c)^{\frac{3}{2}}}{1} \left\{ \left(\frac{\mu_3}{2\alpha} + \mu_3 b + \alpha c + \mu_2 \right) \varepsilon^{2\alpha(y-b)} - \mu_3 y - \frac{\mu_3}{2\alpha} - \mu_2 \right\}^{\frac{1}{2}} \; . . (632);$$

by which expression the variation of the section of the chain of uniform strength is determined.

Differentiating the equation $\dfrac{dy}{dx} = \dfrac{u}{c}$ in respect to x, and substituting for $\dfrac{du}{dx}$ its value from equation (629),

$$c \frac{d^2 y}{dx^2} = \frac{\alpha}{c}(c^2 + u^2) + \mu_2 + \mu_3 y.$$

Substituting for u^2 its value from equation (631),

$$c \frac{d^2 y}{dx^2} = \left(\frac{\mu_3}{2\alpha} + \mu_3 b + \alpha c + \mu_2 \right) \varepsilon^{2\alpha(y-b)} - \frac{\mu_3}{2\alpha}.$$

Multiplying both sides of this equation by $\dfrac{dy}{dx}$, and integrating between the limits b and y, observing that when $y = b$, $\dfrac{dy}{dx} = 0$,

$$\alpha c \left(\frac{dy}{dx} \right)^2 = \left(\frac{\mu_3}{2\alpha} + \mu_3 b + \alpha c + \mu_2 \right) \left(\varepsilon^{2\alpha(y-b)} - 1 \right) - \mu_3 (y - b).$$

Now let it be observed, that the value of τ, being in all practical cases exceedingly great as compared with the values of μ_1 and m, the value of α (equation 630) is exceedingly small; so that we may, without sensible error, assume those terms of the series $\varepsilon^{2\alpha(y-b)}$ which involve powers of $2\alpha(y-b)$ above the first, to vanish as compared with unity. This supposition being made, we have $\varepsilon^{2\alpha(y-b)} - 1 = 2\alpha(y-b)$; whence, by substitution and reduction,

$$c \left(\frac{dy}{dx} \right)^2 = 2(\mu_3 b + \alpha c + \mu_2)(y - b).$$

Extracting the square root of both sides, transposing, and integrating,

$$x^2 = \left(\frac{2c}{\mu_3 b + \alpha c + \mu_2} \right) (y - b) \quad \ldots \ldots \text{(633)};$$

the equation to a parabola whose vertex is in D, and its axis vertical.

The values a and H of x and y at the points of suspension being substituted in this equation, and it being solved in respect to c, we obtain

$$c = \left(\frac{\mu_2 + \mu_3 b}{2H - 2b - \alpha a^2} \right) a^2 \quad \ldots \ldots \text{(634)};$$

by which expression the tension c upon the lowest point of the curve is determined, and thence the length y of the suspending rod at any given distance x from the centre of the span, by equation (633), and the section K of the chain at that point by equation (632), which last equation gives by a reduction similar to the above

$$K = \frac{\alpha c}{\mu_1} \left\{ 2(y - b) \left(\frac{\mu_3 b + \mu_2}{c} + \alpha \right) + 1 \right\}^{\frac{1}{2}} \quad \ldots \ldots \text{(635)}.$$

407. *The section of the chains being of uniform dimensions, as in the common suspension bridge, it is required to determine the conditions of the equilibrium.**

The weight of the suspending rods being neglected, and the same notation being adopted as in the preceding articles, except that μ_1 is taken to represent the weight of one foot in the length of the chains instead of a bar one square inch in section, we have by equation (619), since K is here constant,

$$u = \mu_1 s + \mu_2 x \quad \ldots \ldots \text{(636)}.$$

Differentiating this equation in respect to x, and observing

---

* This problem appears first to have been investigated by Mr. Hodgkinson in the fifth volume of the Manchester Transactions; his investigation extends to the case in which the influence of the weights of the suspending rods is included.

that $\dfrac{ds}{dx} = \left(1 + \dfrac{dy^2}{dx^2}\right)^{\frac{1}{2}} = \left(1 + \dfrac{u^2}{c^2}\right)^{\frac{1}{2}}$ (equation 620), and that

$$\dfrac{du}{dx} = \dfrac{du}{dy}\dfrac{dy}{dx} = \dfrac{du}{dy}\dfrac{u}{c} = \mu_1\left(1 + \dfrac{u^2}{c^2}\right)^{\frac{1}{2}} + \mu_2 ;$$

$$\therefore \quad x = \int_0^u \dfrac{c\,du}{\mu_1(c^2+u^2)^{\frac{1}{2}} + \mu_2 c},\ \ y = \int_0^u \dfrac{u\,du}{\mu_1(c^2+u^2)^{\frac{1}{2}} + \mu_2 c}.$$

The former of these equations may be rationalised by assuming $(c^2+u^2)^{\frac{1}{2}} = c + zu$, and the latter by assuming $(c^2+u^2)^{\frac{1}{2}} = z$; there will thus be obtained by reduction

$$x = 2c\int_0^z \dfrac{(1+z^2)dz}{(1-z^2)\{(\mu_1+\mu_2) + (\mu_1 - \mu_2)z^2\}},\ \ y = \int_c^z \dfrac{z\,dz}{\mu_1 z + \mu_2 c}.$$

The latter equation may be placed under the form

$$y = \dfrac{1}{\mu_1}\int_c^z \left\{ 1 - \dfrac{c\mu_2}{\mu_1 z + \mu_2 c} \right\} dz ;$$

which expression being integrated and its value substituted for z, we obtain

$$y = \dfrac{1}{\mu_1}\left\{ (c^2+u^2)^{\frac{1}{2}} - c - \dfrac{c\mu_2}{\mu_1}\log._\varepsilon \dfrac{\mu_1(c^2+u^2)^{\frac{1}{2}} + \mu_2 c}{(\mu_1+\mu_2)c} \right. \quad .\,. (637).$$

The method of rational fractions (Hymer's *Integ. Calc.* § 2.) being applied to the function under the integral sign in the former equation, it becomes

$$x = \dfrac{2c}{\mu_1}\int_0^z \left\{ \dfrac{1}{1-z^2} - \dfrac{\mu_2}{(\mu_1-\mu_2)z^2 + (\mu_1+\mu_2)} \right\} dz.$$

The integral in the first term in this expression is represented by $\frac{1}{2}\log._\varepsilon\left(\dfrac{1+z}{1-z}\right)$, and that of the second term by

$$\dfrac{\mu_2}{(\mu_1{}^2 - \mu_2{}^2)^{\frac{1}{2}}}\tan.^{-1}\left(\dfrac{\mu_1-\mu_2}{\mu_1+\mu_2}\right)^{\frac{1}{2}}z,\ \text{or}\ \dfrac{\mu_2}{2(\mu_2{}^2-\mu_1{}^2)^{\frac{1}{2}}}\log._\varepsilon\dfrac{(\mu_2+\mu_1)^{\frac{1}{2}} + (\mu_2-\mu_1)^{\frac{1}{2}}z}{(\mu_2+\mu_1)^{\frac{1}{2}} - (\mu_2-\mu_1)^{\frac{1}{2}}z},$$

according as μ_1 is greater or less than μ_2, or according as the weight of each foot in the length of the chains is greater or less than the weight of each foot in the length of the roadway.

Substituting for z its value, we obtain therefore, in the two cases,

$$x = \frac{c}{\mu_1}\left\{ \log. \frac{(u-c)+(u^2+c^2)^{\frac{1}{2}}}{(u+c)-(u^2+c^2)^{\frac{1}{2}}} - \frac{2\mu_2}{(\mu_1{}^2-\mu_2{}^2)^{\frac{1}{2}}} \tan.^{-1}\left(\frac{\mu_1-\mu_2}{\mu_1+\mu_2}\right)^{\frac{1}{2}}\left\{\left(1+\frac{c^2}{u^2}\right)^{\frac{1}{2}}+\frac{c}{u}\right\}\right\}$$

$$x = \frac{c}{\mu_1}\left\{ \log. \frac{(u-c)+(u^2+c^2)^{\frac{1}{2}}}{(u+c)-(u^2+c^2)^{\frac{1}{2}}} - \frac{\mu_2}{(\mu_1{}^2-\mu_2{}^2)^{\frac{1}{2}}} \log. \frac{(\mu_2+\mu_1)^{\frac{1}{2}}u+(\mu_2-\mu_1)^{\frac{1}{2}}\{(u^2+c^2)^{\frac{1}{2}}-c\}}{(\mu_2+\mu_1)^{\frac{1}{2}}u-(\mu_2-\mu_1)^{\frac{1}{2}}\{(u^2+c^2)^{\frac{1}{2}}-c\}}\right\} \quad \cdots (638).$$

If the given values, a and H, of x and y at the points of suspension, be substituted in equations (638) and (637), equations will be obtained, whence the value of the constant c and of u at the points of suspension may be determined by approximation. A series of values of u, diminishing from the value thus found to zero, being substituted in equations (638) and (637), as many corresponding values of x and y will then become known. The curve of the chains may thus be laid down with any required degree of accuracy.

This common method of construction, which assigns a uniform section to the chains, is evidently false in principle; the strength of a bridge, the section of whose chains varied according to the law established in Art. 403. (equation 624), would be far greater, the same quantity of iron being employed in its construction.

Rupture by Compression.

408. It results from the experiments of Mr. Eaton Hodgkinson[*], on the compression of short columns of different heights but of equal sections, first, that after a certain height is passed the crushing pressure remains the same, as the heights are increased, until another height is attained, when they begin to break; not as they have done before, by the sliding of one portion upon a subjacent portion, but by bending. Secondly, that the plane of rupture is always inclined at the same constant angle to the base of the column, when its height is between these limits. These two facts

[*] Seventh Report of the British Association of Science.

explain one another; for if K represent the transverse section of the column in square inches, and α the constant inclination of the plane of rupture to the base, then will K sec. α represent the area of the plane of rupture. So that if γ represent the resistance opposed, by the coherence of the material, to the sliding of one square inch upon the surface of another*, then will γK sec. α represent the resistance which is overcome in the rupture of the column, so long as its height lies between the supposed limits; which resistance being constant, the pressure applied upon the summit of the column to overcome it must evidently be constant. Let this pressure be

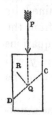

represented by P, and let CD be the plane of rupture. Now it is evident that the inclination of the direction of P to the perpendicular QR to the surface of the plane, or its equal, the inclination α of CD to the base of the column, must be greater than the limiting angle of resistance of the surfaces; if it were not, then would no pressure applied in the direction of P be sufficient to cause the one surface to slide upon the other, even if a separation of the surfaces were produced along that plane.

Let P be resolved into two other pressures, whose directions are perpendicular and parallel to the plane of rupture; the former will be represented by P cos. α, and the friction resulting from it by P cos. α tan. φ; and the latter, represented by P sin. α, will, when rupture is about to take place, be precisely equal to the coherence Kγ sec. α of the plane of rupture increased by its friction P cos. α tan. φ, or P sin. α = Kγ sec. α + P cos. α tan. φ, whence by reduction

$$P = \frac{K\gamma \cos. \varphi}{\sin. (\alpha - \varphi) \cos. \alpha} = \frac{2K\gamma \cos. \varphi}{\sin. (2\alpha - \varphi) - \sin. \varphi} \quad \cdots \cdots (639),$$

It is evident from this expression that if the coherence of the material were the same in all directions, or if the unit of coherence γ opposed to the sliding of one portion of the

---

* The force necessary to overcome a resistance, such as that here spoken of, has been appropriately called by Mr. Hodgkinson the force necessary to *shear* it across.

mass upon another were accurately the same in every direction in which the plane CD may be imagined to intersect the mass, then would the plane of actual rupture be inclined to the base at an angle represented by the formula

$$\alpha = \frac{\pi}{4} + \frac{\varphi}{2} \ \cdots \cdots \ (640);$$

since the value of P would in this case be (equation 639) a minimum when sin. $(2\alpha - \varphi)$ is a maximum, or when $2\alpha - \varphi = \frac{\pi}{2}$, or $\alpha = \frac{\pi}{4} + \frac{\varphi}{2}$; whence it follows that a plane inclined to the base at that angle is that plane along which the rupture will first take place, as P is gradually increased beyond the limits of resistance.

The actual inclination of the plane of rupture was found in the experiments of Mr. Hodgkinson to vary with the material of the column. In cast iron, for instance, it varied according to the quality of the iron from 48° to 58° *, and was different in different species. By this dependence of the angle of rupture upon the nature of the material, it is proved that the value of the modulus of sliding coherence γ is not the same for every direction of the plane of rupture, or that the value of φ varies greatly in different qualities of cast iron.

Solving equation (639) in respect to γ, we obtain

$$\gamma = \frac{P}{K} \sin. (\alpha - \varphi) \cos. \alpha \sec. \varphi \ \cdots \cdots \ (641);$$

from which expression the value of the modulus γ may be determined in respect to any material whose limiting angle of resistance φ is known, the force P producing rupture, under the circumstances supposed, being observed, and also the angle of rupture.†

* Seventh Report of British Association, p. 349.

† A detailed statement of the results obtained in the experiments of Mr. Hodgkinson on this subject is contained in the Appendix to the "Illustrations of Mechanics" by the author of this work.

The Section of Rupture in a Beam.

409. When a beam is deflected under a transverse strain, the material on that side of it on which it sustains the strain is compressed, and the material on the opposite side extended. That imaginary surface which separates the compressed from the extended portion of the material is called its neutral surface (Art. 356.), and its position has been determined under all the ordinary circumstances of flexure. That which constitutes the strength of a beam is the resistance of its material to compression on the one side of its neutral surface, and to extension on the other; so that if either of these yield the beam will be broken.

The *section of rupture* is that transverse section of the beam about which, in its state bordering upon rupture, it is the *most* extended, if it be about to yield by the extension of its material, or the *most* compressed if about to yield by the compression of its material.

In a prismatic beam, or a beam of uniform dimensions, it is evidently that section which passes through the point of greatest curvature of the neutral line, or the point in respect to which the radius of curvature of the neutral line is the least, or its reciprocal the greatest.

General Conditions of the Rupture of a Beam.

410. Let PQ be the section of rupture in a beam sustaining

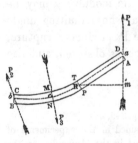

any given pressures, whose resultants are represented, if they be more in number than three, by the three pressures P_1, P_2, P_3. Let the beam be upon the point of breaking by the yielding of its material to extension at the point of greatest extension P; and let R represent, in the state of the beam bordering upon rupture, the intersection of the neutral surface with the section of rupture; which intersection being in the case of rectangular beams a straight line, and being in fact the neutral axis, in that particular position which is assumed by it when the beam is brought into its state bordering upon rupture, may be called the *axis of rupture;* ΔK the area in square inches of any element of the section of rupture, whose perpendicular distance from the axis of rupture R is represented by ρ; S the resistance in pounds opposed to the rupture of each square inch of the section at P; c_1 and c_2 the distances PR and QR in inches.

The forces opposed per square inch to the extension and compression of the material at different points of the section of rupture are to one another as their several perpendicular distances from the axis of rupture, if the elasticity of the material be supposed to remain perfect throughout the section of rupture, up to the period of rupture.

Now at the distance c_1 the force thus opposed to the extension of the material is represented per square inch by S; at the distance ρ the elastic force opposed to the extension or compression of the material (according as that distance is measured on the extended or compressed side), is therefore represented per square inch by $\frac{S}{c_1}\rho$, and the elastic force thus developed upon the element ΔK of the section of rupture by $\frac{S}{c_1}\rho\Delta K$, so that the moment of this elastic force about R is represented by $\frac{S}{c_1}\rho^2\Delta K$, and the sum of the moments of all the elastic forces upon the section of rupture about the axis of rupture by $\frac{S}{c_1}\Sigma\rho^2\Delta K$*; or representing the moment of inertia of the section of rupture about the axis of rupture by I, the sum of the moments of the elastic forces upon the section of rupture about its axis of

---

* It will be observed, as in Art. 360., that the elastic forces of extension and those of compression tend to turn the surface of rupture in the same direction about the axis of rupture.

rupture is represented, at the instant of rupture, by $\dfrac{SI}{c_1}$.*

Now the elastic forces developed upon PQ are in equilibrium with the pressures applied to either of the portions APQD or BPQC, into which the beam is divided by that section; the sum of their moments about the point P is therefore equal to the moment of P_1 about that point. Representing, therefore, by p_1 the perpendicular let fall from the point R upon the direction of P_1, we have

$$P_1 p_1 = \frac{SI}{c_1} \quad \ldots \quad (642).$$

411. If the deflexion be small in the state bordering upon rupture, and the directions of all the deflecting pressures be perpendicular to the surface of the beam, the axis of rupture passes through the centre of gravity of the section, and the value of c_1 is known. Where these conditions do not obtain, the value of c_1 might be determined by the principles laid down in Arts. 357. and 383. This determination would, however, leave the theory of the rupture of beams still incomplete in one important particular. The elasticity of the material has been supposed to remain perfect, at every point of the section of rupture, up to the instant when rupture is about to take place. Now it is to be observed, that by reason of its greater extension about the point P than at any other point of the section of rupture, the elastic limits are there passed before rupture takes place, and before they are attained at points nearer to the axis of rupture; the forces opposed to the extension of the material cannot therefore be assumed to vary, at all points of PR, accurately as their distances from the point R, in that state of the equilibrium of the beam which immediately precedes its rupture; and the sum of their moments cannot therefore be assumed to be accurately represented by the expression $\dfrac{SI}{c_1}$. This remark

---

* This expression is called by the French writers *the moment of rupture*; the beam is of greater or less strength under given circumstances according as it has a greater or less value.

affects, moreover, the determination of the values of h and R (Arts. 357. and 383.), and therefore the value of c_1.

To determine the influence upon the conditions of rupture by transverse strain of that unknown direction of the insistent pressures, and that variation from the law of perfect elasticity which belongs to the state bordering upon rupture, we must fall back upon experiment. From this it has resulted, *in respect to rectangular beams*, that the error produced by these different causes in equation (642) will be corrected if a value be assigned to c_1 bearing, for each given material, a constant ratio to the distance of the point P from the centre of gravity of the section of rupture ; so that c representing the depth of a rectangular beam, the error will be corrected, in respect to a beam of any material, by assigning to c_1 the value $m\frac{1}{2}c$, where m is a certain constant dependent upon the nature of the material. It is evident that this correction is equivalent to assuming $c_1 = \frac{1}{2}c$, and assigning to S the value $\frac{1}{m}$S instead of that which it has hitherto been supposed to represent, viz. the tenacity per square inch of the material of the beam.

It is customary to make this assumption. The values of S corresponding to it have been determined, by experiment, in respect to the materials chiefly used in construction, and will be found in a table at the end of this work. It is to these tables that the values represented by S in all subsequent formulæ are to be referred.

412. From the remarks contained in the preceding article, it is not difficult to conceive the existence of some direct relation between the conditions of rupture by transverse and by longitudinal strain. Such a relation of the simplest kind appears recently to have been discovered by the experiments of Mr. E. Hodgkinson [*], extending to the conditions of rupture by compression, and common to all the different varieties of material included under each of the following great divisions — timber, cast iron, stone, glass.

[*] This discovery was communicated to the British Association of Science at their meeting in 1842; it opens to us a new field of theoretical research.

The following tables contain the summary given by Mr. Hodgkinson of his results : —

Description of Material.	Assumed Crushing Strength per Square Inch.	Mean Tensile Strength per Square Inch.	Mean Transverse Strength of a Bar 1 Inch Square and 1 Foot Long.
Timber - - -	1000	1900	85·1
Cast-iron - -	1000	158	19·8
Stone, including marble - - -	1000	100	9·8
Glass (plate and crown) - - -	1000	123	10·

The following table shows the uniformity of this ratio in respect to different varieties of the same material : —

Description of Material.	Assumed Crushing Strength per Square Inch.	Mean Tensile Strength per Square Inch.	Mean Transverse Strength of a Bar 1 Inch Square and 1 Foot Long.
Black marble - -	1000	143	10·1
Italian marble - -	1000	84	10·6
Rochdale flagstone -	1000	104	9·9
High Moorstone -	1000	100	
Yorkshire flag - -	1000	—	9·5
Stone from Little Hulton, near Bolton -	1000	70	8·8

413. The strongest form of section at any given point in the length of the beam.

Since the extension and the compression of the material are the greatest at those points which are most distant from the neutral axes of the section, it is evident that the material cannot be in the state bordering upon rupture at every point of the section at the same instant (Art. 390.), unless all the material of the compressed side be collected at the *same distance* from the neutral axis, and likewise all the material of the extended side, or unless the material of the extended side and the material of the compressed side be respectively collected into two geometrical lines parallel to the neutral axis ; a distribution manifestly impossible, since it would produce an entire separation of the two sides of the beam.

The nearest practicable approach to this form of section is that represented in the accompanying figure, where the material is shown collected in two thin but wide flanges, but united by a narrow rib.

That which constitutes the strength of the beam being the resistance of its material to compression on the one side of its neutral axis, and its resistance to extension on the other side, it is evidently (Art. 390.) a second condition of the strongest form of any given section that when the beam is about to break across that section by extension on the one side, it may be about to break by compression on the other. So long, therefore, as the distribution of the material is not such as that the compressed and extended sides would yield together, the strongest form of section is not attained. Hence it is apparent that the strongest form of the section collects the greater quantity of the material on the compressed or the extended side of the beam, according as the resistance of the material to compression or to extension is the less. Where the material of the beam is cast iron *, whose resistance to extension is greatly less than its resistance to compression, it is evident that the greater portion of the material must be collected on the extended side.

Thus then it follows, from the preceding condition and this, that the strongest form of section in a cast iron beam is that by which the material is collected into two unequal flanges joined by a rib, the greater flange being on the extended side; and the proportion of this inequality of the flanges being just such as to make up for the inequality of the resistances of the material to rupture by extension and compression respectively.

Mr. Hodgkinson, to whom this suggestion is due, has directed a series of experiments to the determination of that proportion of the flanges by which the strongest form of section is obtained.†

* It is only in cast iron beams that it is customary to seek an economy of the material in the strength of the section of the beam ; the same principle of economy is surely, however, applicable to beams of wood.

† Memoirs of Manchester Philosophical Society, vol. iv. p. 453. Illustrations of Mechanics, Art. 68.

The details of these experiments are found in the following table : —

Number of Experiment.	Ratio of the Sections of the Flanges.	Area of whole Section in Square Inches.	Strength per Square Inch of Section in lbs.
1	1 to 1·	2·82	2368
2	1 to 2·	2·87	2567
3	1 to 4·	3·02	2737
4	1 to 4·5	3·37	3183
5	1 to 5·5	5·0	3346
6	1 to 6·1	6·4	4075

In the first five experiments each beam broke by the tearing asunder of the lower flange. The distribution by which both were about to yield together — that is, the strongest distribution — was not therefore up to that period reached. At length, however, in the last experiment, the beam yielded by the compression of the upper flange. In this experiment, therefore, the *upper* flange was the weakest; in the one before it, the *lower* flange was the weakest. For a form between the two, therefore, the flanges were of equal strength to resist extension and compression respectively; and this was the strongest form of section (Art. 390.).

In this strongest form the lower flange had six times the material of the upper. It is represented in the accompanying figure.

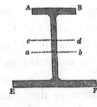

In the best form of cast iron beam or girder used before these experiments, there was never attained a strength of more than 2885 lbs. per square inch of section. There was, therefore, by this form, a gain of 1190 lbs. per square inch of the section, or of $\frac{2}{3}$ths the strength of the beam.

414. THE SECTION OF RUPTURE.

The conditions of rupture being determined in respect to *any* section of the beam by equation (642), it is evident that the particular section across which rupture will actually take place is that in respect to which equation (642) is *first* satisfied, as P_1 is continually increased; or that section in respect to which the formula

$$\frac{I}{p_1 c_1} \quad \ldots \ldots \text{(643)}$$

is the least.

If the beam be loaded along its whole length, and x represent the distance of any section from the extremity at which the load commences, and μ the load on each foot of the length, then (Art. 373.) $P_1 p_1$ is represented by $\frac{1}{2}\mu x^2$. The section of rupture in this case is therefore that section in respect to which μ is first made to satisfy the equation $\frac{1}{2}\mu x^2 = \frac{SI}{c_1}$; or in respect to which the formula

$$\frac{I}{x^2 c^1} \quad \ldots \ldots \text{(644)}$$

is the least.

If the section of the beam be uniform, $\dfrac{I}{c_1}$ is constant; the section of rupture is therefore evidently that which is most distant from the free extremity of the beam.

415. THE BEAM OF GREATEST STRENGTH.

The beam of greatest strength being that (Art. 390.) which presents an equal liability to rupture across every section, or in respect to which every section is brought into the state bordering upon rupture by the same deflecting pressure, is evidently that by which a given value of P is made to satisfy equation (642) for all the possible values of I, p_1, and c_1, or in respect to which the formula

$$\frac{I}{p_1 c_1} \quad \ldots \ldots \text{(645)}$$

is constant.

If the beam be uniformly loaded throughout (Art. 373.), this condition becomes

$$\frac{I}{x^2 c_1} \quad \cdots \cdots (646),$$

or constant, for all points in the length of the beam.

416. ONE EXTREMITY OF A BEAM IS FIRMLY IMBEDDED IN MASONRY, AND A PRESSURE IS APPLIED TO THE OTHER EXTREMITY IN A DIRECTION PERPENDICULAR TO ITS LENGTH: TO DETERMINE THE CONDITIONS OF THE RUPTURE.

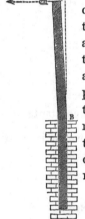

If x represent the distance of any section of the beam from the extremity A to which the load P is applied, and a its whole length, and if the section of the beam be every where the same, then the formula (643) is greatest at the point B, where x is greatest: at this point, therefore, the rupture of the beam will take place. Representing by P the pressure necessary to break the beam, and observing that in this case the perpendicular upon the direction of P from the section of rupture is represented by a, we have (equation 642)

$$P = \frac{SI}{c_1 a} \quad \cdots \cdots (647).$$

If the section of the beam be a rectangle, whose breadth is b and its depth c, then $I = \frac{1}{12} b c^3$, $c_1 = \frac{1}{2} c$.

$$\therefore P = \frac{1}{6} S \frac{bc^2}{a} \quad \cdots \cdots (648).$$

If the section be a solid cylinder, whose radius is c, then (Art. 366.) $I = \frac{1}{4} \pi c^4$, $c_1 = c$.

$$\therefore P = \frac{1}{4} \pi S \frac{c^3}{a} \quad \cdots \cdots (649).$$

If the section be a *hollow* cylinder, whose radii are r_1 and r_2,

$I = \frac{1}{4}\pi \left(r_1{}^4 - r_2{}^4\right)$; which expression may be put under the form $\pi c r \left(r^2 + \frac{1}{4}c^2\right)$ (see Art. 86.), r representing the mean radius of the hollow cylinder, and c its thickness. Also $c_1 = r_1 = r + \frac{1}{2}c$;

$$\therefore P = \pi S \frac{(r^2 + \frac{1}{4}c^2)cr}{(r + \frac{1}{2}c)a} \cdot \cdot \cdot \cdot \cdot (650).$$

417. *The strongest form of beam under the conditions supposed in the last article.*

1st. Let the section of the beam be a rectangle, and let y be the depth of this rectangle at a point whose distance from its extremity A is represented by x, and let its breadth b be the same throughout. In this case $I = \frac{1}{12}by^3$, $c_1 = \frac{1}{2}y$; therefore (equation 642) $P = \frac{SI}{c_1 x} = \frac{1}{6}Sb\frac{y^2}{x}$. If, therefore, P be taken to represent the pressure which the beam is destined just to support, then the form of its section ABC is determined (Art. 415.) by the equation

$$y^2 = \frac{6P}{Sb}x \cdot \cdot \cdot \cdot \cdot (651);$$

it is therefore a parabola, whose vertex is at A.*

If the portion DC of the beam do not rest against masonry at every point, but only at its extremity D, its form should evidently be the same with that of ABC.

2d. Let the section be *a circle*, and let y represent its radius at distance x from its extremity A, then $I = \frac{1}{4}\pi y^4$, $c_1 = y$; therefore $P = \frac{1}{4}\pi S \frac{y^3}{x}$ so that the geometrical form of its longitudinal in section is determined by the equation

* The portion of the beam imbedded in the masonry should have the form described in Art. 419.

$$y^3 = \frac{4P}{\pi S} x \ \cdot \ \cdot \ \cdot \ \cdot \ \cdot \ (652),$$

P representing the greatest pressure to which it is destined to be subjected.

418. THE CONDITIONS OF THE RUPTURE OF A BEAM SUP-
PORTED AT ONE EXTREMITY, AND LOADED THROUGHOUT
ITS WHOLE LENGTH.

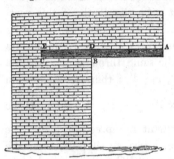

Representing the weight resting upon each inch of its length a by w, and observing that the moment of the weight upon a length x of the beam from A, about the corresponding neutral axis, is represented (Art. 373.) by $\frac{1}{2}\mu x^2$, it is apparent (Art. 414.) that, if the beam be of uniform dimensions, its section of rupture is BD·

Its strength is determined by substituting $\frac{1}{2}\mu a^2$ for $P_1 p_1$ in equation (642), and solving in respect to μ ; we thus obtain

$$\mu = \frac{2SI}{c_1 a^2} \ \cdot \ \cdot \ \cdot \ \cdot \ \cdot \ (653) ;$$

by which equation is determined the uniform load to which the beam may be subjected, on each inch of its length.

For a rectangular beam, whose width is b and its depth c, this expression becomes

$$\mu = \frac{Sbc^2}{3a^2} \ \cdot \ \cdot \ \cdot \ \cdot \ \cdot \ (654).$$

419. To determine the form of greatest strength (Art. 415.)

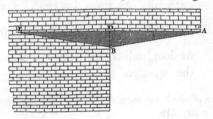

in the case of a beam having a rectangular section of uniform breadth, $\frac{1}{2}\mu x^2$ must be substituted for $P_1 p_1$ in equation (642), and $\frac{1}{12}by^3$ for I, and $\frac{1}{2}y$ for c_1; whence we obtain by reduction

$$y = \left(\frac{3\mu}{Sb}\right)^{\frac{1}{2}} x \ \ . \ . \ . \ . \ (655).$$

The form of greatest strength is therefore, in this case, the straight line joining the points A and B; the distance DB being determined by substituting the distance AD for x in the above equation.

That portion BED of the beam which is imbedded in the masonry should evidently be of the same form with DBA.*

420. If, in addition to the uniform load upon the beam, a given weight W be suspended from A, $\frac{1}{2}\mu x^2 + Wx$ must be

substituted for $P_1 p_1$ in equation (642); we shall thus obtain for the equation to the form of greatest strength

$$y^2 = \frac{3\mu}{Sb}\left\{\frac{2W}{\mu}x + x^2\right\} \ \ . \ . \ . \ . \ (656),$$

which is the equation to an hyperbola having its vertex at A.

---

* It is obvious that in all cases the strength of a beam at each point of its length is dependent upon the dimensions of its cross section at that point, and that its general form may in any way be changed without impairing its strength, provided those dimensions of the section be everywhere preserved.

421. The beam of greatest strength in reference to the form of its section and to the variation of the dimensions of its section, when supported at one extremity in a horizontal position, and loaded uniformly throughout its length.

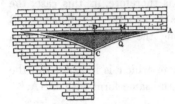

The general form of the section must evidently be that described in Art. 413. Let the same notation be taken as in Art. 367., except that the depth MQ of the plate or rib joining the two flanges is to be represented by y, and its thickness by c, so that $d_3 = y$, and $A_3 = cy$; therefore by equation (508),

$$I = \tfrac{1}{12}(A_1 d_1^2 + A_2 d_2^2 + cy^3) + \tfrac{1}{4} \cdot \left\{ \frac{4A_1 A_2 + (A_1 + A_2)cy}{A_1 + A_2 + cy} \right\} y^2.$$

Also, representing by c_1 the distance of the centre of gravity of the whole section from the upper surface of the beam, since $\tfrac{1}{2}y$ and y may be taken to represent the distances of the centres of gravity of the sections A_3 of the rib and A_1 of the lower flange from the same surface nearly, we have $c_1(A_1 + A_2 + A_3) = \tfrac{1}{2}yA_3 + yA_1$, or $c_1(A_1 + A_2 + cy) = \tfrac{1}{2}cy^2 + A_1 y$. Substituting for I and c_1 in equation (642), and for $P_1 p_1$ its value $\tfrac{1}{2}\mu x^2$, x being taken to represent the distance AM, and μ the load on each inch of that length, we have (Art. 415.)

$$\frac{3\mu}{S} x^2 = \frac{(A_1 d_1^2 + A_2 d_2^2 + cy^3)(A_1 + A_2 + cy) + \{12A_1 A_2 + 3(A_1 + A_2)cy\}y^2}{cy^2 + 2A_1 y} \cdot (657).$$

Let the area cy of the section of the rib now be neglected, as exceedingly small when compared with the areas of the sections of the flanges, an hypothesis which assigns to the beam somewhat less than its actual strength; let also the area of the section of the upper flange be assumed equal to n times that of the lower, or $A_2 = nA_1$,

$$\therefore \frac{6\mu}{SA_1}x^2 = \frac{(1+n)(d_1{}^2+nd_2{}^2)}{y} + 12ny \quad \cdots \cdots \quad (658).$$

If the flanges be exceedingly thin, d_1 and d_2 are exceedingly small; the first term of this expression may therefore be neglected. The equation will then become that to a parabola whose vertex is at A and its axis vertical. This may therefore be assumed as a near approximation to the true form of the curve AQC.

Where the material is cast iron, it appears by Mr. Hodgkinson's experiments (Art. 413.) that n is to be taken $=6$.

422. A BEAM OF UNIFORM SECTION IS SUPPORTED AT ITS EXTREMITIES AND LOADED AT ANY POINT BETWEEN THEM: IT IS REQUIRED TO DETERMINE THE CONDITIONS OF RUPTURE.

The point of rupture in the case of a uniform section is evidently (Art. 414.) the point C, from which the load is suspended; representing AB, AC, BC, by a, a_1, and a_2; and observing that the pressure P_1 upon the point B of the beam $=\dfrac{Wa_1}{a}$, so that the moment of P_1, in respect to the section of rupture $C=\dfrac{Wa_1a_2}{a}$, we have, by equation (642), $\dfrac{Wa_1a_2}{a}=\dfrac{SI}{c_1}$;

$$\therefore W = \frac{SIa}{a_1a_2c_1} \quad \cdots \cdots \quad (659).$$

If the beam be *rectangular*, $I = \frac{1}{12}bc^3$, $c_1 = \frac{1}{2}c$,

$$\therefore W = \frac{S}{6}\frac{bc^2a}{a_1a_2} \quad \cdots \cdots \quad (660);$$

where W represents the breaking weight, S the modulus of rupture, a the length, b the breadth, c the depth, and a_1, a_2 the distances of the point c from the two extremities, all these dimensions being in inches.

If the load be suspended in the middle, $a_1 = a_2 = \frac{1}{2}a$,

$$\therefore W = \frac{2S}{3}\frac{bc^2}{a} \quad \cdots \cdots \quad (661).$$

If the beam be a *solid cylinder*, whose radius $=c$, then $I=\frac{1}{4}\pi c^4$, $c_1=c$; therefore, equation (659),

$$W=\frac{\pi S}{4}\frac{ac^3}{a_1 a^2} \quad \ldots \ldots (662).$$

If the beam be a hollow cylinder, whose mean radius is r, and its thickness c, $I=\pi cr(r^2+\frac{1}{4}c^2)$, $c_1=r+\frac{1}{2}c$; therefore, equation (659),

$$W=\pi S\frac{acr(r^2+\frac{1}{4}c^2)}{a_1 a_2(r+\frac{1}{2}c)} \quad \ldots \ldots (663).$$

If the section of the beam be that represented in Art. 413., being every where of the same dimensions, then, observing that $Ac_1=\frac{1}{2}d_3A_3+d_3A_1$, nearly, we have (equations 508 and 659)

$$W=\frac{Sa}{6}\frac{A(A_1d_1{}^2+A_2d_2{}^2+A_3d_3{}^2)+3(4A_1A_2+A_1A_3+A_2A_3)d_3{}^2}{(2A_1+A_3)a_1 a_2 d_3} \ldots (664) ;$$

where A_1, A_2 represent the areas of the sections of the upper and lower flanges, and A_3 that of the connecting rib or plate, and d_1, d_2, d_3 their respective depths.

423. A BEAM IS SUPPORTED AT ITS EXTREMITIES, AND LOADED AT ANY GIVEN POINT BETWEEN THEM; ITS SEC-TION IS OF A GIVEN GEOMETRICAL FORM, BUT OF VARI-ABLE DIMENSIONS: IT IS REQUIRED TO DETERMINE THE LAW OF THIS VARIATION, SO THAT THE STRENGTH OF THE BEAM MAY BE A MAXIMUM.

W representing the breaking load upon the beam, and a_1, a_2 the distances of its point of suspension C, from A and B, the pressure P_1 upon A is represented by $\dfrac{Wa_2}{a}$. If, there-

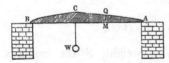

fore (Art. 390.), x represent the horizontal distance of any section MQ from the point of support A, and I its moment of inertia, and c_1 the distance from its centre of gravity to the point where rupture is about to take place (in this case its lowest point); then by equation (642)

$$\frac{Wa_2}{a}x = \frac{SI}{c_1} \quad \ldots \quad (665).$$

1st. Let the section be *rectangular ;* let its breadth b be constant; and let its depth at the distance x from A be represented by y; therefore $I = \frac{1}{12}by^3$, $c_1 = \frac{1}{2}y$. Substituting in the above equation and reducing,

$$y^2 = \frac{6Wa_2}{Sab}x \quad \ldots \quad (666).$$

The curve AC is therefore a parabola, whose vertex is at A, and its axis horizontal. In like manner the curve BC is a parabola, whose equation is identical with the above, except that a_1 is to be substituted in it for a_2.

2d. Let the section of the beam be a circle. Representing the radius of a section at distance x from A by y, we have $I = \frac{1}{4}\pi y^4$, $c_1 = y$; therefore by equation (665)

$$y^3 = \frac{4Wa_2}{S\pi a}x \quad \ldots \quad (667).$$

3d. Let the section of the beam be circular ; but let it be hollow, the thickness of its material being every where the same, and represented by c. If $y =$ mean radius of cylinder at distance x from A, then $I = \pi cy(y^2 + \frac{1}{4}c^2)$, $c_1 = (y + \frac{1}{2}c)$;

$$\therefore \ x = \frac{S\pi acy}{2Wa_2}\left(\frac{4y^2 + c^2}{2y + c}\right) \quad \ldots \quad (668).$$

424. THE BEAM OF GREATEST ABSOLUTE STRENGTH WHEN LOADED AT A GIVEN POINT AND SUPPORTED AT THE EXTREMITIES.

Let the section of the beam be that of greatest strength (Art. 413.). Substituting in equation (665) the value of $\frac{I}{c_1}$, as before in equation (657), and reducing,

$$\frac{6Wa_2x}{Sa} = \frac{(A_1d_1^2 + A_2d_2^2 + cy^3)(A_1 + A_2 + cy) + 12A_1A_2y^2 + 3(A_1 + A_2)cy^3}{cy^2 + 2A_1y}. \ (669).$$

o o 4

If the section cy of the rib be every where exceedingly small as compared with the sections of the flanges, and if $A_2 = nA_1$,

$$\frac{12 W a_2}{S A_1 a} x = \frac{(1+n)(d_1{}^2 + n d_2{}^2)}{y} + 12 ny \ \ . \ . \ . \ . \ (670).$$

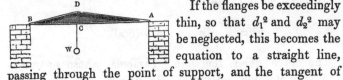

If the flanges be exceedingly thin, so that $d_1{}^2$ and $d_2{}^2$ may be neglected, this becomes the equation to a straight line, passing through the point of support, and the tangent of whose inclination to the horizon is represented by $\dfrac{W a_2}{S A_1 na}$.

If its proper value be assigned to n (Art. 413.), this may be assumed as an approximation to the true form of beam of THE GREATEST ABSOLUTE STRENGTH. When the material is cast iron, it appears by the experiments of Mr. Hodgkinson (Art. 413.) that $n = 6$. A_2 represents in all the above cases the section of the *extended* flange; in this case, therefore, it represents the section of the *lower* flange.

The depth CD at the point of suspension may be determined by substituting a_1 for x in equation (670); its value is thus found to be represented by the formula

$$\overline{CD} = \frac{W a_1 a_2}{S A_1 na} \ \ . \ . \ . \ . \ (671).$$

425. If instead of the depth of the beam being made to vary so as to adapt itself to the condition (Art. 390.) of uniform strength, its breadth b be made thus to vary, the depth c remaining the same; then, assuming the breadth of the upper flange at the distance x from the point of support A to be represented by y, and the section of the lower flange to be n times greater than that of the upper; observing, moreover, that in equation (508) $A_1 = y d_1$, $A_2 = n A_1 = n y d_1$; neglecting also A_3 as exceedingly small when compared with A_1 and A_2, and writing c for d_3, we have by reduction

$$I = \tfrac{1}{12} y (d_1{}^2 + n d_2{}^2) d_1 + y \frac{n c^2 d_1}{n+1}.$$

Also c_1 being the distance of the lower surface of the beam from the common centre of gravity of the sections of the two flanges, we have $c_1(n+1)=c$. Eliminating, therefore, the values of I and c_1 from equation (665),

$$x = \frac{Sa}{Wa_2} \left\{ \tfrac{1}{12}(n+1)(d_1{}^2 + n d_2{}^2)\frac{d_1}{c} + n c d_1 \right\} y \; \dots \; (672),$$

the equation to a straight line. Each flange is therefore in this case a quadrilateral figure, whose dimensions are determined from the greatest breadth; this last being known, for the upper flange, by substituting a_1 for x in the above equation, and solving in respect of y, and for the lower flange from the equation $n b_1 d_1 = b_2 d_2$, in which $b_1\,b_2$ represent the greatest breadths of the two flanges, and $d_1\,d_2$ their depths.

426. A BEAM IS LOADED UNIFORMLY THROUGHOUT ITS WHOLE LENGTH, AND SUPPORTED AT ITS EXTREMITIES: IT IS REQUIRED TO DETERMINE, 1. THE CONDITIONS OF ITS RUPTURE WHEN ITS CROSS SECTION IS UNIFORM THROUGHOUT; 2. THE STRONGEST FORM OF BEAM HAVING EVERY WHERE A RECTANGULAR CROSS SECTION; 3. THE BEAM OF GREATEST STRENGTH IN REFERENCE BOTH TO THE FORM AND THE VARIATION OF ITS CROSS SECTION.

1. If the section of the beam be uniform, its point of rupture
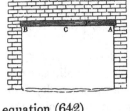
is determined by formula (644) to be its *middle* point. Representing therefore, in this case, the length of the beam by $2a$, the weight on each inch of its length by μ, and its breadth by b; and observing that in this case $P_1 p_1 = \mu a^2 - \tfrac{1}{2}\mu a^2 = \tfrac{1}{2}\mu a^2$, we have by equation (642)

$$\mu = \frac{2SI}{a^2 c_1} \; \dots \; (673),$$

where μ represents the load per inch of the length of the beam necessary to produce rupture. In the case of a rectangular beam, this equation becomes

$$\mu = \frac{Sbc^2}{3a^2} \cdots \cdots (674).$$

2. To determine the form of the beam of greatest strength, having a rectangular section of given breadth b, let y be taken to represent its depth PQ at a point P, and x its horizontal

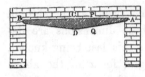

distance from the point A. Then $I = \frac{1}{12}by^3$, $c_1 = \frac{1}{2}y$; also P_1p_1 (equation 642) representing the moment of the resultant of the pressures upon AP about the centre of gravity of $PQ = \mu ax - \frac{1}{2}\mu x^2$; therefore by equation (642) $\mu ax - \frac{1}{2}\mu x^2 = \frac{1}{6}Sby^2$;

$$\therefore \ y^2 = \frac{3\mu}{Sb}\,(2ax - x\ ),$$

the equation to an ellipse, whose vertex is in A, and its centre at C.

3. To determine the beam of absolute maximum strength, let it be assumed, as in Art. 424., that the area of the section of the rib is exceedingly small as compared with the areas of the sections of the flanges; and let the area of the section of the lower or extended flange be n times that of the upper;

then, as in Art. 424., $\dfrac{I}{c_1} = \dfrac{A_1}{12}\left\{ \dfrac{(1+n)(d_1{}^2 + nd_2{}^2)}{y} + 12ny \right\}$;

also $P_1p_1 = \mu ax - \frac{1}{2}\mu x^2$; whence, by equation (642),

$$\mu ax - \tfrac{1}{2}\mu x^2 = \frac{SA_1}{12}\left\{ \frac{(1+n)(d_1{}^2 + nd_2{}^2)}{y} + 12ny \right\} \cdots \cdots (675).$$

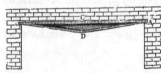

If the depths d_1 and d_2 of the flanges be small, as compared with the whole depth of the beam, at every point, the first term of the second member may be neglected, and we shall have by reduction

$$x(2a-x)=\frac{2nA_1S}{\mu}y \cdot \cdot \cdot \cdot \cdot (676);$$

the equation to a parabola, whose axis is the vertical passing through the centre of the beam whose parameter is $\frac{2nA_1S}{\mu}$, and the position of its vertex D determined by the formula

$$\overline{CD}=\frac{\mu a^2}{2nA_1S} \cdot \cdot \cdot \cdot \cdot (677).$$

4. If it be proposed to make the rib or plate uniting the two flanges every where of the same depth *, and so to vary the breadths of the flanges as to give to the beam a uniform strength at all points under these circumstances; representing by y the breadth of the upper flange at a horizontal distance x from the point of support, we shall obtain, as in Art. 425.,

$$\frac{I}{c_1}=y\frac{(n+1)}{12c}(d_1^2+nd_2^2)d_1+yncd_1.$$

Moreover, $P_1p_1=\mu ax-\frac{1}{2}\mu x^2=\frac{1}{2}\mu x(2a-x)$; whence we obtain by substitution in equation (642), and reduction,

$$x(2a-x)=\left(\frac{Sd_1}{6\mu c}\right)\{(n+1)(d_1^2+nd_2^2)+12nc^2\}y \cdot \cdot \cdot (678);$$

the equation to a parabola, whose axis is in the horizontal line bisecting the flange at right angles, its parameter represented by the coefficient of y in the preceding equation, and half the breadth of the flange in the middle determined by the formula

$$\frac{6ca^2\mu}{\{(n+1)(d_1^2+nd_2^2)+12nc^2\}Sd_1} \cdot \cdot \cdot \cdot \cdot (679).$$

The equation to the lower flange is determined by substituting for y, in equation (678), $\frac{yd_2}{nd_1}$; whence it follows that the breadth of the lower flange in the middle is equal to that of the upper multiplied by the fraction $\frac{yd_2}{nd_1}$.

* As in Mr. Hodgkinson's construction.

427. A RECTANGULAR BEAM OF UNIFORM SECTION, AND UNIFORMLY LOADED THROUGHOUT ITS LENGTH, IS SUPPORTED BY TWO PROPS PLACED AT EQUAL DISTANCES FROM ITS EXTREMITIES: TO DETERMINE THE CONDITIONS OF RUPTURE.

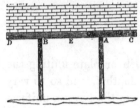

It is evident from formula (644) that the section of rupture of the portion CA of the beam is at A, and therefore that the conditions of its rupture are determined (Art. 418.) by the equation

$$\mu_1 = \frac{Sbc^2}{3a_1^2} \ \cdots \cdots \ (680);$$

where μ_1 represents, as before, the load upon each inch of the length of the beam, b its breadth, c its depth, and a_1 the length of the portion AC.

Again, it is evident that the point of rupture of the portion AB of the beam is at E. Now the value of $P_1 p_1$ (equation 642) is, in respect to the portion AE of the beam, $\mu_2 a (a - a_1) - \frac{1}{2}\mu_2 a^2$; $2a$ representing the whole length of the beam, μ_2 the load upon each inch of the length of the beam which would produce rupture at E, and therefore $\mu_2 a$ the resistance of each prop in the state bordering upon rupture; also $\frac{I}{c_1} = \frac{1}{6}bc^2$. Whence, by equation (642), $\mu_2 a(a - a_1) - \frac{1}{2}\mu_2 a^2 = \mu_2 a(\frac{1}{2}a - a_1) = \frac{1}{6}bc^2 S$;

$$\therefore \ \mu_2 = \frac{Sbc^2}{3a(a - 2a_1)} \ \cdots \cdots \ (681).$$

428. THE BEST POSITIONS OF THE PROPS.

If the load μ be imagined to be continually increased, it is evident that rupture will eventually take place at A or at E according as the limit represented by equation (680), or that represented by equation (681), is first attained, or according as μ_1 or μ_2 is the *less*.

Let μ_1 be conceived to be the less, and let the prop A be moved nearer to the extremity C; a_1 being thus diminished, μ_1 will be increased, and μ_2 diminished. Now if, after this change in the position of the prop, μ_1 still remains less than μ_2, it is evident that the beam will bear a greater load than it would before, and that when by continually increasing the load it is brought into the state bordering upon rupture at A it will not be in the state bordering upon rupture at E. The beam may therefore be strengthened yet further by moving the prop A towards C ; and thus *continually*, so that the beam evidently becomes the strongest when the prop is moved into such a position that μ_1 may just equal μ_2. This position is readily determined from equations (680) and (681) to be that in which

$$a_1 = a(\sqrt{2}-1) = \cdot 44225a \quad \ldots \quad \ldots \quad (682).$$

429. A RECTANGULAR BEAM OF UNIFORM SECTION AND UNIFORMLY LOADED IS SUPPORTED AT ITS EXTREMITIES, AND BY TWO PROPS SITUATED AT EQUAL DISTANCES FROM THEM: TO DETERMINE THE CONDITIONS OF RUPTURE.

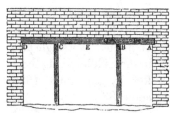

Adopting the same notation as in Art. 376., it appears by equation (548) that the distance x_1 of the point of greatest curvature of the neutral line, and therefore of the section of rupture in AB from A (Art. 409.), is determined by the equation $\mu x_1 = P$, it being observed that, at the section of rupture, the neutral line is *concave* to the axis of x, and therefore the second differential coefficient (equation 548) negative. The value of P is that determined by equation (556); so that

$$x_1 = \tfrac{1}{8}a \cdot \frac{n^3 + 12n^2 - 24n + 8}{n(2n-3)} \quad \ldots \quad \ldots \quad (683),$$

where a represents the distance AE, and na the distance AB.

Let P represent the intersection of the neutral line with the plane of rupture, and μ_1 the load per inch of the whole length of the beam which would produce rupture at P. Now the sum of the moments of the forces impressed on AP (other than the elastic forces on the section of rupture) is represented, in the state bordering upon rupture, by $P_1x_1 - \frac{1}{2}\mu_1x_1^2$; or, since $P_1 = \mu_1x_1$, it is represented by $\frac{1}{2\mu_1}P_1^2$; whence it follows by equation (642) that the conditions of the rupture of the beam between A and B are determined by the equation $\frac{1}{2\mu_1}P_1^2 = \frac{1}{6}Sbc^2$, or

$$P_1^2 = \frac{1}{3}\mu_1 Sbc^2 \ \cdots \ (684).$$

Eliminating the value of P_1 between equations (556) and (684), we obtain

$$\mu_1 = \frac{Sbc^3}{3a^2}\left\{\frac{8n(2n-3)}{n^3 + 12n^2 - 24n + 8}\right\}^2 \ \cdots \ (685).$$

Substituting this value of μ_1 in equation (684), and reducing,

$$P_1 = \frac{8Sbc^2}{3a}\left\{\frac{n(2n-3)}{n^3 + 12n - 24n + 8}\right\} \ \cdots \ (686).$$

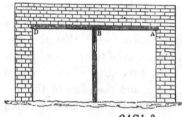

If the points B and C coincide, or the beam be supported by a single prop in the middle, $n = 1$; therefore by equations (685) and (686),

$$\mu_1 = \frac{64Sbc^2}{27a^2} \ \cdots \ (687);$$

$$P_1 = \frac{8Sbc^2}{9a^2} \ \cdots \ (688).$$

Similarly, it appears by equation (552) that the point of greatest curvature between B and C is E; if the rupture of the beam take place first between these points, it will therefore take place in the middle. Let μ_2 represent the load, per inch of the length, which would produce a rupture at E. Now, the sum of the moments about E of the forces impressed upon AE is $P_1 a + P_2(a - na) - \frac{1}{2}\mu_2 a^2 = (P_1 + P_2)a - P_2 na - \frac{1}{2}\mu_2 a^2 = \mu_2 a^2 - (\mu_2 a - P_1)na - \frac{1}{2}\mu_2 a^2$ (since $P_2 + P_2 = \mu_2 a$) $= \frac{1}{2}(1 - 2n)\mu_2 a^2 + P_1 na$. Therefore by equation (642)

$$\frac{1}{2}(1 - 2n)\mu_2 a^2 + P_1 na = \frac{1}{6}Sbc^2 \quad \ldots \ldots (689).$$

Substituting for P_1 its value from equation (556), and solving in respect to μ_2,

$$\mu_2 = \frac{4}{3}\frac{Sbc^2}{a^2}\left\{\frac{2n - 3}{n^3 - 4(1 - n)^2}\right\} \quad \ldots \ldots (690).$$

If the load be continually increased, the beam will break between A and B, or between B and C, according as μ_1 (equation 685) or μ_2 (equation 690) is the *less*.

430. THE BEST POSITIONS OF THE PROPS.

It may be shown, as in Art. 428., that the positions in which the props must be placed so as to cause the beam to bear the greatest possible load distributed uniformly over its whole length, are those by which the values of μ_1 (equation 685) and μ_2 (equation 690) are made equal; the former of these quantities representing the load, per inch of the length, which being uniformly distributed over the whole beam would just produce rupture between A and B, if it did not before take place between B and C; and the latter that which would, under the same circumstances, produce rupture between B and C if it had not before taken place between A and B.

Let, then, *na* represent the distance at which the prop B must be placed from A to produce this equality; and let the value of μ_1 given by equation (684) be substituted for μ_2 in equation (689); we shall thus obtain by reduction

$$P_1^2 a^2 + 2P_1 a \cdot \frac{Sbc^2 n}{3(1 - 2n)} = \frac{(Sbc^2)^2}{9(1 - 2n)}.$$

Solving this quadratic in respect to $P_1 a$,

$$P_1 a = \tfrac{1}{3} S b c^2 \left\{ \frac{\pm(1-n)-n}{1-2n} \right\}$$

The negative sign must be taken in this expression, since the positive would give $P_1 = \mu_1 a$ by equation (684), and corresponds therefore to the case $n=0$. Assuming the negative sign, and reducing, we have $3(2n-1)P_1 a = S b c^2$. Substituting in this expression for P_1 its value from equation (686), and reducing,

$$\frac{8n(2n-1)(2n-3)}{n^3 + 12n^2 - 24n + 8} = 1.$$

The three roots of this equation are 1.57087, $.61078$, and $.26994$. The first and last are inadmissible; the one carrying the point B beyond E, and the other assigning to P_1 a negative value.* The best position of the prop is therefore that which is determined by the value

$$n = .61078 \ldots \ldots (691).$$

431. THE CONDITIONS OF THE RUPTURE OF A RECTANGULAR BEAM LOADED UNIFORMLY THROUGHOUT ITS LENGTH, AND HAVING ITS EXTREMITIES PROLONGED AND FIRMLY IMBEDDED IN MASONRY.

It has been shown (Art. 378.) that the conditions of the deflexion of the beam are, in this case, the same as though its extremities, having been prolonged to a point A (see *fig.* p. 573.) such that AB might equal $.6202$AE, had been supported by a prop at B, and by the resistance of any fixed surface at A. The load which would produce the rupture of the beam is therefore, in this case, the same as that which would produce the rupture of a beam supported by props (Art. 429.) between the props, and is determined by that value of μ_2 (equation 690) which is given by the value $.6202$

---

* We may, nevertheless, suppose the extremity A, instead of being supported from beneath, to be pinned down by a resistance or a pressure acting from above. This case may occur in practice, and the best position of the props corresponding to it is that which is determined by the least root of the equation, viz. $.26994$.

of n. It is, however, to be observed that the symbol a re-

presents in that equation the distance AE (*fig.* Art. 429.); and that if we take it to represent the distance BE in that or the accompanying figure, we must substitute $\frac{a}{1-n}$ for a in equation (690), since $a = \mathrm{BE} = \mathrm{AE} - \mathrm{AB} = (1-n)\mathrm{AE}$; so that $\mathrm{AE} = \frac{a}{1-n}$. This substitution being made, equation (690) becomes

$$\mu_2 = \tfrac{4}{3}\frac{Sbc^2}{a^2}\frac{(2n-3)(1-n)^2}{n^3-4(1-n)^2};$$

and substituting the value 6202 for n, we obtain by reduction

$$\mu_2 = \frac{Sbc^2}{a^2} \quad \cdots \cdots \cdots (692),$$

by which formula the load per inch of the length of the beam necessary to produce rupture is determined.

If the beam had not been prolonged beyond the points of support B and D and imbedded in the masonry, then the load per inch of the length necessary to produce rupture would have been represented by equation (674): elimi-nating between that equation and equation (692), we obtain $\mu_2 = 3\mu$; so that the load per inch of the length neces-sary to produce rupture is 3 times as great, when the extre-mities of the beam are prolonged and firmly imbedded in the masonry, as when they are free; i. e. *the strength of the beam is 3 times as great in the one case as in the other.*

432. THE STRENGTH OF COLUMNS.

For all the knowledge of this subject on which any

P P

reliance can be placed by the engineer he is indebted to experiment.*

The following are the principal results obtained in the valuable series of experimental inquiries recently instituted by Mr. Eaton Hodgkinson. †

FORMULÆ REPRESENTING THE ABSOLUTE STRENGTH OF A CYLINDRICAL COLUMN TO SUSTAIN A PRESSURE IN HE DIRECTION OF ITS LENGTH.

D = external diameter or side of the square of the column in inches.

D_1 = internal diameter of hollow cylinder in inches.

L = length in feet.

W = breaking weight in tons.

Nature of the Column.	Both Ends being rounded, the Length of the Column exceeding 15 times its Diameter.	Both Ends being flat, the Length of the Column exceeding 30 times its Diameter.
Solid cylindrical column of cast iron	$W = 14 \cdot 9 \dfrac{D^{3 \cdot 76}}{L^{1 \cdot 7}}$	$W = 44 \cdot 16 \dfrac{D^{3 \cdot 55}}{L^{1 \cdot 7}}$
Hollow cylindrical column of cast iron	$W = 13 \dfrac{D^{3 \cdot 76} - D_1^{3 \cdot 76}}{L^{1 \cdot 7}}$	$W = 44 \cdot 34 \dfrac{D^{3 \cdot 55} - D_1^{3 \cdot 55}}{L^{1 \cdot 7}}$
Solid cylindrical column of wrought iron	$W = 42 \cdot 8 \dfrac{D^{3 \cdot 76}}{L^2}$	$W = 133 \cdot 75 \dfrac{D^{3 \cdot 55}}{L^2}$
Solid square pillar of Dantzic oak (dry)	- - -	$W = 10 \cdot 95 \dfrac{D^4}{L^2}$
Solid square pillar of red deal (dry)	- - -	$W = 7 \cdot 81 \dfrac{D^4}{L^2}$

* The hypothesis upon which it has been customary to found the theoretical discussion of it, is so obviously insufficient, and the results have been shown by Mr. Hodgkinson to be so little in accordance with those of practice, that the high sanction it has received from labours such as those of Euler, Lagrange, Poisson, and Navier can no longer establish for it a claim to be admitted among the conclusions of science. (See Appendix G.)

† From a paper by Mr. Hodgkinson published in the second part of the Transactions of the Royal Society for 1840, to which the royal medal of the Society was awarded. The experiments were made at the expense of Mr. Fairbairn of Manchester, by whose liberal encouragement the researches of practical science have been in other respects so greatly advanced.

In all cases the strength of a column, one of whose ends was rounded and the other flat, was found to be an arithmetic mean between the strengths of two other columns of the same dimensions, one having both ends rounded and the other having both ends flat.

The above results only apply to the case in which the length of the column is so great that its fracture is produced wholly by the *bending* of its material; this limit is fixed by Mr. Hodgkinson in respect to columns of cast iron at about fifteen times the diameter when the extremities are rounded, and thirty times the diameter when they are flat. In shorter columns fracture takes place partly by the crushing and partly by the bending of the material. To these shorter columns the following rule was found to apply with sufficient accuracy : — " If W_1 represent the weight in tons which would break the column by bending alone (or if it did not crush) as given by the preceding formula, and W_2 the weight in tons which would break the column by crushing alone (or if it did not bend) as determined from the above table, then the actual breaking weight W of the column is represented in tons by the formula

$$W = \frac{W_1 W_2}{W_1 + \frac{3}{4} W_2} \quad \cdot \cdot \cdot \cdot \cdot \quad (693).$$

Columns enlarged in the middle. — It was found that the strengths of columns of cast iron, whose diameters were from one and a half times to twice as great in the middle as at the extremities, were stronger by one seventh than solid columns, containing the same quantity of iron and of the same length, when their extremities were rounded ; and stronger by one eighth or one ninth when their extremities were flat and rendered immoveable by discs.

433. RELATIVE STRENGTH OF LONG COLUMNS OF CAST IRON, WROUGHT IRON, STEEL, AND TIMBER OF THE SAME DIMENSIONS. — Calling the strength of the cast iron column 1000, the strength of the wrought iron column will, according

to these experiments, be 1745, that of the cast steel column 2518, of the column of Dantzic oak 108·8, and of the column of red deal 78·5.

Effect of drying on the strength of columns of timber. — It results from these experiments, that the strength of short columns of wet timber to resist crushing is not *one half* that of columns of the same dimensions of dry timber.

TORSION.

434. *The elasticity of torsion.*

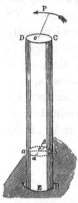

Let ABCD represent a solid cylinder, one of whose transverse sections AEB is immoveably fixed, and every other displaced in its own plane, about its centre, by the action of a pressure P applied, at a given distance *a* from the axis, to the section CD of the cylinder in the plane of that section and round its centre; the cylinder is said, under these circumstances, to be subjected to *torsion*, and the forces opposed to the alteration of its form, and to its rupture, constitute its resistance to torsion.

Let $a\alpha b\beta$ be any section of the cylinder whose distance from the section AEB is represented by x, and let $\alpha\beta$ represent that diameter of the section $a\alpha b\beta$ which was parallel to the diameter AB before the torsion commenced; let ab be the projection of the diameter AB upon the section $a\alpha b\beta$, and let the angle $ac\alpha$ be represented by θ.

Now the elastic forces called into action upon the section $a\alpha b\beta$ are in equilibrium with the pressure P. But these elastic forces result from the *displacement* of the section $a\alpha b\beta$ upon its immediately subjacent section. Moreover, the actual displacement of any small element ΔK of the section $a\alpha b\beta$, upon the subjacent section, evidently depends partly upon the *angular* displacement of the one section upon the other, and partly upon the distance ρ of the element in question from the axis of the cylinder. Now the angle aca or θ is evidently the *sum* of the angular displacements of all the sections between $aab\beta$ and AEB upon their subjacent sections; and the angular displacement of each upon its subjacent section is the same, the circumstances affecting the displacement of each being obviously the same : also the number of these sections varies as x, and the sum of their angular displacements is represented by θ ; therefore the angular displacement of each section upon its subjacent section varies as $\dfrac{\theta}{x}$, and the actual displacement of the small element ΔK of the section $aab\beta$ varies as $\dfrac{\theta}{x}\rho$. Now the material being elastic, the pressure which must be applied to this element in order to keep it in this state of displacement varies as the amount of the displacement (Art. 347.), or as $\dfrac{\theta}{x}\rho$. Let its actual amount, when referred to a unit of surface, be represented by $G\dfrac{\theta}{x}\rho$, where G is a certain constant dependant for its amount on the elastic qualities of the material, and called the modulus of torsion ; then will the force of torsion required to keep the element ΔK in its state of displacement be represented by $G\dfrac{\theta}{x}\rho\Delta K$, and its moment about the axis of the cylinder by $G\dfrac{\theta}{x}\rho^2\Delta K$. So that the sum of the moments of all such forces of torsion in respect to the whole section $aab\beta$ will be represented by $G\dfrac{\theta}{x}\Sigma\rho^2\Delta K$, or by $G\dfrac{\theta}{x}I$, if I represent the moment of inertia of the section about the axis of the cylinder. Now these forces are in equilibrium with P; therefore, by the principle of the equality of moments,

$$Pa = GI\frac{\theta}{x} \ \ . \ . \ . \ . \ (694).$$

If r represent the radius of the cylinder, $I = \frac{1}{2}\pi r^1$ (Art. 85.). Substituting this value, representing by L the whole length of the cylinder, and by Θ the angle through which its extreme section CD is displaced or through which OP is made to revolve, called the *angle of torsion*, and solving in respect to Θ,

$$\Theta = \left(\frac{2a}{\pi G}\right) \cdot \frac{PL}{r^i} \ \ldots \ (695).$$

Thus, then, it appears that when the dimensions of the cylinder are given, the angle of torsion Θ varies directly as the pressure P by which the torsion is produced; whence, also, it follows (Art. 97.) that if the cylinder, after having been deflected through any distance, be set free, it will oscillate isochronously about its position of repose, the time T of each oscillation being represented in seconds (equation 76) by the formula

$$T = \left(\frac{2\pi}{Gg}\right)^{\frac{1}{2}} \cdot \frac{a(WL)^{\frac{1}{2}}}{r^2} \ \ldots \ (696),$$

since by equation (695) $P = \left(\frac{\pi r}{2a^2L}\frac{G}{}\right)(\Theta a)$; in which ex-pression (Θa) represents the length of the path described by the point P from its position of repose, so that the moving force upon the point P, when the pressure producing torsion is removed, varies as the path described by it from its position of repose.

The above is manifestly the theory of Coulomb's Torsion Balance.* W represents in the formula the weight of the mass supposed to be carried round by the point P, and the inertia of the cylinder itself is neglected as exceedingly small when compared with the inertia of this weight.

The torsion of rectangular prisms has been made the subject of the profound investigations of MM. Cauchy†, Lamé et Clapeyron‡, and Poisson.§ It results from these investi-

---

* Illustrations of Mechanics, art. 37.

† Exercises de Mathematique, 4e année.

‡ Crelle's Journal.   § Mémoires de l'Academie, tome viii.

gations* that if b and c be taken to represent the sides of the rectangular section of the prism, and the same notation be adopted in other respects as before, then

$$\Theta = \frac{3PLa(b^2 + c^3)}{Gb^3c^3} \cdot \cdot \cdot \cdot \; (697).$$

M. Cauchy has shown the values of the constant G to be related to those of the modulus of elasticity E by the formula

$$G = \tfrac{2}{5}E \; . \; . \; . \; . \; . \; (698).$$

In using the values of G deduced by this formula from the table of moduli of elasticity, all the dimensions must be taken in inches, and the weights in pounds.

435. ELASTICITY OF TORSION IN A SOLID HAVING A CIRCULAR SECTION OF VARIABLE DIMENSIONS.

Let ab represent an element of the solid contained
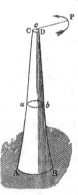
by planes, perpendicular to the axis, whose distance from one another is represented by the exceedingly small increment Δx of the distance x of the section ab from the fixed section AB, and let its radius be represented by y; and suppose the whole of the solid except this single element to become rigid, a supposition by which the conditions of the equilibrium of this particular element will remain unchanged, the pressure P remaining the same, and being that which produces the torsion of this single element. Whence, representing by $\Delta\theta$ the angle of torsion of this element, and considering it a cylinder whose length is Δx, we have by equation (694), substituting for I its value $\tfrac{1}{2}\pi y^4$,

$$Pa = \tfrac{1}{2}G\pi y \frac{\Delta\theta}{\Delta x}.$$

Passing to the limit, and integrating between the limits 0 and

---

* Navier, Resumé des Leçons, &c. Art. 159.

L, observing that at the former limit $\theta = 0$, and at the latter $\theta = \Theta$,

$$\Theta = \frac{2\mathrm{P}a}{\pi\mathrm{G}} \int_0^\mathrm{L} \frac{dx}{y^4} \cdot \cdot \cdot \cdot \cdot (699).$$

If the sides AC and BD of the solid be straight lines, its form being that of a truncated cone, and if r_1 and r_2 represent its diameters AB and CD respectively; then

$$\frac{dx}{dy} = -\frac{\mathrm{L}}{r_1 - r_2}.$$

Also,

$$\int_0^\mathrm{L} y^{-4} dx = \int_{r_1}^{r_2} y^{-4} \frac{dx}{dy} dy = -\frac{\mathrm{L}}{r_1 - r_2} \int_{r_1}^{r_2} y^{-4} dy = \tfrac{1}{3}\frac{\mathrm{L}}{r_1 - r_2}(r_2{}^{-3} - r_1{}^{-3});$$

$$\therefore \ \Theta = \tfrac{2}{3}\frac{\mathrm{P}\mathrm{L}a}{\pi\mathrm{G}} \left(\frac{r_2{}^{-3} - r_1{}^{-3}}{r_1 - r_2} \right) \ \cdot \ \cdot \ \cdot \ \cdot (700).$$

436. The rupture of a cylinder by torsion.

It is evident that rupture will first take place in respect to those elements of the cylinder which are nearest to its surface, the displacement of each section upon its subjacent section being greatest about those points which are nearest to its circumference. If, therefore, we represent by T the pressure per square inch which will cause rupture by the sliding of any section of the mass upon its contiguous section*, then will T represent the resistance of torsion per square inch of the section, at the distance r from the axis, at the instant when rupture is upon the point of taking place, the radius of the cylinder being represented by r. Whence it follows that the displacement, and therefore the resistance to torsion, per square inch of the section, at any other distance ρ from the axis, will be represented at that distance by $\frac{\mathrm{T}\rho}{r}$, the

---

* Or the pressure per square inch necessary to *shear* it across (Art. 408.).

resistance upon any element ΔK by $\dfrac{T}{r}\rho\Delta K$, and the sum of the moments about the axis, of the resistances of all such elements, by $\dfrac{T}{r}\Sigma\rho^2\Delta K$, or by $\dfrac{T}{r}I$, or substituting for I its value (equation 64) by $\frac{1}{2}T\pi r^3$. But these resistances are in equilibrium with the pressure P, which produces torsion, acting at the distance a from the axis;

$$\therefore\ Pa = \tfrac{1}{2}T\pi r^3 \ \dots \ (701).$$

It results from the researches of M. Cauchy, before referred to, that in the case of a rectangular section whose sides are represented by b and c, the conditions of rupture are determined by the equation

$$Pa* = \tfrac{1}{3}T\frac{b^2c^2}{(b^2+c^2)^{\frac{1}{2}}} \ \dots \ (702).$$

The *length* of a prism subjected to torsion does not affect the actual amount of the pressure required to produce rupture, but only the angle of torsion (equation 695) which precedes rupture, and therefore the space through which the pressure must be made to act, and the *amount of* WORK *which must be done* to produce rupture.

According to M. Cauchy, the modulus of rupture by torsion T is connected with that S of rupture by transverse strain by the equation

$$T = \tfrac{4}{3}S \ \dots \ (703).$$

---

* Navier, Resumé d'un Cours, &c. Art. 167.

PART VI.

IMPACT.

437. THE IMPACT OF TWO BODIES WHOSE CENTRES OF GRAVITY MOVE IN THE SAME RIGHT LINE, AND WHOSE POINT OF CONTACT IS IN THAT LINE.

From the period when a body first receives the impact of another, until that period of the impact when both move for an instant with the same velocity, it is evident that the surfaces must have been in a state of continually increasing compression: the instant when they acquire a common velocity is, therefore, that of their *greatest* compression. When this common velocity is attained, their mutual pressures will have ceased if they be inelastic bodies, and they will move with a common motion; if they be elastic, their surfaces will in the act of recovering their forms be mutually repelled, and the velocities will after the impact be different from one another.

438. A BODY WHOSE WEIGHT IS W_1, AND WHICH IS MOVING IN A HORIZONTAL DIRECTION WITH A UNIFORM VELOCITY REPRESENTED BY V_1, IS IMPINGED UPON BY A SECOND BODY WHOSE WEIGHT IS W_2, AND WHICH IS MOVING IN THE SAME STRAIGHT LINE WITH THE VELOCITY V_2: IT IS REQUIRED TO DETERMINE THEIR COMMON VELOCITY V AT THE INSTANT OF GREATEST COMPRESSION.

Let f_1 represent the decrement per second of the velocity of W_1 at any instant of the impact (Art. 94.), or rather the decrement per second which its velocity would experience if the retarding pressure were to remain constant; then will

$\dfrac{W_1}{g} f_1$ represent (Art. 95.) the *effective* force upon W_1; and if f_2 be taken to represent under the same circumstances the increment of velocity received by W_2, then will $\dfrac{W_2}{g} f_2$ represent the effective force upon W_2. Whence it follows, by the principle of D'Alembert (Art. 103.), that if these effective forces be conceived to be applied to the bodies in directions opposite to those in which the corresponding retardation and acceleration take place, they will be in equilibrium with the other forces applied to the bodies. But, by supposition, no other forces than these are applied to the bodies: these are therefore in equilibrium with one another.

$$\therefore \ \frac{W_1}{g} f_1 = \frac{W_2}{g} f_2 \ \ldots \ (704).$$

Let now an exceedingly small increment of the time from the commencement of the impact be represented by Δt, and let Δv_1 and Δv_2 represent the decrement and increment of the velocities of the bodies respectively during that time,

$$\therefore \ (\text{Art. 95.}) \ f_1 \Delta t = \Delta v_1, \ f_2 \Delta t = \Delta v_2;$$
$$\therefore \ (\text{equation 704}) \ W_1 . \ \Delta v_1 = W_2 . \ \Delta v_2;$$

and this equality obtaining throughout that period of the impact which precedes the period of greatest compression, it follows that when the bodies are moving in the same direction

$$W_1(V_1 - V) = W_2(V - V_2) \ \ldots \ (705);$$

since $V_1 - V$ represents the whole velocity lost by W_1 during that period, and $V - V_2$ the whole velocity gained by W_2.

If the bodies be moving in opposite directions, and their common motion at the instant of greatest compression be in the direction of the motion of W_1, then is the velocity lost by W_1 represented as before by $(V_1 - V)$; but the sum of the decrements and increments of velocity communicated to W_2, in order that its velocity V_2 may in the first place be destroyed, and then the velocity V communicated to it in an opposite direction, is represented by $(V_2 + V)$,

$$\therefore \ W_1(V_1 - V) = W_2(V_2 + V).$$

Solving these equations in respect to V, we obtain

$$V = \frac{W_1 V_1 \pm W_2 V_2}{W_1 + W_2} \ \cdot \cdot \cdot \cdot \cdot \ (706);$$

the sign $\pm$ being taken according as the motions of the bodies before impact are both in the same direction or in opposite directions.

If the second body was at rest before impact, $V_2 = 0$, and

$$V = \frac{W_1 V_1}{W_1 + W_2} \ \cdot \cdot \cdot \cdot \cdot \ (707).$$

If the bodies be equal in weight,

$$V = \tfrac{1}{2}(V_1 \pm V_2).$$

The demonstration of this proposition is wholly independent of any hypothesis as to the nature of the impinging bodies or their elastic properties; the proposition is therefore true of all bodies, whatever may be their degrees of hardness or their elasticity, provided only that at the instant of greatest compression every part of each body partakes in the common velocities of the bodies, there being no relative or vibratory motion of the parts of either body among themselves.

439. To DETERMINE THE WORK EXPENDED UPON PRODUCING THE STATE OF THE GREATEST COMPRESSION OF THE SURFACES OF THE BODIES.

The same notation being taken as before, the whole work accumulated in the bodies, before impact, is represented by $\tfrac{1}{2}\dfrac{W_1}{g}V_1{}^2 + \tfrac{1}{2}\dfrac{W_2}{g}V_2{}^2$; and the work accumulated in them at the period of greatest compression, when they move with the common velocity V, is represented by $\tfrac{1}{2}\dfrac{W_1 + W_2}{g}V^2$. Now the difference between the amounts of work accumulated in the bodies in these two states of their motion has been expended in producing their compression; if, therefore, the amount of work thus expended be represented by u, we have

$$u = \tfrac{1}{2}\frac{W_1}{g}V_1{}^2 + \tfrac{1}{2}\frac{W_2}{g}V_2{}^2 - \tfrac{1}{2}\frac{W_1 + W_2}{g}V^2 \; ;$$

or substituting for V its value from equation (706), and reducing,

$$u = \frac{1}{2g}\left(\frac{W_1 W_2}{W_1 + W_2}\right)(V_1 \mp V_2)^2 \;\; \ldots \;\; (708).$$

This expression represents the amount of work *permanently* lost in the impact of two *inelastic* bodies, their common velocity after impact being represented by equation (706). If W_2 be exceedingly great as compared with W_1,

$$u = \frac{W_1}{2g}(V_1 \mp V_2)^2 \;\; \ldots \;\; (709).$$

440. Two ELASTIC BODIES IMPINGE UPON ONE ANOTHER : IT IS REQUIRED TO DETERMINE THE VELOCITY AFTER IMPACT.

If the impinging bodies be perfectly elastic, it is evident that after the period of their greatest compression is passed, they will, in the act of expanding their surfaces, exert mutual pressures upon one another, which are, in corresponding positions of the surfaces, precisely the same with those which they sustained whilst in the act of compression; whence it follows that the decrements of velocity experienced by that body whose motion is retarded by this expansion of the surfaces, and the increments acquired by that whose velocity is accelerated, will be equal to those before received in passing through corresponding positions, and therefore the whole decrements and increments thus received during the whole expansion equal to those received during the whole compression.

Now the velocity lost by W_1 during the compression is represented by $(V_1 - V)$; that lost by it during the expansion, or from the period of greatest compression to that when the bodies separate from one another, is therefore represented by the same quantity. But at the instant of greatest compression both bodies had the velocity V; the velocity v_1 of W_1 at the instant of separation is therefore $V-(V_1-V)$, or $2V-V_1$. In like manner, the velocity gained by W_2 during compression, and therefore during expansion, being represented by $(V \mp V_2)$, and its velocity at the instant of greatest compression by V, its velocity v_2 at the instant of separation is represented by $V+(V \mp V_2)$, or by $2V \mp V_2$, the sign $\mp$ being taken according as the motion of the bodies before impact was in the same or opposite directions.

Substituting for V its value in these expressions (equation 706), and reducing, we obtain

$$v_1 = \frac{(W_1-W_2)V_1 \pm 2W_2V_2}{W_1+W_2} \ \dots \ (710);$$

$$v_2 = \frac{\mp(W_1-W_2)V_2 + 2W_1V_1}{W_1+W_2} \ \dots (711).$$

If the bodies be perfectly elastic and equal in weight, $v_1=V_2$, $v_2=V_1$: they therefore, in this case, *interchange* their velocities by impact; and if either was at rest before impact, the other will be at rest after impact.

If W_2 be exceedingly great as compared with W_1, $v_1 = -V_1 \pm 2V_2$, $v_2 = \pm V_2$. In this case v_1 is negative, or the motion of the lesser body alters its direction after impact, when their motions before impact were in opposite directions; or when they were in the same direction, provided that $2V_2$ be not greater than V_1.

441. If the elasticities of the balls be imperfect, the force with which they tend to separate at any given point of the expansion is different from that at the corresponding point of the compression; the decrements and increments of the velocities, produced during given corresponding periods of the compression and expansion, are therefore different; whence it follows that the whole amounts of velocity, lost by the one and gained by the other during the two periods, are different:

let them bear to one another the ratio of 1 to e. Now the velocity lost during compression by W_1 is under all circumstances represented by (V_1-V); that lost during expansion is therefore represented, in this case, by $e(V_1-V)$; therefore, $v_1 = V - e(V_1 - V) = (1+e)V - eV_1$. In like manner, the velocity gained by W_2 during compression is in all cases represented by $(V \mp V_1)$: that gained during expansion is therefore represented by $e(V \mp V_2)$; therefore, $v_2 = V + e(V \mp V_2) = (1+e)V \mp eV_2$. Substituting for V, and reducing,

$$v_1 = \frac{(W_1 - eW_2)V_1 \pm (1+e)W_2V_2}{W_1 + W_2} \ \dots \ (712);$$

$$v_2 = \frac{\mp (W_1 - W_2)eV_1 + (1+e)W_1V_1}{W_1 + W_2} \ \dots \ (713).$$

442. In the impact of two elastic bodies, to determine the accumulated work, or one half the vis viva, lost by the one and gained by the other.

The vis viva lost by W_1 during the impact is evidently represented by $\frac{W_1}{g}V_1^2 - \frac{W_1}{g}v_1^2 = \frac{W_1}{g}(V_1^2 - v_1^2) =$

$\frac{W_1}{g}\left\{ V_1^2 - \{(1+e)V - eV_1\}^2 \right\} = \frac{W_1}{g}\{(1-e^2)V_1^2 + 2e(1+e)VV_1$

$-(1+e)^2V^2\} = \frac{W_1}{g}(1+e)(V_1-V)\{V_1(1-e)+V(1+e)\}$.

Substituting in this expression its value for V (equation 706), reducing, and representing by u_1 one half the vis viva lost by W_1 in its impact, or the amount by which its accumulated work is diminished by the impact (Art. 67.),

$$u_1 = \frac{(1+e)W_1W_2(V_1 \mp V_2)}{2g(W_1+W_2)^2}\{2W_1V_1 + (1-e)W_2V_1 \pm (1+e)W_2V_2\} \ \dots \ (714).$$

Similarly, if u_2 be taken to represent one half the vis viva gained by W_2, or the amount by which its accumulated work is increased by the impact, then

$$u_2 = \pm\frac{(1+e)W_1W_2(V_1 \mp V_2)}{2g(W_1+W_2)^2}\{2W_2V_2 + (1-e)W_1V_2 \pm (1+e)W_1V_1\} \ \dots \ (715).$$

443. Let u be taken to represent the whole amount of the work accumulated in the two bodies before their impact, which is lost during their impact. This amount of work is evidently equal to the difference between that gained by the one body and lost by the other ; so that $u = u_1 - u_2$. Substituting the values of u_1 and u_2 from the preceding equations, and reducing, we obtain

$$u = \frac{(1 - e^2) W_1 W_2 (V_1 \mp V_2)^2}{2g(W_1 + W_2)} \quad \ldots \quad (716).$$

This expression is equal to one half the vis viva lost during the impact of the bodies. If the bodies be perfectly elastic, $e = 1$, and $u = 0$. In this case there is no real loss of vis viva in the impact, all that which the one body yields, during the impact, being taken up by the other. *

444. In the preceding propositions it has been supposed that the motions of the impinging body, and the body impinged upon, are opposed by no resistance whatever during the period of the impact. There is no practical case in which this condition obtains accurately. If, nevertheless, the resistance opposed to the motion of each body be small, as compared with the pressure exerted by each upon the other, at any period of the impact, then it is evident that all the circumstances of the impact as it proceeds, and the motion of each body at the instant when it ceases, will be very nearly the same as though no resistance were opposed to the motion of either. †

---

* It has been customary, nevertheless, to speak of a *loss* of vis viva in the impact of perfectly elastic bodies. This loss is in all such cases to be understood only as a loss experienced by *one* of the bodies, and not as an absolute loss. When the impinging bodies are perfectly elastic, it is evident that the one flies away with all the vis viva which is lost in the impact by the other.

† Let P_1 and P_2 represent resistances opposed to the motions of two impinging bodies whose weights are W_1 and W_2 ; also let $\dfrac{W_1}{g} f_1$, and

445. As an illustration of the principle established in the last article, let it be required to determine the space through which a nail will be driven by the blow of a hammer; and let it be supposed that the resistance opposed to the driving of the nail is partly a constant resistance overcome at its point, and partly a resistance opposed by the friction of the mass into which it is driven upon its sides, varying in amount directly with the length of it x, at any time imbedded in the wood. Let this resistance be represented by $\alpha + \beta x$; then will the work which must be expended in driving it to a depth D be represented (Art. 51.) by

$$\int_0^D (\alpha + \beta x) dx, \text{ or by } (\alpha D + \tfrac{1}{2}\beta D^2).$$

Let W_2 represent the weight of the nail, and V the velocity with which a hammer whose weight is W_1 must impinge upon it to drive it to this depth, and let the surfaces of the nail and hammer both be supposed inelastic; then will the work

$\dfrac{W_2}{g} f_2$ represent the effective forces upon the two bodies at any period of the impact; then, by D'Alembert's principle,

$$\frac{W_1}{g} f_1 - P_1 - \frac{W_2}{g} f_2 - P_2 = 0;$$

or representing by t the time occupied in the impact, up to the period of greatest compression, by V their common velocity at that period, and by v_1 and v_2 their velocities at any period of the impact, and substituting for f_1 and f_2 their values (equation 72),

$$\frac{W_1}{g} \frac{dv_1}{dt} - P_1 - \frac{W_2}{g} \frac{dv_2}{dt} - P_2 = 0.$$

Transposing and integrating between the limits 0 and t,

$$\frac{W_1}{g}(V_1 - V) = \frac{W_2}{g}(V - V_1) + \int_0^t (P_1 + P_2) dt.$$

Now if P_1 and P_2 be not exceedingly great, the integral in the second member of the equation is exceedingly small as compared with its other terms, and may be neglected; the above equation will then become identical with equation (705).

accumulated in the hammer before impact be represented by $\frac{1}{2}\frac{W_1}{g}V^2$, and the amount of this work lost during the impact by the compression of the surfaces of contact will be represented (equation 716) by $\frac{1}{2g}\left(\frac{W_1 W_2}{W_1 + W_2}\right)V^2$. The work remaining, and effective to drive the nail, is therefore represented by the difference of these quantities; and this work being that actually expended in driving the nail, we have

$$\frac{1}{2g}W_1 V^2 - \frac{1}{2g}\left(\frac{W_1 W_2}{W_1 + W_2}\right)V^2 = \alpha D + \tfrac{1}{2}\beta D^2 ;$$

$$\frac{1}{g}\frac{V^2 W_1^2}{W_1 + W_2} = 2\alpha D + \beta D^2 \ . \ . \ . \ . \ (717) ;$$

by the solution of which quadratic equation, D may be determined.

446. TWO SOLID PRISMS HAVE A COMMON AXIS; THE EX-
TREMITY OF ONE OF THEM RESTS AGAINST A FIXED SURFACE,
AND ITS OPPOSITE EXTREMITY RECEIVES THE IMPACT, IN A
HORIZONTAL DIRECTION, OF THE OTHER PRISM: IT IS RE-
QUIRED TO DETERMINE THE COMPRESSION OF EACH PRISM,
THE LIMITS OF PERFECT ELASTICITY NOT BEING PASSED IN
THE IMPACT.

Let W represent the weight of the impinging prism, and V its velocity before impact; L_1 and L_2 the lengths of the prisms before compression; E_1 and E_2 their moduli of elasticity; K_1 and K_2 their sections; l_1 and l_2 the greatest compressions produced in them respectively by the impact; then will the amounts of work which must have been done upon the prisms to produce these compressions be represented (equation 491) by the formulæ

$$\tfrac{1}{2}\frac{K_1 E_1 l_1^2}{L_1} \text{ and } \tfrac{1}{2}\frac{K_2 E_2 l_2^2}{L_2},$$

and the whole work thus expended by

$$\tfrac{1}{2}\left(\frac{K_1E_1{l_1}^2}{L_1}+\frac{K_2E_2{l_2}^2}{L_2}\right).$$

But this work has been done by the work $\tfrac{1}{2}\dfrac{W}{g}V^2$, accumulated (Art. 66.) before impact in the impinging body, and that work has been exhausted in doing it ;

$$\therefore\ \tfrac{1}{2}\left(\frac{K_1E_1{l_1}^2}{L_1}+\frac{K_2E_2{l_2}^2}{L_2}\right)=\tfrac{1}{2}\frac{W}{g}V^2.$$

Moreover, the mutual pressures upon the surfaces of contact are at every period of the impact equal, and at the instant of greatest compression they are represented respectively (equation 490) by $\dfrac{K_1E_1l_1}{L_1}$ and $\dfrac{K_2E_2l_2}{L_2}$;

$$\therefore\ \frac{K_1E_1l_1}{L_1}=\frac{K_2E_2l_2}{L_2}=P\ \dots\ (718).$$

Eliminating l_2 between this equation and the preceding, and reducing,

$$l_1=\frac{L_1V}{K_1E}\left\{\left(\frac{L_1}{K_1E_1}+\frac{L_2}{K_2E_2}\right)g\right\}^{-\frac{1}{2}}\sqrt{W}\ \dots\ (719);$$

$$P=V\left\{\left(\frac{L_1}{K_1E_1}+\frac{L_2}{K_2E_2}\right)g\right\}^{-\frac{1}{2}}\sqrt{W}\ \dots\ (720);$$

in which expressions l_1 represents the greatest compression of the prism whose section is K_1, and P the driving pressure at the instant of greatest compression.

447. *The mutual pressures* P *of the surfaces of contact at any period of the impact.*

Let l represent the space described by that extremity of the impinging prism, by which it does not impinge: it is evident that this space is made up of the two corresponding compressions of the surfaces of impact of the prisms ; so that if these be represented by l_1 and l_2, then $l=l_1+l_2$. But (equation 718) $l_1 = \dfrac{PL_1}{K_1E_1}$, $l_2 = \dfrac{PL_2}{K_2E_2}$; therefore $l = P\Big(\dfrac{L_1}{K_1E} +$

$\dfrac{L_2}{K_2E_2}\Big)$;

$$\therefore \; P = l\Big(\dfrac{L_1}{K_1E_1} + \dfrac{L_2}{K_2E_2}\Big)^{-1} \; . \; . \; . \; . \; (721).$$

448. *A measure of the compressibility of the prisms.*

If λ be taken to represent the space through which that extremity of the impinging prism by which it does *not* impinge will have moved when the mutual pressure of the surfaces of contact is 1 lb.; or, in other words, if λ represent the aggregate space through which the prisms would be compressed by a pressure of 1 lb.; then, by the preceding equation,

$$\lambda = \dfrac{L_1}{K_1E_1} + \dfrac{L_2}{K_2E_2} \; . \; . \; . \; . \; (722).$$

λ may be taken as *a measure of the aggregate compressibility of the prisms, being the space through which their opposite extremities would be made to approach one another by a pressure of 1 lb. applied in the direction of their length.*

If λ_1 and λ_2 represent the spaces through which the prisms would *severally* be compressed by pressures of 1 lb. applied to each, then $\lambda_1 = \dfrac{L_1}{K_1E_1}$, $\lambda_2 = \dfrac{L_2}{K_2E_2}$; therefore $\lambda = \lambda_1 + \lambda_2$, or the aggregate compressibility of the two prisms is equal to the sum of their separate compressibilities.

449. *The work u expended upon the compression of the prisms at any period of the impact.*

The work expended upon the compression l_1 is represented by $\frac{1}{2}\dfrac{K_1E_1}{L_1}l_1{}^2$; or, substituting its value for l_1 (equation 718), it is represented by $\frac{1}{2}\dfrac{L_1}{K_1E_1}P^2$. And, similarly, the work expended on the compression l_2 is represented by $\frac{1}{2}\frac{L_2}{K_2 E_2}P^2$;

therefore $u = \frac{1}{2}\left(\frac{L_1}{K_1 E_1} + \frac{L_2}{K_2 E_2}\right)P^2$; or substituting for P its value from equation (721),

$$u = \frac{1}{2}l^2\left(\frac{L_1}{K_1 E_1} + \frac{L_2}{K_2 E_2}\right)^{-1} = \frac{1}{2}\frac{l^2}{\lambda} \ \ldots \ (723).$$

450. *The velocity of the impinging body at any period of the impact, the impact being supposed to take place vertically.*

It is evident that at any period of the impact, when the velocity of the impinging body is represented by v, there will have been expended, upon the compression of the two bodies, an amount of work which is represented by the work accumulated in the impinging body before impact, increased by the work done upon it by gravity during the impact, and diminished by that which still remains accumulated in it, or by $\frac{1}{2}\frac{W}{g}V^2 + Wl - \frac{1}{2}\frac{W}{g}v^2$.

Representing, therefore, by u the work expended upon the compression of the bodies, we have $\frac{1}{2}\frac{W}{g}V^2 + Wl - \frac{1}{2}\frac{W}{g}v^2 = u$.

Substituting, therefore, for u its value from equation (723),

$$\frac{1}{2}\frac{W}{g}V^2 + Wl - \frac{1}{2}\frac{W}{g}v^2 = \frac{1}{2}l^2\left(\frac{L_1}{K E_1} + \frac{L_2}{K_2 E_2}\right)^{-1} ;$$

$$\therefore \ v^2 = V^2 + 2lg - \frac{l^2 g}{W}\left(\frac{L_1}{K_1 E_1} + \frac{L_2}{K_2 E_2}\right)^{-1} \ \ldots \ (724).$$

Or substituting for l its value in terms of P (equation 721),

$$v^2 = V^2 + g\left(\frac{L_1}{K_1 E_1} + \frac{L_2}{K_2 E_2}\right)\left(2P - \frac{P^2}{W}\right) \ \ldots \ (725).$$

THE PILE DRIVER.

451. It is evident that the pile will not begin to be driven until a period of the impact is attained, when the pressure of the ram upon its head, together with the weight of the pile, exceeds the resistance opposed to its motion by the coherence and the friction of the mass into which it is driven. Let this resistance be represented by P; let V represent the velocity of the ram at the instant of impact, and v its velocity at the instant when the pile begins to move, and W_1, W_2 the weights of the ram and pile; then, since the pile will have been at rest during the whole of the intervening period of the impact, since moreover the mutual pressures Q of the surfaces of contact are, at the instant of motion, represented by $P - W_2$, we have by equation (725)

$$v^2 = V^2 - g\left(\frac{L_1}{K_1 E_1} + \frac{L_2}{K_2 E_2}\right)\left\{\frac{(P - W_2)^2}{W_1} - 2(P - W_2)\right\} \cdots (726).$$

If the value of v determined by this equation be not a possible quantity, no motion can be communicated to the pile by the impact of the ram: the following inequality is therefore a condition necessary to the driving of the pile,

$$V^2 > g\left(\frac{L_1}{K_1 E_1} + \frac{L_2}{K_2 E_2}\right)\left\{\frac{(P - W_2)^2}{W_1} - 2(P - W_2)\right\} \cdots (727).$$

After the pile has moved through any given distance, one portion of the work accumulated in the ram before its impact will have been expended in overcoming, through that distance, the resistance opposed to the motion of the pile; another portion will have been expended upon the compression of the surfaces of the ram and pile; and the remainder will be accumulated in the moving masses of the ram and pile. The motion of the pile cannot cease until after the period of the greatest compression of the ram and pile is attained; since the

reaction of the surface of the pile upon the ram, and therefore the driving pressure upon the pile, increases continually
with the compression. If the surfaces be inelastic, having no
tendency to recover the forms they may have received at the
instant of greatest compression, they will move on afterwards
with a common velocity, and come to rest together; so that the
whole work expended prejudicially during the impact will be
that expended upon the compression of the inelastic surfaces
of the ram and pile. If, however, both surfaces be elastic,
that of the ram will return from its position of greatest compression, and the ram will thus acquire a velocity relatively
to the pile, in a direction opposite to the motion of the pile.
Until it has thus reached the position in respect to the pile
in which it first began to drive it, their mutual reaction Q
will exceed the resistance P, and the pile will continue to be
driven. After the ram has, in its return, passed this point,
the pile will still continue to be driven through a certain
space, by the work which has been accumulating in it during
the period in which Q has been in excess of P. When the
motion of the pile ceases, the ram on its return will thus have
passed the point at which it first began to drive the pile: if it
has not also then passed the point at which its weight is just
balanced by the elasticity of the surfaces, it will have been
continually acquiring velocity relatively to the pile from the
period of greatest compression; it will thus have a certain
velocity, and a certain amount of work will be accumulated
in it when the motion of the pile ceases: this amount of
work, together with that which must have been done to produce that compression which the surfaces of contact retain at
that instant, will in no respect have contributed to the driving
of the pile, and will have been expended uselessly. If the
ram in its return has, at the instant when the motion of the
pile ceases, passed the point at which its weight would just
be balanced by the elasticity of the surfaces of contact, its
velocity relatively to the pile will be in the act of diminishing;
or it may, for an instant, cease at the instant when the
pile ceases to move. In this last case, the pile and ram, for
an instant, coming to rest together, the whole work accumu-

lated in the impinging ram will have been usefully expended
in driving the pile, excepting only that by which the remaining compression of the surfaces has been produced;
which compression is less than that due to the weight of the
ram. This, therefore, may be considered the case in which a
maximum useful effect is produced by the ram. The following article contains an analytical discussion of these conditions
under their most general form.

452. *A prism impinged upon by another is moveable in the direction of its axis, and its motion is opposed by a constant
pressure* P: *it is required to determine the conditions of the
motion during the period of impact, the circumstances of the
impact being in other respects the same as in Article* 450.

Let f_1 and f_2 represent the additional velocities which
would be lost and acquired per second (see Art. 95.) by the
impinging prism and the prism impinged upon, if the pressures, at any instant operating upon them, were to remain
from that instant constant; then will $\dfrac{W_1}{g}f_1$, $\dfrac{W_2}{g}f_2$ represent the effective forces upon the two bodies (Art. 103.) or
the pressures which would, by the principle of D'Alembert,
be in equilibrium with the unbalanced pressures upon them,
if applied in opposite directions.

Now the unbalanced pressure upon the system
BP composed of the two prisms is represented by
$(W_1 + W_2 - P)$,

$$\therefore \frac{W_1}{g}f_1 + \frac{W_2}{g}f_2 = W_1 + W_2 - P \ldots \ldots (728);$$

also the unbalanced pressure upon the prism PQ
$= W_2 + Q - P$, where Q represents the mutual pressure of the prisms at Q;

$$\therefore \frac{W_2}{g}f_2 = W_2 + Q - P \ldots \ldots (729).$$

Let A have been the position of the extremity B of the

impinging prism at the instant of impact; and let x_1 represent the space through which the aggregate length BP of the two prisms has been diminished since that period of the impact, and x_2 the space through which the point P has moved; then (equation 721)

$$Q = x_1 \left(\frac{L_1}{K_1 E_1} + \frac{L_2}{K_2 E_2} \right)^{-1} = \frac{x_1}{\lambda} \quad \cdots \quad (730).$$

Also $AB = x_1 + x_2$; therefore velocity of point $B = \dfrac{d(x_1 + x_2)}{dt}$,

(Art. 96.) ; therefore $f_1 = \dfrac{d^2 x_1}{dt} + \dfrac{d^2 x_2}{dt^2} = \dfrac{d^2 x_1}{dt^2} + f_2$.

Substituting these values of f_1 and Q in equations (728) and (729), and eliminating f_2 between the resulting equations,

$$\frac{d^2 x_1}{dt^2} = -\frac{g}{\lambda} \left(\frac{1}{W_1} + \frac{1}{W_2} \right) x_1 + \frac{Pg}{W_2} \quad \cdots \quad (731).$$

Integrating this equation by the known rules, we obtain

$$x_1 = A \sin. \gamma t + B \cos. \gamma t + \frac{Pg}{\gamma^2 W_2} \quad \cdots \quad (732);$$

in which expression the value of γ is determined by the equation

$$\gamma^2 = \frac{g}{\lambda} \left(\frac{1}{W_1} + \frac{1}{W_2} \right) = g \left\{ \frac{W_1^{-1} + W_2^{-1}}{L_1 (K_1 E_1)^{-1} + L_2 (K_2 E_2)^{-1}} \right\} \quad \cdots \quad (733);$$

and A and B are certain constants to be determined by the conditions of the question. Substituting in equation (729) the value of Q from equation (730), and solving in respect to f_2,

$$f_2 = \frac{g}{W_2 \lambda} x_1 + \left(1 - \frac{P}{W_2} \right) g \quad \cdots \quad (734).$$

Substituting for x_1 its value from equation (732), and for f_2 its value $\dfrac{d^2 x_2}{dt^2}$, and reducing,

$$\frac{d' x_2}{dt'} = \frac{Ag}{W_2 \lambda} \sin. \gamma t + \frac{Bg}{W_2 \lambda} \cos. \gamma t + \left(1 - \frac{P}{W_1 + W_2} \right) g.$$

Integrating between the limits 0 and t, and observing that when $t=0$, $\dfrac{dx_2}{dt}=0$; the time being supposed to commence with the motion of the prism PQ;

$$\frac{dx_2}{dt}=\frac{Ag}{W_2\lambda\gamma}(1-\cos.\,\gamma t)+\frac{Bg}{W_2\lambda\gamma}\sin.\,\gamma t+\left(1-\frac{P}{W_1+W_2}\right)gt.$$

Integrating a second time between the same limits,

$$x_2=\frac{Ag}{W_2\lambda\gamma^2}(\gamma t-\sin.\,\gamma t)+\frac{Bg}{W_2\lambda\gamma^2}(1-\cos.\,\gamma t)+\tfrac{1}{2}\left(1-\frac{P}{W_1+W_2}\right)gt^2..(735).$$

Now when the motion of the second prism ceases $\dfrac{dx_2}{dt}=0$; whence, if the corresponding value of t be represented by T,

$$A(1-\cos.\,\gamma\,T)+B\sin.\,\gamma T+\left(1-\frac{P}{W_1+W_2}\right)W_2\lambda\gamma T=0\;\;.\;.\;.\;(736).$$

To determine the constants A and B, let it be observed that the motion of the prism QP cannot commence until the pressure Q of the impinging prism upon it, added to its own weight W_2, is equal to the resistance P opposed to its motion. So that if c be taken to represent the value of x_1 (*i. e.* the aggregate compression of the two prisms) at that instant, then, substituting for Q its value from equation (730), $\dfrac{c}{\lambda}+W_2=P$;

$$\therefore\; c=(P-W_2)\lambda=(P-W_2)\left(\frac{L_1}{K_1E_1}+\frac{L_2}{K_2E_2}\right)\;.\;.\;.\;.\;(737).$$

Now since the time t is supposed to commence at the instant when this compression is attained, and the prism PQ is upon the point of moving, substituting the above value of c for x_1 in equation (732), and observing that when $x=c$, $t=0$, we obtain $(P-W_2)\lambda=B+\dfrac{Pg}{\gamma^2W_2}$; whence by substitution from equation (733), and reduction,

$$B=\frac{(P-W_1-W_2)g}{\gamma^2W_1}=\lambda\left(\frac{P}{W_1+W_2}-1\right)W_2\;.\;.\;.\;.\;.\;(738).$$

So long as the extremity P, of the prism impinged upon, is at rest, the whole motion of the point B arises from the compression of the two prisms, and is represented by $\frac{dx_1}{dt}$. The

value of $\frac{dx_1}{dt}$, when $t=0$, is represented therefore by v (equation 726). Differentiating, therefore, equation (732), assuming $t=0$, and substituting v for $\frac{dx_1}{dt}$, we obtain $v=\gamma A$; whence it appears that the value of A is determined by dividing the square root of the second member of equation (726) by γ.

Substituting for A and B their values in equation (736),

$$\frac{v}{g}(1-\cos.\ \gamma T)+\lambda W_2\left(\frac{P}{W_1+W_2}-1\right)\sin.\gamma T+\left(1-\frac{P}{W_1+W_2}\right)W_2\lambda\gamma T=$$

Reducing, and dividing by the common factor of the two last terms,

$$\frac{v(1-\cos.\gamma T)}{g\lambda W_2\{P(W_1+W_2)^{-1}-1\}}+\sin.\gamma T-\gamma T=0\ \ldots\ (739).$$

Substituting for A and B their values in equation (735), and representing by D the value of x_2, when $t=T$,

$$D=\frac{vg}{W_2\lambda\gamma^3}(\gamma T-\sin.\gamma T)+g\left(\frac{P}{W_1+W_2}-1\right)\left(\frac{\text{vers.}\,\gamma T}{\gamma^2}-\tfrac{1}{2}T^2\right)\ldots\ (740).$$

The value of T determined by equation (739) being substituted in equation (740), an expression is obtained for the whole space through which the second prism is driven by the impact of the first.*

---

* The method of the above investigation is, from equation (731), nearly the same with that given by Dr. Whewell, in the last edition of his Mechanics, the principle of the investigation appears to be due to Mr. Airey. If the value of γ, as determined by equation (739), were not exceedingly great, then, since the value of T is in all practical cases exceedingly small, the value of γT would in all cases be exceedingly small, and we might approximate to the value of T in equation (740), by substituting for cos. γT and sin. γT, the two first terms of the expansions of those functions, in terms of γT.

APPENDIX.

NOTE A.

THEOREM. — *The definite integral* $\int_a^b fx\,dx$ *is the limit of the sums of the values severally assumed by the product* fx . Δx, *as* x *is made to vary by successive equal increments of* Δx, *from* a *to* b, *and as each such equal increment is continually and infinitely diminished, and their number therefore continually and infinitely increased.*

To prove this, let the general integral be represented by Fx ; let us suppose that fx does not become infinite for àny value of x between a and b, and let any two such values be x and $x+\Delta x$; therefore, by Taylor's theorem, $F(x+\Delta x)=Fx+\Delta x fx+(\Delta x)^{1+\lambda}M$, where the exponent $1+\lambda$ is given to the third term of the expansion instead of the exponent 2, that the case may be included in which the second differential coefficient of Fx, $\dfrac{dfx}{dx}$, is infinite, and in which the exponent of Δx in that term is therefore a fraction less than 2.

Let the difference between a and b be divided into n equal parts ; and let each be represented by Δx, so that

$$\frac{b-a}{n}=\Delta x.$$

Giving to x, then, the successive values $a,\ a+\Delta x,\ a+2\Delta x\ ..\ a+(n-1)\Delta x$, and adding,

$$F(a+n\Delta x)=Fa+\Delta x\Sigma_1^n f\{a+(n-1)\Delta x\}+(\Delta x)^{1+\lambda}\Sigma M_n,$$

$$\therefore\ Fb-Fa=\Delta x\Sigma_1^n f\{a+(n-1)\Delta x\}+(\Delta x)^{1+\lambda}\Sigma M_n.$$

Now none of the values of M are infinite, since for none of these values is fx infinite. If, therefore, M be the greatest of these values, then is ΣM_n less than nM ; and therefore

$$Fb-Fa-\Delta x\Sigma_1^n f\{a+(n-1)\Delta x\}<(b-a)M(\Delta x)^\lambda.$$

The difference of the definite integral $Fb-Fa$, and the sum $\Sigma_1^n(\Delta x)f\{a+(n-1)\Delta x\}$ is always, therefore, less than $(b-a)M(\Delta x)^\lambda$. Now M is *finite*, and $(b-a)$ is *given*, and as n is increased Δx is diminished continually ; and therefore $(\Delta x)^\lambda$ is diminished continually, λ being positive.

Thus by increasing n indefinitely, the difference of the definite integral and the sum may be diminished indefinitely, and therefore, in the limit, the definite integral is equal to the sum (*i. e.*)

$$Fb - Fa = \text{limit } \Sigma_1''(\Delta x) \cdot f\{a + (n-1)\Delta x\} \; ;$$

or, *interpreting* this formula, $Fb - Fa$ is the sum of the values of $\Delta x \quad fx$, when x is made to pass by infinitesimal increments, each represented by Δx, from a to b.

NOTE B.
PONCELET'S FIRST THEOREM.

* The values of a and b in the radical $\sqrt{a^2 + b^2}$ being linear and rational, let it be required to determine the values of two indeterminate quantities α and β, such that the errors which result from assuming $\sqrt{a^2 + b^2} = \alpha a + \beta b$, through a given range of the values of the ratio $\left(\dfrac{a}{b}\right)$, may be the least possible in reference to the true value of the radical; or that $\dfrac{\alpha a + \beta b - \sqrt{a^2 + b^2}}{\sqrt{a^2 + b^2}}$, or $\dfrac{\alpha a + \beta b}{\sqrt{a^2 + b^2}} - 1$, may be the least possible in respect to all that range of values which this formula may be made to assume between two given extreme values of the ratio $\dfrac{a}{b}$. Let these extreme values of the ratio $\dfrac{a}{b}$ be represented by cot. ψ_1 and cot. ψ_2, and any other value by cot. ψ. Substituting cot. ψ for $\dfrac{a}{b}$ in the preceding formula, and observing that $\sqrt{a^2 + b^2} = \sqrt{b^2 \cot.^2\psi + b^2} = b$ cosec. ψ, also that $\alpha a + \beta b = \alpha b \cot.\psi + \beta b = (\alpha \cos.\psi + \beta \sin.\psi)b$ cosec. ψ, the corresponding error is represented by

$$\alpha \cos. \psi + \beta \sin. \psi - 1 \quad \ldots \ldots \quad (1);$$

which expression is evidently a maximum for that value ψ_3 of ψ which is determined by the equation

$$\cot. \psi_3 = \frac{\alpha}{\beta} \quad \cdots \cdots \quad (2);$$

so that its maximum value is

$$\sqrt{\alpha^2 + \beta^2} - 1 \quad \ldots \ldots \quad (3).$$

Moreover, the function admits of no other maximum value, nor of any minimum value. The values of α and β being arbitrary, let them be assumed to be such that $\dfrac{\alpha}{\beta}$ or cot. ψ_3 may be less than cot. ψ_1, and greater

---

* The method of this investigation is not the same as that adopted by M. Poncelet; the principle is the same.

than cot. ψ_2. Now, so long as all the values of the error (formula 1) remain positive, between the proposed limits, they are all manifestly diminished by diminishing a and β ; but when by this diminution the error is at length rendered negative in respect to one or both of the extreme values ψ_1, or ψ_2 of ψ, and to others adjacent to them, then do these negative errors continually *increase*, as a and β are yet farther diminished, whilst the positive maximum error (formula 3) continually *diminishes*. Now the most favourable condition, in respect to the whole range of the errors between the proposed limits of variation, will manifestly be attained when, by thus diminishing the positive and thereby increasing the negative errors, the greatest positive error is rendered equal to each of the two negative errors; a condition which will be found to be consistent with that before made in respect to the arbitrary values of a and β, and which supposes that the values of the error (formula 1) corresponding to the values ψ_1 and ψ_2 are each equal, when taken negatively, to the maximum error represented by formula 3, or that the constants a and β are taken so as to satisfy the two following equations.

$$1-(a \cos. \psi_1 + \beta \sin. \psi_1) = \sqrt{a^2+\beta^2} - 1.$$
$$1-(a \cos. \psi_1 + \beta \sin. \psi_1) = 1-(a \cos. \psi_2 - \beta \sin. \psi_2).$$

The last equation gives us by reduction

$$a \cos. \psi_1 + \beta \sin. \psi_1 = \beta \frac{\cos. \frac{1}{2}(\psi_1-\psi_2)}{\sin. \frac{1}{2}(\psi_1+\psi_2)},$$
$$\text{and } a = \beta \cot. \tfrac{1}{2}(\psi_1+\psi_2).$$

Substituting these values in the first equation, and reducing,

$$\beta = \frac{2 \sin. \frac{1}{2}(\psi_1+\psi_2)}{1+\cos. \frac{1}{2}(\psi_1-\psi_2)} = \frac{\sin. \frac{1}{2}(\psi_1+\psi_2)}{\cos.^2 \frac{1}{4}(\psi_1-\psi_2)} \quad \ldots \ldots (4) ;$$

$$\therefore a = \frac{2 \cos. \frac{1}{2}(\psi_1+\psi_2)}{1+\cos. \frac{1}{2}(\psi_1-\psi_2)} = \frac{\cos. \frac{1}{2}(\psi_1+\psi_2)}{\cos.^2 \frac{1}{4}(\psi_1-\psi_2)} \quad \ldots \ldots (5).$$

These values of a and β give for the maximum error (formula 3) the expression

$$\tan.^2 \tfrac{1}{4}(\psi_1-\psi_2) \, \ldots \ldots (6).$$

Thus, then, it appears that the value of the radical $\sqrt{a^2+b^2}$ is represented, in respect to all those values of $\frac{a}{b}$ which are included between the limits cot. ψ_1 and cot. ψ_2, by the formula

$$\frac{\cos. \frac{1}{2}(\psi_1+\psi_2)}{\cos.^2 \frac{1}{4}(\psi_1-\psi_2)} + b \frac{\sin. \frac{1}{2}(\psi_1+\psi_2)}{\cos.^2 \frac{1}{4}(\psi_1-\psi_2)} \quad \ldots \ldots (7),$$

with a degree of approximation which is determined by the value of $\tan.^2 \tfrac{1}{4}(\psi_1-\psi_2)$.

If in the proposed radical the value of a admits of being increased infinitely in respect to b, or the value of b infinitely diminished in respect to a, then cot. $\psi_1 =$ infinity ; therefore $\psi_1 = 0$. In this case the formula of approximation becomes

$$a\left(1-\tan.^2\tfrac{1}{4}\psi_2\right)+2b\tan.\tfrac{1}{4}\psi_2 \quad\ldots\ldots\ (8)\ ;$$

and the maximum error

$$\tan.^2\tfrac{1}{4}\psi_2 \quad\ldots\ldots\ (9).$$

If the values of a and b are wholly unlimited, so that a may be infinitely small or infinitely great as compared with b, then cot. $\psi_1 =$ infinity, cot. $\psi_2 = 0$;

therefore $\psi_1 = 0$, $\psi_2 = \dfrac{\pi}{2}$. Substituting these values, the formula of approximation becomes

$$\cdot 8284a + \cdot 8284b \quad\ldots\ldots\ (10)\ ;$$

and the maximum error

$$\cdot 1716,\ \text{or}\ \tfrac{1}{6}\text{th nearly.}$$

If b is essentially less than a, but may be of *any* value less than it, so that $\dfrac{a}{b}$ is always greater than unity, but may be infinite, then cot. $\psi_1 =$ infinity, cot. $\psi_2 = 1$; therefore $\psi_1 = 0$, $\psi_2 = \dfrac{\pi}{2}$. Substituting these values in the formula of approximation, and reducing, it becomes

$$\cdot 96046a + \cdot 39783b \quad\ldots\ldots\ (11)\ ;$$

and the maximum error

$$\cdot 03945,\ \text{or}\ \tfrac{1}{25}\text{th nearly.}$$

It is in its application to this case that the formula has been employed in the preceding pages of this work.

The following table, calculated by M. Gosselin, contains the values of the coefficients α and β for a series of values of the inferior limit cot. ψ_2 the superior limit being in every case *infinity*.

Relation of a to b.	Value of Cot. ψ_2.	Value of α.	Value of β.	Maximum Error.	Aproximate Value of $\sqrt{a^2+b^2}$.
a and b any whatever	0	0·82840	0·82840	0·17160 or $\tfrac{1}{6}$	0·8284 $(a+b)$
$a>b$	1	0·96046	0·39783	0·03954 or $\tfrac{1}{25}$	·96046a + ·39783b
$a>2b$	2	0·98592	0·23270	0·01408 or $\tfrac{1}{71}$	·98592a + ·23270b
$a>3b$	3	0·99350	0·16123	0·00650 or $\tfrac{1}{154}$	·99350a + ·16123b
$a>4b$	4	0·99625	0·12260	0·00375 or $\tfrac{1}{266}$	·99625a + ·12260b
$a>5b$	5	0·99757	0·09878	0·00243 or $\tfrac{1}{417}$	·99757a + ·09878b
$a>6b$	6	0·99826	0·08261	0·00174 or $\tfrac{1}{589}$	·99826a + ·08261b
$a>7b$	7	0·99875	0·07098	0·00125 or $\tfrac{1}{805}$	·99875a + ·07098b
$a>8b$	8	0·99905	0·06220	0·00095 or $\tfrac{1}{1049}$	·99905a + ·06220b
$a>9b$	9	0·99930	0·05535	0·00070 or $\tfrac{1}{1428}$	·99930a + ·05535b
$a>10b$	10	0·99935	0·04984	0·00065 or $\tfrac{1}{1538}$	·99935a + ·04984b

PONCELET'S SECOND THEOREM.

To approximate to the value of $\sqrt{a^2-b^2}$, let $aa-\beta b$ be the formula of approximation, then will the relative error be represented by

$$\frac{\sqrt{a^2-b^2}-(aa-\beta b)}{\sqrt{a^2-b^2}}, \text{ or by } 1-\frac{\left(a\dfrac{a}{b}-\beta\right)}{\sqrt{\dfrac{a^2}{b^2}-1}}.$$

Now, let it be observed that a^2 being essentially greater than b^2, $\dfrac{a}{b}>1$; let $\dfrac{a}{b}$, therefore, be represented by cosec. ψ, then will the relative error be represented by $1-\dfrac{(a\,\text{cosec.}\,\psi-\beta)}{\sqrt{\text{cosec.}\,^2\psi-1}}$, or by

$$1-a\,\text{sec.}\,\psi+\beta\,\text{tan.}\,\psi \ \ldots\ldots (12),$$

which function attains its maximum when sin. $\psi=\dfrac{\beta}{a}$. Substituting this value in the preceding formula, and observing that $-a\,\text{sec.}\,\psi+\beta\,\text{tan.}\,\psi=$

$$-\text{sec.}\,\psi(a-\beta\,\text{sin.}\,\psi)=-\frac{\left(a-\dfrac{\beta^2}{a}\right)}{\sqrt{1-\dfrac{\beta^2}{a^2}}}=-\sqrt{a^2-\beta^2}, \text{ we obtain for the maxi-}$$

mum error the expression

$$1-\sqrt{a^2-\beta^2} \ \ldots\ldots (13).$$

Assuming ψ_1 and ψ_2 to represent the values of ψ, corresponding to the greatest and least values of $\dfrac{a}{b}$, and observing that in this case, as in the preceding, the values of a and β, which satisfy the conditions of the question, are those which render the values of the error corresponding to these limits equal, when taken with contrary signs, to the maximum error, we have

$$-1+a\,\text{sec.}\,\psi_1-\beta\,\text{tan.}\,\psi_1=1-\sqrt{a^2-\beta^2} \ \ldots (14).$$

$$1-a\,\text{sec.}\,\psi_1+\beta\,\text{tan.}\,\psi_1=1-a\,\text{sec.}\,\psi_2+\beta\,\text{tan.}\,\psi_2 \ \ldots (15).$$

The latter equation gives, by reduction,

$$a=\beta\frac{\cos.\,\frac{1}{2}(\psi_1-\psi_2)}{\sin.\,\frac{1}{2}(\psi_1+\psi_2)} \ \ldots (16).$$

$$a^2-\beta^2=\beta^2\left\{\frac{\cos.^2\,\frac{1}{2}(\psi_1-\psi_2)}{\sin.^2\,\frac{1}{2}(\psi_1+\psi_2)}-1\right\}=\beta^2\frac{\cos.\,\psi_1\cos.\,\psi_2}{\sin.^2\,\frac{1}{2}(\psi_1+\psi_2)}.$$

And $a\,\text{sec.}\,\psi_1-\beta\,\text{tan.}\,\psi_1=\beta\,\cot.\,\frac{1}{2}(\psi_1+\psi_2) \ \ldots (17).$

Substituting these values in equation (14), and solving in respect to β,

R R

$$\beta = \frac{2 \sin. \frac{1}{2}(\psi_1 + \psi_2)}{\cos. \frac{1}{2}(\psi_1 + \psi_2) + \sqrt{\cos. \psi_1 \cos. \psi_2}} \dots \text{(18)}.$$

$$a = \frac{2 \cos. \frac{1}{2}(\psi_1 - \psi_2)}{\cos. \frac{1}{2}(\psi_1 + \psi_2) + \sqrt{\cos. \psi_1 \cos. \psi_2}} \dots \text{(19)}.$$

The maximum error is represented by the formula

$$1 - \frac{2\sqrt{\cos. \psi_1 \cos. \psi_2}}{\cos. \frac{1}{2}(\psi_1 + \psi_2) + \sqrt{\cos. \psi_1 \cos. \psi_2}} \dots \text{(20)}.$$

These formulæ will be adapted to logarithmic calculation, if we assume $\frac{1}{2}(\psi_1 + \psi_2) = \Psi_1$, and $\frac{\cos. \frac{1}{2}(\psi_1 - \psi_2)}{\sin. \frac{1}{2}(\psi_1 + \psi_2)} = \text{cosec.} \Psi_2$; we shall thus obtain from equations (16) and (17) $a = \beta$ cosec. Ψ_2, $\sqrt{a^2 - \beta^2} = \beta$ cot. Ψ_2, and a sec. ψ_1' $-\beta$ tan. $\psi_1 = \beta$ cot. Ψ_1; therefore, by equation (14),

$$\left. \begin{array}{l} \beta = \dfrac{2}{\cot. \Psi_1 + \cot. \Psi_2} = \dfrac{2 \sin. \Psi_1 \sin. \Psi_2}{\sin. (\Psi_1 + \Psi_2)} \\[2mm] a = \dfrac{2 \text{ cosec. } \Psi_2}{\cot. \Psi_1 + \cot. \Psi_2} = \dfrac{2 \sin. \Psi_1}{\sin. (\Psi_1 + \Psi_2)} \end{array} \right\} \dots \text{(21)}.$$

$$\text{Maximum error} = \frac{\sin. (\Psi_1 - \Psi_2)}{\sin. (\Psi_1 + \Psi_2)} \dots \text{(22)}.$$

The form under which this theorem has been given by M. Poncelet is different from the above. Assuming, as in the previous case, the limiting values of $\frac{a}{b}$ to be represented by cot. ψ_1 and cot. ψ_2, and proceeding by a geometrical method of investigation, he has shown that if we assume tan. $\psi_1 = \cos. \omega_1$, tan. $\psi_2 = \cos. \omega_2$, $\omega_1 + \omega_2 = 2\gamma_1$, $\omega_1 - \omega_2 = 2\delta$, and cos. $\gamma_2 = \frac{\cos. \gamma_1}{\cos. \delta}$; then

$$a = \frac{2 \cos. \gamma_1}{\sin. (\gamma_1 + \gamma_2)}, \beta = \frac{2 \cos.^2 \gamma_1}{\sin. (\gamma_1 + \gamma_2) \cos. \delta}, \text{maximum error} = \frac{\sin. (\gamma_1 - \gamma_2)}{\sin. (\gamma_1 + \gamma_2)}.$$

If the least possible value of a be $1\frac{1}{10}b$, and its greatest possible value be infinite as compared with b, M. Poncelet has shown the formula of approximation to become

$$\sqrt{a^2 - b^2} = 1 \cdot 1319 a - 0 \cdot 72636 b \dots \text{(23)},$$

with a possible error of $0 \cdot 1319$ or $\frac{1}{7}$th nearly.

If the least possible value of a be $2b$, and its greatest possible value infinite compared with b; then

$$\sqrt{a^2 - b^2} = 1 \cdot 018623 a - 0 \cdot 272944 b \dots \text{(24)},$$

with a possible error of $\cdot 0186$ or $\frac{1}{52}$d nearly.

NOTE C.

The following is the general theorem expressing the relation between any number of pressures P_1, P_2, P_3, &c., applied to a body moveable about a cylindrical axis, in its state bordering upon motion. It is demonstrated in a " Memoir upon the Theory of Machines," printed in the Second Part of the *Transactions of the Royal Society* for 1841. — Let ι_{12}, ι_{13}, ι_{14}, ι_{23}, &c. be taken to represent the inclinations of the directions of the pressures to one another; a_1, a_2, a_3, &c. the perpendiculars upon them, severally, from the centre of the axis; ρ the radius of the axis; and ϕ the limiting angle of rèsistance; then

$$P_1 = -\frac{P_2 a_2 + P_3 a_3 + \ldots}{a_1} + \frac{\rho\sin.\phi}{a_1^2}\left\{\begin{matrix}P_2^2 L_{12}^2 + P_3^2 L_{13}^2 + P_4^2 L_{14}^2 + \ldots \\ + 2 P_2 P_3 M_{23} + 2 P_2 P_4 M_{24} + \ldots\end{matrix}\right\}^{\frac{1}{2}};$$

in which expression L_{12}, L_{13}, &c. are taken to represent the lines which join the foot of the perpendicular a_1 let fall upon the pressure P_1, with the feet of the perpendiculars a_2, a_3, a_4, &c. let fall upon the other pressures of the system; and in which M_{23}, M_{24}, &c. are taken to represent the different values assumed by the function

$$\{a_2 a_3 - a_1(a_1 \cos. \iota_{23} + a_2 \cos. \iota_{13} + a_3 \cos. \iota_{12})\}.$$

It is evident that equation (161) is but a particular case of this more general theorem.

NOTE D.

THE BEST DIMENSIONS OF A BUTTRESS.

If m_1 (Art. 301.) represent the modulus of stability of the portion AG of the wall, it may be shown, as before, that

$$P\{(h_1 - h_2)\sin. a - (l - a_2 - m_1)\cos. a\} = (\tfrac{1}{2}a_1 - m_1)(h_1 - h_2)a_1 \mu;$$

$$\therefore P\{(h_1 - h_2)\sin. a - (l - a_2)\cos. a\} = \tfrac{1}{2}(h_1 - h_2)a_1^2 \mu - m_1\{P\cos. a + (h_1 - h_2)a_1\mu\} \ldots (25).$$

If $m_1 = m$, the stability of the *portion* AG of the structure is the same with that of the *whole* AC; an arrangement by which the greatest strength is obtained with a given quantity of material (see Art. 390.). This supposition being made, and m eliminated between the above equation and equation (393), that relation between the dimensions of the buttress and those of the wall which is consistent with the greatest economy of the material used will be determined. The following is that relation : —

$$\frac{\left(a_1^2 h_1 + 2 a_1 a_2 h_1 + \frac{1}{n}a_2^2 h_2\right) - P(h_1 \sin. a - l\cos. a)}{P\cos. a + \mu\left(a_1 h_1 + \frac{1}{n}a_2 h_2\right)} = \frac{\tfrac{1}{2}\mu(h_1 - h_2)a_1^2 - P\{(h_1 - h_2)\sin. a - (l - a_2)\cos. a\}}{P\cos. a + \mu a_1(h_1 - h_2)} \quad \ldots (26).$$

It is necessary to the greatest economy of the material of the Gothic buttress (Art. 303.) that the stability of the portions Qa and Qb, upon their respective bases ac and be, should be the same with that of the whole buttress on its base EC. If, in the preceding equation (26), h_1-h_3 be substituted for h_1, and h_2-h_3 for h_2, the resulting equation, together with that deduced as explained in the conclusion of Art. 303., will determine this condition, and will establish those relations between the dimensions of the several portions of the buttress which are consistent with the greatest economy of the material, or which yield the greatest strength to the structure from the use of a given quantity of material.

NOTE E.

Dimensions of the Teeth of Wheels.

The following rules are extracted from the work of M. Morin, entitled *Aide Memoire de Mecanique Pratique:* — If we represent by a the width in parts of a foot of the tooth measured parallel to the axis of the wheel, and by b its breadth or thickness measured parallel to the plane of rotation upon the pitch circle, then, the teeth being constantly greased, the relation of a and b should be expressed, when the velocity of the pitch circle does not exceed 5 feet per second, by $a=4b$; when it exceeds 5 feet per second, by $a=5b$; if the wheels are constantly exposed to wet, by $a=6b$.

These relations being established, the width or thickness of the tooth will be determined by the formulæ contained in the columns of the following table : —

Material.	French Measures, Cents. and Kils.	English Measures, Feet and Pounds.
Cast iron - -	$b = \cdot105\sqrt{P}$	$b = \cdot002319\sqrt{P}$
Brass - - -	$b = \cdot131\sqrt{P}$	$b = \cdot002894\sqrt{P}$
Hard wood - -	$b = \cdot145\sqrt{P}$	$b = \cdot003203\sqrt{P}$

Assuming that when the teeth are carefully executed the space between the teeth should be $\frac{1}{15}$th greater than their thickness, and $\frac{1}{10}$th greater when the least labour is bestowed on them, the values of the pitch T will in these two cases be represented by $b(2+\frac{1}{15})$ and $b(2+\frac{1}{10})$, or by $2\cdot067b$ and $2\cdot1b$. Substituting in these expressions the values of b given by the formulæ of the preceding table, then determining from the resulting values of c (see equation 238) the corresponding values of the coefficient C (see equation 239), the following table is obtained : —

Material.	Value of c (equation 238).		Value of C (equation 239).	
	For Teeth of the best Workmanship.	For Teeth of inferior Workmanship.	For Teeth of the best Workmanship.	For Teeth of inferior Workmanship.
Cast iron -	·004795	·004870	0·912	0·922
Brass - -	·005982	·006077	1·057	1·068
Hard wood -	·006621	·006726	1·131	1·143

The following are the pitches commonly in use among mechanics : —

$$1, \quad 1\tfrac{1}{8}, \quad 1\tfrac{1}{4}, \quad 1\tfrac{1}{2}, \quad 2, \quad 2\tfrac{1}{2}, \quad 3.$$
(in. in. in. in. in. in. in.)

Prof. Willis considers the following to be sufficient below inch pitch : —

$$\tfrac{1}{4}, \quad \tfrac{3}{8}, \quad \tfrac{1}{2}, \quad \tfrac{5}{8}, \quad \tfrac{3}{4}.$$
(in. in. in. in. in.)

Having, therefore, determined the proper pitch to be given to the tooth from formula 239, the nearest pitch is to be taken from the above series to that thus determined.

NOTE F.

EXPERIMENTS OF M. MORIN ON THE TRACTION OF CARRIAGES.

The following are among the general results deduced by M. Morin from his experiments : —

1. The traction is directly proportional to the load, and inversely proportional to the diameter of the wheel.
2. Upon a paved or a hard Macadamized road, the resistance is independent of the width of the tire when it exceeds from 3 to 4 inches.
3. At a walking pace the traction is the same, under the same circumstances, for carriages with springs and without them.
4. Upon hard Macadamized and upon paved roads, the traction increases with the velocity ; the increments of traction being directly proportional to the increments of velocity above the velocity 3·28 feet per second, or about 2¼ miles per hour. The equal increment of traction thus due to each equal increment of velocity is less as the road is more smooth, and the carriage less rigid or better hung.
4. Upon soft roads of earth, or sand or turf, or roads fresh and thickly gravelled, the traction is independent of the velocity.
5. Upon a well-made and compact pavement of hewn stones, the traction at a walking pace is not more than three fourths of that upon the best Macadamized road under similar circumstances; at a trotting pace it is equal to it.
6. The destruction of the road is in all cases greater as the diameters of the wheels are less, and it is greater in carriages without than with springs.

NOTE G.

ON THE STRENGTH OF COLUMNS.

Mr. Hodgkinson has obligingly communicated the following observations on Art. 432. : —

1. The reader must be made to understand that the rounding of the ends of the pillars is to make them moveable there, as if they turned by means of an universal joint ; and the flat-ended pillars are conceived to be supported in every part of the ends by means of flat surfaces or otherwise, rendering the ends perfectly immoveable.

2. The coefficient (13) for hollow columns with rounded ends is deduced from the whole of the experiments first made, including some which were very defective on account of the difficulty experienced in the earlier attempts to cast good hollow columns so small as were wanted. The first castings were made lying on their side ; and this, notwithstanding every effort, prevented the core being in the middle : some of the columns were reduced, too, in thickness, half way between the middle and the ends, and near to the ends, and this slightly reduced the strength. These causes of weakness existed much more among the pillars with rounded ends than those with flat ones ; they are alluded to in the paper (Art. 47.). Had it not been for them, the coefficient (13) would, I conceive, have been equal to that for solid pillars (or 14·9).

3. The fact of long pillars with flat ends being about three times as strong as those of the same dimensions with rounded ends is, I conceive, well made out, in cast iron, wrought iron, and timber ; you have, however, omitted it, being perhaps led to do it through the low value of the coefficient (13) above mentioned.

The same may be mentioned with respect to the near approach in strength of long pillars with flat ends, and those of half the length with rounded ends. It may be said that the law of the 1·7 power of the length would nearly indicate the latter ; but this last, and the other powers 3·76 and 3·55, are only approximations, and not exactly constant, though nearly so, and I do not know whether the other equal quantities are not, with some slight modifications, physical facts.

4. The strength of pillars of *similar form* and of the same materials varies as the 1·865 power, or nearly as the square of their like linear dimensions, or as the area of their cross section.

TABLE I.

The Numerical Values of COMPLETE *Elliptic Functions of the* FIRST *and* SECOND *Orders for Values of the Modulus* k *corresponding to each Degree of the Angle* sin.^{-1}k.

Sin.^{-1}k.	F_1.	E_1.	Sin.^{-1}k.	F_1.	E_1.
0°	1·57079	1·57079	46	1·86914	1·34180
1	1·57091	1·57067	47	1·88480	1·33286
2	1·57127	1·57031	48	1·90108	1·32384
3	1·57187	1·56972	49	1·91799	1·31472
4	1·57271	1·56888	50	1·93558	1·30553
5	1·57379	1·56780	51	1·95386	1·29627
6	1·57511	1·56649	52	1·97288	1·28695
7	1·57667	1·56494	53	1·99266	1·27757
8	1·57848	1·56316	54	2·01326	1·26814
9	1·58054	1·56114	55	2·03471	1·25867
10	1·58284	1·55888	56	2·05706	1·24918
11	1·58539	1·55639	57	2·08035	1·23966
12	1·58819	1·55368	58	2·10465	1·23012
13	1·59125	1·55073	59	2·13002	1·22058
14	1·59456	1·54755	60	2·15651	1·21105
15	1·59814	1·54415	61	2·18421	1·20153
16	1·60197	1·54052	62	2·21319	1·19204
17	1·60608	1·53666	63	2·24354	1·18258
18	1·61045	1·53259	64	2·27537	1·17317
19	1·61510	1·52830	65	2·30878	1·16382
20	1·62002	1·52379	66	2·34390	1·15454
21	1·62523	1·51907	67	2·38087	1·14534
22	1·63072	1·51414	68	2·41984	1·13624
23	1·63651	1·50900	69	2·46099	1·12724
24	1·64260	1·50366	70	2·50455	1·11837
25	1·64899	1·49811	71	2·55073	1·10964
26	1·65569	1·49236	72	2·59981	1·10106
27	1·66271	1·48642	73	2·65213	1·09265
28	1·67005	1·48029	74	2·70806	1·08442
29	1·67773	1·47396	75	2·76806	1·07640
30	1·68575	1·46746	76	2·83267	1·06860
31	1·69411	1·46077	77	2·90256	1·06105
32	1·70283	1·45390	78	2·97856	1·05377
33	1·71192	1·44686	79	3·06172	1·04678
34	1·72139	1·43966	80	3·15338	1·04011
35	1·73124	1·43229	81	3·25530	1·03378
36	1·74149	1·42476	82	3·36986	1·02784
37	1·75216	1·41707	83	3·50042	1·02231
38	1·76325	1·40923	84	3·65185	1·01723
39	1·77478	1·40125	85	3·83174	1·01266
40	1·78676	1·39314	86	4·05275	1·00864
41	1·79922	1·38488	87	4·33865	1·00525
42	1·81215	1·37650	88	4·74271	1·00258
43	1·82560	1·36799	89	5·43490	1·00075
44	1·83956	1·35937			
45	1·85407	1·35064			

The Tables of M. Garidel.

TABLE II.

Showing the Angle of Rupture Ψ *of an Arch whose Loading is of the same Material with its Voussoirs, and whose Extrados is inclined at a given Angle to the Horizon.* (See Art. 346.)

a = ratio of lengths of voussoirs to radius of intrados.

c = ratio of depth of load over crown to radius of intrados, so that $c=\beta(1+a)$. (Art. 340.)

ι = inclination of extrados to horizon.

$\iota = 0.$

a	$c=0$	$c=0.1$	$c=0.2$	$c=0.3$	$c=0.4$	$c=0.5$	$c=1.0$
0·05	68·0°	59·19°	54·04°	51·15°	49·35°	48·20°	45·74°
0·10	65·4	60·48	57·70	56·01	54 93	54·17	52·34
0·15	64·0	61·3	59·7	58·69	58·0	57·49	56·21
0·20	63·1	61·7	60·88	60·30	59·90	59·60	58·80
0·25	62·24	61·76	61·44	61·22	61·05	60·94	60·59
0·30	61·3	61·42	61·54	61·60	61·66	61·67	61·81
0·35	60·17	60·80	61·21	61·54	61·78	61·98	62·56
0·40	58·8	59·8	60·52	61·05	61·48	61·67	62·9
0·45	57·32	58·53	59·45	60·19	60·80	61·28	62·85
0·50	55·63	56·97	58·09	58·98	59·72	60·34	62·40

$\iota = 7° \ 30'.$

a	$c=0$	$c=0.1$	$c=0.2$	$c=0.3$	$c=0.4$	$c=0.5$	$c=1.0$
0·05	68·3°	57·3°	51·69°	48·61°	47·84°	46·11°	44·85°
0·10	64·3	58·68	55·95	54·52	53·64	53·03	51·68
0·15	62·43	59·67	58·33	57·55	57·00	56·61	55·66
0·20	61·48	60·42	59·72	59·35	59·07	58·87	58·29
0·25	60·75	60·55	60·44	60·38	60·33	60·21	60·17
0·30	60·09	60·49	60·77	60·95	61·08	61·18	61·48
0·35	59 27	60·12	60·62	61·02	61·33	61·59	62·31
0·40	58·25	59·33	60·11	60·72	61·18	61·57	62·7
0·45	57·11	58·35	59·29	60·05	60·67	61·16	62·78
0·50	55·82	57·13	58·21	59·08	59·81	60·41	62·45

$\iota=15°$.

α	c=0	c=0·1	c=0·2	c=0·3	c=0·4	c=0·5	c=1·0
0·05	64·8°	50·5°	46·95°	45·69°	45·03°	44·67°	43·9°
0·10	59·3	55·07	53·34	52·47	51·99	51·69	50·93
0·15	59·08	57·32	56.65	56·05	55·75	55·55	55·05
0·20	59·06	58·60	58·35	58·20	58·10	58·02	57·84
0·25	59·05	59·28	59·42	59·53	59·60	59·65	59·79
0·30	58·90	59·57	59·98	60·26	60·48	60·66	61·15
0·35	58·53	59·41	60·09	60·57	60·93	61·17	62·0
0·40	57·99	59·08	59·87	60·48	60·95	61·36	62·6
0·45	57·26	58·43	59·34	60·06	60·67	61·15	62·7
0·50	56·38	57·61	58·58	59·36	60·06	60·64	62·5

$\iota=22° \ 30'$.

α	c=0	c=0·1	c=0·2	c=0·3	c=0·4	c=0·5	c=1·0
0·05	36·1°	41·2°	42·0°	42·3°	42·6°	42·7°	42·9°
0·10	50·5	50·3	50·19	50·17	50·14	50·13	50·11
0·15	54·25	54·31	54·35	54·35	54·36	54·36	54·38
0·20	56·17	56·60	56·82	56·95	57·04	57·11	57·28
0·25	57·27	57·93	58·33	58·61	58·79	58·95	59·33
0·30	57·85	58·68	59·23	59·60	59·93	60·16	60·83
0·35	58·07	59·01	59·70	60·21	60·61	60·91	61·85
0·40	58·02	59·02	59·79	60·38	60·87	61·25	62·2
0·45	57·74	58·78	59·60	60.26	60·82	61·27	62·7
0·50	57·30	58·31	59·16	59·88	60·47	61·00	62·9

$\iota=30°$.

α	c=0	c=0·1	c=0·2	c=0·3	c=0·4	c=0·5	c=1·0
0·05	31·3°	36·2°	38·4°	39·57°	40·28°	40·77°	41·9°
0·10	43·3	46·06	47·25	47·90	48·30	48·59	49·24
0·15	50·07	51·46	52·18	52·63	52·94	53·14	53·68
0·20	53·66	54·69	55·27	55·67	55·96	56·16	56·72
0·25	55·80	56·72	57·30	57·72	58·01	58·23	58·89
0·30	57·13	58·01	58·62	59·06	59·40	59·69	60·48
0·35	57·93	58·80	59·43	59·94	60·33	60·66	61·64
0·40	58·33	59·20	59·89	60·42	60·87	61·23	62·39
0·45	58.47	59·33	60·03	60·61	61·08	61·48	62·87
0·50	58·38	59·22	59·93	60·53	61·03	61·47	63·0

$$\iota = 37^\circ\ 30\ .$$

α	$c=0$	$c=0\cdot1$	$c=0\cdot2$	$c=0\cdot3$	$c=0\cdot4$	$c=0\cdot5$	$c=1\cdot0$
0·05	31·1°	34·3°	36·28°	37·59°	38·48°	39·16°	40·82°
0·10	40·98	43·59	45·09	46·01	46·67	47·14	48·35
0·15	47·71	49·40	50·43	51·12	51·61	51·96	52·93
0·20	52·01	53·23	54·01	54·54	54·94	55·24	56·10
0·25	54·87	55·80	56·45	56·94	57·29	57·59	58·41
0·30	56·77	57·58	58·16	58·62	58·98	59·26	60·16
0·35	58·04	58·78	59·34	59·81	60·17	60.47	61·45
0·40	58·89	59·58	60·13	60·60	60·97	61·30	62·4
0·45	59·38	60·06	60·62	61·07	61·47	61·83	63·0
0·50	56·69	60·29	60·84	61·26	61·72	62·07	63·3

$$\iota = 45^\circ.$$

α	$c=0$	$c=0\cdot1$	$c=0\cdot2$	$c=0\cdot3$	$c=0\cdot4$	$c=0\cdot5$	$c=1\cdot0$
0·05	31·3°	33·68°	35·46°	36·36°	37·22°	38·0°	39·9⁶
0·10	40·6	42·4	43·7	44·64	45·35	45·92	47·45
0·15	46·77	48·20	49·18	49·93	50·47	50·92	52·15
0·20	51·23	52·27	53·05	53·64	54·07	54·42	55·47
0·25	54·42	55·22	55·84	56·31	56·70	57·01	57·97
0·30	56·72	57·38	57·90	58·30	58·65	58·94	59·85
0·35	58·35	58·94	59·40	59·79	60·11	60·38	61·30
0·40	59·56	60·09	60·52	60·89	61·19	61·46	62·4
0·45	60·40	60·89	61·29	61·67	61·97	62·24	63·2
0·50	60·99	61·43	61·8	62·2	62·5	62·8	63·8

THE TABLES OF M. GARIDEL.

TABLE III.

Showing the Horizontal Thrust of an Arch, the Radius of whose Intrados is Unity, and the Weight of each Cubic Foot of its Material and of that of its Loading, Unity. (See Art. 346.)

N.B. To find the horizontal thrust of any other arch, multiply that given in the table by the square of the radius of the intrados and by the weight of a cubic foot of the material.

$\iota = 0.$

α	$c=0$ $\frac{P}{r^2}$	$c=0\cdot1$ $\frac{P}{r^2}$	$c=0\cdot2$ $\frac{P}{r^2}$	$c=0\cdot3$ $\frac{P}{r^2}$	$c=0\cdot4$ $\frac{P}{r^2}$	$c=0\cdot5$ $\frac{P}{r^2}$	$c=1\cdot0$ $\frac{P}{r^2}$
0·05	0·08174	0·14797	0·21762	0·28877	0·36060	0·43277	0·79541
0·10	0·10279	0·16370	0·22588	0·28862	0·35164	0·41481	0·73161
0·15	0·11894	0·17480	0·23111	0·28764	0·34429	0·40100	0·68504
0·20	0·13073	0·18191	0·23322	0·28460	0·33603	0·38747	0·64488
0·25	0·13871	0·18553	0·23237	0·27922	0·32607	0·37293	0·60727
0·30	0·14333	0·18604	0·22874	0·27145	0·31416	0·35687	0·57041
0·35	0·14504	0·18379	0·22258	0·26140	0·30023	0·33907	0·53335
0·40	0·14422	0·17913	0·21415	0·24924	0·28437	0·31953	0·49560
0·45	0·14124	0·17240	0·20374	0·23520	0·26674	0·29835	0·45693
0·50	0·13649	0·16396	0·19168	0·21957	0·24760	0·27573	0·41728

$\iota = 7° \ 30'.$

α	$c=0$ $\frac{P}{r^2}$	$c=0\cdot1$ $\frac{P}{r^2}$	$c=0\cdot2$ $\frac{P}{r^2}$	$c=0\cdot3$ $\frac{P}{r^2}$	$c=0\cdot4$ $\frac{P}{r^2}$	$c=0\cdot5$ $\frac{P}{r^2}$	$c=1\cdot0$ $\frac{P}{r^2}$
0·05	0·06180	0·12867	0·19937	0·27125	0·34356	0·41606	0·77944
0·10	0·08514	0·14666	0·20930	0·27237	0·33561	0·39895	0·71618
0·15	0·10380	0·16001	0·21657	0·27326	0·33003	0·38683	0·67110
0·20	0·11813	0·16948	0·22089	0·27237	0·32384	0·37533	0·63286
0·25	0·12870	0·17557	0·22244	0·26932	0·31619	0·36306	0·59743
0·30	0·13598	0·17866	0·22134	0·26403	0·30673	0·34943	0·56295
0·35	0·14040	0·17909	0·21783	0·25661	0·29542	0·33424	0·52846
0·40	0·14234	0·17718	0·21215	0·24720	0·28230	0·31744	0·49344
0·45	0·14211	0·17323	0·20454	0·23598	0·26751	0·29910	0·45763
0·50	0·14003	0·16753	0·19528	0·22319	0·25124	0·27938	0·42096

$$\iota = 15°.$$

α	$c=0$ $\dfrac{P}{r^2}$	$c=0.1$ $\dfrac{P}{r^2}$	$c=0.2$ $\dfrac{P}{r^2}$	$c=0.3$ $\dfrac{P}{r^2}$	$c=0.4$ $\dfrac{P}{r^2}$	$c=0.5$ $\dfrac{P}{r^2}$	$c=1.0$ $\dfrac{P}{r^2}$
0·05	0·05310	0·12265	0·19488	0·26748	0·34018	0·41293	0·77681
0·10	0·07903	0·14170	0·20493	0·26832	0·33176	0·39524	0·71277
0·15	0·09990	0·15658	0·21336	0·27022	0·32708	0·38395	0·66840
0·20	0·11631	0·16781	0·21931	0·27083	0·32234	0·37386	0·63145
0·25	0·12894	0·17582	0·22268	0·26955	0·31643	0·36330	0·59767
0·30	0·13835	0·18096	0·23361	0·26627	0·30895	0·35163	0·56510
0·35	0·14494	0·18355	0·22224	0·26098	0·29976	0·33855	0·53271
0·40	0·14905	0·18384	0·21878	0·25380	0·28888	0·32399	0·49995
0·45	0·15097	0·18212	0·21344	0·24488	0·27641	0·30800	0·46652
0·50	0·15099	0·17860	0·20642	0·23439	0·26247	0·29065	0·43232

$$\iota = 22° \; 30'.$$

α	$c=0$ $\dfrac{P}{r^2}$	$c=0.1$ $\dfrac{P}{r^2}$	$c=0.2$ $\dfrac{P}{r^2}$	$c=0.3$ $\dfrac{P}{r^2}$	$c=0.4$ $\dfrac{P}{r^2}$	$c=0.5$ $\dfrac{P}{r^2}$	$c=1.0$ $\dfrac{P}{r^2}$
0·05	0·06102	0·13346	0·20621	0·27899	0·35178	0·42458	0·78857
0·10	0·08700	0·15053	0·21407	0·27760	0·34113	0·40466	0·72233
0·15	0·10877	0·16567	0·22257	0·27947	0·33638	0·39328	0·67778
0·20	0·12635	0·17785	0·22936	0·28087	0·33239	0 38391	0·64150
0·25	0·14037	0·18716	0·23399	0·28082	0·32767	0·37453	0·60886
0·30	0·15129	0·19381	0·23640	0·27902	0·32166	0·36432	0·57773
0·35	0·15948	0·19804	0·23669	0·27540	0·31415	0·35292	0·54700
0·40	0·16525	0·20005	0·23497	0·26999	0·30506	0·34017	0·51608
0·45	0·16883	0·20005	0.23141	0·26289	0·29444	0·32604	0·48460
0·50	0·17047	0·19824	0·22617	0·25423	0·28238	0·51060	0·45241

$$\iota = 30°.$$

α	$c=0$ $\dfrac{P}{r^2}$	$c=0.1$ $\dfrac{P}{r^2}$	$c=0.2$ $\dfrac{P}{r^2}$	$c=0.3$ $\dfrac{P}{r^2}$	$c=0.4$ $\dfrac{P}{r^2}$	$c=0.5$ $\dfrac{P}{r^2}$	$c=1.0$ $\dfrac{P}{r^2}$
0·05	0·09355	0·16408	0·23605	0·30845	0·38101	0·45365	0·81731
0·10	0·11297	0·17592	0·23922	0·30263	0·36609	0·42957	0·74711
0·15	0·13295	0·18962	0·24640	0·30323	0·36009	0·41696	0·70138
0·20	0·15038	0·20172	0·25314	0·30459	0·35606	0·40755	0·66506
0·25	0·16493	0·21160	0·25834	0·30513	0·35193	0·39876	0·63299
0·30	0·17673	0·21917	0·26170	0·30427	0·34688	0·38951	0·60282
0·35	0·18599	0·22452	0·26314	0·30182	0·34055	0·37930	0·57332
0·40	0·19293	0·22777	0·26271	0·29773	0·33280	0·36791	0·54380
0·45	0·19774	0·22906	0·26050	0·29202	0·32361	0·35524	0·51385
0·50	0·20060	0·22854	0·25661	0·28476	0 31299	0·34128	0·48327

$\iota = 37° \ 30'.$

α	$c=0$ $\dfrac{P}{r^2}$	$c=0\cdot1$ $\dfrac{P}{r^2}$	$c=0\cdot2$ $\dfrac{P}{r^2}$	$c=0\cdot3$ $\dfrac{P}{r^2}$	$c=0\cdot4$ $\dfrac{P}{r^2}$	$c=0\cdot5$ $\dfrac{P}{r^2}$	$c=1\cdot0$ $\dfrac{P}{r^2}$
0·05	0·14749	0·21733	0·28854	0·36038	0·43255	0·50490	0·86784
0·10	0·15949	0·22174	0·28457	0·34768	0·41093	0·47426	0·79141
0·15	0·17605	0·23233	0·28886	0·34553	0·40226	0·45904	0·74322
0·20	0·19209	0·24321	0·29448	0·34583	0·39722	0·44865	0·70598
0·25	0·20627	0·25282	0·29948	0·34619	0·39294	0·43972	0·67382
0·30	0·21827	0·26066	0·30314	0·34568	0·38825	0·43085	0·64406
0·35	0·22805	0·26659	0·30521	0·34388	0·38259	0·42133	0·61529
0·40	0·23570	0·27060	0·30558	0·34062	0·37571	0·41083	0·58673
0·45	0·24130	0·27275	0·30427	0·33586	0·36749	0·39916	0·55787
0·50	0·24499	0·27312	0 30132	0·32958	0·35789	0·38625	0·52845

$\iota = 45°.$

α	$c=0$ $\dfrac{P}{r^2}$	$c=0\cdot1$ $\dfrac{P}{r^2}$	$c=0\cdot2$ $\dfrac{P}{r^2}$	$c=0\cdot3$ $\dfrac{P}{r^2}$	$c=0\cdot4$ $\dfrac{P}{r^2}$	$c=0\cdot5$ $\dfrac{P}{r^2}$	$c=1\cdot0$ $\dfrac{P}{r^2}$
0·05	0·23105	0·30081	0·37162	0·44305	0·51485	0·58688	0·94881
0·10	0·23318	0·29507	0·35754	0·42034	0·48333	0·54646	0·86800
0·15	0·24478	0·30079	0·35708	0·41355	0·47013	0·52678	0·81059
0·20	0·25819	0·30915	0·36028	0·41151	0·46281	0·51415	0·77124
0·25	0·27104	0·31752	0·36410	0·41074	0·45744	0·50417	0·73809
0·30	0·28248	0·32486	0·36731	0·40981	0·45235	0·49493	0·70803
0·35	0·29216	0·33073	0·36935	0·40803	0·44674	0·48547	0·67939
0·40	0·29997	0·33494	0·36998	0·40506	0·44016	0·47530	0·65123
0·45	0·30589	0·33745	0·36907	0·40072	0·43240	0·46412	0·62294
0·50	0·30996	0·33824	0·36657	0·39494	0·42334	0·45177	0·59419

TABLE IV.

Mechanical Properties of the Materials of Construction.

Note. — The capitals affixed to the numbers in this table refer to the following authorities : —

B.	Barlow, *Report to the Commissioners of the Navy, &c.*
Be.	Bevan.
Br.	Belidor, *Arch. Hydr.*
Bru.	Brunel.
C.	Couch.
D. W.	Daniell and Wheatstone, *Report on the Stone for the Houses of Parliament.*
F.	Fairbairn.
H.	Hodgkinson, *Report to the British Association of Science, &c.*
K.	Kirwan.
La.	Lamé.
M.	Muschenbroek, *Introd. ad Phil. Nat.* i.
Mi.	Mitis.
Mt.	Mushet.
Pa.	Colonel Paseley.
R.	Rondelet. *L'Art de Batir,* iv.
Re.	Rennie, *Phil. Trans., &c.*
T.	Thompson, *Chemistry.*
Te.	Telford.
Tr.	Tredgold, *Essay on the Strength of Cast Iron.*
W.	Watson.

Names of Materials.	Specific Gravity.	Weight of 1 Cubic Foot in lbs.	Tenacity per Square Inch in lbs.	Crushing Force per Square Inch in lbs.	Modulus of Elasticity E.	Modulus of Rupture S.
Acacia (English growth)	·710 B.	44·37	16000 Be.	- -	1152000 B.	11202 B.
Air (atmospheric) -	·001228	0·0768				
Alabaster (yellow Malta)	2·699	168·68				
do. (stained brown) -	2·744	171·50				
do. (oriental white) -	2·730	170·62				
Alder - - - -	·800 M.	50·00	14186 M.	6895 H.		
Antimony, (cast) - -	4·500 M.	281·25	1066 M.			
Apple-tree - - -	·793 M.	49·56	19500 Be.			
Ash - - - - {	·690 to ·845	43·12 53·81	} 17207 B.	{ 8683 H. 9363 H.	} 1644800 B.	12156 B.
Bay-tree - - -	·822 M.	51·37	12396	7158 H.		
Bean (Tonquin) - -	1·080	67·50	- -	- -	2601600 B.	20886 B.
Beech - - - - {	·854 to ·690	53·37 43·12	15784 B. 17850 B.	7733 H. 9363 dry. H.	} 1353600 B.	9336 B.
Birch (common) - -	·792 B.	49·50	15000	{ 4533 H. 640 2dry. H.	} 1562400 B.	10920 B.
do. (American) - -	·648 B.	40·50	-	11663 H.	1257600 B.	9624 B.
Bismuth (cast) - -	9·810 M.	613·87	3250 M.			
Bone of an ox - -	1·656 M.	103·50	5265			
Box (dry) - - -	·960 B.	60·00	19891 B.	10299 H.		
Brass (cast) - - -	8·399	525·00	17968 Re.	10304 Re.	8930000	
do. (wire-drawn) -	8·544	534·00				
Brick (red) - - -	2·168 Re.	135·50	280	807 Re.		
do. (pale red) - -	2·085 Re.	130·31	300	562 Re.		
Brick-work - - -	1·800	112·50				
Bullet-tree (Berbice) -	1·029 B.	64·31	-	- -	2601600 B.	15636 B.
Cane - - - -	0·400	25·00	6300 Be.			
Cedar (Canadian, fresh)	0·909 C.	56·81	11400 Be.	5674 H.		
do. (seasoned) - -	0·753	47·06	- -	4912 H.		
Chalk - - - {	2·784 to 1·869	174·00 116·81	}- -	334 Re.		
Chestnut - - -	0·657 R.	41·06	13300 R.			
Clay (common) - -	1·919 Br.	119·93				
Coal (Welsh furnace) -	1·337 Mt.	83·56				
do. (coke) - - -	1·000 Mt.	62·50				
do. (Alfreton) - -	1·235 Mt.	77·18				
do. (Butterly) - -	1·264 Mt.	79·00				
do. (coke) - - -	1·100 Mt.	68·75				
do. (Welsh stone) -	1·358 Mt.	85·50				
do. (coke) - - -	1·390 Mt.	86·87				
do. (Welsh slaty) -	1·409 Mt.	88·06				
do. (Derbyshire cannel)	1·278 Mt.	79·87				
do. (Kilkenny) - -	1·602 Mt.	100·12				
do. (coke) - - -	1·657 Mt.	103·56				
do. (slaty) - - -	1·443 Mt.	90·18				

Names of Materials.	Specific Gravity.	Weight of 1 Cubic Foot in lbs.	Tenacity per Square Inch in lbs.	Crushing Force per Square Inch in lbs.	Modulus of Elasticity E.	Modulus of Rupture S.
Coal (Bonlavooneen) -	1·436 Mt.	89·75				
do. (coke) - - -	1·596 Mt.	99·75				
do. (Corgee) - -	1·403 Mt.	87·68				
do. (coke) - - -	1·656 Mt.	103·50				
do. (Staffordshire) -	1·240	78·12				
do. (Swansea) - -	1·357 K.	84·81				
do. (Wigan) - -	1·268 K.	79·25				
do. (Glasgow) - -	1·290	80·62				
do. (Newcastle) -	1·257 K.	78·56				
do. (common cannel)	1·232 K.	77·00				
do. (slaty cannel) -	1·426 K.	89·12				
Copper (cast) - -	8·607	537·03	19072			
do. (sheet) - -	8·785	549·06				
do. (wire-drawn) -	8·878	560·00	61228			
do. (in bolts) - -	- -	- -	48000			
Crab-tree - - -	0·765	47·80	- -	6499 H.		
Deal (Christiania middle) - - - -	0·698 B.	43·62	12400	- -	1672000 B.	9864 B.
do. (Memel middle) -	0·590 B.	36·87	- - -	- -	1535200 B.	10386 B.
do. (Norway spruce) -	0·340	21·25	17600			
do. (English) - -	0·470	29·37	7000			
Earth, (rammed) - -	1·584 Pa.	99·00				
Elder - - - -	0·695 M.	43·43	10230	8467 H.		
Elm (seasoned) - -	0·588 C.	36·75	13489 M.	10331 H.	699840 B.	6078 B.
Fir (New England) -	0·553 B.	34·56	- -	- -	2191200 B.	6612 B.
do. (Riga) - -	0·753 B.	47·06	{ 11549 B. to 12857 B.	5748 H. to 6586 H.	1328800 B.	6648 B.
do. (Mar Forest) -	0·693 B.	43·31			869600 B.	7572 B.
Flint - - - -	2·630 T.	164·37				
Glass (plate) - - -	2·453	153·31	9420			
Gravel - - - -	1·920	120·00				
Granite (Aberdeen) -	2·625	164·00				
do. (Cornish) - -	2·662	166·30				
do. (red Egyptian) -	2·654	165·80				
Hawthorn - - -	0·91 Be.	38·12	10500 Be.			
Hazel - - - -	0·86 Be.	53·75	18000 Be.			
Holly - - - -	0·76 Be.	47·5	16000 Be.			
Horn of an ox - -	1·689 M.	105·56	8949			
Hornbeam (dry) - -	0·760 R.	47·50	20240 Be.	7289 H.		
Iron (wrought English)	7·700	481·20	25¼ tons, La.			
do. (in bars) - -	{ ·7600 to 7·800	475·50 487·00	25¼ tons, La.			
do. (hammered) - -	- -	- -	30 tons, Bru.			
do. (Russian), in bars	- -	- -	27 tons, La.			
do. (Swedish) in bars	- -	- -	32 tons, R.			
do. (English) in wire 1-10th inch diameter	}- -	- -	{ 36 to 43 tons, Te.			
do. (Russian) in wire 1-20th to 1-30th inch diameter - - -	}- -	- -	{ 60 to 91 tons, La.			
do. rolled in sheets and cut lengthwise -	- -	- -	14 tons, Mi.			
do. cut crosswise - -	- -	- -	18 tons, Mi.			
do. in chains, oval links, 6 inches clear, iron ½ inch diameter	- -	- -	21½ tons, Br.			
do. (Brunton's) with stay across link -	- -	- -	25 tons, B.			
Iron, cast (old Park) -	- -	- -	- -	- -	18014400 T.	48240 T.
do. (Adelphi) - -	- -	- -	- -	- -	18353600 T.	45360 T.
do. (Alfreton) - -	- -	- -	- -	- -	17686400 T.	44046 T.
do. (scrap) - -	- -	- -	- -	- -	18032000 T.	45828 T.
do. (Carron, No. 2. cold blast) - -	7·066 H.	441·62	16683 H.	106375 H.	17270500 H.	38556 H.*
do. (hot blast) -	7·046 H.	440·37	13505 H.	108540 H.	16085000 H.	37503 H.*

* The numbers marked with an asterisk are calculated from the experiments of Messrs. Hodgkinson and Fairbairn.

Names of Materials.	Specific Gravity.	Weight of 1 Cubic Foot in lbs.	Tenacity per Square Inch in lbs.	Crushing Force per Square Inch in lbs.	Modulus of Elasticity E.	Modulus of Rupture S.
Iron, cast (do. No. 3. cold blast)	7·094 F.	443·37	14200 H.	115442 H.	16246966 F.	35980 F.*
do. (hot blast)	7·056 F.	441·00	17755 H.	133440 H.	17873100 F.	42120 F.*
do. (Devon, No. 3. cold blast)	7·295 H.	455·93	-	-	22907700 H.	36288 H.*
do. (hot blast)	7·229 H.	451·81	21907 H.	145435 H.	22473650 H.	43497 H.*
do. (Buffery, No. 1. cold blast)	7·079 H.	442·43	17466 H.	93366 H.	15381200 H.	37503 H.*
do. (hot blast)	6·998 H.	437·37	13434 H.	86397 H.	13730500 H.	35316 H.*
do. (Coed Talon, No. 2. cold blast)	6·955 F.	434·06	18855 H.	81770 H.	14313500 F.	33104*F.*
do. (hot blast)	6·968 F.	435·50	16676 H.	82739 H.	14322500 F.	33145,F.*
do. (Coed Talon, No. 3. cold blast)	7·194 F.	449·62	-	-	17102000 F.	43541 F.*
do. (hot blast)	6·970 F.	435·62	-	-	14707900 F.	40159 F.*
do. (Elsicar, No 1. cold blast)	7·030 F.	439·37	-	-	13981000 F.	34862 F.*
do. (Milton, No. 1. hot blast)	6·976 F.	436·00	-	-	11974500 F.	28552 F.*
do. (Muirkirk, No. 1. cold blast)	7·113 F.	444·56	-	-	14003550 F.	35923 F.*
do (hot blast)	6·953 F.	434·56	-	-	13294400 F.	33850 F.*
Ivory	1·826 P.	114·12	16·626			
Laburnum	0·92 Be.	57·50	10500 Be.			
Lance-wood	1·022	63·87	24696			
Larch	0·522 B.	32·62	10220 R.	3201 H. (green)	897600 B.	4992 B.
do. (second specimen)	0·560 B.	35·00	8900 B.	5568 H. (dry)	1052800 B.	6894 B.
Lead (cast English)	11·446 M.	717·45	1824 Re.	-	720000 Tr.	
do. (milled sheet)	11·407 T.	712·93	3328 Tr.			
do. (wire)	11·317 T.	705·12	2581 M.			
Lignum vitæ	1·220	76·25	11800 M.			
Limestone (arenaceous)	2·742	171·37				
do. (foliated)	2·837	177·31				
do. (white fluor)	3·156	197·25				
do. (green)	3·182	198·87				
Lime-tree	0·760	47·50	23500 Be.			
Lime (quick)	0·483 Br.	52·68				
Mahogany (Spanish)	0·800	50·00	16500	8198 H.		
Maple (Norway)	0·793	49·56	10584			
Marble (white Italian)	2·638 H.	164·87	-	-	2520000 T.	1062
do. (black Galway)	2·695 H.	168·25	-	-	-	2664
Mercury (at 32°)	13·619	851·18				
do. (at 60°)	13·580	848·75				
Marl	1·600 to 2·877 T.	100·00 / 118·31				
Mortar	1·751 Br.	107·18	50			
Oak (English)	0·934 B.	58·37	17·300 M.	4684 H. / 9509 H. (dry)	1451200 B.	10032 B.
do. (Canadian)	0·872 B.	54·50	10·253	4231 H. / 9509 H. (dry)	2148800 B.	10596 B.
do. (Dantzic)	0·756 B.	47·24	12·780	-	1191200 B.	8742 B.
do. (Adriatic)	0·993 B.	62·06	-	-	974400 B.	8298 B.
do. (African middle)	0·972 B.	60·75	-	-	2283200 B.	13566 B.
Pear-tree	0·661 M.	41·31	-	7518 H.	-	
Pine (pitch)	0·660 B.	41·25	7818 M.	-	1225600 B.	9792 B.
do. (red)	0·657 B.	41·06	-	5375 H.	1840000 B.	8946 B.
do. (American yellow)	0·461 C.	28·81	-	5445 H.	1600000 Tr.	
Plane-tree	0·64 Be.	40·00	11700 Bc.			
Plum-tree	0·785 M.	49·06	11351	9367 H. / 3657 H. (wet)		
Poplar	0·383 M.	23·93	7200 Be.	3107 H. / 5124 H. (dry)		
Pozzolano	2·677 K.	169·37				
Sand (river)	1·886	117·87				
Serpentine (green)	2·574 K.	163·87				

Names of Materials.	Specific Gravity.	Weight of 1 Cubic Foot in lbs.	Tenacity per Square Inch in lbs.	Crushing Force per Square Inch in lbs.	Modulus of Elasticity E.	Modulus of Rupture S.
Shingle - - - -	1·424 Pa.	89·00				
Silver (standard) - -	10·312 T.	644·50	40902 M.			
Slate (Welsh) - -	2·888	180·50	12800	- -	15800000 Tr.	11766 Re.
do. (Westmoreland)	2·791 W.	174·43	- -	- -	12900000 Tr.	
do. (Valentia) -	2·880 Re.	180·00	- -	- -	- -	5226 Re.
do. (Scotch) - -	-	-	9600	- -	15700000 Tr.	
Steel (soft) - - -	7·780	486·25	120000			
do. (razor-tempered) -	7·840	490·00	150000	- -	29000000 Y.	
Stone (Ancaster) - -	2·182 D.W.	136·37				
do. Barnack - -	2·090 D.W.	130·62				
do. Binnie - - -	2·194 D.W.	137·12				
do. Bolsover - -	2·316 D.W.	144·75				
do. Box - - -	1·839 D.W.	114·93				
do. Bramham Moor -	2·008 D.W.	125·50				
do. Brodsworth - -	2 093 D.W.	130·81				
do. Cadeby - -	1·951 D.W.	121·93				
do. Caithness - -	2·764 Re.	172·75	- -	- -	- -	5142 Re.
do. Craigleith - -	2·266 D.W.	141·62				
do. Chilmark (A) -	2·366 D.W.	147·87				
do. Chilmark (B) -	2·383 D.W.	148·93				
do. Chilmark (C) -	2·481 D.W.	155·06				
do. Darby Dale (Stancliffe) - - -	2·628 D.W.	164·25				
do. Giffneuk - -	2·230 D.W.	139·37				
do. Gunbarrel(Stanley)	2·260 D.W.	141·25				
do. Ham Hill - -	2·260 D.W.	141·25				
do. Haydon - -	2·040 D.W.	127·50				
do. Heddon - -	2·229 D.W.	139·31				
do. Hildenly - -	2·098 D.W.	131·12				
do. Hookstone - -	2·253 D.W.	140·81				
do. Huddlestone -	2·147 D.W.	134·18				
do. Little Hulton -	2·357 H.	147·31	- -	- -	- -	774 H.
do. Kenton - -	2·247 D.W.	140·43				
do. Ketton - -	2·045 D.W.	127·81				
do. Ketton rag - -	2·490 D.W.	155·62				
do. Mansfield, or Lindley's red - - -	2·338 D.W.	146·12				
do. white - - -	2·277 D.W.	142·31				
do. Morley Moor -	2·053 D.W.	128·31				
do. Park Nook - -	2·138 D.W.	133·62				
do. Park Spring -	2·321 D.W.	145·06				
do. Portland (Waycroft Quarry) - - -	2·145 D.W.	134·06				
do. Redgate - -	2·239 D.W.	139·93				
do. Roach Abbey -	2·134 D.W.	133·37				
do. Rochdale - -	2·577 H.	161·06	- -	- -	2358 H.	
do. Stanley - -	2·227 D.W.	139·18				
do. Taynton - -	2·103 D.W.	131·43				
do. Totterwhoe - -	2·891 D.W.	118·18				
do. Jackdaw crag -	2·070 D.W.	129·37				
do Yorkshire flag -	2·320 H.	145·00	- -	- -	- -	1116 H.
do. green moor - -	2·534 Re.	158·37	- -	- -	- -	2010 Re.
Sycamore - - -	0·69 Be.	43·12	13000 Be.			
Teak (dry) - - -	0·657 C.	41·06	15000 B.	12101 H.	2414400 B.	14772 B.
Tile (common) - -	1·815 Br.	113·43				
Tin (cast) - - -	7·291 Tr.	455·68	5322 M.	- -	4608000 Tr.	
Water (sea) - - -	1·027 T.	64·18				
do. (rain) - -	1·000	62·50				
Walnut - - - -	0·671 M.	41·93	8130 M.	6645 H.		
Whalebone - - -	-	-	7667	- -	820000 Tr.	
Willow (dry) - -	0·390	24·37	14000 Be.			
Yew (Spanish) - -	0·807 M.	50·43	8000 Be.			
Zinc - - - -	7·028 W.	439·25	- -	- -	13680000 Tr.	

Note.—The experiments of Mr. Hodgkinson, from which the moduli of rupture of stones contained in the last column of the above table have been calculated, together with those detailed in Art. 412., form part of a more extended inquiry, which, when completed, will be laid before the Royal Society. The crushing forces given in the above tables are in every case determined by the compression of prisms of such a height as to allow the fracture to take place by the sliding of the upper portion freely off, along its plane of separation from the lower (see Art 408.). The experiments of Mr. Hodgkinson have shown that all results obtained without reference to this circumstance are erroneous. (See *Illustrations of Mechanics*, p. 402.)

TABLE V.

Useful Numbers.

$\pi \ldots\ldots = 3 \cdot 1415927$

Log. $\pi = 0 \cdot 4971499$

Log.$_\varepsilon \pi = 1 \cdot 1447299$

$\dfrac{1}{\pi} \ldots\ldots = 0 \cdot 3183099$

$\pi^2 \ldots\ldots = 9 \cdot 8696044$

$\dfrac{1}{\pi^2} \ldots\ldots = 0 \cdot 1013212$

$\sqrt{\pi} \ldots = 1 \cdot 7724538$

$\dfrac{1}{\sqrt{\pi}} \ldots = 0 \cdot 5641896$

$\sqrt{2} \ldots = 1 \cdot 4142136$

$\dfrac{1}{\sqrt{2}} \ldots = 0 \cdot 7071068$

$\pi \sqrt{2} \ldots = 4 \cdot 4428829$

$\dfrac{\pi}{\sqrt{2}} \ldots = 2 \cdot 2214415$

$\dfrac{\sqrt{2}}{\pi} \ldots\ldots = 0 \cdot 4501582$

$\sqrt{\dfrac{\pi}{2}} \ldots\ldots 1 \cdot 2533141$

$\sqrt{\dfrac{2}{\pi}} \ldots\ldots 0 \cdot 7978846$

$\varepsilon \ldots\ldots\ldots = 2 \cdot 7182818$

Log. $\varepsilon \ldots\ldots\ldots = 0 \cdot 4342945$

Modulus of common logarithms $\ldots\ldots = \cdot 434294482$

Log. of ditto $\ldots\ldots = 9 \cdot 6377843$

$g \ldots\ldots\ldots = 32 \cdot 19084$

$\sqrt{g} \ldots\ldots\ldots = 5 \cdot 67363$

Log. $g \ldots\ldots = 1 \cdot 5077222$

Inches in a French métre $\ldots\ldots = 39 \cdot 37079$

Log. of ditto $\ldots\ldots = 1 \cdot 5951741$

Feet in ditto $\ldots\ldots = 3 \cdot 2808992$

Log. of ditto $\ldots\ldots = 0 \cdot 5159929$

Square feet in the square métre $\ldots\ldots = 10 \cdot 764297$

Acres in the Are $\ldots\ldots = 0 \cdot 024711$

Lbs. in a kilogramme $\ldots\ldots = 2 \cdot 20548$

Log. of ditto $\ldots\ldots = 0 \cdot 3435031$

Imperial gallons in a litre $\ldots\ldots = 0 \cdot 2200967$

Lbs. per square inch in 1 kilogramme per square millimetre $= 1422$

Cwts. ditto, ditto $\ldots\ldots = 12 \cdot 7$

Volume of a sphere whose diameter is 1 $\ldots\ldots = 0 \cdot 5235988$

Arc of $1°$ to rad. 1 $\ldots\ldots = 0 \cdot 017453293$

Arc of $1'$ to rad. 1 $\ldots\ldots = 0 \cdot 000290888$

Arc of $1''$ to rad. 1 $\ldots\ldots = 0 \cdot 000004848$

Degrees in an arc whose length is 1 $\ldots\ldots = 57 \cdot 295780°$

Grains in 1 oz. avoirdupois $\ldots\ldots = 437\frac{1}{2}$

Grains in 1 lb. ditto...=7000
Grains in a cubic inch of distilled water, Bar. 30 in., Th. 62°=252·458
Cubic inches in an ounce of water..=1·73298
Cubic inches in the Imperial gallon..=277·276
Feet in a geographical mile..=6075·6
Log. of ditto...=3·7835892
Feet in a statute mile..=5280
Log. of ditto...=3·7226339
Length of seconds' pendulum in inches.................................=39·19084
Cubic inches in 1 cwt. of cast iron..=430·25
   —   Bar iron.................................=397·60
   —   Cast brass..............................=368·88
   —   Cast copper............................=352·41
   —   Cast lead................................=272·80
Cubic feet in 1 ton of paving stone..=14·835
   —   Granite...................................=13·505
   —   Marble....................................=13·070
   —   Chalk.......................................=12·874
   —   Limestone................................=11·273
   —   Elm..=64·460
   —   Honduras mahogany..................=64·000
   —   Mar Forest fir...........................=51·650
   —   Beech......................................=51·494
   —   Riga fir....................................=47·762
   —   Ash and Dantzic oak..................=47·158
   —   Spanish mahogany.....................=42·066
   —   English oak...............................=36·205
To find the weight in lbs. of 1 foot of common rope, multi-
  ply the square of its circumference in inches by......·044 to ·046
Ditto for a cable...·027

Note. — The numerical values of the functions of π in this table were calculated by Mr. Goodwin. These, together with the numbers of cubic inches and
feet per cwt. or ton of different materials, are taken from the late Dr. Gregory's
excellent treatise, entitled *Mechanics for Practical Men.* The other numbers of
the table are principally taken from Mr. Babbage's *Tables of Logarithms* and the
Aide Memoire of M. Morin.

THE END.

LONDON:
Printed by A. SPOTTISWOODE,
New-Street-Square.